Advances in Cybersecurity: Challenges and Solutions

Advances in Cybersecurity: Challenges and Solutions

Editors

Peter R. J. Trim
Yang-Im Lee

MDPI

Basel • Beijing • Wuhan • Barcelona • Belgrade • Novi Sad • Cluj • Manchester

Editors

Peter R. J. Trim
Birkbeck, University of London
London
UK

Yang-Im Lee
University of Westminster
London
UK

Editorial Office
MDPI AG
Grosspeteranlage 5
4052 Basel, Switzerland

This is a reprint of articles from the Special Issue published online in the open access journal *Applied Sciences* (ISSN 2076-3417) (available at: https://www.mdpi.com/journal/applsci/special_issues/Cybersecurity_Challenges_Solutions).

For citation purposes, cite each article independently as indicated on the article page online and as indicated below:

Lastname, A.A.; Lastname, B.B. Article Title. *Journal Name* **Year**, *Volume Number*, Page Range.

ISBN 978-3-7258-1897-6 (Hbk)
ISBN 978-3-7258-1898-3 (PDF)
doi.org/10.3390/books978-3-7258-1898-3

Contents

About the Editors

Peter R. J. Trim

Peter R. J. Trim is a Reader in Marketing and Security Management at Birkbeck, University of London and holds degrees from various institutions including City University (City, University of London), Cranfield Institute of Technology (Cranfield University), and the University of Cambridge. He is a Fellow of the Higher Education Academy and the Royal Society of Arts. Peter has published 69 academic journal articles and 13 books and is the co-editor of two reports for the government. He has also edited a journal special section and is the co-editor of four journal Special Issues. In addition, he has authored 67 chapters in books and delivered 79 conference papers. Peter has been involved in a number of funded research projects involving government, industry, and academia, has worked in several industries, and has overseas work experience. Peter was also a founding member of the academic liaison panel of the Information Assurance Advisory Council and attended regular meetings in London. He is the co-author, with Yang-Im Lee, of a book entitled Strategic Cyber Security Management, published by Routledge, which draws on a social science perspective in order to link cyber security management with resilience and business continuity planning, for example. Currently, Peter is involved in various aspects of research involving cyber security management and online marketing, and he remains actively involved in organizing cyber security management research workshops and conferences. He is also involved in various international cyber security network initiatives.

Yang-Im Lee

Yang-Im has studied and worked in Korea, Japan, and the UK. She undertook her postgraduate studies at the School of Oriental and African Studies, University of London, and was awarded a scholarship by Stirling University to undertake a PhD at that institution. Yang-Im is currently a Senior Lecturer in Marketing at Westminster Business School, University of Westminster, where she teaches various aspects of marketing. Yang-Im has published 35 articles in a range of academic journals and contributed 14 book chapters. She has co-authored books and presented 45 conference papers. She is also the co-editor of three Special Issues. Yang-Im is a Fellow of the Royal Society of Arts and has been a Visiting Fellow at Birkbeck, University of London. Yang-Im has a deep interest in education and the use of technology and, in the past, provided support for the Information Assurance Advisory Council, assuming the role of academic liaison panel co-ordinator for a number of years. Yang-Im has been involved in a number of funded research projects in the UK and is currently undertaking research into online marketing and cyber security management.

Preface

The papers in this reprint bear witness to the fact that the body of cyber security knowledge is continuing to evolve, and they will, we are sure, prove inspirational to those aspiring to undertake cyber security-related studies and/or research. What is evident is the holistic nature of the subject and the fact that there are many issues and challenges to be addressed, which remain ongoing. We are indebted to the authors of the papers who have provided material for inclusion in this reprint, and we take much pleasure knowing that the community of interest that is emerging is devoting time and effort to put in place countermeasures that will prevent innocent people falling victim to cyber-attacks. We also recognize the intellectual challenges to be confronted and consider that there are many more challenges ahead, which will require teamwork and international cooperation.

Peter R. J. Trim and Yang-Im Lee
Editors

applied sciences

MDPI

Editorial

Advances in Cybersecurity: Challenges and Solutions

Peter R. J. Trim [1,*] and Yang-Im Lee [2,*]

1 Birkbeck Business School, Birkbeck, University of London, Malet Street, London WC1E 7HX, UK
2 Westminster Business School, University of Westminster, 35 Marylebone Road, London NW1 5LS, UK
* Correspondence: p.trim@bbk.ac.uk (P.R.J.T.); y.lee@westminster.ac.uk (Y.-I.L.)

1. Introduction

Cyberattacks have increased in intensity and sophistication in recent years, resulting in defensive actions to safeguard company assets and vulnerable people. Research undertaken into various forms of cyberattacks has introduced a number of methods and approaches to help counteract the actions of those responsible for attacks of this nature [1]. However, fully understanding the factors involved requires an in-depth appreciation of the type of attack and the possible impact on an organization should it be successful in penetrating the organization's defences. Indeed, as change occurs in society and people adapt accordingly, new vulnerabilities emerge. For example, although remote working is perceived as beneficial from a cost effectiveness perspective, it can be argued that the benefits afforded to employees, which include the opportunity to work flexibly and utilize their own personal device(s) to access organizational computer systems when working from home, need to be weighed against the possible risks involved. It must be highlighted that the Bring Your Own Device (BYOD) model [2] can put both the employee and the organization at risk if the employee does not follow the security guidance provided. Hence, monitoring and organizational control are important factors in terms of ensuring that the use of a personal device does not prove problematic.

With the advances in artificial intelligence (AI) and the anticipated advantages and threats associated with it, it is pleasing to note that work that is being carried out to ensure that networking platforms are safer, is focused on making networks more robust. Studies relating to data-driven edge intelligence vis-à-vis robust network anomaly detection will contribute significantly to making data secure, and the benefits associated with network anomaly detection [3] will help security provision. The evolving nature of cyber threats has resulted in various initiatives involving corporate, government and academic researchers, all of whom have contributed to the defence of society. In the process, cooperation involving institutions and cybersecurity experts, has witnessed cross border initiatives that have helped preserve the quality of life. Future technological collaboration and knowledge transfer between cybersecurity researchers will do much to speed up the process of developing new technological solutions to combat innovative practices emanating from cyber criminals.

Cyberattacks have an international dimension; as such, cybersecurity researchers need to find ways to collaborate, which requires clear leadership. However, although new technologies are emerging and being approved and are partially funded by the government, smart cities will be at risk if the technologies that underpin critical infrastructure are deficient. It has been suggested [4] that attention needs to be paid to SCADA systems, and in particular power grid subsystems. With cybersecurity threat detection remaining high on the agenda of cybersecurity researchers and senior managers, it can be expected that more attention will be given to establishing how the Internet of Things (IoT) will be prone to malware attacks [5]. Much is known about such attacks, but the perpetrators of such attacks are increasingly seeking to exploit new vulnerabilities and will continue to do so for a considerable time. Whether their motivation is associated with financial gain or attributed to a desire to cause disruption and gain publicity is of interest to cybersecurity researchers.

Citation: Trim, P.R.J.; Lee, Y.-I. Advances in Cybersecurity: Challenges and Solutions. *Appl. Sci.* **2024**, *14*, 4300. https://doi.org/10.3390/app14104300

Received: 30 April 2024
Accepted: 17 May 2024
Published: 19 May 2024

Clearly, international cooperation to counteract the actions of cyber criminals and threat agents will continue to be the focus of policy makers, highlighting the importance of identifying solutions to recurring threats. Acknowledging that cybersecurity needs to be properly managed and resourced focuses attention on various research initiatives, both present and evolving, that will help identify solutions and make organizations less vulnerable to attacks. Therefore, building a practical environment in which cybersecurity training and weapon system test evaluations [6] can be undertaken is essential. Acknowledging that cybercrime is also associated with acts of cyber war and cyber terrorism, provides policy makers with the grounds to regulate more widely to prevent the evolution of more advanced forms of cyberattacks. Advances in artificial intelligence (AI) will be a game changer and require more investment in order to better understand how to defend against AI-orchestrated attacks. However, the advances made in technology will not distract from the fact that managers in both the public and private sectors need to ensure that staff are compliant and comply with security practices [7]. To ensure that this happens, appropriate governance framework(s) and mechanism(s) need to be put in place.

To solve the underlying root of recurring cybersecurity threats and issues, cybersecurity researchers need to implement cybersecurity policy and strategy initiatives that will help counteract the effort of those intent on destabilizing society and causing untold damage for their own gain. Hence, this Special Issue is dedicated to developments in cybersecurity from an interdisciplinary and multidisciplinary perspective, and the collection of papers focus on the challenges confronting companies, governments and society. The topics covered establish the ways in which technology and human–technology interactions are enhancing cybersecurity provision. By adopting a holistic view of cybersecurity and outlining the strategies to implement cybersecurity solutions, it is possible for society to be better-protected and more able to withstand sustained cyberattacks. A broad range of papers are included in the Special Issue, and various methodological approaches are represented that help us understand how cybersecurity theory and practice are linked and how we can devise and implement effective cybersecurity solutions.

2. An Overview of the Published Articles

The range of topics covered and knowledge accumulated by the authors can be considered inspirational, setting the scene for future research into cyber security and the related areas of study. Indeed, Ayedh et al. pay attention to an important but under-researched topic, Bring Your Own Device (BYOD), referring to the relevant security and privacy requirements. As well as covering BYOD security policies, reference is made to state-of-the art security policy technologies, technology trends and the measures employed to enhance security.

Another area of increased attention is the need for maintaining a secure system by acquiring necessary learning data. In their paper, Cha et al. make reference to a digital twin environment and focus on the need to ensure that systems and data in the genuine system are safeguarded. One of the benefits of this approach is that new malware is generated through image conversion and an adversarial generative neural network, which has the benefit of predicting and preventing the generation of malware in the future.

Regarding the detection of anomalies in data streams, Demertzis et al. establish a cross-modal dynamic attention neural architecture (CM-DANA), which represents a dynamic attention mechanism that can be trained through harnessing multimodal learning tasks. The data are derived from different cyber modalities and have the benefit of being able to detect suspicious abnormal behaviour.

Mejjaouli and Guizani propose a model based on the fuzzy unordered rule induction algorithm (FURIA), which detects malware associated with portable document format (PDF) malware. A comparative analysis is made of various machine learning models using standard assessment measures. The FURIA-based model was found to outperform other machine learning models.

Considering the problems created by malware and the need to adequately classify viruses, Wu et al. offer guidance on detection rates, for example, and clarify how a static classification model encompassing a malicious code fused with TCN and BiGRU can both extract and integrate the opcode features and the byte features of a malicious code.

Early threat detection has occupied the minds of researchers for some time and López-Vizcaíno et al. focus attention on the time-aware F-score (TaF) metric for early detection, as it considers the number of items/individual elements processed in relation to establishing if an element is an anomaly to be detected or not relevant for detection. The results are validated via an operative system (OS) scan attack. It was concluded that the TaF metric is adequate in terms of a time-sensitive detection system.

Zhang et al. pay attention to detecting phishing scams on Ethereum, and the bagging multiedge graph convolutional network (BM-GCN) scheme is proposed. The BM-GCN (0.877 AUC) scheme was found to outperform other baseline classification methods.

Regarding the unbalanced intrusion detection data vis-à-vis a multi-class classification problem, Bacevicius and Paulauskaite-Taraseviciene evaluate the performance of multi-class classification for network intrusions and utilize the CIC-IDS2017 and CSE-CIC-IDS2018 datasets. The classification performance of six machine learning models was compared, and it was discovered that decision trees using the CART algorithm outperformed the other machine learning models by achieving an average macro $F1$-score of 0.96878.

Supervision control and data acquisition (SCADA) systems are open to attack and can be subject to much disruption. In this context, Söğüt and Erdem carried out research involving five attack scenarios vis-à-vis DDos attacks. By monitoring the SCADA system networks, various models were applied to the obtained data, and it was discovered that the hybrid model and the decision tree were the most suitable and could be used in harmony on real field systems.

Huang et al. focus on cyber mimic defence, and with the need to partition complex networks, multidimensional evaluation metrics were established to assess the effectiveness of cyber mimic defence technology.

Regarding the use of a cyber range to effectively integrate a number of factors in relation to a battlefield environment, Park et al. explain how a multi-cyber range can benefit those engaged in a training environment. There are several advantages: the impacts associated with DDos attacks are highlighted and the interoperability between systems is maintained.

In relation to the security of database management systems (DBMSs) and grey-box fuzzing activity, Wen et al. implement Squill, a grey-box fuzzer, in order to address the challenges associated with DBMS fuzzing. In their study, 30 bugs were found in MySQL, 27 were found in MariaDB and 6 were unearthed in OceanBase, with 9 CVEs assigned. As a consequence, it was proven that Squill was able to locate more bugs in DBMSs as opposed to other known tools.

Additional insights into grey-box fuzzing were provided by Xie et al. Their aim was to rectify the inefficiencies associated with traditional seed scheduling strategies by advocating a seed scheduling strategy guided by untouched edges. As such, a new instrumentation method was put forward. The prototype UntouchFuzz was used to evaluate the experiments against seed scheduling strategies, and 13 vulnerabilities were discovered in the open-source projects and 7 of these had assigned CVEs.

Ransomware attacks are common, and Al-Awadi et al. pay specific attention to evaluating the effectiveness of Windows 11 Pro in relation to its capability to counteract ransomware attacks. A dual examination revealed that Windows 11 Pro does have formidable defences. Recommendations that will benefit technology developers and end-users are provided, which makes an important contribution to cybersecurity knowledge enhancement.

Pan et al. outline a scheme for encrypting linear controllers, the objective of which is to remove security risks and improve security in relation to networked control systems.

The authors use precomputation vis-à-vis data encryption and demonstrate how security can be improved.

With reference to essential cybersecurity control (ECC), Alfaadhel et al. advocate for a comprehensive and customized risk-based cybersecurity compliance assessment system. RC2AS helps staff identify current weaknesses and formalize planning. In addition, the assessment results appear in dashboards. RC2AS can be used to calculate the overall compliance score, which can be considered highly beneficial.

3. Conclusions

As can be deduced from the above, the scope and depth of the knowledge encompassed by the papers that make up this Special Issue will do much to underpin the advancement of cybersecurity, further focusing the minds of senior managers, policy makers and researchers on cyber threat detection and prevention. Indeed, those involved in cybersecurity research are very much involved in defencive actions, and it is hoped that the work of the experts outlined herewith will do much to inspire people to learn more about cybersecurity and engage in cybersecurity research. Guidance is provided in terms of what needs to be achieved to counteract the various types of cyberattack that have proliferated in recent years, and this can be considered beneficial in terms of the issues and challenges that have emerged and are continuing to emerge. The research findings encourage the cooperative spirit of the researchers, and we thank them for sharing their knowledge with us and providing insights that can be drawn upon by a wide audience. It is pleasing to note that those involved in cyber security research are working hard to expand the theoretical base of cybersecurity, which is evolving as an established and distinct body of knowledge.

Conflicts of Interest: The authors declare no conflicts of interest.

List of Contributions:

1. Ayedh M.A.T.; Wahab, A.W.A.; Idris, M.Y.I. Systematic literature review on security access control policies and techniques based on privacy requirements in a BYOD environment: State of the art and future directions. *Appl. Sci.* **2023**, *13*, 8048. https://doi.org/10.3390/app13148048.
2. Cha, H.-J.; Yang, H.-K.; Song, Y.-J.; Kang, A.R. Intelligent anomaly detection system through malware image augmentation in IIoT environment based on digital twin. *Appl. Sci.* **2023**, *13*, 10196. https://doi.org/10.3390/app131810196.
3. Demertzis, K.; Rantos, K.; Magafas, L.; Iliadis, L. A cross-modal dynamic attention neural architecture to detect anomalies in data streams from smart communication environments. *Appl. Sci.* **2023**, *13*, 9648. https://doi.org/10.3390/app13179648.
4. Mejjaouli, S.; Guizani, S. PDF malware detection based on fuzzy unordered rule induction algorithm (FURIA). *Appl. Sci.* **2023**, *13*, 3980. https://doi.org/10.3390/app13063980.
5. Wu, X.; Song, Y.; Hou, X.; Ma, Z.; Chen, C. Deep learning model with sequential features for malware classification. *Appl. Sci.* **2022**, *12*, 9994. https://doi.org/10.3390/app12199994.
6. López-Vizcaíno, M.; Nóvoa, F.J.; Fernández, D.; Cacheda, F. Time aware F-score for cybersecurity early detection evaluation. *Appl. Sci.* **2024**, *14*, 574. https://doi.org/10.3390/app14020574.
7. Zhang, Z.; He, T.; Chen, K.; Zhang, B.; Wang, Q.; Yuan, L. Phishing node detection in ethereum transaction network using graph convolutional networks. *Appl. Sci.* **2023**, *13*, 6430. https://doi.org/10.3390/app13116430.
8. Bacevicius, M.; Paulauskaite-Taraseviciene, A. Machine learning algorithms for raw and unbalanced intrusion detection data in a multi-class classification problem. *Appl. Sci.* **2023**, *13*, 7328. https://doi.org/10.3390/app13127328.
9. Söğüt, E.; Erdem, O.A. A multi-model proposal for classification and detection of DDoS attacks on SCADA systems. *Appl. Sci.* **2023**, *13*, 5993. https://doi.org/10.3390/app13105993.
10. Huang, Z.; Yuan, Y.; Fu, J.; He, J.; Zhu, H.; Cheng, G. Location-aware measurement for cyber mimic defense: You cannot improve what you cannot measure. *Appl. Sci.* **2023**, *13*, 9213. https://doi.org/10.3390/app13169213.
11. Park, M.; Lee, H.; Kim, Y.; Kim, K.; Shin, D. Design and implementation of multi-cyber range for cyber training and testing. *Appl. Sci.* **2022**, *12*, 12546. https://doi.org/10.3390/app122412546.
12. Wen, S.; Jia, P.; Yang, P.; Hu, C. Squill: Testing DBMS with correctness feedback and accurate instantiation. *Appl. Sci.* **2023**, *13*, 2519. https://doi.org/10.3390/app13042519.

13. Xie, C.; Jia, P.; Yang, P.; Hu, C.; Kuang, H.; Ye, G.; Hong, X. Not all seeds are important: Fuzzing guided by untouched edges. *Appl. Sci.* **2023**, *13*, 13172. https://doi.org/10.3390/app132413172.
14. Al-Awadi, Y.M.; Baydoun, A.; Ur Rehman, H. Can Windows 11 stop well-known ransomware variants? An examination of its built-in security features. *Appl. Sci.* **2024**, *14*, 3520. https://doi.org/10.3390/app14083520.
15. Pan, J.; Sui, T.; Liu, W.; Wang, J.; Kong, L.; Zhao, Y.; Wei, Z. Secure control of linear controllers using fully homomorphic encryption. *Appl. Sci.* **2023**, *13*, 13071. https://doi.org/10.3390/app132413071.
16. Alfaadhel, A.; Almomani, I.; Ahmed, M. Risk-based cybersecurity compliance assessment system (RC2AS). *Appl. Sci.* **2023**, *13*, 6145. https://doi.org/10.3390/app13106145.

References

1. Li, Y.; Liu, Q. A comprehensive review study of cyber-attacks and cyber security: Emerging trends and recent development. *Energy Rep.* **2021**, *7*, 8176–8186. [CrossRef]
2. Lee, J.; Warkentin, M.; Crossler, R.E.; Otondo, R.F. Implications of monitoring mechanisms on Bring Your Own Devise adoption. *J. Comput. Inf. Syst.* **2017**, *57*, 309–318. [CrossRef]
3. Xu, S.; Qian, Y.; Hu, R.Q. Data-driven edge intelligence for robust network anomaly detection. *IEEE Trans. Netw. Sci. Eng.* **2019**, *7*, 1481–1492. [CrossRef]
4. Ma, C. Smart city and cyber-security; technologies used, leading challenges and future recommendations. *Energy Rep.* **2021**, *7*, 7999–8012. [CrossRef]
5. Ullah, F.; Naeem, H.; Jabbar, S.; Khalid, S.; Latif, M.A.; Al-Turjman, F.; Mostarda, L. Cyber security threats detection in Internet of Things using deep learning approach. *IEEE Access* **2019**, *7*, 124379–124389. [CrossRef]
6. Park, M.; Lee, H.; Kim, Y.; Kim, K.; Shin, D. Design and implementation of multi-cyber range for cyber training and testing. *Appl. Sci.* **2022**, *12*, 12546. [CrossRef]
7. Donalds, C.; Osei-Bryson, K.-M. Cybersecurity compliance behavior: Exploring the influences of individual decision style and other antecedents. *Int. J. Inf. Manag.* **2020**, *51*, 102056. [CrossRef]

applied sciences

MDPI

Review

Systematic Literature Review on Security Access Control Policies and Techniques Based on Privacy Requirements in a BYOD Environment: State of the Art and Future Directions

Aljuaid Turkea Ayedh M [1,2], Ainuddin Wahid Abdul Wahab [1,*] and Mohd Yamani Idna Idris [1,3]

1 Faculty of Computer Science and Information Technology, Universiti Malaya, Kuala Lumpur 50603, Malaysia; taljuaid@su.edu.sa (A.T.A.M.)
2 Faculty of Computing and Information Technology, Shaqra University, Shaqra 11961, Saudi Arabia
3 Center for Mobile Cloud Computing, Faculty of Computer Science and Information Technology, Universiti Malaya, Kuala Lumpur 50603, Malaysia
* Correspondence: ainuddin@um.edu.my; Tel.: +60-3-7967-6383

Abstract: The number of devices connected within organisational networks through "Bring Your Own Device" (BYOD) initiatives has steadily increased. BYOD security risks have resulted in significant privacy and security issues impacting organisational security. Many researchers have reviewed security and privacy issues in BYOD policies. However, not all of them have fully investigated security and privacy requirements. In addition to describing a system's capabilities and functions, these requirements also reflect the system's ability to eliminate various threats. This paper aims to conduct a comprehensive review of privacy and security criteria in BYOD security policies, as well as the various technical policy methods used to mitigate these threats, to identify future research opportunities. This study reviews existing research and highlights the following points: (1) classification of privacy and security requirements in the context of BYOD policies; (2) comprehensive analyses of proposed state-of-the-art security policy technologies based on three layers of security BYOD policies, followed by analyses of these technologies in terms of the privacy requirements they satisfy; (3) technological trends; (4) measures employed to assess the efficacy of techniques to enhance privacy and security; and (5) future research in the area of BYOD security and privacy.

Keywords: access control policies; security techniques; BYOD security layers; risk access control; onboarding access control; authentication; attack detection; privacy and security requirements; BYOD environment

Citation: Ayedh M, A.T.; Wahab, A.W.A.; Idris, M.Y.I. Systematic Literature Review on Security Access Control Policies and Techniques Based on Privacy Requirements in a BYOD Environment: State of the Art and Future Directions. *Appl. Sci.* **2023**, *13*, 8048. https://doi.org/10.3390/app13148048

Academic Editors: Antonio Fernández-Caballero, Peter R. J. Trim, Yang-Im Lee and Luis Javier García Villalba

Received: 24 February 2023
Revised: 15 June 2023
Accepted: 4 July 2023
Published: 10 July 2023

1. Introduction

The bring your own device (BYOD) paradigm, which allows employees to connect their mobile devices to the organization's network, rapidly changes organisational operations by enhancing flexibility, productivity, and effectiveness [1]. Despite these advantages, security concerns continue to affect organisational environments [2] and introduce security challenges and significant security risks [2]. One of the primary concerns with BYOD is the need for more control over employee devices. Personal devices have different security measures and software updates from company-issued devices. This disparity exposes vulnerabilities that attackers could exploit, resulting in data breaches or unauthorized access to sensitive information [3]. In addition, the variety of devices in a BYOD environment complicates the implementation of standard security policies. Different operating systems, versions, and security configurations must be considered by organizations, making it challenging to implement uniform security measures across all devices [4]. This variation heightens the possibility that cybercriminals will exploit security gaps or obsolete security software.

Furthermore, when personal devices are used for work-related purposes, the possibility of mixing personal and business information increases. This mixing of data poses the risk of data leakage or inadvertent disclosure. It becomes more difficult for organizations on these devices to ensure that sensitive information is adequately protected and segregated from personal files. The BYOD direction also raises concerns about the possibility of device loss or theft. If an employee's device containing sensitive company information is lost or stolen, the risk of unauthorised access to that information increases [5].

To address these risks, organisations can effectively manage BYOD usage by implementing security access control and security policy technologies that address these vulnerabilities and obstacles [6]. However, there are significant gaps between the security offered by current BYOD access control policies and the desired outcomes [7]. Rhee et al. [8] provide fundamental access control policies that address security and privacy issues in three primary categories: authenticity, confidentiality, and integrity. These policies encompass network access control policies, mobile access control policies, mobile information management policies, mobile application management policies, and enterprise mobility management policies. Initially, these policies assisted organisations in managing and governing BYOD devices effectively. Nonetheless, the increasing complexity of attacks targeting BYOD devices and networks [8–10] has rendered these policies and the security requirements they fulfil insufficient [11,12]. To adequately meet the security and privacy requirements of BYOD, it is essential to adopt integrated and comprehensive BYOD security policies, emphasising the implementation of three-tiered policies and a thorough understanding of security and privacy requirements, as mentioned by Bello et al. [13] in their work on consumerization.

There have been few systematic reviews of the security of BYOD, as seen in Table 1. However, most earlier research has systematically ignored examining security and privacy needs in the BYOD context. Additionally, previous survey studies have yet to investigate the most commonly used security policy techniques based on a three-tiered BYOD policy architecture with the appropriate technology to fulfill security and privacy requirements. For instance, in [14], Oktavia et al. presented a survey on privacy concerns and BYOD challenges. This study examined these issues in depth. However, the analysis of policy mechanisms based on security and privacy requirements was not included and was limited to raising concerns.

Similarly, Jamal et al. [15] surveyed BYOD authentication techniques, focusing on authenticity criteria while giving less attention to other security and privacy criteria. The term "other security and privacy criteria*" indicates additional privacy requirements in BYOD security, such as confidentiality, integrity, availability, authenticity, privacy preservation, non-repudiation, and attack detection. Furthermore, Palanisamy et al. [16] presented a thorough review of compliance theories that are used to interpret and predict security practices in the Bring Your Own Device (BYOD) sector. However, they did not conduct an assessment of technologies in relation to security and privacy standards. Instead, their primary focus was largely on the theoretical aspects of the subject. Additionally, Wani et al. [17] highlighted significant security problems associated with hospital BYOD practices but did not extensively address the identified concerns by analyzing technologies. While several survey studies have contributed to understanding BYOD's privacy and security challenges, most of them have provided limited information on the inherent privacy and security concerns of BYOD. Furthermore, many of these studies examined only a subset of the problem or conducted a review during the early stages of BYOD adoption. In [18], the researchers conducted a comprehensive review of attack detection strategies that utilize machine learning. However, they did not delve into other aspects related to privacy needs. To put it another way, while they extensively studied how machine learning can be used to identify cyber attacks, they did not explore other crucial components of privacy, such as data protection, anonymity, and user consent. Table 1 provides an overview of the main focus and limitations of some of the earliest (pre-2023) literature on BYOD security. Overall, while these studies have made strides in offering access control solutions and risk analyses, they exhibit limitations. Particularly, they do not critically evaluate the studies in terms of their contribution to satisfying the privacy requirements essential for BYOD systems.

Table 1. Focus and limitations of some of the key older (pre-2023) publications.

Ref	Concentrate on	Limitations
[14,17,19,20]	Focused on discussing risks and security issues related to BYOD.	Not comprehensive for all security and privacy requirements and access control based on three layers.
[15]	Focused on techniques connected to the authenticity criteria.	Other * security and privacy criteria received less attention.
[16]	Overview of the BYOD compliance theories.	Not analysing technologies based on privacy criteria.
[21]	Classification scheme for proposed solutions based on identified security issues.	Not analysing technologies based on privacy criteria.
[18]	Mapping review of attack detection strategy based on machine learning.	Other * security and privacy requirements were ignored.

* indicates additional privacy requirements in BYOD security.

Therefore, this paper extensively reviews security policies and access control in three security policy layers. It evaluates the effectiveness of existing security policies and access control mechanisms in meeting privacy requirements. The study contributes to understanding privacy-focused security measures in BYOD environments and informs future research and improvements in security policies and access control strategies. To ensure a systematic approach, we developed a review protocol that outlines the critical phases necessary to achieve the objectives of this study. The paper focuses on seven critical security and privacy criteria within the three layers of secure BYOD control policies designed for enterprises adopting BYOD practices. These criteria include confidentiality, integrity, availability, authenticity, privacy preservation, non-repudiation, and attack detection. Achieving these criteria ensures the system can eliminate potential privacy and security vulnerabilities and comply with regulatory guidelines [5,22]. The process of this study will ensure unbiased data retrieval and thorough search procedures. The contributions of this study to the overall review can be summarized as follows:

- Identifies privacy and security criteria needed in the BYOD policy setting.
- Analyses existing policy techniques based on privacy and security requirements in three security policy layers.
- Introduces a novel taxonomy that categorizes policy techniques into three layers according to their alignment with privacy and security requirements. This taxonomy provides a structured framework for understanding and organizing policy techniques, contributing to the existing knowledge in the field.
- Identifies and discusses the current trends in technology related to policy techniques by examining the technological advancements within each layer.
- Addresses the measures used to evaluate the effectiveness of policy techniques, mainly through performance analysis, by discussing these evaluation measures.
- Presents a comprehensive evaluation of policy techniques' technical advantages and limitations by highlighting the strengths and weaknesses of each technique.
- Identifies potential areas for future research and improvement in policy techniques. By pointing out the gaps and limitations in the existing techniques, the paper stimulates further exploration and encourages researchers to develop innovative approaches to address the identified challenges.

The remainder of the article is divided into the following sections: Section 2 presents the background information, while Section 3 compares the conventional BYOD security approach with the desired state and examines BYOD security policies across the three layers of BYOD architecture policy. Section 4 provides an overview of the privacy and security requirements for the development of security policies. The methodology for conducting a systematic literature review is presented in Section 5. The data analysis process is detailed in Section 6, followed by a discussion of critical findings in Section 7. Section 8 highlights open research issues, and Section 9 concludes the paper.

2. Background and Related Concepts

This section examines the security and privacy challenges posed by the bring your own device (BYOD) environment and introduces the fundamental concepts that will be investigated in this study.

2.1. BYOD Risks

The expanding adoption of BYOD raises notable privacy and security concerns. According to a report by Hewlett-Packard [13], employees now utilize multiple mobile devices at work, creating a situation where their activities are invisible and untraceable to the IT department. Consequently, this trend poses various challenges and security threats for enterprises.

Firstly, the need for clear security expectations is a primary challenge, resulting in costly consequences, particularly when inexperienced personnel are entrusted with data security. Moreover, phishing attacks pose a serious risk by compromising employee devices and company information. Using unsecured Wi-Fi networks outside the office further amplifies security risks, as highlighted by Kaspersky Lab [23]. Additionally, malware infections and unauthorized app installations pose additional threats [23,24]. Lastly, the BYOD trend raises concerns about employees accessing social media during work hours, potentially violating company policies.

These risks associated with BYOD significantly impact organizational security, giving rise to various challenges. These challenges include implementing security policies across different BYOD operating systems and devices, effectively managing numerous BYOD devices within the organization, securing BYOD devices, tracking their activities, and monitoring their usage outside of work hours [24]. To address the above challenges, access control policies and technologies play a crucial role in mitigating risks and meeting the privacy requirements of the BYOD environment.

As a result, this study aims to investigate appropriate access control solutions and privacy requirements within the three layers of the BYOD security architecture. The objective is to effectively tackle the privacy and security issues associated with BYOD.

2.2. Security Challenges

The term "security", in the context of an organization, pertains to the safeguarding of valuable assets, including information, resources, processes, and records [13]. Companies that prioritise security allocate significant resources to establish an effective information security system to protect their data. However, despite these efforts, they remain a target for various threats [25].

Managing security poses challenges for large corporations with multiple departments or small branches sharing the same network. In the latter scenario, this structure makes it easier for cybercriminals to target small branches before moving on to headquarters. Security plays a crucial role in supporting organizational operational processes and methods. Therefore, organizations must protect confidential information from potential threats or harm resulting in damage, loss, modification, or unauthorized disclosure.

According to Whitman and Mattord [26], an organisation's security should be a complex system that includes computer systems supporting the organisational environment, software applications and databases securing data, as well as policies, procedures, training, and other human-reliant components. The primary objective of securing an organization's critical information resources and assets is to implement security policies across three layers: onboarding access control, authentication access control, and risk access control policies utilizing related technologies [13].

Whitman and Mattord further suggest that the objectives of security policies should encompass the protection of confidentiality, integrity, and availability of information-reliant entities and information-delivering systems, ultimately fulfilling privacy requirements [26].

Hence, confidentiality, integrity, and availability objectives are essential for accomplishing security policy objectives.

2.3. Privacy Challenges

Privacy is the behaviour or attitude of a company toward protecting its information resources and the personally identifiable information of its customers [27]. Organisations are increasingly leaking personal information, either deliberately or due to compromised information systems. Users of BYOD devices express concerns about their ability to manage personal data and trust organizations to protect their users. According to Johnston and Anna [28], there has been an ongoing debate on whether individuals should sign contracts with organizations to manage and safeguard their information or if it should be the responsibility of the enterprise to protect individual privacy. According to various publications, organisations devote regular financial resources to implementing safeguards and information privacy programmes to protect information from leaks and other threats. However, they fail to effectively utilise these protections and programmes [29]. This strategy can easily result in privacy breaches, which have historically been a significant concern. The primary objective of these hacking attacks is to steal, delete, or alter sensitive information. Organisations must recognise the significance and necessity of protecting personal data because it presents various privacy issues, risks, and other data breaches. New, difficult-to-detect vulnerabilities are generated for hackers as technology progresses and becomes more sophisticated [30]. BYOD has and will continue to raise concerns about data and user privacy.

2.4. Security Policies

Security policies refer to the rules established by organizational leaders to ensure the appropriate level of security is aligned with the organization's needs. The access control policy is crucial in safeguarding the organization's data and resources against internal and external threats, reducing vulnerability to cyber and physical attacks. Each access control type employs specific security techniques to meet the organization's security and privacy requirements, which will be discussed in detail in Section 3.2.

2.5. Security Technologies

Security technologies encompass control policies across various technological components to fulfil privacy and security requirements within the BYOD organizational environment. These technologies cover devices, information, applications, and communication. Each layer of the BYOD security architecture incorporates access policies and technologies that address specific security and privacy needs. For instance, the identification access control policy employs a variety of mechanisms, such as authentication algorithms or other technologies, to carry out authentication and authorization functions. Section 3.2 will comprehensively explain these techniques, where the control policies and techniques associated with each access control policy will be discussed.

2.6. Privacy and Security Requirements

Privacy and security requirements define the security standards that policies and technologies in the BYOD context should meet. Standard terms include confidentiality, integrity, availability, non-repudiation, authentication, privacy preservation, and attack detection. The rationale for selecting these requirements and the underlying concepts will be discussed in Section 4.

3. Comparison Traditional Access Control vs. Security Access Control Based on Three Layers

This section will highlight the primary differentiation between traditional BYOD security policies and contemporary approaches to BYOD security, as depicted in Figure 1. According to Macaraeg [31], traditional security policies primarily focus on protecting devices, networks, data, and applications. These policies encompass network access control, mobile device management, mobile application management, and enterprise mobility management policies, as described in more detail in Section 3.1. While these access control policies may provide security mechanisms and solutions to protect the company's network, data, and devices, they often overlook risk control access policies. Consequently, they fail to meet the requirement for attack detection. To address this limitation, Bello et al. [1] proposed an enhanced BYOD access control approach based on three security layers, aiming to comprehensively address security and privacy requirements that fulfill organizational security needs and user privacy. This three-layered protection approach, as described in Section 3.2, includes an access control policy and corresponding mechanism within each layer. A comparison between this security approach and traditional security policies and their respective shortcomings is outlined in Table 2:

- Traditional security approaches are often deemed insufficient and potentially vulnerable in providing the required level of protection [1,31]. For example, network access control policies in traditional security heavily rely on mobile device management (MDM) technology for authenticating and managing BYOD devices within the organization. In contrast, ideal BYOD security emphasizes identification access control methods, such as biometrics and two-factor authentication. Additionally, ideal security policies for information protection encompass advanced measures like communication access control, encryption, virtual private networks (VPNs), and data wiping. On the other hand, traditional security approaches utilize mobile information management (MIM) to enforce information management policies.
- Traditional security policies lack the inclusion of risk access control policies, despite their importance in BYOD security, as recommended by Bello et al. [1].
- Traditional security policies fall short in adequately addressing all privacy requirements due to limitations in policy techniques and defense mechanisms [31]. These limitations serve as a motivation to address the challenges surrounding privacy and security and enhance security policies through the implementation of three layers of security policies, thus meeting the privacy requirements of BYOD security.

Figure 1. Comparison between traditional security policy and BYOD security.

Table 2. Comparison between traditional security policy and BYOD security.

Security Policy Approaches	Traditional Security	BYOD Security That Should Be Implemented
Mobile Device Management Policy.	✓	✓
Application Management Policy.	✓	✓
Information Management Policy.	✓	✓
Network Access Control Policy.	✓	✓
Enterprise Mobility Management Policy.	✓	✓
Security Communication Policy and Data Protection.	✗	✓
Private Network (VPN), Data Wiping and Data Backup.	✗	✓
Identification Access Control Policy.	✗	✓
Risk Access control policy.	✗	✓
Information Security Policy Compliance.	✗	✓
Comprehensive for the privacy requirements that the BYOD needs.	✗	✓

3.1. *Traditional Security-Based Fundamental Access Control Policy Approaches*

This section highlights and discusses several traditional management approaches and techniques utilised to control the security of organisational resources and avoid security risks to manage employee devices at work effectively. These management approaches can assist as security mechanisms, or solutions, that protect corporate networks, data, and devices while considering employees' privacy rights. The fundamental security policy approach for BYOD includes network access control policies, mobile device management policies, mobile application management policies, and enterprise mobility management policies. These access control policies may provide security mechanisms or solutions. Their shortcomings are detailed below.

3.1.1. Network Access Control Policy (NAC)

This technique focuses on the management and control of access to enterprise networks. NAC controls the devices that access the corporate network and provides secure and regulated network access to various devices from various locations. In addition, NAC can implement authentication and encryption security controls and integrate MDM tools into the network infrastructure to manage network services and resources and monitor the entire network. This strategy uses virtual local area network (LANs)to reduce network traffic by classifying users according to access control policies or functions [13]. Some network BYOD solutions, such as those developed by Cisco and Meru Networks, recommend BYOD management through network methods [13]. However, NAC is subject to the following limitations:

- Managing and accessing rich media material can contribute to network congestion.
- Malicious devices linked to the network can contaminate it.
- Malicious or infected devices can infect others on the same virtual local area network (VLAN).

3.1.2. Mobile Device Management Policy (MDM)

MDM, which manages and controls mobile devices, is based on a product or software platform [32]. MDM controls apps, cameras and the cloud on staff devices. Google Device Manager, Apple Profile Manager and Microsoft Exchange ActiveSync are among MDM technologies [13,33]. Organisations may use MDM to monitor, manage and secure mobile devices in the workplace by enforcing security requirements and ensuring devices conform to these regulations. The MDM approach can manage desktops, mobile devices and servers using the same tools. In addition, it can enforce device access regulations for all devices attached to the MDM platform. However, it has certain limitations, including:

- There is a limit on the number of devices and operating systems.
- Personal and corporate data might be combined.
- Third-party apps are required to utilise MDM functionalities fully.

3.1.3. Mobile Application Management Policy (MAM)

This differs from the MDM approach in that it controls, manages, and secures only specific enterprise software instead of the entire device. Consequently, a corporation may utilise MAM to protect and control email apps and other corporate applications on the mobile devices of its workers [33]. For instance, ZixOne is a mobile application that offers a BYOD solution, providing management access to business email via secure email encryption features [34].

3.1.4. Enterprise Mobility Management Policy (EMM)

This integrates MDM, MAM, and MIM capabilities. It is a solution for BYOD security that handles all devices, apps and data [34]. Enterprise mobility management EMM distinctive characteristics include separating work and personal data on the same device, managing threats proactively, and providing an application store for business apps. However, the EMM strategy has drawbacks such as:

- It combines all of the limitations and challenges of the strategies previously discussed.
- The user experience and satisfaction with BYOD may be compromised by this solution.
- By employing data separation techniques such as containers, corporate data can be vulnerable to security threats.

3.1.5. Mobile Information Management Policy (MIM)

MIM focused on managing BYOD-based data and documents that synchronise across many devices [33]. Mobile information management (MIM) is a device-independent security strategy that encrypts sensitive data and permits only authorised applications to access or transmit it. Mobile information management faces significant obstacles in enterprise mobility management.

3.2. Security Access Control Based on Three Layers (Ideal Security Policy)

This section will introduce the architecture of security policy layers and the control mechanisms present at each layer. According to Bello et al. [1], the security policy of BYOD is divided into three layers that organisations can manage to support comprehensive BYOD device management and security. These layers consist of the operational, tactical, and strategic layers. Figure 2 shows the three-layer BYOD policy. Each layer has a security control function and works with the other layers to manage BYOD information security and privacy. In addition, the protection function of each layer has many security control mechanisms. Bello et al. confirm in [1] that the three layers should be considered when implementing access control solutions between an organization's resources and BYOD devices in order to protect the environment from security and privacy threats, where each layer function is complementary to the function of the other layer to obtain optimal and comprehensive security, in addition to achieving the privacy and security requirements that the BYOD strategy needs, which include confidentiality, integrity, availability, authenticity, privacy preservation, non-repudiation, and attack detection. The operational layer is the primary layer, which focuses on the service level agreement (SLA) that the system owner proposes for the agreement's policies; also, BYOD users register their devices as an initial step. Following that, secure BYOD access control should be added to the tactical layer of policies concerned with authentication and confidentiality of the communication channel between the organization's resources and devices. Finally, applications should be subject to safe control policies. In addition, the strategic layer, an essential addition to the previous two, is responsible for detecting and monitoring employee-device-based attacks. Therefore, if organisations adopt a three-layer policy approach, they will achieve a more secure, competitive environment than traditional policies.

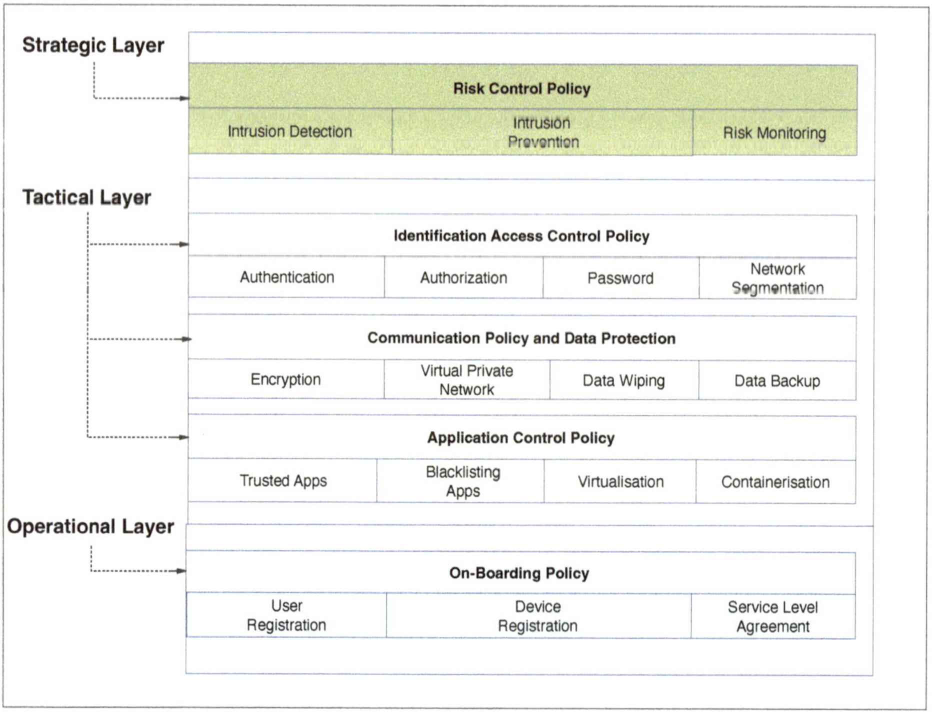

Figure 2. Secure Three-Layer Architecture for BYOD Access Control Policy [1].

3.2.1. Operational Layer

This layer encompasses onboarding access control, which facilitates the identification of authorized users who can access the organization's resources and network through their BYOD devices [1]. Within this layer, two key security functions are performed. Firstly, device account registration control allows users to create and register their accounts, requiring verifiable information such as employee ID, job role, and assigned services. Secondly, a service level agreement (SLA) is established, requiring BYOD users to agree to the terms of use and assume responsibility for the information and services provided by the corporate system.

3.2.2. Tactical Layer

The second layer, known as the tactical layer, encompasses a BYOD policy that focuses on authentication through access control, consisting of three key functions [1]. Firstly, the identification access control policy establishes access controls for BYOD devices to safeguard an organization's information services and resources from unauthorized access. Within BYOD ecosystems, activities such as illegal use or inappropriate communication in information-sensitive applications, including password authentication, authorization, and network segmentation, are strictly prohibited.

The second function is the security communication policy, also known as data protection. Its objective is to ensure the protection of confidential data while enabling secure and protected access, sharing, and transfer between BYOD devices and the organizational infrastructure. Various security control mechanisms, such as encryption, virtual private networks (VPN), data wiping, and data backup, are employed to achieve data confidentiality.

Thirdly, the application control policy is responsible for safeguarding organizations that adopt BYOD from malicious applications. The application management policy aims to prevent any confusion between personal and business data, allowing the exchange of data between personal and company applications on BYOD devices. Control mechanisms such as virtualization, licensing, application blocklisting, application whitelisting, and containerization support this layer.

Overall, the tactical layer addresses multiple security requirements, including confidentiality, integrity, and prevention of unauthorized access.

3.2.3. Strategic Layer

The third layer, referred to as the strategic layer, encompasses a risk control policy known as risk-based access control. This policy focuses on safeguarding both BYOD devices and organizational resources from malware attacks, unauthorized access, and attacks originating from or transmitted through BYOD devices. It plays a critical role in detecting, preventing, and monitoring risks. Intrusion prevention systems (IPS) and intrusion detection systems (IDS) are examples of systems utilized within this layer to achieve these objectives [1].

4. Overview of the Privacy and Security Requirements in a BYOD Environment

Privacy requirements are the set of security and privacy requirements that BYOD security policies should achieve to provide sufficient security. Every decision made by BYOD users within an organisation is influenced by security and privacy, including permitting email on a personal device. However, this could expose the organisation to numerous risks and data loss. Employees are also concerned about how much personal data employers can access and use to control their devices, although enterprises are authorised to protect company data. Therefore, privacy requirements are insufficient for the organization's and BYOD users' security. The security access control in BYOD should consider the privacy and security requirements of both the organisation and BYOD users. Several security models have been proposed to overcome these challenges and concerns, including the CIA triad, which refers to confidentiality, integrity, and availability and is designed to guide information security policies that include confidentiality, integrity, and availability within an organisation. Also, the IAS-Octave has confidentiality, integrity, availability, accountability, auditability, authenticity, trustworthiness, non-repudiation, and privacy preservation.

The conventional CIA triad framework is inadequate for addressing emerging threats in shared environments such as BYOD, as evidenced by Mosenia and Ioannis's research [35]. The introduction of BYOD exposes organizations to various security risks, including email phishing attacks, embedded viruses in applications, and denial-of-service incidents. Consequently, the CIA triad framework fails to meet evolving security and privacy requirements, necessitating an upgrade to the more comprehensive IAS-Octave standard. Yahuza et al. [22] propose a combined approach that integrates the privacy requirements of the CIA triad model and the IAS-Octave model, as depicted in Figure 3. This upgraded model incorporates the IAS-Octave security requirements and introduces additional attack detection requirements. Muktar et al. [22] further suggest enhancing security and privacy requirements in edge computing, comprising eight essential privacy requirements, including the CIA and IAS-Octave standards, attack detection, and reliability. Adopting this same model while excluding the reliability requirement is advisable for BYOD security, as previous studies have not emphasized its significance, prioritizing other security and privacy requirements [1]. In the forthcoming sections, we will investigate each component of this model to improve privacy requirements and investigate their relevance to BYOD security.

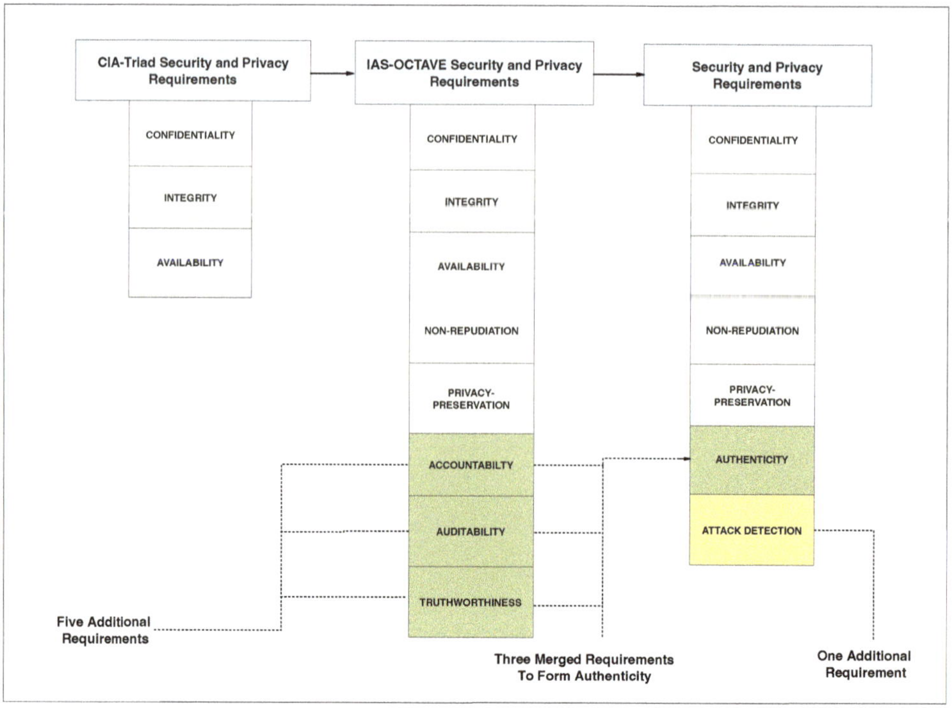

Figure 3. Development of BYOD security and privacy requirements [22].

4.1. CIA Triad Criteria Security and Privacy Requirements

The CIA triad, which encompasses the principles of confidentiality, integrity, and availability, serves as a fundamental model for the development of security systems [16]. These three categories, confidentiality, integrity, and availability, have traditionally formed a classification system for security and privacy requirements [16]. For several reasons, it is essential to include these requirements within the privacy guidelines for BYOD. Firstly, confidentiality plays a critical role in ensuring security and privacy [36]. Its primary objective is to prevent unauthorized access to sensitive company information. Cybersecurity issues may arise because BYOD devices and enterprises are connected to the internet. For instance, the man-in-the-middle (MITM) attack represents one of the threats targeting the security of BYOD users. This attack aims to intercept and steal information exchanged between two parties. Secondly, the integrity requirement ensures that data is only accessible to authorized BYOD devices and remains unmodified. Preserving data integrity becomes challenging in BYOD scenarios, as the organization loses visibility and control when a BYOD device operates outside its network, leading to potential data corruption or loss [37]. Therefore, integrity is a fundamental security requirement for organizations employing BYOD devices [38]. Lastly, availability emphasizes the continuous accessibility of devices, data, and resources, ensuring that BYOD devices always have secure access to services. In cases of an unexpected surge in data traffic volume, client-server communications may suffer from data loss [39].

4.2. IAS-Octave Security Criteria and Privacy Requirements

The IAS-Octave classification extends the CIA triad framework by introducing additional security requirements: accountability, auditability, trustworthiness, non-repudiation, and privacy preservation [22]. Accountability and auditability contribute to establishing trustworthiness, which is closely associated with the authenticity requirement. Alternatively, it has been proposed that these three requirements can be combined to form the concept of authenticity [22]. Therefore, authenticity is essential to the standard set of privacy and security requirements for BYOD. It ensures accurate monitoring and verification of BYOD users' identities, establishing trust between BYOD devices, organizations, and their resources. Additionally, including privacy-preservation and non-repudiation requirements is crucial as part of the BYOD security framework. Privacy preservation ensures the security and monitoring of all information, end users, networks, and resources involved in the organization's BYOD implementation. On the other hand, non-repudiation guarantees that a BYOD device cannot later deny its signature's validity or an event that occurred within the organization. This requirement aids in tracing the origin of any BYOD device that poses security issues [40]. Thus, the privacy requirements for BYOD are derived from the CIA triad model and supplemented with three additional standards from the IAS-Octave framework: privacy preservation, non-repudiation, and authenticity.

4.3. Addition of Attack Detection Requirements

In the context of BYOD devices, "attack detection" refers to identifying and mitigating potential threats. When employees bring their devices to the office and connect them to the wireless network, there is a risk of theft and other malicious activities. Security threats such as phishing, malware, and potentially spreading infected devices through BYOD can compromise a company's security [41]. Hence, the security and privacy requirements for BYOD should include provisions for attack detection. Attack detection requirements are necessary to meet the required level of security by identifying and preventing attacks before they occur [42]. Also, Bello et al. [1] highlight the importance of incorporating attack detection requirements into BYOD security policies, particularly concerning risk access control policies discussed in a previous section. Attack detection requirements are intrinsically linked to risk-based access control policies, as illustrated in Figure 3. The attack detection criteria should be integrated into the existing privacy requirements and are crucial as they address the negative aspects of implementing the BYOD concept in organizations. This helps mitigate significant challenges such as attacks, risks, unauthorized access, data leakage, and user privacy concerns. By incorporating attack detection along with privacy and security requirements, organizations can effectively tackle these challenges and enhance the security of their BYOD environments. This study examines access control policies and techniques in the context of BYOD based on the privacy requirements that have been met. The aim is to determine the privacy requirements that have been thoroughly examined and to encourage researchers to enhance and propose techniques that align with privacy requirements that may need more attention and require further investigation. Table 3 comprehensively describes the privacy criteria that will be evaluated during the review process. Furthermore, Figure 3 illustrates the advancements in privacy and security requirements for BYOD security.

Table 3. Security and privacy requirements in the BYOD environment [22].

Requirement	Description
Confidentiality	Ensures that unauthorized individuals are prevented from accessing shared data within BYOD-enabled organizations.
Integrity	Ensures that data is delivered exclusively to authorized BYOD devices without any unauthorized modifications.
Authenticity	Ensures accurate monitoring and verification of BYOD users' identities, fostering trust between the BYOD devices, organizations, and their resources.
Nonrepudiation	Ensures accurate monitoring and verification of BYOD users' identities, fostering trust between the BYOD devices, organizations, and their resources.
Privacy-Preservation	Ensures the secure and monitored storage of all confidential information related to BYOD devices, including end users.
Attack Detection	ensures the timely identification and effective mitigation of any security breaches or threats targeting BYOD devices.

5. SLR Methodology

This study has conducted a systemic literature review (SLR) following Kitchenham's recommendations [43]. The literature analysis process consisted of five steps before the review:

Step 1: Define research questions and objectives to give the study a broad scope.
Step 2: Determine a search strategy for published research in available digital libraries.
Step 3: Employ a screening process that uses inclusion and exclusion criteria to decide which studies to include.
Step 4: Perform classification and data extraction, aided by keywording.
Step 5: Extract and map data.

In addition, the preparation, collection, retrieval and implementation processes were conducted to identify any study discrepancies in the previous literature and thereby contribute to the subsequent study. The primary purpose of this search was to find publications that investigated BYOD security policy techniques based on security and privacy requirements. Figure 4 illustrates the measures taken in the methodology of the analysis work.

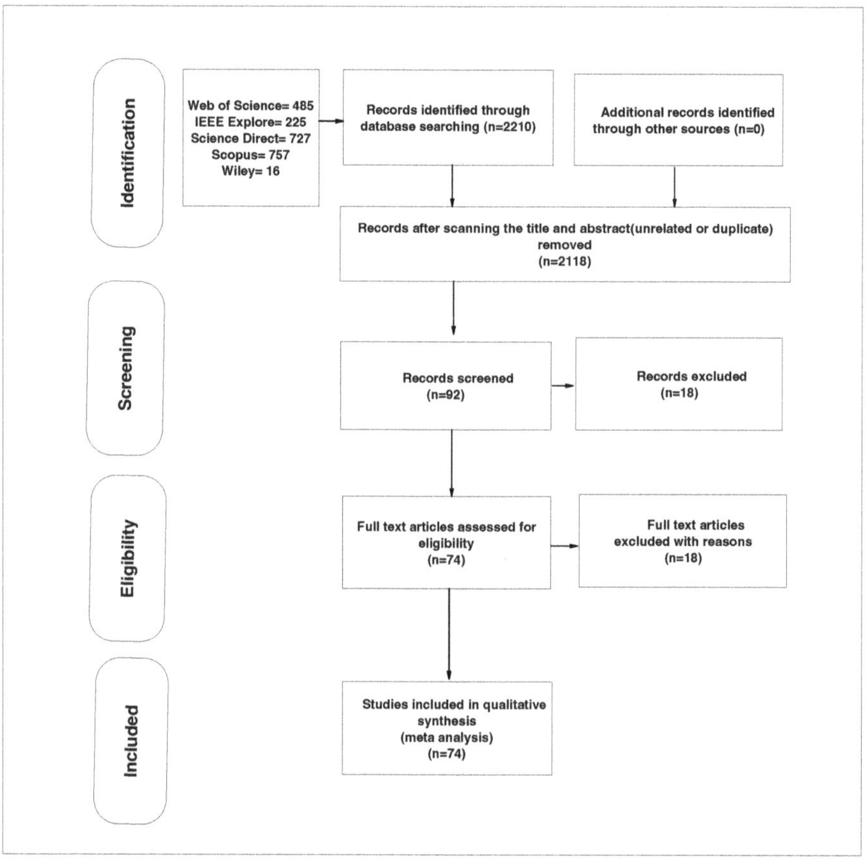

Figure 4. Flowchart of the systematic review process.

5.1. Research Questions and Objectives

This study aims to highlight the results of existing primary studies published on security policy techniques and privacy in the BYOD environment to identify current trends and open issues in the domain. Table 4 shows our research questions and the objectives of each research question.

Table 4. Research questions and objectives.

Research Question	Research Objective
RQ1: For BYOD environments, what is the classification of privacy and security criteria?	To define the security and privacy requirements to ensure the highest level of security for the data of enterprises and BYOD users.
RQ2: What policy techniques are employed to ensure security and privacy requirements have been identified?	To analyse existing solutions to security policy techniques in terms of the three layers of BYOD security policy that are used to achieve specific security and privacy requirements.
RQ3: What are the trends in technological methods used by the identified techniques?	To identify trends in technological approaches used by the indicated methodology.
RQ4: What evaluation procedures should use to evaluate the performance measurement of technologies?	To determine the appropriate evaluation metrics used in evaluating the performance of technologies.
RQ5: What future research opportunities and gaps exist in the security policy and privacy field in BYOD for researchers?	To identify the currently open issues for privacy and policy issues in BYOD.

5.2. Data Search Strategy

All defense policy research, including analysis and technical studies, was thoroughly searched in the BYOD environment. Five major electronic databases, including the Web of Science, IEEE Explore, Wiley, Science Direct and Scopus, were used in the research. The quest was limited to technology, computer technology, informatics, and engineering. In addition, the boundaries of the analysis were limited by the subject areas. The first search was conducted by screening conference papers and journals published between 2015 and June 2022. The search was restricted to the five online electronic databases. To build a search query, specific keywords with similar meanings were used. The queries were then followed up in three phases: title, abstract scanning and reading of the full text. The three phases of an inquiry are detailed below:

- "Bring your own device" OR "BYOD" AND "Security Policy" OR "Policy" AND "Security and Privacy"
- "Secure*" AND "Policy Techniques " AND "Bring your own device" OR "BYOD"
- "Access control" AND "BYOD" OR "Bring your own device" AND "BYOD" OR "Bring your own device"

5.3. Criteria for Study Selection

The inclusion criteria were established to ensure well-defined boundaries of the review topics and to facilitate article selection. The search criteria included collecting applicable data from journal articles and conference papers published in public databases from 2015 to June 2022 (see Table 5). There were a total of 2208 posts found initially. These were from Science (485), IEEE Explore (225), Science Direct (727), Scopus (757) and Wiley (16). Next, the title and abstract of the papers were scanned. Following the scan, 2118 papers were found to be outside the scope of the review and were excluded. The remaining 90 papers were subsequently selected through inclusion and exclusion procedures. If an article met the criteria of inclusion set out in Table 5, it was eligible for inclusion. Otherwise, it was excluded. The remaining 92 publications were scrutinised after the full-text review. Several articles were subsequently removed, leaving 74 articles for inclusion. The reason for including these articles was that they addressed the study's objectives, which were access control and policy techniques in terms of three layers and privacy requirements, and they met the established inclusion criteria.

Table 5. Inclusion and exclusion criteria.

Inclusion	Exclusion
IC1: Papers related to research questions.	**EC1:** The papers do not address security policies based on BYOD.
IC2: Papers from journals or conferences.	**EC2:** Techniques and models used in the security policy are not addressed.
IC3: Papers are written in English only.	**EC3:** Duplicate papers.
IC4: Papers published between 2015 and 2022.	**EC4:** Full text is unavailable.
IC5: The full text is available.	**EC5:** Non-English.
IC6: Articles that present techniques and models of security policy in a BYOD environment.	**EC6:** White papers, chapters in books and magazines.

5.4. Data Extraction

The 74 papers that met the inclusion criteria were reviewed to summarise relevant data that addressed the research questions. As a result, the following are documented: the authors, the publication year, the type of article, the policy technique under a specific category of security and privacy requirement, the category of technological approaches used and the performance metrics used in evaluating the proposed technique's performance. Additionally, the flaws of each recognised technique provided research opportunities.

6. Data Analysis

This section examines all the research that fulfilled the inclusion requirements. Section 6.1 describes general data analysis. In addition, Section 6.2 conducts an analysis to address the research questions. According to RQ1,the classification of privacy and security criteria is examined and discussed in this Section 4. This study found that seven security and privacy requirements identified to meet the needs of the BYOD organization's security and privacy users include confidentiality, integrity, availability, non-repudiation, authentication, privacy preservation, and attack detection. The RQ2 is related to analyzing access control techniques based on security and privacy requirements. These were then divided into three categories. The first are techniques that meet one of the security and privacy requirements; the second are techniques that meet more than one requirement; and the third are techniques that do not meet any requirement. In addition to RQ3, which identified policy techniques used to ensure security and privacy requirements that achieved RQ4, evaluate performance measurement, and RQ5, related to performance evaluation metrics, are discussed in detail.

6.1. Analysis of General Data

The review included 74 papers from various journals and conference proceedings indexed in five electronic databases. The percentage of papers published from 2015 to 2022 is shown in Figure 5. According to the review results, research on security policy techniques in the BYOD environment only gained popularity in 2017. Fifteen percent and 17 percent of all publications in 2017 and 2018, respectively, were found in journals. Journal articles accounted for 17 percent of all publications identified from review efforts in 2020. From 2021 to December 2022, this rose to 20 percent, suggesting more researchers had become interested in the field. Figure 6 illustrates the distribution of publications throughout the various databases covering a wide range of subjects. WOS provided the majority of the articles, followed by the IEEE database. Scopus was the second most popular database, followed by Science Direct and Wiley, with the lowest percentages.

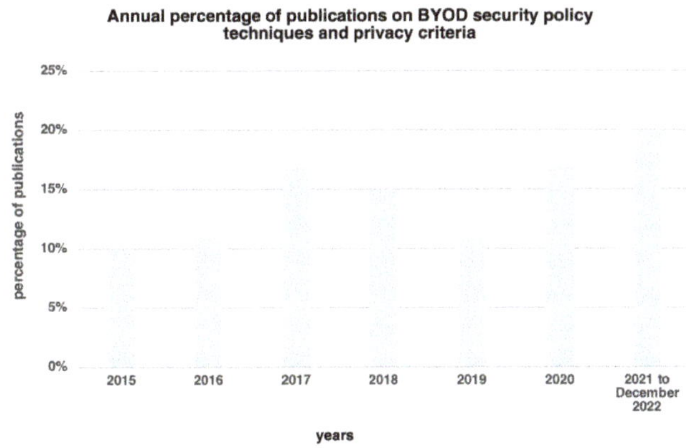

Figure 5. Percentage of publications related to BYOD security policy techniques per year.

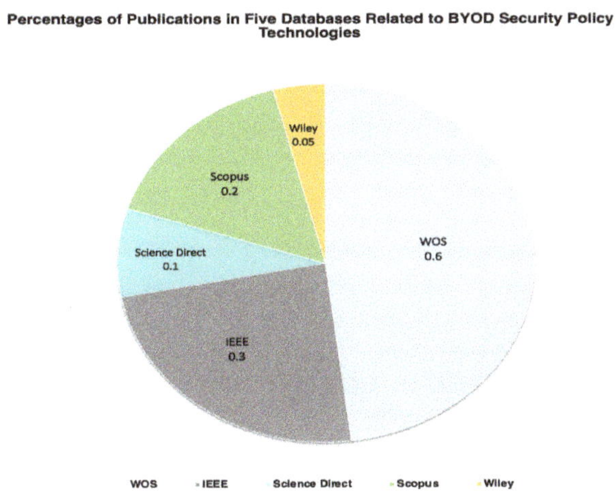

Figure 6. Percentages of publications in five databases relevant to security policy in BYOD.

6.2. Analysis of BYOD Security Policy Techniques Based on Privacy and Security Requirements

The purpose of this section is to present an analysis of the results. Firstly, we analysed existing policy techniques based on privacy and security requirements to ensure specific criteria were met. Then, we examined the trends in technical approaches used by the methodologies identified. According to the prior studies of policy techniques in BYOD security, the analysis was divided into three categories based on the privacy requirements they fulfilled. The first group includes technologies that meet one of the privacy requirements, such as techniques that achieve authenticity, confidentiality, and attack detection, as discussed in Section 6.2.1. In the second group, some techniques address more than one requirement, such as confidentiality and authentication, confidentiality and attack detection, and others that combine the requirements for authentication, attack detection, and confidentiality as explained in Section 6.2.2. The third group relates to security policy methods that did not address any of the suggested privacy requirements as stated in Section 6.2.3. Finally, our analysis identifies the performance measures and metrics that evaluate the effectiveness of the techniques discussed in Section 6.2.4. In addition, the tables summarize techniques within specific categories of privacy and security requirements, providing a brief overview of the methodology, the technology used, the performance assessment analysis, the main advantages and the limitations. Table 6 summarises the techniques that address the authenticity requirement. Moreover, the techniques that address confidentiality requirements based on cryptography are summarised in Table 7. Table 8 summarises the techniques that address confidentiality requirements based on isolation. The techniques that address the attack detection requirement are investigated in Tables 9 and 10. Table 11 focuses on approaches that address more than one criterion. Table 12 summarises the technologies that do not meet the privacy as mentioned above and security standards. Finally, Tables 13 and 14 summarises the performance measures and evaluation metrics used to evaluate the effectiveness of the techniques mentioned.

Table 6. Summary of authentication techniques.

Ref	Technology Employed	Performance Analysis	Advantage	Limitation
[44]	MDM-based model.	Prototype implementation.	It can identify whether the BYOD device is disabled or enabled.	Only a limited number of devices and operating systems.
[45]	Pattern lock method and four-digit PIN(CirclePIN).	Experiment analysis and real dataset.	Protect against side-channel, shoulder-surfing and single-recording threats.	Less secure technologies.
[46]	Based on WPA2-Enterprise.	Experimental analyses, case study.	It eliminates shared password risks.	Increase processing power.
[47]	Set policies based on IEEE 802.1X/Certificated.	Case study of a Greek school.	Introduces security issues and their resolution.	Secure but vulnerable if authentication policies are simple.
[48]	Co-proximity authentication protocol(fingerprint sensors and biometric).	Prototype implementation and behavioural, biometric dataset.	Context-aware authentication.	Complex and time-intensive.
[49]	EZ-Net system.	Experimental analysis, Case study on campus.	Low-cost, high-performance.	Each user's monthly authentication time is limited.
[50]	Lightweight network access control (NAC) module.	Prototype implementation, OpenWrt, NAC module.	Improves administrator management and OpenWrt is a free wireless router.	Unsuitable for all systems.
[51]	Pseudo-code-based two-factor authentication.	Mathematical analysis.	Simple to implement and inexpensive.	It is difficult to do while utilising a mobile phone rather than a laptop because the keyboard is different.
[52]	NFC technology.	Simulation (computer simulation experimental).	More secure, fast and convenient authentication.	Limited number of nodes.
[53]	AppShield scheme, certificate-based authentication.	Prototype implementation, synthetic dataset (1000 data access operations).	Most secure BYOD infrastructure control.	Complicated and time-consuming.
[54]	Fine-grained security policies (set policy as an individual, group for user and device).	Prototype implementation.	Very low cost.	Network address interpretation is weak.

Table 7. Summary of cryptographic techniques.

Ref	Technology Employed	Performance Analysis	Advantage	Limitation
[55]	Cryptography-based algorithm (self-encryption).	Prototype implementation, (Encryption and decryption for Several files).	Effective in data storage units in enterprises that use BYOD.	Limitation in the execution time, compression.
[56]	RSA-based algorithm, property-based token attestation (PTA).	Mathematical analyses, a scyther tool for verification.	Secure enterprise network access.	Cloudlet-based BYOD models require modification of this PTA protocol.
[57]	MAC-based algorithm (symmetric key cryptographic), file-grained data.	Prototype implementation.	Sets policy rules in the endpoint and achieves confidentiality.	Makes use of the shared key.
[58]	ABE scheme-based algorithm.	Algorithmic proof.	Confidentiality.	Time increase due to sensor data collection and processing to determine attributes.
[59]	Based on the algorithm (RSA-Tokens).	Prototype implementation and Private cloud.	It reduces authentication time and information exchange from 3000 ms to 2000 ms using TLS.	It transfers data slowly.
[60]	Based on symmetric algorithm.	Prototype implementation.	BYODENCE is low-cost and delivers high accuracy and speed.	More complex.

Table 8. Summary of isolation techniques.

Ref	Technology Employed	Performance Analysis	Advantage	Limitation
[61]	VNF scheme-based virtualization technique.	Prototype implementation and testing in real word network.	Improved security, mobility, and response time with virtual and dynamic networks.	Synchronization of physical and virtual security.
[62]	Remote mobile screen (RMS) system based virtualization method.	Experimental analysis.	RMS ensures data confidentiality, policy compliance and space isolation.	It poses numerous security risks.
[63]	vNative-based virtualization method.	Prototype implementation, evaluation testbed configuration.	Data confidentiality and isolation.	Limited availability of BYOD/mobile devices.
[64]	Multi-level architecture for isolation.	Experimental analyses.	Provided privacy for android end-user.	The solution is only for android.
[65]	Brahma-based virtualization method.	Prototype implementation (KVM Module, Zenfone).	Privacy.	Less security.
[66]	EMM, SDN and NFV.	Prototype implementation.	SDN can enhance NFV performance with virtualization.	SSDN and NFV are independent.
[67]	MSS system-based virtualisation method.	Experimental analysis.	MSS secures sensitive data and security activities in a separate domain.	The MSS is only for one environment.
[68]	MSS system.	Experimental analyses.	Confidentiality.	Low cost.

Table 9. Summary of the attack detection techniques.

Ref	Technology Employed	Performance Analysis	Advantage	Limitation
[69]	Risk access scheme, NGFW technique based deep packet inspection firewall.	Experiment analyses and Case study (Tokyo University).	Reduces costs while increasing security.	It is weaker because package content should be verified instead of only filtered network traffic.
[70]	Forensics investigation method.	Experiment analyses, real dataset (Android log).	Works well to detect and analyse BYOD malware.	Only deal with log files without applications.
[71]	Forensic investigation model.	Simulation performed in three stages (computer, simulation, and experimental)	tracked BYOD user's traffic.	Forensic investigation requires an end-to-end ecosystem.
[72]	Clustering algorithms	Simulation, public malware dataset, branchy model.	High precision and lower traffic.	long training time.
[73]	Risk-detection-based ML random forest algorithm.	Experimental analysis, tested by comparing infected, unaffected android, public malware App dataset.	The method examines the whole issue to see if BYOD is legal.	Less efficient.
[74]	Risk access control model based on a risk estimation algorithm (fuzzy model)	Prototype implementation and Smart contracts dataset.	Making access decisions based on anomalies.	The method requires further development.
[75]	Andrologger tool.	Experimental analyses, Real dataset.	Automatically sending BYOD data and user actions to the company's server for analysis.	The method only works on android phones.
[76]	OPPRIM-based risk policy model	simulations (AnyLogic), Mathematical Analysis.	Adaptability, cost reduction.	Simulating risk will result in conservative behaviours from attackers.
[77]	Neural network-based algorithm to detect HTTP botnets.	Simulations (Anylogic), Drebin dataset, PRISM model checker and Mathematical analysis (Correlation)˙	High accuracy.	Complexity.
[78]	IDS model-based ML algorithm.	Prototype implementation, NSL KDD dataset.	Able to detect DoS, probing and torrent traffic.	The dataset quantity is large.
[79]	Behaviour-based abnormality detection model, Pattern Analysis (data mining).	Mathematical analysis.	Analyzing user behaviour to detect abnormal behaviour.	High false alarm rate.
[80]	System based SDN.	Simulation (NS2), attacks dataset (SYN attack, ICMP flood, Dos attack).	Self-adaptive network can defend against internal threats and reduce attack reaction time.	Should be extended and improved to detect sophisticated APTs.
[81]	Malware identification scheme based supervised classification algorithms/ML.	Prototype implemented-android applications dataset.	High accuracy.	The decision limit may be overtrained.
[82]	MUSES framework based fuzzy Logic, ML.	Prototype implementation.	Enhanced security and provided intrusion detection systems.	Difficult interpretation.

Table 10. Summary of the attack detection techniques.

Ref	Technology Employed	Performance Analysis	Advantage	Limitation
[83]	ANN algorithm (deep learning).	Experimental analysis, Real data from the Media Lab of MIT dataset (incoming, outgoing, type of call).	Precision and accuracy of over 99%.	ANN has difficulty converting data into numerical values.
[84]	Border patrol system.	Prototype implementation, 2000 apps of Google Play an android emulator.	Introduced new policies for corporate networks to manage organised activities.	Limited ability to block malicious apps.
[85]	Anomaly detection method based ML algorithms.	Prototype implementation, a proof-of-concept and Spark dataset.	Detecting intrusions and anomalies.	Need more verification.
[86]	CVSS system-based decision engine logic algorithm.	Simulation (Computer simulation and experimental).	Privacy for the infrastructure layer.	The assessment time is still a major concern.
[87]	IFT technique based on a clustering algorithm/ML.	Experimental analyses, Public data set containing packet IAT features.	Employed the packet inter-arrival time feature to detect abnormal behaviour.	Need to improve the algorithm within large datasets.
[88]	Real-time traffic classification system based on ML.	Experimental analyses, proof of concept and real dataset.	Employed a fine-grained, real-time traffic classification.	Required a change in the entire network infrastructure to implement SDN protocol.
[89]	Detection technique-based novel algorithm.	Prototype implementation.	Reduced network risk and increased BYOD infrastructure security.	The analysis should include traffic instead of just the login log.
[90]	Other methods (white box method).	Prototype implementation.	Benefits both end-users and organisational use.	Exploiting apps makes security testing difficult.
[41,91]	DFRM model based honeypot technology (digital forensic readiness).	Prototype implementation.	Efficient method.	limited in digital evidence.
[92]	DFR framework-based honeypot technology.	Prototype implementation.	DFR improves BYOD security and reduces issues.	Honeypots only gather data when attacked.
[93]	Roving proxy server framework.	Prototype implementation and SMS spam dataset.	Efficient with a small dataset.	Need more work on a massive dataset.
[94]	Classification framework.	Correlation analysis, Real malicious traffic logs.	Useful for the network administrator.	Only classifies cyber-attack patterns and not types.
[95]	Based on dynamic decision tree algorithm (ML).	Experimentation and IBM's internal network dataset.	Reducing enterprise risk.	Limited application installation.
[96]	Network scanning technique-based algorithm.	Proof-of-concept, Experimental analysis and Public dataset(Attack)	Preventing network eavesdropping and spoofing.	It should be implemented on the switch for better security.
[97]	SIDD system.	Prototype implementation, network attacks dataset (e.g., zero-day, worms, DoS).	Validate up to 99% and help to detect zero-day malware.	Should be applied to various attacks to ensure effectiveness.
[98]	ARANAC, a Novel access control	Experimental analysis, Case study using more than 80 Android devices at a university campus	ARANAC has monitoring, risk estimation, and attack detection modules.	Need to extend the model to estimate risk values for the remaining features.

Table 11. Summary of techniques based on multiple security and privacy requirements.

Privacy Requirements	Ref	Technology Employed	Performance Analysis	Advantage	Limitation
Confidentiality and Authentication.	[99]	AES,HMAC algorithm	Prototype implementation.	Authentication-based cryptography secures networks and users.	Certificate management challenges.
Confidentiality and Authentication.	[100]	Optimizing and encrypting Schema-Based Access Control.	Prototype implementation.	Protects data from key leakage.	Secure but complicated.
Confidentiality and Authentication.	[101]	Biohashing technique.	Mathematical analysis, Experimental.	Increases security and lowers error rates.	Data increases collision probability.
Attack Detection, Authentication, and Confidentiality.	[102]	SDN and 2FA (TOTP)-based virtualization method.	Prototype implementation, Proof of concept.	Security and assault reduction.	ARP spoofing is the only attack allowed with this method.
Confidentiality and attack detection.	[103]	PHE encryption, tracing, and revoking (tracing algorithm, public revocation algorithm).	Simulation, experimental analysis (an RBAC simulation system with ten classes and around ten users per class).	Tracking and key service interruptions provide safety.	It needs big master public keys.

Table 12. Summary of the techniques that did not consider any of the proposed requirements.

Ref	Technology Employed	Performance Analysis	Advantage	Limitation
[104]	Enforce access control policy architecture.	NA	Simple policy.	It is only a preliminary proposal that needs implementation
[105]	STRIDE-based BYOD threat model.	NA	Help understand how BYOD concerns affect corporate data	Only provides an initial model.
[106]	An investigation framework.	Safe-Logic experts evaluated based on particular criteria.	Excellent awareness of BYOD threats and solutions.	Only provides an initial model.
[107]	Proactive approach based GQM (goal, question, metric).	NA	Creating benchmarks for BYOD security protocols.	Only identified security metric.
[108]	OPPRIM model.	Mathematical analysis and quantitative model.	Integrating trust and risk management mechanisms with threat analysis	Only provides an initial model.
[109]	Poise policy for device, app.	Prototype implementation.	It is network analysis.	It should build and test a complete prototype.
[110]	Fine-grained policies (BYODroid model).	Case Study.	Apply policy on the infrastructure layer.	Implementation is needed to verify rules.
[111,112]	Security architecture.	Case Study.	Flexible and adaptable to various business fields.	Only provides a starting point.
[113]	Metamodeling techniques.	Case study (Interview domain experts at the Ottawa Hospital to validate).	Identify key BYOD risk assessment metamodel concepts.	Initial and limited architecture.
[114,115]	Other methods (middleboxes and alert systems).	Prototype implementation.	Accurate transparency, usability, and performance with HTTPS security.	It is only a preliminary proposal that needs implementation.
[116]	Plugin Framework Based fine-grained security policy.	Prototype implementation.	Allows correct security policy execution on any device connection network organisation.	No other systems (only Android).
[117]	BYOD security framework	Case study (Australian, questionnaire instrument)	Building new framework to improve attack and threat defence.	Discusses only the theoretical aspect.

Table 13. Summary of evaluation metrics.

Ref	Requirement under Consideration	Purpose for the Evaluation	Evaluation Metrics
[44,46,47,50,51,54]	Authenticity.	To evaluate the functionalities of the authentication policy.	Access Response (Deny or Permit).
[45]	Authenticity.	To evaluate the possibility of an authentication policy for preventing unauthorized access.	Authentication Time, Access Response and Error Rate.
[48]	Authenticity.	To evaluate the performance of biometric authentication.	Accuracy and Frequency Detection Latency.
[49]	Authenticity.	To evaluate the functionalities of the authentication policy.	Response time and Cost.
[52,53]	Authenticity.	To evaluate the performance of the authentication system.	Performance, Memory Consumption, CPU consumption, Time overhead and Code Size
[55]	Confidentiality.	To reduce the time when switching the key.	Execution Time.
[67,68]	Confidentiality.	To evaluate the performance measurement of the MSS system.	Time and Performance Measurement.
[69]	Attack detection.	To evaluate the possibility of reducing management costs for Wireless Network Monitoring.	Cost.
[72]	Attack detection.	To evaluate the possibility of reducing costs and improving detection accuracy and efficiency.	Detection Accuracy and Cost.
[73,81]	Attack detection.	To evaluate the accuracy of detecting a specific malware and abnormal application class.	Accuracy.
[74]	Attack detection.	To evaluate the risk value associated with each access request.	Risk Value, Scalability and Context Awareness
[76]	Attack detection.	To determine the impact of threat and opportunity estimations on the risk.	threat probability.
[77]	Attack detection.	To evaluate the rate of detection and accuracy.	Accuracy and rate of detection
[78]	Attack detection.	To evaluate the model's accuracy for detecting peer-to-peer traffic from torrent clients.	Accuracy and usability.
[79]	Attack detection.	To evaluate the ability to classify abnormal behavior.	Behaviour occurrence probability.
[80]	Attack detection.	To evaluate response time against internal threats.	Delay time.
[83]	Attack detection.	To evaluate the system's ability to detect unauthorized access.	Precision (PPV), recall (TPR) and accuracy (ACC).
[84]	Attack detection.	To evaluate the ability to block unwanted application functions selectively.	Overhead and latency
[85]	Attack detection.	To evaluate RAM and CPU consumption over time.	RAM usage over time. CPU usage over time.
[86]	Attack detection.	To evaluate time duration and vulnerability assessment against the cyber threat.	Time and score assessment.

Table 14. Summary of evaluation metrics.

Ref	Requirement under Consideration	Purpose for the Evaluation	Evaluation Metrics
[56]	Confidentiality.	To evaluate the performance of PTA protocol.	Computation cost and performance consumption.
[57]	Confidentiality.	To evaluate the response time for each subsequent request.	Execution time, complexity of the access policy and block size.
[58]	Confidentiality.	Protecting BYOD users and network privacy with practical effect on communication performance.	Time complexity, communication costs and communications complexity.
[59]	Confidentiality.	To ensure the proper handshaking time between clients and serves.	Cost and exchange key time.
[61,64]	Confidentiality.	To reduce the delay in access time and run time.	Access time, power consumption and switching cost.
[62]	Confidentiality.	To evaluate the scalability of the approach.	CPU usage, memory usage and boot time.
[65]	Confidentiality.	To evaluate memory resource usage and power consumption due to virtualization.	Run-time memory usage and power consumption.

6.2.1. Techniques that Satisfy Only One of the Privacy and Security Requirements

This section describes the findings of previous studies related to proposed policy technologies based on how well they meet privacy and security criteria. As stated in Section 3.2, the requirements for BYOD security were classified into seven categories.

These are confidentiality, integrity, availability, non-repudiation, authentication, privacy preservation, and attack detection. These requirements contributed to the creation of the review-related taxonomy. Figure 7 illustrates the taxonomy of security policies and privacy techniques based on the security requirements in BYOD environments, which include the seven categories described above. It can be seen that some techniques address one aspect of privacy and security criteria. For instance, confidentiality requirements have been met through cryptographic techniques and isolation techniques. Authentication algorithms can address authenticity requirements. Machine learning algorithms, deep learning algorithms, and SDN techniques have accomplished attack detection requirements. According to the review's findings, a technique that satisfies integrity, privacy, availability and non-repudiation requirements alone cannot be produced. As seen in Figure 7, we labelled this with the symbol NA, indicating non-availability. After analysing and categorising previous research results, we failed to identify any research that addressed just these categories. Other researchers have focused on the privacy and security needs of authenticity, confidentiality, and attack detection but failed to consider all the privacy and security requirements that BYOD demands. In the following section, we will examine security policy strategies that only address a single privacy and security criterion and summarize previous research.

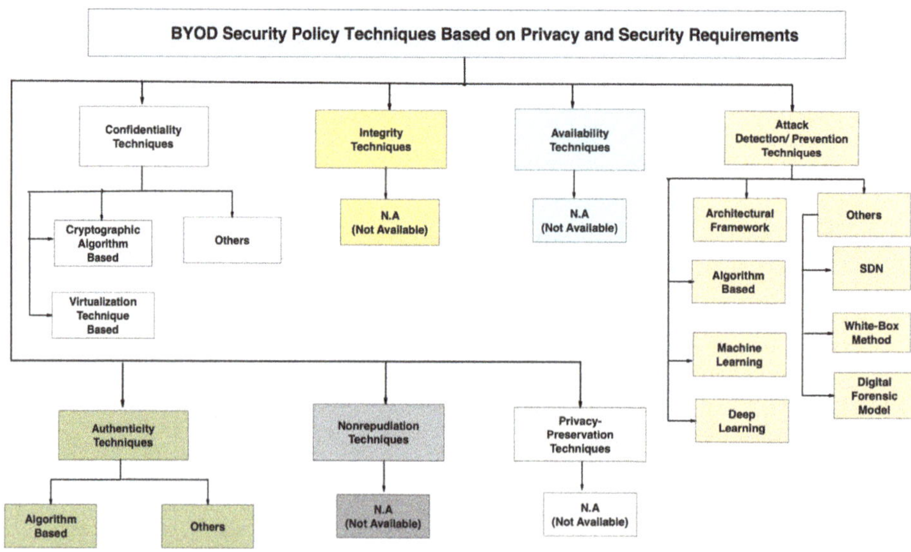

Figure 7. The taxonomy security policy and technological trends based on privacy requirements.

A. Techniques Based on Authenticity Requirements:

Authenticity requirements refer to verifying the identification of a user, process, or BYOD device and are frequently required before granting access to resources in an environment [118]. Many companies believe that the BYOD concept benefits both individuals and enterprises. However, the primary security issue is how to restrict BYOD device access to organisational information. As a result, when a BYOD client accesses a company network via their device, the device must be authenticated in some way. Lee et al. [44] proposed a secure authentication for BYOD devices that would be able to select disablers and enablers, determining if a BYOD device is either a disabler or an enabler. The proposed solution involved a mobile application acting as an MDM client, connecting to a server-side MDM server. However, the proposed method is only compatible with a limited number of devices and a specific operating system. Guerar et al. [45] presented CirclePIN, a novel authentication technique geared towards BYOD, particularly smart watches, that is both resilient to most common threats and was highly rated in actual tests with real users. It enhanced authentication between BYOD and the company network and gave protection against unauthorised access. However, it still performed at a similar level to other less secure alternatives. The review also showed that previous researchers had put much effort into achieving authenticity based on authentication techniques. In Table 6, a summary of the technology used, performance analysis, advantages and limitations is presented.

B. Cryptographic Techniques Based on Confidentiality Requirements:

Confidentiality requirement refers to the protection of organisational data from unauthorised parties' access to records. In other words, it ensures that no unauthorised users can access the data shared in BYOD-enabled enterprises. Therefore, in BYOD situations, encryption mechanisms should be seen as the first line of defence that must be in place to protect company data. Previous researchers have studied confidentiality requirements based on cryptographic techniques. For example, Vinh et al. [56] implemented property-based token attestation (PTA) to protect users and company networks in BYOD environments. These data are processed as binary and property-based attestation and the method is effective in data storage units in enterprises. Nevertheless, the PTA protocol must be changed to make a more secure BYOD model based on the Cloudlet idea. Rahardjo et al. [55] implemented the self encryption algorithm into various applications. However, the proposed method has limitations in terms of execution time and compression. Catuogno et al. [57] addressed the problem of preserving data access control over a mobile device's storage space while running different and distinct third-party applications. To that end, they proposed a general-purpose protected file system capable of providing fine-grained data protection at the operating system level. Facilities for trusted execution environments (TEE) are used for data encryption, critical protection, and policy compliance, providing safe access to corporate networks. Table 7 represents additional efforts made by researchers to accomplish confidentiality requirements using cryptographic techniques.

C. Isolation Techniques Based on Confidentiality Requirements:

There are numerous methods for achieving confidentiality, such as encryption, which we discussed previously, and isolation strategies. The difference is that cryptographic techniques based on confidentiality aim to protect the confidentiality of data stored in systems or transmitted over the organisation's network and BYOD. However, isolation techniques based on confidentiality are utilised to isolate corporate space, enforce security policies, and protect corporate data when using BYOD. In addition, security isolation methods allow users to separate personal data from corporate space. In these various ways, cryptographic and isolation techniques constitute a requirement for confidentiality.

In a BYOD scenario, personal and company data are stored on the same device. There are numerous benefits to integrating BYOD devices into the company infrastructure. However, this causes safety issues such as space isolation, data confidentiality, policy compliance and vulnerabilities since small and medium-sized enterprises cannot afford a suitable product solution that allows personal and professional data to coexist securely on employees' devices. Previous research has suggested isolation-based methods. Ocano et al. [62] proposed a remote mobile screen (RMS) based on the virtualization method to solve these problems. RMS allows for data confidentiality and space isolation. However, the complexity of managing physical roles is challenging. In addition, an essential real-world scenario in BYOD requires individuals to be isolated from the enterprise's privileged information access, according to Kim et al. [61]. They proposed an architecture called vNative that builds one foreground virtual machine (FVM), providing data confidentiality and isolation. However, the proposed solution is restricted to a few mobile intelligent devices. The proposed virtual network function (VNF) scheme is described in [63].

The suggested approach enables users to take advantage of the NFV server's more powerful corporate resources and increases service quality regarding security, mobility, and response time. It features a prototype technique [64] that takes advantage of virtualization characteristics common in today's mobile processors to fulfill this isolation demand.

In addition, this solution offers typical Android users security and privacy features. Table 8 summarises the isolation strategies discussed in previous studies.

D. Technique-Based Attack Detection Requirement:

According to previous research, there are numerous approaches for identifying attacks and preventing and monitoring risks. Kim and Lee [70] suggested a network-based risk identification tool based on the ML algorithm. The goal was to detect and evaluate malware on infected mobile devices under BYOD. However, this strategy cannot be employed for applications using log files. Aldini et al. [76] improved the method based on a machine learning algorithm to detect denial of service assaults, probing attacks, and torrent traffic in BYOD. The proposed approach is adaptable and flexible, lowering costs. Ammar et al. [80] proposed a supervised classification-based malware detection technique for Android BYOD devices. It provides a self-adaptive network capable of protecting critical services and defending against internal attacks and should be upgraded and expanded to detect advanced APTs. Ref. [72] proposed method based on clustering algorithms that may improve malware detection in BYOD was used with a simulation data set and a public dataset and achieved high precision. Ref. [69] employed a risk access scheme based on NGFW (next-generation firewall) technology, and a deep packet inspection firewall is used to protect a campus network. This method tracks and categorises risk packages. The experiment was analysed using a case study (Tokyo University). It reduced costs while increasing security. However, it is weaker because package content needs to be verified for network traffic filtering.

Furthermore, Ali et al. [71] proposed a risk-based access control model based on a dynamic risk estimation method, using features to analyse each request's security risk. The algorithm evaluates access based on user context, resource sensitivity, action severity, and risk history to represent an end-to-end ecosystem. Zungur et al. [84] presented a framework for detecting anomalous behaviour and unauthorised access to BYOD devices using artificial neural networks (ANN and data mining. A real dataset used in the evaluation had a precision of over 99 percent. One of the study's goals by [90] was to identify anomalous behaviour in BYOD devices connected to a corporate platform and then submit those devices to access control system management (ACSM) for platform access using an intelligent filtering technique. This method can set end-user access control policies and protect against malicious mobile apps, but security testing becomes difficult when apps are constantly attacked. The author in [79] proposed a behaviour-based abnormality detection method based on data mining that examines vulnerabilities and patterns in diverse information use contexts. Finally, ref. [83] provided a methodology using ANN techniques for detecting unusual behaviour and identifying unauthorised access to BYOD devices. It performed well on a real-world dataset, with a precision and accuracy of more than 99%. The experimental analysis assisted organisations in addressing three types of legitimate user behaviour. ANN, on the other hand, has difficulty converting data into numerical values. Previous studies show that many techniques have been developed for detecting risk in BYOD. Some techniques, such as fuzzy models and neural networks, are based on machine learning and deep learning. Tables 9 and 10 summarise previous research on BYOD attack detection techniques.

6.2.2. Techniques Based on More Than One Security and Privacy Requirement

Some techniques have satisfied more than one of the requirements, as seen in Figure 8, which shows the classification of policy technologies that consider multiple requirements. For example, some technologies fulfil multiple security and privacy requirements, according to previous studies. Firstly, some methods address both confidentiality and authentication. Many businesses have the necessary security access for BYOD to wireless networks. The network enterprise delivers WPA2 based on the 802.1X standard [46,47]. It is insufficient and should be strengthened. Pomak and Limpiyakom, proposed encryption-based authentication in [99]. They proposed employing near field communication multi-factor authentication (NFC) for secure authentication in combination with hybrid cryptosystems. Chen et al. [100] proposed an authentication scheme for cloud data in IoT/BYOD. Participants can employ ciphertext policy attribute-based encryption to construct fine-grained access control rules (CP-ABE). The approach uses hybrid cloud infrastructure to offload expensive CP-ABE activities while protecting privacy. This way, encryption and optimization methods protect critical data at the item level, preventing leakage. Secondly, some procedures fulfil the criteria for confidentiality and detection of attacks. Zhu et al. [103] employed a new model combining anomaly detection, tracing, and revocation procedures. Partially ordered hierarchical encryption (PHE) is a novel threshold public key-based cryptosystem that implements a partial-order key hierarchy similar to RBAC roles. Tracking and large service interruptions are critical security measures. Thirdly, some techniques combine authentication, attack detection and confidentiality requirements. For instance, the technique proposed by Geber et al. [102] can enable SDN-based authentication in a BYOD setting using OpenFlo-based infrastructure. It uses SDN characteristics to create virtual networks for monitoring dynamically selected BYOD users. By employing the portal and two-factor authentication, the user is provided with a safe and convenient means for enabling access to critical applications only and barring unauthorised requests, thereby limiting the services that a potentially attacking device can access. Changes to switching streaming rules may be deployed quickly, ensuring that users always have the most up-to-date access to the network with attack predictions. Three security and privacy requirements have been fulfilled: the authentication of BYOD devices used two-factor authentication, attack detection was achieved via continuous risk monitoring, and confidentiality was based on virtual networks. Table 11 depicts those techniques that have employed more than one security and privacy requirement in previous studies in the field.

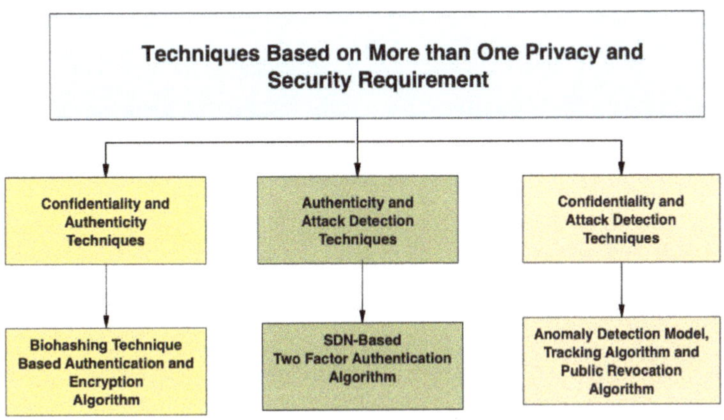

Figure 8. Techniques based on more than one privacy requirement.

6.2.3. Techniques That Did Not Fulfil Any of the Classified Security and Privacy Requirements

Figure 9 shows methods that did not identify any of the suggested requirements. Most studies in this section focus on enforcing simple policies or security architectures to improve overall security for organisations that support BYOD. These methods are preliminary models or policies, not technologies that address the identified requirements for privacy and security. For example, Selviandro et al. [104] suggested an architectural framework for enforced access control policies to reduce BYOD vulnerabilities. while Armando et al. introduced a reliable and policy-aware architecture for enforcing fine-grained security requirements on BYOD devices [105]. They implemented a user and device security policy based on existing NATO CIA guidelines (NCI Agency). Aldini et al. [108] presented an opportunity-enabled risk management (OPPRIM) approach that aimed to balance understanding primary threats with an appreciation of the increased opportunities that may emerge from BYOD. OPPRIM combines risk estimation methods with trust and threat metrics, and the OPPRIM policy and metric formulation paradigm is formalised in this study using logic. The model was checked using qualitative tools. The next Table 12 summarises the studies that address the security of BYOD without focusing on a specific security and privacy requirement.

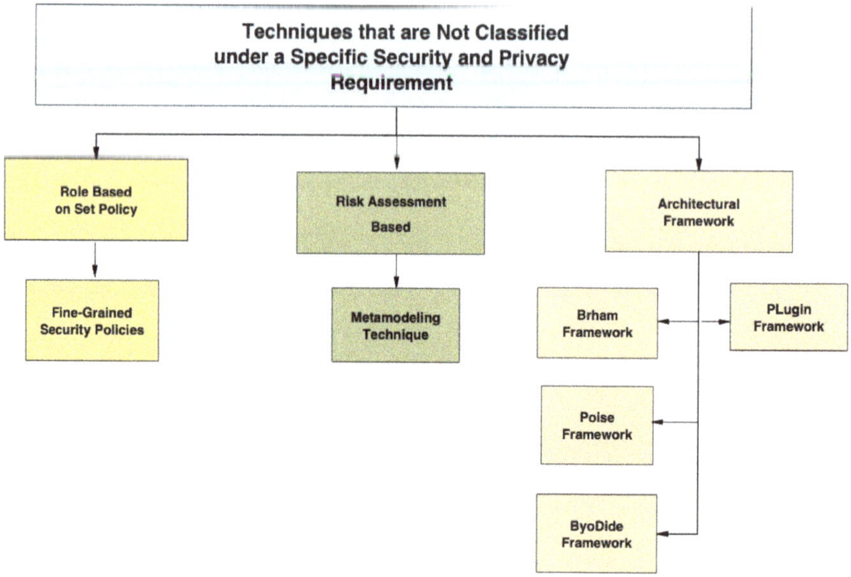

Figure 9. Techniques that did not satisfy classified privacy requirements.

6.2.4. Performance Evaluation and Metrics Employed by the Techniques

This section reviews the performance evaluation and the metrics used to evaluate the techniques. To categorize the performance evaluation analysis methods, the review distinguishes between techniques with tools (methods utilizing software or hardware in the assessment process) and methods without tools (methods relying on mathematical analysis). Figure 10 illustrates the methods using tools, including case studies, simulations (e.g., MATLAB, NS2, computer simulation experiment, and Anylogic), datasets (e.g., Public, Real, and Synthetic), formal security proofs (e.g., Scyther), and prototype implementations. Conversely, analysis without tools relies solely on mathematical analysis.

Previous research has identified specific evaluation metrics for each security and privacy requirement in technology. These metrics have been replicated in other studies to assess various security requirements. For instance, the authenticity requirement is evaluated based on response time, access response (deny or permit), and cost. The confidentiality requirement is measured using metrics such as execution time, communication costs, communication complexity, exchange key time, CPU usage, memory usage, and boot time. Attack detection requirements are assessed through metrics like accuracy, precision (PPV), recall (TPR), detection rate, RAM usage over time, CPU usage over time, overhead, and time cost. However, it is important to note that performance evaluation matrices may not adequately cover all the security and privacy requirements outlined in BYOD policies, highlighting the need for further exploration in this field.

Overall, performance evaluation and metrics provide invaluable insight into the efficacy, efficiency, and vulnerabilities of security access control systems. By employing these evaluation techniques, organizations can strengthen their access control measures, mitigate security risks, and protect sensitive resources and data. Figure 10 illustrates the classification of the performance evaluation analysis methods adopted by the techniques examined in this review. Furthermore, Tables 13 and 14 provide a summary of the metrics used by the techniques and their respective purposes.

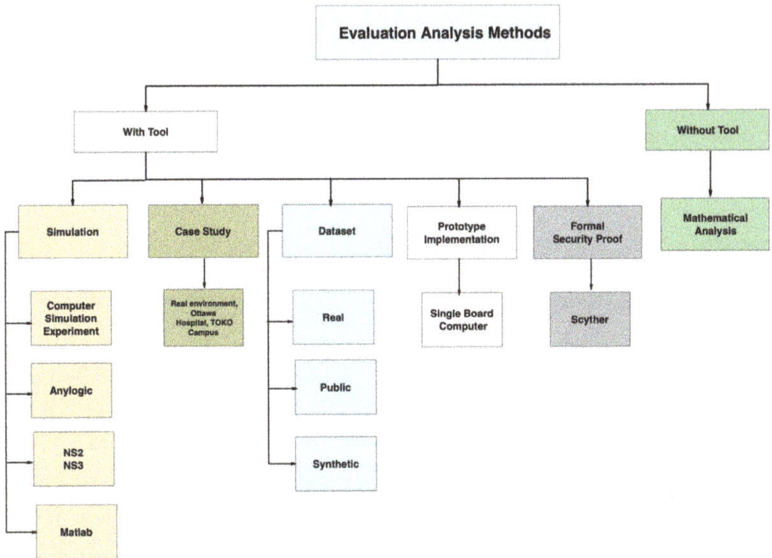

Figure 10. Classification of the performance analysis methods.

7. Discussion

The core security policy for BYOD encompasses various policies such as Network Access Control (NAC), Mobile Device Management (MDM), Mobile Application Management (MAM), Mobile Information Management (MIM), and Enterprise Mobility Management (EMM). However, the increasing number of BYOD attacks indicates that these security policies must be revised to develop BYOD security while failing to meet privacy requirements. Instead, a three-tiered security policy framework consisting of operational, tactical, and strategic layers should be implemented, working in harmony. This framework requires critical policies, including on-boarding access control policy, authentication access control policy, communication policy, application control policy, and risk control policy. These policies encompass BYOD devices and the organization's network, resources, communications, and applications while addressing seven essential security and privacy requirements: confidentiality, integrity, availability, non-repudiation, authentication, privacy preservation, and attack detection. Despite previous reviews providing foundational knowledge on the security and privacy challenges associated with BYOD policies, many still need to examine security policy techniques and privacy requirements thoroughly. Additionally, the exploration of technological methods employed to ensure compliance with these requirements still needs to be completed. This review adopts a systematic approach to comprehensively understand security policy techniques and privacy requirements within the context of a three-layered security policy architecture, which is considered ideal for optimizing BYOD security.

The study formulated and addressed five research questions (RQ1–RQ5) to accomplish its objectives. RQ1 focused on identifying the primary criteria that meet the security requirements of bring your own device (BYOD), namely confidentiality, integrity, availability, non-repudiation, authentication, privacy preservation, and attack detection. The results pertaining to RQ1 are discussed in Section 4. Similarly, the findings for RQ2 revealed the effectiveness of security policy techniques in meeting the security and privacy requirements. This analysis is presented in Section 6. The reviewed studies highlighted that among the various criteria, the most attention was given to attack detection, with 30 proposed techniques, followed by authenticity with 11 recommended techniques. Eight techniques focused on confidentiality using isolation techniques, while six techniques centered on confidentiality using cryptography. Additionally, 13 studies did not fall under specific requirements as they presented initial suggestions within the policy architecture framework. Furthermore, five proposed techniques addressed a combination of two or more different requirements, with only three techniques addressing both authenticity and confidentiality, and one technique covering authentication, attack detection, and confidentiality requirements. Additionally, one technique fulfilled the confidentiality and attack detection requirements. However, the review study noted a lack of techniques addressing the integrity, availability, non-repudiation, and privacy preservation requirements. This indicates the need for future research to focus on these aspects. Figure 11 illustrates the distribution of techniques based on the analysed requirements.

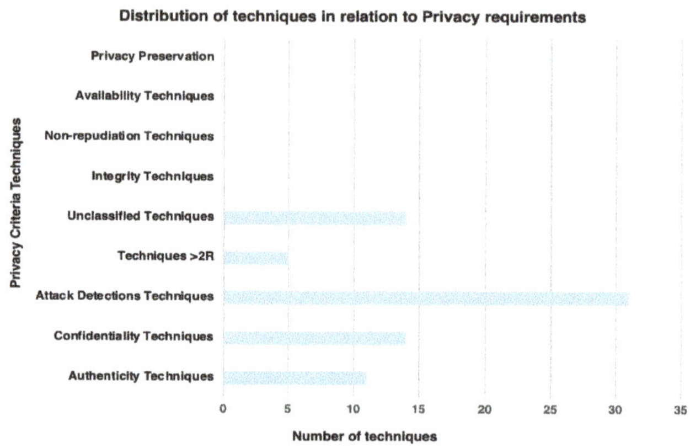

Figure 11. Distribution of techniques in relation to Privacy requirements.

Following the research review conducted under RQ3, the identified methodologies were further categorised based on their technological approaches. These approaches encompassed software-defined networking (SDN), machine learning (ML) algorithms, digital forensic models, and mobile security solution (MSS) systems. Among these, the most commonly employed technology was cryptography, specifically based on the RSA algorithm. Furthermore, the evaluation of technique effectiveness involved the examination of performance metrics, as explored under RQ4. Each category of requirements employed specific metrics to assess the performance of the techniques. Tables 13 and 14 highlight the purpose of evaluating techniques using these specific metrics. These tables provide valuable guidance for future researchers regarding the appropriate metrics for measuring technique performance. The review indicates that simulation experiments were the most frequently utilized approach for performance analysis in response to RQ4. Among the studies that employed datasets, a particular dataset emerged as the most commonly used. Prototype implementations predominantly employed embedded devices (single-board computers), while only one study utilized the Scyther tool for formal security analysis. Informal security analysis was the prevailing method utilized. Finally, based on the analysis and evaluation conducted, the study also identified significant challenges faced by the field. These challenges hold relevance for future researchers and will be expounded upon in the subsequent section.

8. Open Issues

This section aims to respond to RQ5 by discussing pertinent open research concerns regarding security policy techniques and privacy requirements. It intends to highlight opportunities for future research in this field. The following are the main open issues that merit further investigation:

8.1. Blockchain-Based Authentication Techniques

Academic researchers have proposed a variety of authentication approaches and techniques. These approaches can assist organisations in developing more secure access control between their networks, resources and BYOD. However, most techniques are designed to target either the user or the device individually. BYOD requires both the device and the user to be trusted with access to the organisation's resources. Knowledge-based authentication, possession-based authentication, biometric-based authentication and multi-factor authentication are all vulnerable to attack. As a result, more robust authentication measures are required. Most authentication systems require storing authentication information (password, ID, public key). The issue with consolidating storage is that it might become a single failure point. The initial BYOD solutions proposed by earlier academics are insufficient to secure the BYOD environment. Future research should focus on developing BYOD access control, which will address most of the issues in the BYOD environment and employ blockchain technology for security.

8.2. Secure Two-Way Communication

It is challenging to set up secure two-way communication in a BYOD environment. As a result, lightweight key exchange algorithms that suit access control in BYOD should be devised in the future to provide safe two-way communication between companies and BYOD devices.

8.3. Real-Time Authorization

BYOD encompasses a wide range of smart devices that contact the organisation's server and depend on providing real-time data access. Therefore, a delay in procuring an access decision within a BYOD environment may lead to unauthorised access to sensitive data, resulting in vulnerabilities to the system. In the scenario mentioned earlier, it can be inferred that access decision-making must remain close to the origin of access requests, which can provide the advantage of real-time data for the access control system, thus enabling access decisions with a minimal end-to-end delay. Moreover, authorised access must be ensured throughout the session, not only at the time of the request. In addition, real-time authorization means continuous access decisions throughout the session, not only when requesting access.

8.4. Context-Aware Role-Based Access Control

In security access control, many methods target communication between the organisation's platform and BYOD devices. The method of access control proposed [109] is not sensitive to contexts such as user status, location of the BYOD user, or time. According to the literature, these factors are context-aware elements. These factors are essential because they help identify BYOD devices inside the organisation. Therefore, access control based on context awareness should be given more attention to ensure the privacy of the BYOD environment. This is a major challenge and represents a promising field of study, but changes in context with access control must be adapted iteratively to avoid any risks.

8.5. Risk-Adaptive Policy Adjustments

Researchers in the field of BYOD security are interested in detecting and preventing insider attacks. Numerous methods based on behavioural models and anomaly detection approaches for addressing insider risks in BYOD environments have developed, such as [81,82,88,89]. While these methodologies do not focus on adapting access control policies to avoid insider assaults, they provide valuable insights into insider threat detection in BYOD situations. However, adaptive policy adjustment approaches pose a significant challenge and are an unexplored study area.

8.6. Evaluation Metrics Employed by the Techniques

According to the evaluation metrics, cost reduction concerning policy procedures remains a significant issue, particularly for small and growing businesses. Furthermore, it should be noted that most attack detection techniques [72,73,93] evaluate attack detection accuracy but not adaptive policy adjustment accuracy. As a result, two evaluation metrics, time and adaptive policy adjustment accuracy, were required to evaluate adaptive policy adjustment techniques. Because there has been little research in this area, other security and privacy requirements such as integrity, availability, privacy preservation and non-repudiation should be measured using evaluation metrics.

8.7. More Effort Is Required to Fulfill Specific Privacy and Security Criteria

As previously stated, several security and privacy requirements summarised above either need to explore particular technologies fully or include them. For instance, integrity, availability, non-repudiation and privacy-preservation requirements are not taken into account by any technique on a stand-alone basis but integrated with other criteria. Therefore, future research should focus on developing techniques that consider these considerations.

9. Conclusions and Future Work

BYOD security requires applying access control policies at all three security levels: operational, tactical and strategic, considering privacy requirements. This study is the first systematic review of security policies based on the privacy criteria that they meet. It also examines current research and focuses on the following aspects: (1) categorising privacy and security requirements within bring your own device (BYOD) policies; (2) analysing advanced security policy technologies for BYOD, considering three layers of security, and assessing their alignment with privacy requirements; (3) identifying technological trends in the field; (4) evaluating the effectiveness of techniques to enhance privacy and security through various measures; and (5) identifying potential areas for future BYOD security and privacy research.

In total, 74 articles were analysed based on SLR standard procedures to extract the key findings that underpin this study. Firstly, a taxonomy of security policy techniques and privacy requirements was introduced. According to the survey, BYOD security and privacy requirements can be divided into seven categories: confidentiality, integrity, availability, non-repudiation, authentication, privacy preservation, and attack detection. Secondly, the survey found that every criterion can be fulfilled with a specific technique, except for integrity, availability, non-repudiation, and privacy-preservation requirements, which were combined with other requirements. Third, having identified contemporary trends relevant to BYOD security, the techniques identified were classified according to their associated technological methods. Fourth, there was a discussion of the performance criteria used to evaluate the effectiveness of each technique. The review effort also showed the limitations of each of the methodologies. Lastly, for the benefit of academics interested in working on BYOD security and privacy, potential research areas were identified for future studies. For future studies, researchers are encouraged to focus on developing access control to address most issues in the BYOD environment. Utilising blockchain technology for security and emphasising context-aware access control to protect the privacy of the BYOD environment are recommended. The study also emphasises the need to consider privacy requirements such as integrity, availability, non-repudiation, and privacy preservation.

Author Contributions: Conceptualization, A.T.A.M.; methodology, A.T.A.M.; validation, A.T.A.M.; formal analysis, A.T.A.M.; investigation, A.T.A.M.; resources, A.T.A.M.; data curation, A.T.A.M.; writing—original draft preparation, A.T.A.M.; writing—review and editing, A.T.A.M.; visualization, A.T.A.M.; supervision, A.W.A.W. and M.Y.I.I. All authors have read and agreed to the published version of the manuscript.

Funding: This work was partly supported by the University of Malaya Impact Oriented Interdisciplinary Research Grant under Grant IIRG008 (A, B, C)-19IISS.

Institutional Review Board Statement: Not applicable.

Informed Consent Statement: Not applicable.

Acknowledgments: The authors of this paper appreciate the reviewers' observations, comments and suggestions for improving the manuscript's content.

Conflicts of Interest: The authors declare no conflicts of interest.

References

1. Bello, A.G.; Murray, D.; Armarego, J. A systematic approach to investigating how information security and privacy can be achieved in BYOD environments. *Inf. Comput. Secur.* **2017**, *25*, 475–492. [CrossRef]
2. Agrawal, A.; Pandey, A.K.; Baz, A.; Alhakami, H.; Alhakami, W.; Kumar, R.; Khan, R.A. Evaluating the security impact of healthcare Web applications through fuzzy based hybrid approach of multi-criteria decision-making analysis. *IEEE Access* **2020**, *8*, 135770–135783. [CrossRef]
3. Beckett, P. BYOD–popular and problematic. *Netw. Secur.* **2014**, *2014*, 7–9. [CrossRef]
4. Njuguna, D.; Kanyi, W. An evaluation of BYOD integration cybersecurity concerns: A case study. *Int. J. Recent Res. Math. Comput. Sci. Inf. Technol.* **2023**, *9*, 80–91.
5. Conteh, N.Y.; Schmick, P.J. Cybersecurity: risks, vulnerabilities and countermeasures to prevent social engineering attacks. *Int. J. Adv. Comput. Res.* **2016**, *6*, 31. [CrossRef]
6. Clarke, J.; Hidalgo, M.G.; Lioy, A.; Petkovic, M.; Vishik, C.; Ward, J. Consumerization of IT: Top risks and opportunities. In *ENISA Deliverables*; European Network and Information Security Agency (ENISA) Report; European Network and Information Security Agency (ENISA): Athens, Greece, 2012.
7. Utter, C.J.; Rea, A. The" Bring your own device" conundrum for organizations and investigators: An examination of the policy and legal concerns in light of investigatory challenges. *J. Digit. Forensics Secur. Law* **2015**, *10*, 4. [CrossRef]
8. Rhee, K.; Won, D.; Jang, S.W.; Chae, S.; Park, S. Threat modeling of a mobile device management system for secure smart work. *Electron. Commer. Res.* **2013**, *13*, 243–256. [CrossRef]
9. Morrow, B. BYOD security challenges: Control and protect your most sensitive data. *Netw. Secur.* **2012**, *2012*, 5–8. [CrossRef]
10. La Polla, M.; Martinelli, F.; Sgandurra, D. A survey on security for mobile devices. *IEEE Commun. Surv. Tutor.* **2012**, *15*, 446–471. [CrossRef]
11. Kok, J.; Kurz, B. Analysis of the botnet ecosystem. In Proceedings of the 10th Conference of Telecommunication, Media and Internet Techno-Economics (CTTE), Berlin, Germany, 16–18 May 2011; VDE: Frankfurt am Main, Germany, 2011; pp. 1–10.
12. Niehaves, B.; Köffer, S.; Ortbach, K. IT consumerization—A theory and practice review. In Proceedings of the 18th Americas Conference on Information Systems (AMCIS 2012), Seattle, WA, USA, 9–11 August 2012.
13. Garba, A.B.; Armarego, J.; Murray, D.; Kenworthy, W. Review of the information security and privacy challenges in Bring Your Own Device (BYOD) environments. *J. Inf. Priv. Secur.* **2015**, *11*, 38–54. [CrossRef]
14. Oktavia, T.; Yanti; Prabowo, H.; Meyliana. Security and privacy challenge in Bring Your Own Device environment: A systematic literature review. In Proceedings of the 2016 International Conference on Information Management and Technology (ICIMTech), Bandung, Indonesia, 16–18 November 2016; IEEE: Piscataway, NJ, USA, 2016; pp. 194–199.
15. Jamal, F.; Taufik, M.; Abdullah, A.A.; Hanapi, Z.M. A Systematic Review Of Bring Your Own Device (BYOD) Authentication Technique. *J. Phys. Conf. Ser.* **2020**, *1529*, 042071. [CrossRef]
16. Palanisamy, R.; Norman, A.A.; Kiah, M.L.M. Compliance with bring your own device security policies in organizations: A systematic literature review. *Comput. Secur.* **2020**, *98*, 101998. [CrossRef]
17. Wani, T.A.; Mendoza, A.; Gray, K. Hospital bring-your-own-device security challenges and solutions: Systematic review of gray literature. *JMIR mHealth uHealth* **2020**, *8*, e18175. [CrossRef] [PubMed]
18. Eke, C.I.; Norman, A.A.; Mulenga, M. Machine learning approach for detecting and combating bring your own device (BYOD) security threats and attacks: A systematic mapping review. *Artif. Intell. Rev.* **2023**, *56*, 8815–8858. [CrossRef]
19. AL-Azazi, O.A.A.S.; Norman, A.A.; Ghani, N.B.A. BrA Systematic Literature Review and Bibliometric Analysis (2017–2022) Your Own Device Information Security Policy Compliance Framework. In Proceedings of the 2022 International Conference on Cyber Resilience (ICCR), Dubai, United Arab Emirates, 6–7 October 2022; IEEE: Piscataway, NJ, USA, 2022; pp. 1–8.
20. Soubhagyalakshmi, P.; Reddy, K.S. An efficient security analysis of bring your own device. *IAES Int. J. Artif. Intell.* **2023**, *12*, 696. [CrossRef]
21. Ratchford, M.; El-Gayar, O.; Noteboom, C.; Wang, Y. BYOD security issues: A systematic literature review. *Inf. Secur. J. Glob. Perspect.* **2022**, *31*, 253–273. [CrossRef]
22. Yahuza, M.; Idris, M.Y.I.B.; Wahab, A.W.B.A.; Ho, A.T.; Khan, S.; Musa, S.N.B.; Taha, A.Z.B. Systematic review on security and privacy requirements in edge computing: State of the art and future research opportunities. *IEEE Access* **2020**, *8*, 76541–76567. [CrossRef]
23. kaspersky. Available online: https://www.kaspersky.com (accessed on 9 May 2023).
24. Batool, H.; Masood, A. Enterprise mobile device management requirements and features. In Proceedings of the IEEE INFOCOM 2020-IEEE Conference on Computer Communications Workshops (INFOCOM WKSHPS), Toronto, ON, Canada, 6–9 July 2020; IEEE: Piscataway, NJ, USA, 2020; pp. 109–114.
25. Da Veiga, A.; Astakhova, L.V.; Botha, A.; Herselman, M. Defining organisational information security culture—Perspectives from academia and industry. *Comput. Secur.* **2020**, *92*, 101713. [CrossRef]
26. Whitman, M.E.; Mattord, H.J. *Principles of Information Security*; Cengage Learning: Boston, MA, USA, 2021.

27. Mohsin, M.K.A.; Ab Hamid, Z. Bring Your Own Device (BYOD): Legal Protection of The Employee in Malaysia. *Malays. J. Soc. Sci. Humanit. (MJSSH)* **2022**, *7*, e001609.
28. Johnston, Z.A. Exploring Privacy Concern Effect on Organizational BYOD Policies and Security Measures Compliancy. Ph.D. Thesis, Capella University, Minneapolis, MN, USA, 2022.
29. Véliz, C. Privacy and digital ethics after the pandemic. *Nat. Electron.* **2021**, *4*, 10–11. [CrossRef]
30. White, B. The Influence of BYOD Security Risk on SME Information Security Effectiveness. Ph.D. Thesis, Capella University, Minneapolis, MN, USA, 2022.
31. Macaraeg, T.A., Jr. *Bring-Your-Own-Device (BYOD): Issues and Implementation in Local Colleges and Universities in the Philippines*; ResearchGate: Berlin, Germany, 2013.
32. Herrera, A.V.; Ron, M.; Rabadão, C. National cyber-security policies oriented to BYOD (bring your own device): Systematic review. In Proceedings of the 2017 12th Iberian Conference on Information Systems and Technologies (CISTI), Lisbon, Portugal, 21–24 June 2017; IEEE: Piscataway, NJ, USA, 2017; pp. 1–4.
33. Scarfo, A. New security perspectives around BYOD. In Proceedings of the 2012 Seventh International Conference on Broadband, Wireless Computing, Communication and Applications, Victoria, BC, Canada, 12–14 November 2012; IEEE: Piscataway, NJ, USA, 2012, pp. 446–451.
34. Alotaibi, B.; Almagwashi, H. A review of BYOD security challenges, solutions and policy best practices. In Proceedings of the 2018 1st International Conference on Computer Applications & Information Security (ICCAIS), Riyadh, Saudi Arabia, 4–6 April 2018; IEEE: Piscataway, NJ, USA, 2018; pp. 1–6.
35. Mosenia, A.; Jha, N.K. A comprehensive study of security of internet-of-things. *IEEE Trans. Emerg. Top. Comput.* **2016**, *5*, 586–602. [CrossRef]
36. Karimi, K.; Krit, S. Smart home-smartphone systems: Threats, security requirements and open research challenges. In Proceedings of the 2019 International Conference of Computer Science and Renewable Energies (ICCSRE), Agadir, Morocco, 22–24 July 2019; IEEE: Piscataway, NJ, USA, 2019; pp. 1–5.
37. Rodríguez, N.R.; Murazzo, M.A.; Chávez, S.B.; Valenzuela, F.A.; Martín, A.E.; Villafane, D.A. Key aspects for the development of applications for Mobile Cloud Computing. *J. Comput. Sci. Technol.* **2013**, *13*, 143–148.
38. Downer, K.; Bhattacharya, M. BYOD security: A new business challenge. In Proceedings of the 2015 IEEE International Conference on Smart City/SocialCom/SustainCom (SmartCity), Chengdu, China, 19–21 December 2015; IEEE: Piscataway, NJ, USA, 2015; pp. 1128–1133.
39. Doh, I.; Lim, J.; Chae, K. Secure authentication for structured smart grid system. In Proceedings of the 2015 9th International Conference on Innovative Mobile and Internet Services in Ubiquitous Computing, Santa Catarina, Brazil, 8–10 July 2015; IEEE: Piscataway, NJ, USA, 2015; pp. 200–204.
40. Almarhabi, K.; Jambi, K.; Eassa, F.; Batarfi, O. Survey on access control and management issues in cloud and BYOD environment. *Int. J. Comput. Sci. Mob. Comput.* **2017**, *6*, 44–54.
41. Ali, M.I.; Kaur, S. Next-generation digital forensic readiness BYOD framework. *Secur. Commun. Netw.* **2021**, *2021*, 6664426. [CrossRef]
42. Sushil, G.S.; Deshmuk, R.K.; Junnarkar, A.A. Security Challenges and Cyber Forensics For IoT Driven BYOD Systems. In Proceedings of the 2022 IEEE 7th International conference for Convergence in Technology (I2CT), Mumbai, India, 7–9 April 2022; IEEE: Piscataway, NJ, USA, 2022; pp. 1–7.
43. Kitchenham, B.; Brereton, P. A systematic review of systematic review process research in software engineering. *Inf. Softw. Technol.* **2013**, *55*, 2049–2075. [CrossRef]
44. Lee, J.E.; Park, S.H.; Yoon, H. Security policy based device management for supporting various mobile os. In Proceedings of the 2015 Second International Conference on Computing Technology and Information Management (ICCTIM), Johor, Malaysia, 21–23 April 2015; IEEE: Piscataway, NJ, USA, 2015; pp. 156–161.
45. Guerar, M.; Verderame, L.; Merlo, A.; Palmieri, F.; Migliardi, M.; Vallerini, L. CirclePIN: A novel authentication mechanism for smartwatches to prevent unauthorized access to IoT devices. *ACM Trans.-Cyber-Phys. Syst.* **2020**, *4*, 1–19. [CrossRef]
46. Yanson, K. Results of implementing WPA2-enterprise in educational institution. In Proceedings of the 2016 IEEE 10th International Conference on Application of Information and Communication Technologies (AICT), Baku, Azerbaijan, 12–14 October 2016; IEEE: Piscataway, NJ, USA, 2016; pp. 1–4.
47. Gkamas, V.; Paraskevas, M.; Varvarigos, E. Design of a secure BYOD policy for the Greek School Network: a Case Study. In Proceedings of the 2016 IEEE International Conference on Computational Science and Engineering (CSE) and IEEE International Conference on Embedded and Ubiquitous Computing (EUC) and 15th International Symposium on Distributed Computing and Applications for Business Engineering (DCABES), Paris, France, 24–26 August 2016; IEEE: Piscataway, NJ, USA, 2016; pp. 557–560.
48. Oluwatimi, O.; Damiani, M.L.; Bertino, E. A context-aware system to secure enterprise content: Incorporating reliability specifiers. *Comput. Secur.* **2018**, *77*, 162–178. [CrossRef]
49. Kao, Y.C.; Chang, Y.C.; Chang, R.S. EZ-Net BYOD service management in campus wireless networks. *J. Internet Technol.* **2017**, *18*, 907–917.
50. Heo, H.; Ryou, J. Design and implementation of lightweight network access control technique on wireless router. *Int. J. Serv. Technol. Manag.* **2017**, *23*, 101–116. [CrossRef]
51. Jaha, F.; Kartit, A. Pseudo code of two-factor authentication for BYOD. In Proceedings of the 2017 International Conference on Electrical and Information Technologies (ICEIT), Rabat, Morocco, 15–18 November 2017; IEEE: Piscataway, NJ, USA, 2017; pp. 1–7.

52. Cai, C.; Weng, J.; Liu, J. Mobile authentication system based on national regulation and NFC technology. In Proceedings of the 2016 IEEE First International Conference on Data Science in Cyberspace (DSC), Changsha, China, 13–16 June 2016; IEEE: Piscataway, NJ, USA, 2016; pp. 590–595.
53. Deng, R.; Weng, J.; Ren, K.; Yegneswaran, V. Security and privacy in communication networks. In Proceedings of the Security and Privacy in Communication Networks: 12th International Conference (SecureComm 2016), Guangzhou, China, 10–12 October 2016; Springer: Berlin/Heidelberg, Germany, 2017; Volume 198.
54. Seneviratne, B.; Senaratne, S. Integrated Corporate Network Service Architecture for Bring Your Own Device (BYOD) Policy. In Proceedings of the 2018 3rd International Conference on Information Technology Research (ICITR), Moratuwa, Sri Lanka, 5–7 December 2018; IEEE: Piscataway, NJ, USA, 2018; pp. 1–6.
55. Rahardjo, M.R.D.; Shidik, G.F. Design and implementation of self encryption method on file security. In Proceedings of the 2017 International Seminar on Application for Technology of Information and Communication (iSemantic), Semarang, Indonesia, 7–8 October 2017; IEEE: Piscataway, NJ, USA, 2017; pp. 181–186.
56. Vinh, T.L.; Cagnon, H.; Bouzefrane, S.; Banerjee, S. Property-based token attestation in mobile computing. *Concurr. Comput. Pract. Exp.* **2020**, *32*, e4350. [CrossRef]
57. Catuogno, L.; Galdi, C. A Fine-grained General Purpose Secure Storage Facility for Trusted Execution Environment. In Proceedings of the International Conference on Information Systems Security and Privacy (ICISSP 2019), Prague, Czech Republic, 23–25 February 2019; pp. 588–595.
58. Li, F.; Rahulamathavan, Y.; Conti, M.; Rajarajan, M. Robust access control framework for mobile cloud computing network. *Comput. Commun.* **2015**, *68*, 61–72. [CrossRef]
59. Gupta, S. Single Sign-On beyond Corporate Boundaries. In Proceedings of the 2018 8th International Conference on Intelligent Systems, Modelling and Simulation (ISMS), Kuala Lumpur, Malaysia, 8–10 May 2018; IEEE: Piscataway, NJ, USA, 2018; pp. 38–42.
60. Abisheka, P.C.; Azra, M.F.; Poobalan, A.; Wijekoon, J.; Yapa, K.; Murthaja, M. An Automated Solution For Securing Confidential Documents in a BYOD Environment. In Proceedings of the 2021 3rd International Conference on Advancements in Computing (ICAC), Colombo, Sri Lanka, 9–11 December 2021; IEEE: Piscataway, NJ, USA, 2021; pp. 61–66.
61. Kim, J.; Kim, T.Y.; Kim, D. Network based vByod scheme in NFV platform. In Proceedings of the 2017 International Conference on Information and Communication Technology Convergence (ICTC), Jeju, Republic of Korea, 18–20 October 2017; IEEE: Piscataway, NJ, USA, 2017; pp. 1143–1145.
62. Ocano, S.G.; Ramamurthy, B.; Wang, Y. Remote mobile screen (RMS): An approach for secure BYOD environments. In Proceedings of the 2015 International Conference on Computing, Networking and Communications (ICNC), Garden Grove, CA, USA, 16–19 February 2015; IEEE: Piscataway, NJ, USA, 2015; pp. 52–56.
63. Dong, Y.; Mao, J.; Guan, H.; Li, J.; Chen, Y. A virtualization solution for BYOD with dynamic platform context switching. *IEEE Micro* **2015**, *35*, 34–43. [CrossRef]
64. Averlant, G. Multi-level isolation for android applications. In Proceedings of the 2017 IEEE International Symposium on Software Reliability Engineering Workshops (ISSREW), Toulouse, France, 23–26 October 2017; IEEE: Piscataway, NJ, USA, 2017; pp. 128–131.
65. Chiueh, T.C.; Lin, H.; Chao, A.; Wu, T.G.; Wang, C.M.; Wu, Y.S. Smartphone virtualization. In Proceedings of the 2016 IEEE 22nd International Conference on Parallel and Distributed Systems (ICPADS), Wuhan, China, 13–16 December 2016; IEEE: Piscataway, NJ, USA, 2016; pp. 141–150.
66. Ketel, M. Enhancing BYOD security through SDN. In Proceedings of the SoutheastCon 2018, St. Petersburg, FL, USA, 19–22 April 2018; IEEE: Piscataway, NJ, USA, 2018; pp. 1–2.
67. Kim, G.; Jeon, Y.; Kim, J. Secure mobile device management based on domain separation. In Proceedings of the 2016 International Conference on Information and Communication Technology Convergence (ICTC), Jeju, Republic of Korea, 19–21 October 2016; IEEE: Piscataway, NJ, USA, 2016; pp. 918–920.
68. Kim, G.; Kim, J. Secure voice communication service based on security platform for mobile devices. In Proceedings of the 2017 International Conference on Information and Communication Technology Convergence (ICTC), Jeju, Republic of Korea, 18–20 October 2017; IEEE: Piscataway, NJ, USA, 2017; pp. 1190–1192.
69. Mishima, K.; Sakurada, T.; Hagiwara, Y.; Tsujisawa, T. Secure Campus Network System with Automatic Isolation of High Security Risk Device. In Proceedings of the 2018 ACM SIGUCCS Annual Conference, Orlando, FL, USA, 7–10 October 2018; pp. 107–110.
70. Kim, D.; Lee, S. Study of identifying and managing the potential evidence for effective Android forensics. *Forensic Sci. Int. Digit. Investig.* **2020**, *33*, 200897. [CrossRef]
71. Ali, M.I.; Kaur, S.; Khamparia, A.; Gupta, D.; Kumar, S.; Khanna, A.; Al-Turjman, F. Security challenges and cyber forensic ecosystem in IOT driven BYOD environment. *IEEE Access* **2020**, *8*, 172770–172782. [CrossRef]
72. Tan, X.; Li, H.; Wang, L.; Xu, Z. End-Edge Coordinated Inference for Real-Time BYOD Malware Detection using Deep Learning. In Proceedings of the 2020 IEEE Wireless Communications and Networking Conference (WCNC), Seoul, Republic of Korea, 25–28 May 2020; IEEE: Piscataway, NJ, USA, 2020; pp. 1–6.
73. Watkins, L.; Kalathummarath, A.L.; Robinson, W.H. Network-based detection of mobile malware exhibiting obfuscated or silent network behavior. In Proceedings of the 2018 15th IEEE Annual Consumer Communications & Networking Conference (CCNC), Las Vegas, NV, USA, 12–15 January 2018; IEEE: Piscataway, NJ, USA, 2018; pp. 1–4.

74. Atlam, H.F.; Alenezi, A.; Walters, R.J.; Wills, G.B.; Daniel, J. Developing an adaptive Risk-based access control model for the Internet of Things. In Proceedings of the 2017 IEEE International Conference on Internet of Things (iThings) and IEEE Green Computing and Communications (GreenCom) and IEEE Cyber, Physical and Social Computing (CPSCom) and IEEE Smart Data (SmartData), Exeter, UK, 21–23 June 2017; IEEE: Piscataway, NJ, USA, 2017; pp. 655–661.
75. Tiwari, P.K.; Velayutham, T. Androloggor: Collecting and correlating events to identify suspicious activities in android. In Proceedings of the 2017 8th International Conference on Computing, Communication and Networking Technologies (ICCCNT), Delhi, India, 3–5 July 2017; IEEE: Piscataway, NJ, USA, 2017; pp. 1–7.
76. Aldini, A.; Seigneur, J.M.; Lafuente, C.B.; Titi, X.; Guislain, J. Design and validation of a trust-based opportunity-enabled risk management system. *Inf. Comput. Secur.* **2017**. [CrossRef]
77. Eslahi, M.; Yousefi, M.; Naseri, M.V.; Yussof, Y.; Tahir, N.; Hashim, H. Mobile botnet detection model based on retrospective pattern recognition. *Int. J. Secur. Appl.* **2016**, *10*, 39–44. [CrossRef]
78. Joshi, P.; Jindal, C.; Chowkwale, M.; Shethia, R.; Shaikh, S.A.; Ved, D. Protego: A passive intrusion detection system for android smartphones. In Proceedings of the 2016 International Conference on Computing, Analytics and Security Trends (CAST), Pune, India, 19–21 December 2016; IEEE: Piscataway, NJ, USA, 2016; pp. 232–237.
79. Kim, T. A Study on the Detection of Abnormal Behavior and Vulnerability Analysis in BYOD. In Proceedings of the International Internet of Things Summit; Springer: Berlin/Heidelberg, Germany, 2015; pp. 162–167.
80. Ammar, M.; Rizk, M.; Abdel-Hamid, A.; Aboul-Seoud, A.K. A framework for security enhancement in SDN-based datacenters. In Proceedings of the 2016 8th IFIP international conference on new technologies, Mobility and security (NTMS), Larnaca, Cyprus, 21–23 November 2016; IEEE: Piscataway, NJ, USA, 2016; pp. 1–4.
81. Akhuseyinoglu, N.B.; Akhuseyinoglu, K. AntiWare: An automated Android malware detection tool based on machine learning approach and official market metadata. In Proceedings of the 2016 IEEE 7th Annual Ubiquitous Computing, Electronics & Mobile Communication Conference (UEMCON), New York, NY, USA, 20–22 October 2016; IEEE: Piscataway, NJ, USA, 2016; pp. 1–7.
82. De las Cuevas, P.; Mora, A.; Merelo, J.J.; Castillo, P.A.; Garcia-Sanchez, P.; Fernandez-Ares, A. Corporate security solutions for BYOD: A novel user-centric and self-adaptive system. *Comput. Commun.* **2015**, *68*, 83–95. [CrossRef]
83. Petrov, D.; Znati, T. Context-aware deep learning-driven framework for mitigation of security risks in BYOD-enabled environments. In Proceedings of the 2018 IEEE 4th International Conference on Collaboration and Internet Computing (CIC), Philadelphia, PA, USA, 18–20 October 2018; IEEE: Piscataway, NJ, USA, 2018; pp. 166–175.
84. Zungur, O.; Suarez-Tangil, G.; Stringhini, G.; Egele, M. Borderpatrol: Securing byod using fine-grained contextual information. In Proceedings of the 2019 49th Annual IEEE/IFIP International Conference on Dependable Systems and Networks (DSN), Portland, OR, USA, 24–27 June 2019; IEEE: Piscataway, NJ, USA, 2019; pp. 460–472.
85. Lima, A.; Rosa, L.; Cruz, T.; Simões, P. A Security Monitoring Framework for Mobile Devices. *Electronics* **2020**, *9*, 1197. [CrossRef]
86. Nikoloudakis, Y.; Pallis, E.; Mastorakis, G.; Mavromoustakis, C.X.; Skianis, C.; Markakis, E.K. Vulnerability assessment as a service for fog-centric ICT ecosystems: A healthcare use case. *Peer-to-Peer Netw. Appl.* **2019**, *12*, 1216–1224. [CrossRef]
87. Muhammad, M.A.; Ayesh, A.; Zadeh, P.B. Developing an intelligent filtering technique for bring your own device network access control. In Proceedings of the International Conference on Future Networks and Distributed Systems, Cambridge UK, 19–20 July 2017; pp. 1–8.
88. Uddin, M.; Nadeem, T. TrafficVision: A case for pushing software defined networks to wireless edges. In Proceedings of the 2016 IEEE 13th International Conference on Mobile Ad Hoc and Sensor Systems (MASS), Brasilia, Brazil, 10–13 October 2016; IEEE: Piscataway, NJ, USA, 2016; pp. 37–46.
89. Ali, M.I.; Kaur, S. BYOD Cyber Threat Detection and Protection Model. In Proceedings of the 2021 International Conference on Computing, Communication, and Intelligent Systems (ICCCIS), Greater Noida, India, 19–20 February 2021; IEEE: Piscataway, NJ, USA, 2021; pp. 211–218.
90. Alghamdi, A.M.; Almarhabi, K. A Proposed Framework for the Automated Authorization Testing of Mobile Applications. *Int. J. Comput. Sci. Netw. Secur.* **2021**, *21*, 217–221.
91. Kebande, V.R.; Karie, N.M.; Venter, H. A generic Digital Forensic Readiness model for BYOD using honeypot technology. In Proceedings of the 2016 IST-Africa Week Conference, Durban, South Africa, 11–13 May 2016; IEEE: Piscataway, NJ, USA, 2016; pp. 1–12.
92. Asante, A.; Amankona, V. Digital Forensic Readiness Framework Based on Honeypot Technology for BYOD. *J. Digit. Forensics Secur. Law* **2021**, *16*, 1–17.
93. Eshmawi, A.; Nair, S. The Roving Proxy Framewrok for SMS Spam and Phishing Detection. In Proceedings of the 2019 2nd International Conference on Computer Applications & Information Security (ICCAIS), Riyadh, Saudi Arabia, 1–3 May 2019; IEEE: Piscataway, NJ, USA, 2019; pp. 1–6.
94. Awan, M.S.; AlGhamdi, M.; AlMotiri, S.; Burnap, P.; Rana, O. A classification framework for distinct cyber-attacks based on occurrence patterns. In Proceedings of the 8th International Conference on Security of Information and Networks, Sochi, Russia 8–10 September 2015; pp. 165–168.
95. Stoecklin, M.P.; Singh, K.; Koved, L.; Hu, X.; Chari, S.N.; Rao, J.R.; Cheng, P.C.; Christodorescu, M.; Sailer, R.; Schales, D.L. Passive security intelligence to analyze the security risks of mobile/BYOD activities. *IBM J. Res. Dev.* **2016**, *60*, 9–1. [CrossRef]
96. Chen, Y.; Hu, H.c.; Cheng, G.z. Design and implementation of a novel enterprise network defense system bymaneuveringmulti-dimensional network properties. *Front. Inf. Technol. Electron. Eng.* **2019**, *20*, 238–252. [CrossRef]

97. Yang, C.; Hong-Chao, H.; Guo-Zhen, C. A software-defined intranet dynamic defense system. In Proceedings of the 2018 IEEE 18th International Conference on Communication Technology (ICCT), Chongqing, China, 8–11 October 2018; IEEE: Piscataway, NJ, USA, 2018; pp. 849–854.
98. Gómez-Hernández, J.A.; Camacho, J.; Holgado-Terriza, J.A.; García-Teodoro, P.; Maciá-Fernández, G. ARANAC: A Bring-Your-Own-Permissions Network Access Control Methodology for Android Devices. *IEEE Access* **2021**, *9*, 101321–101334. [CrossRef]
99. Pomak, W.; Limpiyakom, Y. Enterprise WiFi Hotspot Authentication with Hybrid Encryption on NFC-Enabled Smartphones. In Proceedings of the 2018 8th International Conference on Electronics Information and Emergency Communication (ICEIEC), Beijing, China, 15–17 June 2018; IEEE: Piscataway, NJ, USA, 2018, pp. 247–250.
100. Qi, S.; Lu, Y.; Wei, W.; Chen, X. Efficient data access control with fine-grained data protection in cloud-assisted IIoT. *IEEE Internet Things J.* **2020**, *8*, 2886–2899. [CrossRef]
101. Zheng, Y.; Cao, Y.; Chang, C.H. Facial biohashing based user-device physical unclonable function for bring your own device security. In Proceedings of the 2018 IEEE International Conference on Consumer Electronics (ICCE), Las Vegas, NV, USA, 12–14 January 2018; IEEE: Piscataway, NJ, USA, 2018; pp. 1–6.
102. Gebert, S.; Zinner, T.; Gray, N.; Durner, R.; Lorenz, C.; Lange, S. Demonstrating a personalized secure-by-default bring your own device solution based on software defined networking. In Proceedings of the 2016 28th International Teletraffic Congress (ITC 28), Würzburg, Germany, 12–16 September 2016; IEEE: Piscataway, NJ, USA, 2016; Volume 1, pp. 197–200.
103. Zhu, Y.; Gan, G.; Guo, R.; Huang, D. PHE: An efficient traitor tracing and revocation for encrypted file syncing-and-sharing in cloud. *IEEE Trans. Cloud Comput.* **2016**, *6*, 1110–1124. [CrossRef]
104. Selviandro, N.; Wisudiawan, G.; Puspitasari, S.; Adrian, M. Preliminary study for determining bring your own device implementation framework based on organizational culture analysis enhanced by cloud management control. In Proceedings of the 2015 3rd International Conference on Information and Communication Technology (ICoICT), Nusa Dua, Bali, Indonesia, 27–29 May 2015; IEEE: Piscataway, NJ, USA, 2015; pp. 113–118.
105. Flores, D.A.; Qazi, F.; Jhumka, A. Bring your own disclosure: analysing BYOD threats to corporate information. In Proceedings of the 2016 IEEE Trustcom/BigDataSE/ISPA, Tianjin, China, 23–26 August 2016; IEEE: Piscataway, NJ, USA, 2016, pp. 1008–1015.
106. Zulkefli, Z.; Singh, M.M.; Malim, N.H.A.H. Advanced persistent threat mitigation using multi level security–access control framework. In Proceedings of the International Conference on Computational Science and Its Applications; Springer: Berlin/Heidelberg, Germany, 2015; pp. 90–105.
107. Hajdarevic, K.; Allen, P.; Spremic, M. Proactive security metrics for bring your own device (byod) in iso 27001 supported environments. In Proceedings of the 2016 24th Telecommunications Forum (TELFOR), Belgrade, Serbia, 22–23 November 2016; IEEE: Piscataway, NJ, USA, 2016; pp. 1–4.
108. Aldini, A.; Seigneur, J.M.; Lafuente, C.B.; Titi, X.; Guislain, J. Formal modeling and verification of opportunity-enabled risk management. In Proceedings of the 2015 IEEE Trustcom/BigDataSE/ISPA, Helsinki, Finland, 20–22 August 2015; IEEE: Piscataway, NJ, USA, 2015; Volume 1, pp. 676–684.
109. Morrison, A.; Xue, L.; Chen, A.; Luo, X. Enforcing Context-Aware BYOD Policies with In-Network Security. In Proceedings of the 10th USENIX Workshop on Hot Topics in Cloud Computing (HotCloud 18), Boston, MA, USA, 8–9 July 2018.
110. Armando, A.; Costa, G.; Merlo, A.; Verderame, L.; Wrona, K. Developing a NATO BYOD security policy. In Proceedings of the 2016 International Conference on Military Communications and Information Systems (ICMCIS), Brussels, Belgium, 23–24 May 2016; IEEE: Piscataway, NJ, USA, 2016; pp. 1–6.
111. Samaras, V.; Daskapan, S.; Ahmad, R.; Ray, S.K. An enterprise security architecture for accessing SaaS cloud services with BYOD. In Proceedings of the 2014 Australasian Telecommunication Networks and Applications Conference (ATNAC), Southbank, VIC, Australia, 26–28 November 2014; IEEE: Piscataway, NJ, USA, 2014; pp. 129–134.
112. Perini, V.L.; de Fátima Webber do Prado Lima, M. BYOD Manager Kit: Integration of Administration and Security Tools BYOD. In Proceedings of the XIV Brazilian Symposium on Information Systems, Caxias do Sul, Brazil, 4–8 June 2018; pp. 1–9.
113. Zain, Z.M.; Othman, S.H.; Kadir, R. Security-Based BYOD Risk Assessment Metamodelling Approach. In Proceedings of the 21st Pacific Asia Conference on Information Systems (PACIS 2017), Langkawi, Malaysia, 16–20 July 2017.
114. Liu, X.; Qian, F.; Qian, Z. Selective HTTPS traffic manipulation at middleboxes for BYOD devices. In Proceedings of the 2017 IEEE 25th International Conference on Network Protocols (ICNP), Toronto, ON, Canada, 10–13 October 2017; IEEE: Piscataway, NJ, USA, 2017; pp. 1–10.
115. Koesyairy, A.A.; Kurniawan, A.; Hidayanto, A.N.; Budi, N.F.A.; Samik-Ibrahim, R.M. Mapping Internal Control of Data Security Issues of BYOD Program in Indonesian Banking Sector. In Proceedings of the 2019 5th International Conference on Computing Engineering and Design (ICCED), Singapore, 11–13 April 2019; IEEE: Piscataway, NJ, USA, 2019; pp. 1–5.
116. Chu, P.Y.; Lu, W.H.; Lin, J.W.; Wu, Y.S. Enforcing enterprise mobile application security policy with plugin framework. In Proceedings of the 2018 IEEE 23rd Pacific Rim International Symposium on Dependable Computing (PRDC), Taipei, Taiwan, 4–7 December 2018; IEEE: Piscataway, NJ, USA, 2018; pp. 263–268.

117. Downer, K.; Bhattacharya, M. BYOD security: A study of human dimensions. *Informatics* **2022**, *9*, 16. [CrossRef]
118. Ali, S.; Qureshi, M.N.; Abbasi, A.G. Analysis of BYOD security frameworks. In Proceedings of the 2015 Conference on Information Assurance and Cyber Security (CIACS), Rawalpindi, Pakistan, 18 December 2015; IEEE: Piscataway, NJ, USA, 2015; pp. 56–61.

applied sciences

MDPI

Article

Intelligent Anomaly Detection System through Malware Image Augmentation in IIoT Environment Based on Digital Twin

Hyun-Jong Cha [1], Ho-Kyung Yang [2], You-Jin Song [3] and Ah Reum Kang [4,*]

[1] Department of AI Software Engineering, Pai Chai University, Daejeon 35345, Republic of Korea; hjcha@pcu.ac.kr
[2] Department of Defense Acquisition Program, Kwangwoon University, Seoul 01897, Republic of Korea; porori0421@naver.com
[3] Department of Information Management, Dongguk University WISE Campus, Gyeongju-si 38066, Republic of Korea; song@dongguk.ac.kr
[4] Department of Information Security, Pai Chai University, Daejeon 35345, Republic of Korea
* Correspondence: armk@pcu.ac.kr; Tel.: +82-42-722-2522

Abstract: Due to the recent rapid development of the ICT (Information and Communications Technology) field, the industrial sector is also experiencing rapid informatization. As a result, malware targeting information leakage and financial gain are increasingly found within IIoT (the Industrial Internet of Things). Moreover, the number of malware variants is rapidly increasing. Therefore, there is a pressing need for a safe and preemptive malware detection method capable of responding to these rapid changes. The existing malware detection method relies on specific byte sequence inclusion in a binary file. However, this method faces challenges in impacting the system or detecting variant malware. In this paper, we propose a data augmentation method based on an adversarial generative neural network to maintain a secure system and acquire necessary learning data. Specifically, we introduce a digital twin environment to safeguard systems and data. The proposed system creates fixed-size images from malware binaries in the virtual environment of the digital twin. Additionally, it generates new malware through an adversarial generative neural network. The image information produced in this manner is then employed for malware detection through deep learning. As a result, the detection performance, in preparation for the emergence of new malware, demonstrated high accuracy, exceeding 97%.

Keywords: digital twin; IIoT; malware; generative adversarial network; image interpolation

Citation: Cha, H.-J.; Yang, H.-K.; Song, Y.-J.; Kang, A.R. Intelligent Anomaly Detection System through Malware Image Augmentation in IIoT Environment Based on Digital Twin. *Appl. Sci.* **2023**, *13*, 10196. https://doi.org/10.3390/app131810196

Academic Editors: Peter R.J. Trim and Yang-Im Lee

Received: 26 July 2023
Revised: 6 September 2023
Accepted: 7 September 2023
Published: 11 September 2023

1. Introduction

IT (Information Technology) technology combined with the Internet is evolving into a national infrastructure. Notably, the field of factory automation is experiencing significant advancements with the emergence of the IIoT, which aims to digitalize all manufacturing processes beyond conventional process automation [1]. Simultaneously, the rapid development of the 4th industrial revolution, artificial intelligence, IoT (Internet of Things), and big data has introduced new challenges, including information theft, hacking, and eavesdropping. Consequently, the importance of cybersecurity has grown significantly [2].

Malware refers to any software that interferes with or adversely affects normal operation. New types of malware emerge daily. As the number of new malware continues to rise, security threats rapidly expand into the automation of industrial processes. Recently, there has been a growing interest in technology that detects files suspected of being malware by executing them directly in a virtual machine [3]. This interest is fueled by the development of IoT technology and virtualization technology, which has led to the emergence of the concept of a digital twin. The digital twin structure replicates the same virtual system as the corresponding real physical system. It achieves near real-time synchronization of certain value changes in the real world and virtually simulates the dynamic aspects of real objects [4].

Among the technologies using virtual machines, one method for detecting malware involves executing a target file directly in a virtual machine to analyze its behavior. By employing the digital twin in this manner, the code operates in an isolated space, independent of the actual system. In other words, utilizing the digital twin environment does not have any adverse effects on the actual system. Furthermore, even if the system becomes infected with malware, it can be quickly reset to its initial state. This capability enables the efficient examination of multiple codes in succession.

Machine learning technology is also applied to the detection of malware. As a result, researchers are exploring the use of image recognition technology to convert executable files into images and subsequently determine or classify them as malicious [5]. Deep learning eliminates the need for preprocessing to extract features from images, making it an efficient approach for detecting and analyzing malware.

Improving the performance of deep learning models requires a significant amount of data. Obtaining sufficient data that aligns with the learning goals of the desired model, especially in the case of analyzing malware, poses a challenge. Insufficient data hinders the effective training of malware classification and detection models. To overcome this difficulty, data augmentation can be applied. Data augmentation techniques are widely used across different fields, leading to a variety of methods [6,7].

In this paper, we propose a data augmentation method for malware using image conversion and adversarial generative neural networks in a digital twin environment. The proposed method consists of two parts: the image conversion method and data generation. The image conversion method suggests suitable quality and size for detection using deep learning. Subsequently, an adversarial generative neural network is employed to augment the missing data. Additionally, the generated images are trained with CNN to assess their performance. This generative network predicts and prevents the generation of malware in the future through digital twins. By utilizing real data, a digital twin can proactively block or defend against potential damage by simulating the virtual world. This approach enables the detection of combinations that could potentially harm the system through advanced simulations. Malware often involves partial modifications of existing code. Through an adversarial generative neural network, future malware are generated, and then a detection model is created using CNN learning. This model demonstrates robust performance in effectively protecting the system in real-life scenarios.

The thesis comprises a total of eight sections. Section 2 provides a detailed description of each element technology that forms the foundation of the proposed method. Section 3 examines existing research that is comparable to the proposed system. In Section 4, the configuration of the proposed system is explained. Moving on to Section 5, experiments are conducted to validate and assess the performance of the proposed system. In Section 6, the experimental results are evaluated to prove the superiority of the proposed system. Section 7 provides appropriate insights through a summary of the research results and limitations of the proposed system. Finally, Section 8 presents the paper's future research directions.

2. Background

2.1. Digital Twin

In this paper, we utilize a digital twin to create a virtual environment for the detection of dynamic and insecure malware. Our objective is to establish a framework and architecture that aligns with our research goals. Subsequently, we conduct experiments within this virtual world, utilizing the primary functionalities of the digital twin. By doing so, we effectively mitigate physical risks and reduce associated costs, making our approach more efficient and feasible.

The concept of the digital twin was originally proposed by Michael Grieves of the University of Michigan in 2002, defining it as "a digital product corresponding to a physical product" [8]. Subsequently, NASA (National Aeronautics and Space Administration) further refined the digital twin concept in 2012 [9]. As General Electric of the United States introduced an industrial cloud-based open platform [10], the concept of 'digitizing a real object, utilizing it, and reflecting the result in reality' became more refined. Through extensive research and development, it has been elaborated as 'technology that enables real-time prediction, optimization, monitoring, control, and decision support' [11]. Figure 1 depicts a technical, conceptual diagram of the digital twin.

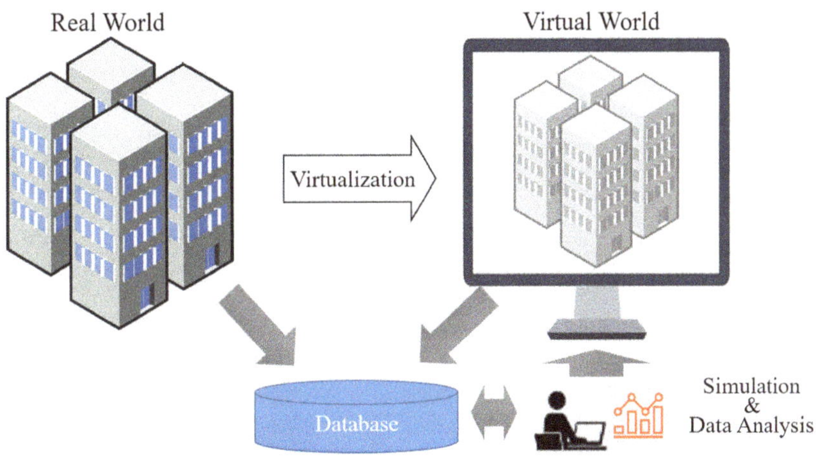

Figure 1. Technical concept of digital twin.

Essentially, before a malware is activated in the physical world, it is analyzed and prevented from causing system malfunctions. Repeatedly implementing a safe physical world within the virtual environment allows for iterative testing and improvement.

In this study, we propose a method to analyze and defend against new types of malware using digital twins.

2.2. Malware

Malware refers to any software that can interfere with or adversely affect the normal operation of an electronic device. As the number of malware rapidly increases, the types of attacks are becoming more diverse and intelligent [12]. According to the 2019 McAfee Labs Threats Report, new malware continues to be discovered, with tens of millions of malware being identified each quarter [13]. Table 1 provides a description of the types of representative malware known so far, along with their characteristics.

Table 1. Types and characteristics of malware.

Category	Characteristic
Worm	Infects itself through the network
Worm Virus	Equipped with both worm and virus infection methods
Trojan Horse	No self-replicating ability
Spyware	Steals user information
Adware	Displays ads automatically
Hijacker	Redirects to unintended sites or opens pop-up windows
Ransomware	Makes files unusable and demands money for recovery
Keylogger	Logs user's keyboard input to uncover sensitive information

2.3. Image Interpolation

Interpolation is a technique used to enlarge or reduce an image. When an image is enlarged, additional pixels must be added between the pixels of the original image. When an image is reduced, one pixel must replace a certain number of pixels in the existing image. Interpolation is distinguished by how pixels are added and how many pixels are mapped to a single pixel.

In this paper, three image interpolation methods were tested: nearest neighbor, bicubic, and bilinear. The detection performance was measured by CNN by adjusting the image size using each interpolation method. In the final proposed system, a method with high performance and short processing time was applied.

The nearest neighbor method is the simplest among interpolation methods and has the shortest processing time. This technique applies the value of the supplemented or replaced pixel to the value of the pixel in the existing image closest to that location. That is, if a two dimensional image is magnified twice, one pixel is enlarged to 2 × 2 pixels, and the same value as the existing pixel is applied to the value of the 2 × 2 pixel. When an image is reduced by a factor of two, 2 × 2 pixels are replaced with 1 pixel, and the value of one of the existing pixels is applied to the value of the replaced pixel [14]. The bilinear technique considers the four adjacent pixels of a new pixel. The value of the new pixel is determined based on the weighted average of the distances to them. Four pixels are considered, but they are most affected by pixel values of closer distances [15]. The bicubic method considers the 16 adjacent pixels of a new pixel and determines the value of the new pixel based on the weighted average of the distances to them. This technique provides a smoother result and is often preferred for image resizing due to its better preservation of image details and reduced artifacts [16].

2.4. Generative Adversarial Network

GAN (generative adversarial network) [17,18] consists of a generator/generative model and a discriminator/discriminator model.

Figure 2 shows the learning structure in a GAN. GANs are often described with examples of banknote counterfeiters and the police. The generator is likened to a banknote counterfeiter, while the discriminator is compared to the police. In this analogy, the criminal (generator) attempts to trick the police (discriminator) with fake banknotes, and the police strive to identify the authenticity of the counterfeit bills. Through this competitive process, each model improves its ability to distinguish between genuine and counterfeit banknotes. As a result, the discriminator (D) tries to make accurate judgments using original data, while the generator (G) generates fake data to prevent the discriminator (D) from making accurate determinations. Through this process of adversarial learning, both models iteratively improve their performance [19].

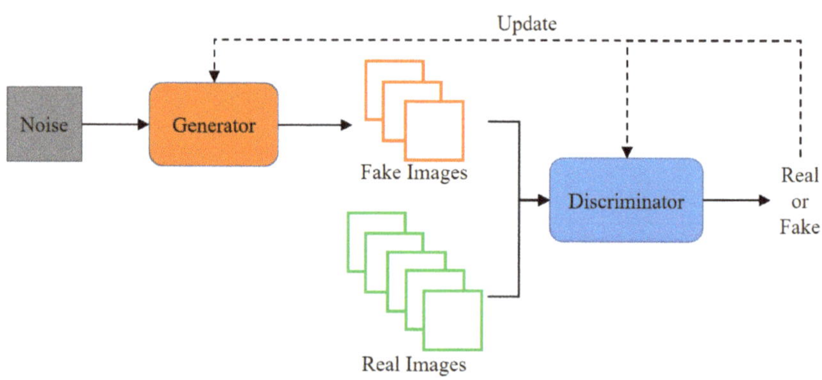

Figure 2. Learning structure of GAN.

Eventually, the discriminator (D) becomes less able to distinguish between the original data and the fake data generated by the generator (G).

Since GANs are unsupervised, they do not require manual labeling, and by learning the data, they can acquire the internal representation of the data. GANs are capable of generating audio, video, text, and image data that is indistinguishable from real data. As a versatile technology, they find applications in marketing, e-commerce, gaming, advertising, and various other industries.

3. Literature Review

3.1. Digital Twin Application Research

Pokhrel et al. [20] provide a comprehensive overview of research on predicting cybersecurity incidents using digital twin technology. The paper covers the integration of digital twin models and cybersecurity frameworks, machine learning, and data analysis techniques for accident prediction. It also emphasizes the importance of real-time monitoring and anomaly detection in the digital twin context.

Eckhart and Ekelhart [21] conducted a study focusing on the application of digital twins to enhance the security of CPS (cyber-physical systems). The paper reviews the concept of digital twins in the context of CPS security and covers various aspects, such as modeling and simulation capabilities, real-time monitoring, and anomaly detection and analysis.

The proposed technique learns real-time data collected from the physical world of digital twins. Therefore, it provides an environment that can be analyzed safely without affecting systems in the physical world.

3.2. Malware Detection through Imaging

Traditional machine learning techniques such as KNN (k-nearest neighbor) and SVM (support vector machine) were used [22]. However, in recent years, studies that apply deep learning are the main focus. Additionally, the malware detection performance using deep learning is generally superior to that using machine learning techniques [23]. In 2022, Atitallah et al. [24]. proposed detection and multiple classification of malware converted into images. The proposed approach combines two machine learning techniques: fine-tuned CNN and random forest voting. This approach takes advantage of CNN's ability to capture complex patterns in data and random forests, which reduce overfitting and enhance generalization.

However, malware sample images of different classes to which the same packing technique is applied may appear similar. Moreover, limitations arise in terms of generalization error and data imbalance due to the size and diversity of the dataset [25].

In this paper, the malware is expressed in grayscale and converted into an image. Additionally, image interpolation is considered to convert the image into a suitable size for analysis.

3.3. Malware Detection through Machine Learning

In 2009, M. Zubair Shafiq et al. [26] proposed PE-Miner (Portable Executable-Miner), a framework for detecting malware following the PE format (Portable Executable Format) in RAID (Redundant Array of Independent Disks). PE Miner extracts information from PE files and creates vector values. Then, a classification model was proposed through a preprocessing process. For more than 1000 malware, the classification time was reduced to 0.244 s, and a high performance of 99% detection rate was shown.

In 2018, Anderson and Roth [27]. proposed Ember, a machine learning model based on machine learning static analysis. Ember extracts information from PE files through static analysis and then creates vector values. A detection model was created using a LightGBM (Light Gradient Boosting Machine) model, which utilizes a boosting technique, a kind of ensemble technique. Ember showed a high detection performance of 92% for more than 800,000 malware.

In 2020, Hojjat Aghakhani et al. [28]. published a study on classifying packed malware using static analysis-based features and machine learning-based classifiers. They then created a classification model using a random forest. The study utilized a real dataset consisting of 4396 unpacked normal files, 12,647 packed normal files, and 33,681 packed malicious files. The results showed 41.84% false negatives and 7.27% false positives for the packed dataset. However, it was found to be difficult to accurately classify packed files based solely on static analysis features.

3.4. Malware Detection through Deep Learning

Saxe and Berlin [29]. proposed a DNN based PE file malware detection model. The vector created by analyzing the PE file is used as an input for a deep learning model with two hidden layers. Additionally, the Bayesian model is used to provide the probability of being malicious in addition to detection. The proposed model shows a high detection rate of 95%.

MalConv, proposed in 2017 by Edward Raff et al. [30], is a CNN model that learns without separate feature extraction and preprocessing from PE files. To validate the model, it was verified using two datasets and showed a high accuracy of 93%.

Mahmoud Kalash et al. [31]. proposed the M-CNN model at the 2018. M-CNN converts PE files into images and classifies them using convolutional neural networks. The resulting image is classified using the VGG-16 (Visual Geometry Group-16) model. The model was validated using 8394 data points, achieving an accuracy of over 96%.

The proposed system in this paper applies a CNN model through digital twin to safely execute the malware. It converts the binary of a file to a grayscale image. Afterwards, the detection performance is measured for each size and interpolation method to select an appropriate size and interpolation method. Additionally, the detection performance of data generated through adversarial generative neural networks is measured.

4. Materials and Methods

The proposed system is built upon a digital twin-based malware detection system. Figure 3 shows the architecture of the proposed system.

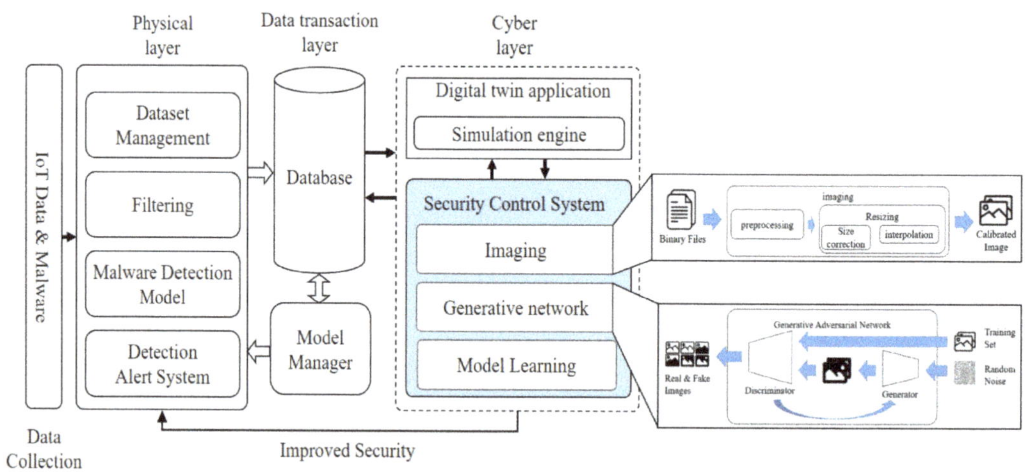

Figure 3. Digital twin-based conceptual diagram of the proposed system.

The architecture is divided into three layer: physical, data, and cyber. The physical layer represents the anomaly detection system for actual malware, consisting of data collection, analysis, and system management components. The data layer includes a database for storing data and learning models, as well as a system for managing these models. The database stores data collected from the physical layer, malware information, and a learning model that serves as the standard for filtering. The model manager is responsible for managing the learning model based on the stored data. The cyber layer encompasses digital twin applications and security control systems. The digital twin application replicates the actual malware detection system in the virtual world. The security control system is responsible for creating and learning about new types of malware. The learning process is conducted based on information received from the data layer, and the model is updated through interactions with the simulation in the digital twin application, thereby improving the system's detection capabilities.

The proposed system collects IoT data and classifies it through a malware detection model. While the system can effectively filter known malware in the physical world using pretrained models, it faces limitations in detecting new types of malware that have not been previously learned. To address this challenge, the proposed system leverages the digital world by creating and learning new types of malware based on existing ones. By generating and learning these new malware variants in the digital twin environment, the system enhances its detection capabilities and applies the knowledge gained to the physical world's detection system. As a result, the proposed system is capable of preemptively detecting emerging threats, surpassing the capabilities of existing methods.

Among the proposed systems, the security control system of the cyber layer is composed of three main components: image generation, new malware image generation, and model learning. Figure 4 illustrates the pre-processing and learning process of the security control system.

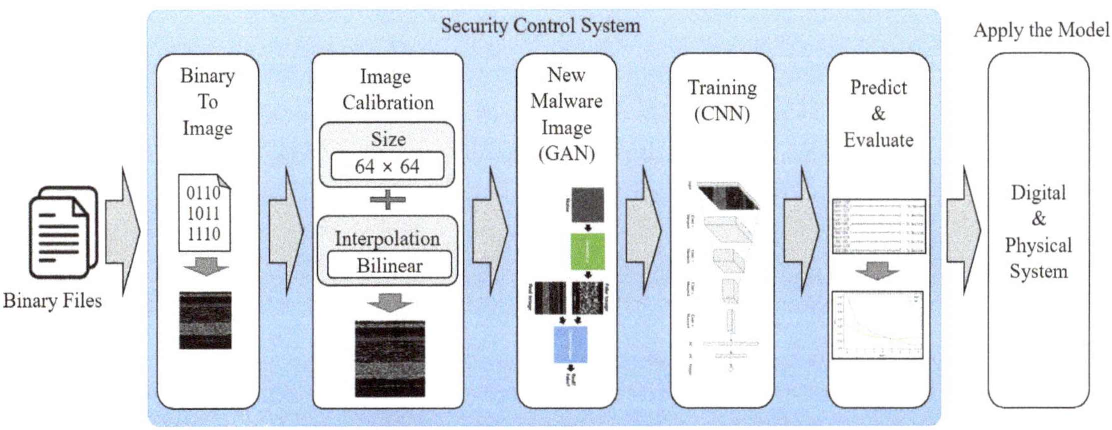

Figure 4. Pre-processing and learning process of security control system.

The image creation process involves converting an input executable file into an image. To achieve this, the input binary information is digitized and converted into pixel values. Additionally, the image size must be standardized for effective learning. Image correction techniques are employed to prevent information loss during image resizing. In the proposed system, the image size is set to 64 × 64 pixels, and the bilinear method is used for image correction. To determine the optimal image size and correction method, three image sizes (32 × 32, 64 × 64, and 128 × 128) and three correction methods (nearest neighbor, bilinear, and bicubic) are combined. The CNN classification performance is then evaluated to select the most efficient configuration. Next, the new malware image creation process utilizes DCGAN, a generative artificial intelligence technique. This method generates new malware images by extracting characteristics from existing malware images. The generated images are then trained using a convolutional neural network, creating a model capable of detecting malware. This approach enables the system to proactively detect and defend against new types of malware through simulated detection in the virtual environment.

4.1. Imaging of Malware

In Figure 5, the process of converting a binary file into an image is depicted. This conversion enables the smooth analysis of malware through image processing. During this process, images of all malware are used as training data after being adjusted to the same size. This step ensures consistency and uniformity in the training data, which is crucial for accurate and effective learning in the subsequent stages of the proposed system.

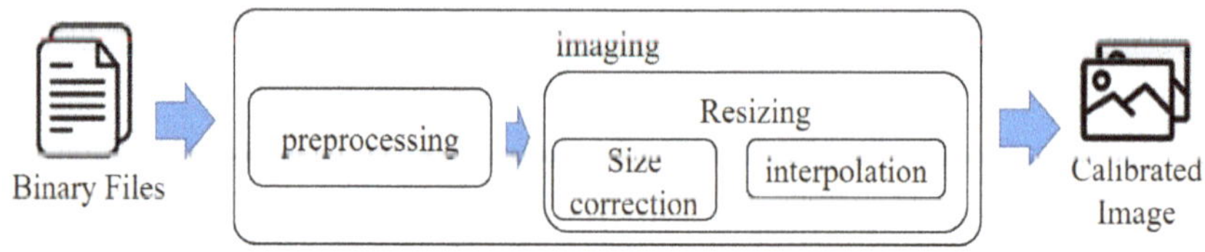

Figure 5. Imaging process of malware.

Figure 6 depicts the pre-processing process of converting a malware into an image. In this process, the size of the image is not a primary concern. The binary data of the malware file is read as a vector of 8-bit unsigned integers. Subsequently, each byte is converted into one pixel of the image, with values ranging from 0 to 255, creating a grayscale image. The size of the image is determined based on the file size. Therefore, the width and height of the image are calculated as the square root of the total number of bytes [32]. As the file size varies, the size of the converted image will also differ accordingly.

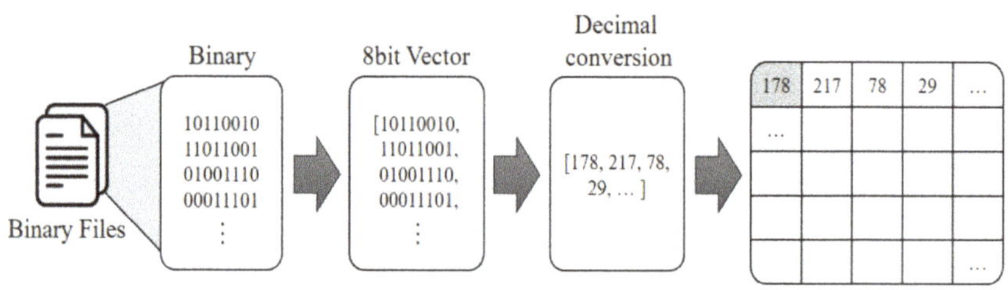

Figure 6. Imaging pre-process of malware.

For training purposes, all images used as inputs to the deep learning model must be of the same size. Therefore, the malware image is resized to a specific size after the pre-processing phase. During this process, it is crucial to adjust the size while preserving the essential characteristics of the image as much as possible. Interpolation is employed to resize the image while maintaining its overall shape. Depending on the interpolation method applied and the desired size for conversion, the shape of the image may slightly vary. In the proposed system, an experiment was conducted to assess the detection performance based on different image sizes and interpolation methods. The image size options included 32×32, 64×64, and 128×128, while interpolation methods considered were nearest neighbor, bilinear, and bicubic. Larger image sizes generally lead to better detection performance, but it comes at the cost of increased processing time. Hence, in the proposed system, the image size was standardized to 64×64 using bilinear interpolation, striking a balance between performance and efficiency.

4.2. Data Augmentation through Generative Adversarial Networks

The generative model of the proposed system uses DCGAN. The process of creating a new malware image is shown in Figure 7. In the generative model of the proposed system, there is one generator and one discriminator. The generator considers as input an image made of random noise, while the discriminator receives the malware image and the image created by the generator to determine their authenticity. Malware images are generated by converting binary files into images, and then resizing them to 64×64 size.

Among these images, we divide them into a training set and a test set to evaluate the system's performance.

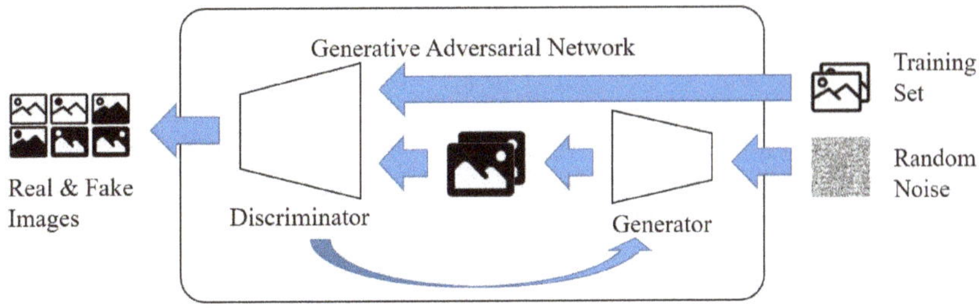

Figure 7. Process of creating malware images.

The discriminator's goal is to predict whether an image is real or fake. This is similar to an image classification problem in supervised learning. Therefore, it is possible to use a network structure in which convolutional layers are stacked, and the output layer serves as the connection layer. The structure of the discriminator model in the proposed system is illustrated in Figure 8.

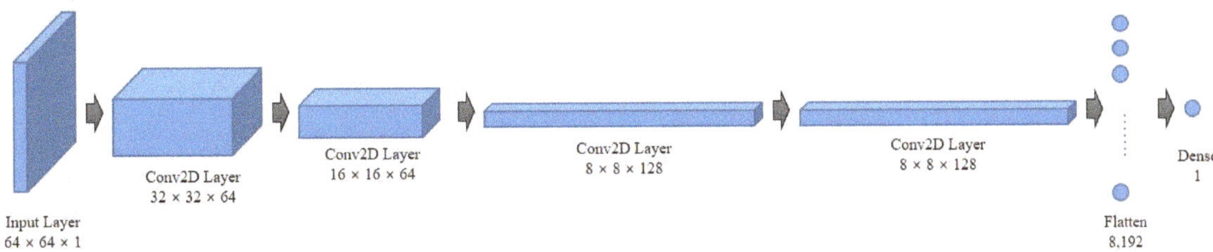

Figure 8. Composition of the discriminator model of the proposed system.

The input is an image of size 64 × 64. The first convolutional layer, Conv2D, employs 64 filters, reducing the feature map size to 32 × 32. In the second convolutional layer, 64 filters are used, further reducing the feature map to 16 × 16. Subsequently, the third convolutional layer utilizes 128 filters, resulting in an 8 × 8 feature map. The last convolutional layer also employs 128 filters, preserving the 8 × 8 feature map size. The feature map is then flattened and converted into a vector. Passing through a dense layer with a single unit, the output is transformed into a value between 0 and 1. This model considers an image as input and outputs a number that distinguishes real from fake. Next, we consider the number of parameters in the discriminator. The first convolutional layer contains 64 filters, resulting in 1664 parameters. The second convolutional layer, with 64 filters, contributes 102,464 parameters. The third convolutional layer employs 128 filters and adds 204,928 parameters. Finally, the last convolutional layer has 128 filters and 409,728 parameters. When these parameters are flattened and connected as a single unit, the count becomes 8193. The total number of discriminator parameters sums up to 726,977.

The generator typically considers a vector sampled from a multivariate standard normal distribution as input and produces an image of the same dimensions as the original training data. Its role is to transform latent space vectors into images. The structure of the proposed system's generator model is illustrated in Figure 9.

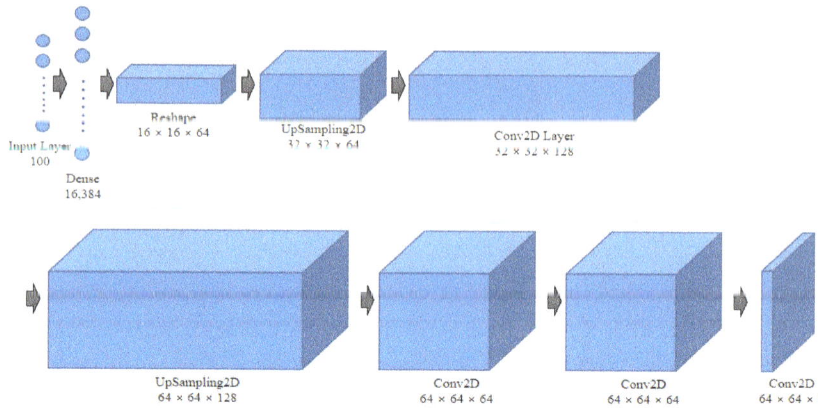

Figure 9. Configuration of the generator model of the proposed system.

The constructor considers a vector of length 100 as input and passes it through a dense layer with 16,384 units. Batch normalization and ReLU activation functions are applied to transform the feature map into a size of 16 × 16 with 64 filters. Subsequently, an upsampling layer increases the feature map to 32 × 32. The first convolutional layer uses 128 filters. Another upsampling layer increases the feature map to 64 × 64. Through two additional convolutional layers with 64 filters, a final 64 × 64 feature map is obtained, which matches the size of the original image. This model considers a vector of length 100 as input and outputs a 64 × 64 image. Next, let us consider the parameter sizes of the constructor model. The initial input of 100 vectors is connected to 16,384 neurons, resulting in 1,654,784 parameters. After applying batch normalization, there are 65,536 parameters. The first convolutional layer has 204,928 parameters, and after applying batch normalization, there are 512 parameters. The second convolutional layer contains 204,864 parameters, and batch normalization contributes 256 parameters. The third convolutional layer has 102,464 parameters. Lastly, the final convolutional layer adds 1601 parameters. The total number of parameters is 2,235,201. Among them, 33,280 parameters are unlearned, leaving 2,201,921 parameters used for learning.

4.3. Deep Learning Model for Malware Detection

The proposed system's malware detection model utilizes a convolutional neural network (CNN) architecture. The model's structure is illustrated in Figure 10. It consists of a series of layers, including a convolutional layer, a pooling layer, and a Rectified Linear Unit (ReLU) activation function.

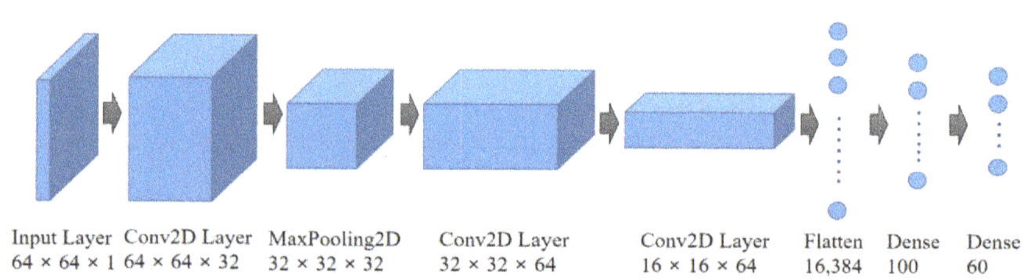

Figure 10. Composition of malicious code detection model.

In the convolutional layer, the model performs convolutions on the input data to extract relevant features. The pooling layer reduces the spatial dimensions of the feature maps, aiding in feature extraction and dimensionality reduction. Additionally, the ReLU activation function introduces non-linearity to the model, enhancing its capacity to capture complex patterns and relationships in the data.

The model receives an input of size 64 × 64. The first convolutional layer, Conv2D, employs 32 filters with the ReLU activation function. Subsequently, a pooling layer is added, reducing the feature map size to 32 × 32. Moving on, the second convolutional layer and pooling layer use 64 filters, resulting in a feature map size of 16 × 16. The 3D feature maps are then flattened to facilitate feature classification. To prevent overfitting, a dense hidden layer and an output layer are added, enabling multi-classification into 60 categories. For the experiments using the Mal60 dataset, the model classifies 60 types of malware. In the case of Malimg, the classification is adjusted to 25 types. To calculate the number of parameters in the model, the first convolutional layer with 32 filters has 320 parameters. The second convolutional layer with 64 filters contributes 18,496 parameters. Flattening the feature maps and connecting them with 100 neurons results in 1,638,500 parameters. The total number of parameters in the model sums up to 1,663,376.

5. Experiment

5.1. Experimental Data

The proposed model was validated using three datasets. The first dataset is the Malimg dataset [33], which was also utilized in previous studies by Kamundala and Kim [34], AlGarni et al. [35], Go et al. [36], and Bhodia et al. [37]. The second dataset used is the Mal60 dataset [38], as studied by Kang and Kim [39]. The last dataset was VXHeaven's 2010 virus collection dataset [40].

The Malimg dataset used in malware imaging research is categorized into 25 families. The dataset comprises a total of 9458 malware samples. However, for the purpose of this paper, which focuses on classification with a limited number of samples, only 20 samples were randomly selected from each family, resulting in a total of 500 samples in the composed dataset.

The Mal60 dataset used in this research is classified into 60 families, each containing a total of 20 malware samples, resulting in a dataset size of 1200 samples. Notably, each family represents a unique category of malware. In some cases, different security companies may assign different names to the same family of malware based on their naming criteria. However, despite these naming differences, the code samples are classified under the same group as they exhibit similar malicious behavior or characteristics. Additionally, certain malware samples in Kaspersky and Bitdefender products may have diagnosis names that cannot be confirmed or are classified as heuristic diagnosis names. Specifically, two examples of such samples are identified as belonging to the Akdoor and Rifdoor families, targeting specific areas.

The VXHeaven dataset comprises 270 k malware executable files, but it does not include malware family information. To determine the family of each sample in the VXHeaven dataset, we added family labels through Virustotal [41], a malware analysis site. When a suspected malware file is uploaded to Virustotal, the site analyzes it for family prediction using various antivirus engines. The malware family label is assigned based on the results obtained from the Microsoft Antivirus engine.

A total of 70% of each dataset is used for training, and the remaining 30% for testing.

5.2. Experiment Environment

The proposed system underwent testing by dividing it into two stages: the imaging stage and the data generation stage. In the imaging stage, the detection performance was measured for each image size and correction method. Additionally, in the data generation stage, data was generated, and the detection performance of the generated data was measured.

To ensure a robust and efficient experimental environment, the experiments were conducted using Google Colaboratory's GPU environment, which helped overcome the limitations of local environments and enabled consistent research in the same environment. The hardware environment and experimental setup used for learning and implementing the model are summarized in Table 2.

Table 2. Experiment environment.

Experimental Elements		Element Value
Local System	CPU	Intel® Core™ i5-12500 3.00 GHz
	Memory	32 GB
	Main Storage	Samsung SSD 256 G
	Support Storage	Seagate HDD 1 TB
Google Colaboratory	Engine	Python 3 Google Compute Engine
	RAM	13 GB
	CPU type	NVIDIA T4
	GPU RAM	15 GB
	Storage	78.2 GB
Model Learning	DCGAN epochs	20,000
	CNN epochs	200
	Batch size	128

5.3. Experimental Setup of Imaging

The experimental procedure, as depicted in Figure 11, involves the following steps: Step 1. Data Preparation: The training data and test data are adjusted to create nine different cases based on the image size (32 × 32, 64 × 64, 128 × 128) and interpolation method (nearest neighbor, bilinear, and bicubic).

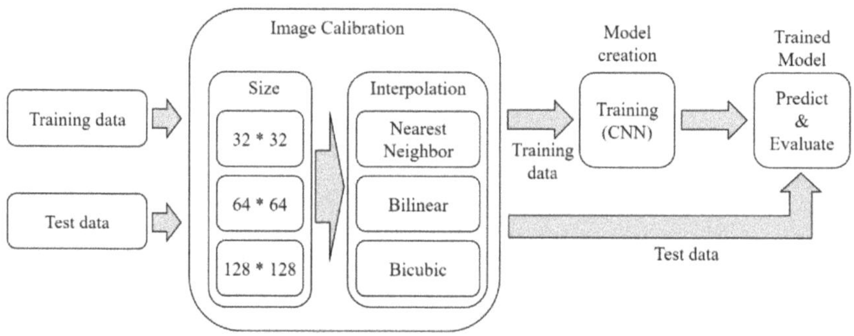

Figure 11. Experimental process of imaging.

Step 2. Model Training: The adjusted training data is used to train a CNN-based deep learning model. As a result, nine different models are generated, each corresponding to a specific combination of image size and interpolation method. During the training process, a batch size of 128 is used, and the learning is repeated for 200 epochs.

Step 3. Model Evaluation: Evaluation of the models is performed separately for each image size. First, a sample image is tested to verify the model's performance. Then, the test is conducted using malware samples. By following this experimental procedure, we can assess the detection performance of the CNN-based models under various settings, such as different image sizes and interpolation methods.

5.4. Experimental Setup of Data Augmentation through Generative Adversarial Networks

The experimental procedure is shown in Figure 12. The image required for the experiment determines the size and correction method based on the results of the imaging experiment. Therefore, the experiment is conducted using a 64 × 64 size image with good performance results, corrected by the bilinear method. The experiment proceeds in two stages. The first step involves the similarity evaluation for fake images generated using GAN. Step 2 is the CNN performance evaluation for both existing data and generated data.

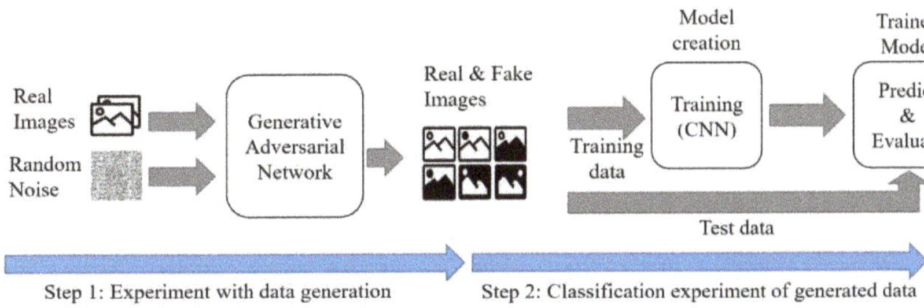

Figure 12. Data generation experiment process.

Step 1 involves creating a new image using DCGAN. The image created in the preprocessing process has a size of 64 × 64 pixels, and the image generated by DCGAN is also 64 × 64 pixels. The time taken to create the image increases with higher resolution. The evaluation assesses the similarity between the real and generated images, using factors such as FID (Fréchet Inception Distance) and the cross-correlation coefficient. Additionally, the loss value is checked to ensure that mode collapses are minimized during creation. The batch size was set to 128, and the learning process was repeated 20,000 times.

In the second step, the detection performance of the generated images and the existing images, which are the results of the first step, is measured. The detection process uses the same CNN model as used in the imaging experiment. The experiment evaluates the detection performance of the generated images. The training and test images are classified in a ratio of 8:2. Evaluation factors include precision, recall, f1-score, and accuracy. The batch size is set to 128, and the learning process is repeated 200 times.

6. Performance Evaluation

6.1. Performance Evaluation of Imaging

When a malicious code file is expressed as an image, the width or length of the image is usually fixed to a certain size. Since the file size of the malicious code is different, the size of the converted image also appears in various ways. Figure 13 shows only a portion of the results of converting malicious code into an image. Malicious code images are converted into images with a fixed length. In other words, it can be seen that Adialer.C malware is the largest. And Agent.FYI malware has the smallest size.

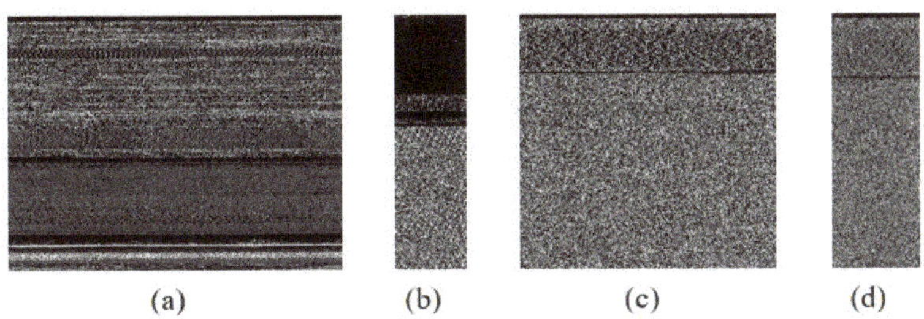

Figure 13. Sample images from the Malimg dataset. (**a**) Adialer.C, (**b**) Agent.FYI, (**c**) Allaple.A, and (**d**) Allaple.L.

For training purposes, all images utilized as inputs for the deep learning model need to share the same dimensions. Consequently, the malicious code images are transformed to a predetermined size. During this process, interpolation is employed to adjust the size while preserving the image's characteristics to the greatest extent possible. Nearest neighbor, bilinear, and bicubic techniques are employed for image refinement. Nearest neighbor, although simple and swift, sacrifices visual quality, resulting in jagged edges, particularly during magnification. This technique is straightforward to implement and comprehend, suitable when prioritizing speed and accepting some degree of quality loss.

The bilinear technique yields smoother outcomes but may lack sharpness, potentially leading to slightly blurred edges. It is slower than nearest neighbor but remains a fast method, delivering superior quality compared to the former and serving well for general-purpose scaling. On the other hand, the bicubic technique affords the highest quality but comes with increased computational complexity. This involves intricate calculations, typically considering 16 adjacent pixels. Bicubic produces considerably smoother images than bilinear and is the preferred choice for high-quality resizing, especially when maintaining sharpness is crucial.

Table 3 showcases the results of time measurements required for image conversion. The time needed was gauged with respect to each interpolation method. A total of 2210 malware images were utilized for time assessment, with time measured in seconds. The results of the measurements were presented with two decimal places. The outcomes illustrate that computation time diminishes in the sequence of nearest neighbor, bilinear, and bicubic techniques.

Table 3. Time required for each interpolation method.

	Nearest Neighbor	Bilinear	Bicubic
32 × 32	5.03	5.50	6.37
64 × 64	20.11	22.02	25.47
128 × 128	80.44	88.08	101.88

Table 4 present the f1-score results for imaging. Overall, the detection performance is relatively better when bilinear or bicubic interpolation methods are applied than when nearest neighbor is used. Additionally, the larger the size of the image, the better the detection performance. However, selecting the optimal interpolation method and image size to be applied to the model should not be solely based on detection performance. It is essential to consider other factors, such as computation requirements. As the size of the image increases, the amount of computation also increases, with bilinear and bicubic methods requiring more computation than nearest neighbor. Therefore, depending on the type of service to be implemented, the interpolation method and image size must be selected, taking into consideration both detection performance and processing time. In this paper, a data generation experiment was conducted using 64 × 64 size images and the bilinear interpolation method. The average accuracy of the entire experiment was 98.2%, the average precision was 96.5%, and the average recall was 97.5%.

Table 4. F1-score analysis results.

	Nearest Neighbor	Bilinear	Bicubic
32 × 32	0.9254	0.9799	0.9784
64 × 64	0.9766	0.9763	0.9762
128 × 128	0.9899	0.9971	0.9948
Mean of f1-score	0.9654	0.9872	0.9855

Figures 14–16 are some of the results of creating an image and adjusting its size using a correction method.

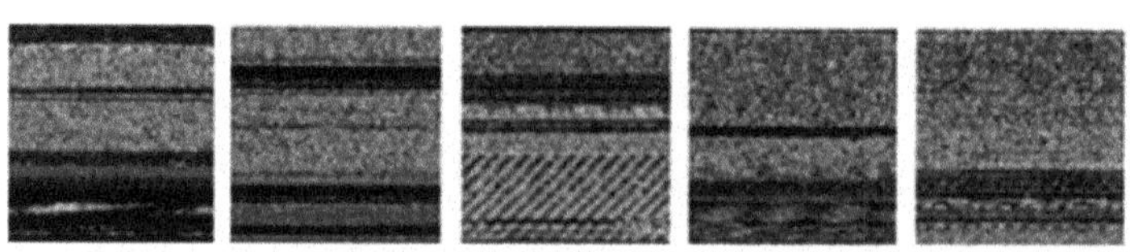

(a) (b) (c) (d) (e)

Figure 14. Sample image from Mal60 dataset. (**a**) Abnores, (**b**) Adposhel, (**c**) Akdoor, (**d**) CRyptXXX, and (**e**) Downloader.

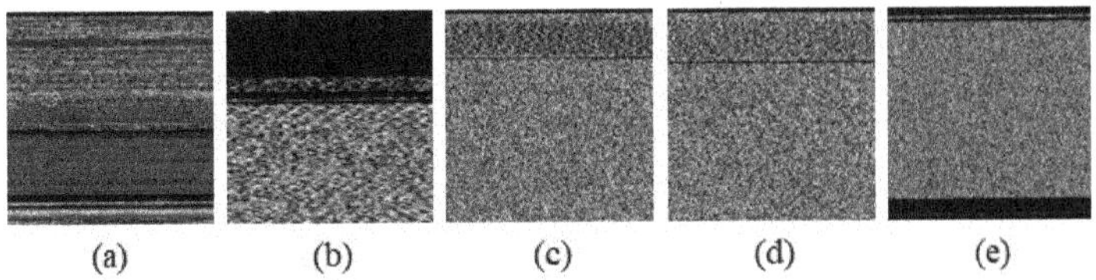

Figure 15. Sample image from Malimg dataset. (**a**) Adialer.C, (**b**) Agent.FYI, (**c**) Allaple.A, (**d**) Allaple.L, and (**e**) Alueron.gen!J.

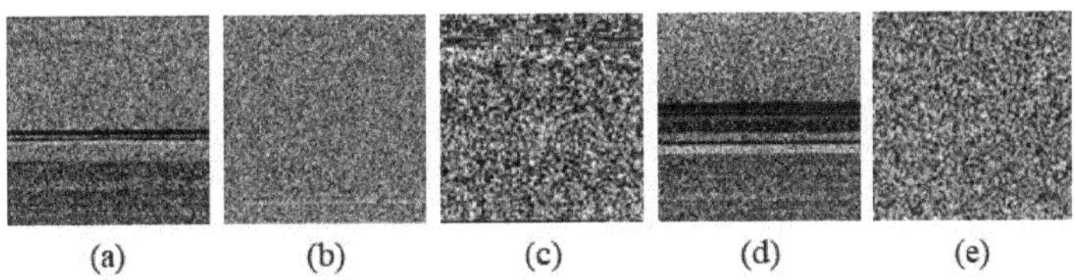

Figure 16. Sample image from VXHeavens dataset. (**a**) C2Lop.A, (**b**) Helpud.A, (**c**) Treemz.gen!A, (**d**) Seimon.D, and (**e**) Storark.A.

6.2. Performance Evaluation for Data Generation

For data generation, training was performed with 20,000 epochs, and this study was repeated four times. Figure 17 shows the progression of fake images generated from the generative network during the learning process. Figure 17a represents the real image of the Abnores type among malware. Subsequent images, Figure 17b–d, depict the results at 1000, 10,000, and 20,000 epochs, respectively.

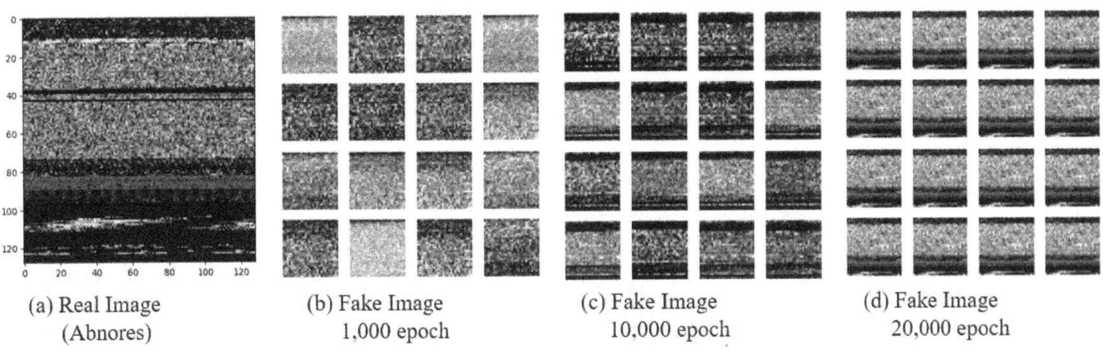

Figure 17. Fake image according to epochs.

At 1000 epochs, the generated fake images lack proper features and exhibit a noticeable presence of different noise patterns compared to the real images. As training progresses to 10,000 epochs, the ambient noise in the generated images starts to somewhat resemble the real images. Finally, at 20,000 epochs, we can observe a significant improvement, with the generated images exhibiting a closer resemblance to the real images. This iterative learning process demonstrates that with sufficient training, the generative model can achieve images that are more similar to the real ones.

Figure 18 depicts the loss values of the generator and discriminator during the data generation process. In both the first and second experiments, we observe that the generator loss value stabilizes around 3000 epochs. This indicates that the model is progressing without experiencing mode collapse during the generation of data. Mode collapse occurs when the generator fails to produce diverse samples and is stuck generating only a limited set of outputs. The stable loss values suggest that the training process is robust, ensuring that the generator continues to generate diverse and meaningful data throughout the epochs.

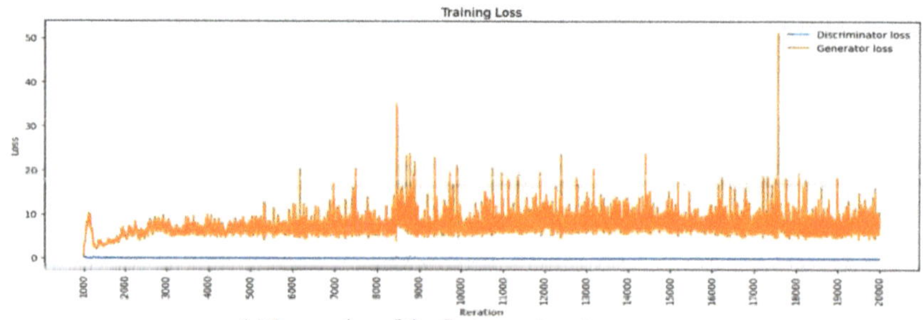

(a) Loss value of the first experiment

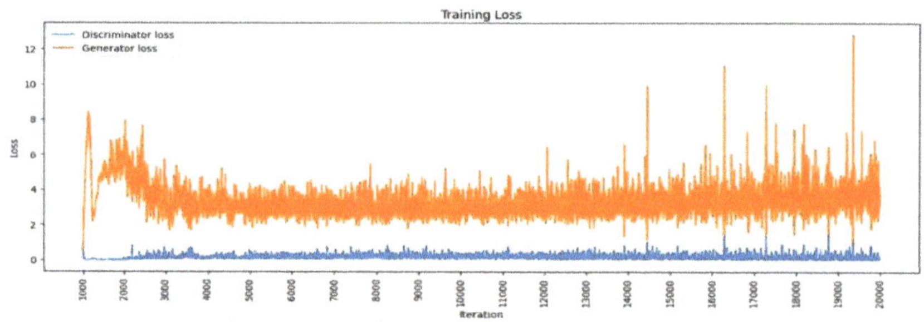

(b) Loss value of the second experiment

Figure 18. Loss value of generative model.

Table 5 presents the FID and cross-correlation values for each count. A smaller FID value indicates that the model generates data more similar to the original data. The average FID index obtained is 5.4603, which suggests that the generated data is relatively similar to the original data. Additionally, the average correlation coefficient is 0.8898, indicating a high degree of similarity between the original data and the generated data. These results, along with the FID, demonstrate that the generative model has successfully produced data that closely resembles the characteristics of the original data.

Table 5. FID and cross-correlation for each count.

Count	FID	Cross-Correlation
1	1.6798	0.9845
2	8.2314	0.7722
3	0.7558	0.9513
4	11.1742	0.8512
Average	5.4603	0.8898

6.3. Detection Performance Evaluation of Generated Data

CNN was used to verify the usability of the data generated using the generative model. Training was conducted by constructing generated images and noise images. The test for verification consisted of images not used in the generative model, generated images, and noise images. The CNN model for learning the generated image used the same model as Figure 11. This is the same model that tested the detection performance of the image size.

The model's performance is presented in Table 6, using evaluation indices such as accuracy, precision, recall, and f1-score. The training results demonstrate high performance, achieving an accuracy of approximately 0.98, a precision of 0.94, a recall of 0.96, and an f1-score value of 0.95 when training was conducted solely with generated images. However, the detection performance using the test set appears to be relatively lower. This can be attributed to the fact that the model was well-trained to distinguish the characteristics of the images during the training stage. Moreover, the inclusion of both malware and normal images in the practice may lead to slightly lower performance for detecting normal images. Nevertheless, the overall training results indicate promising performance, showcasing the model's ability to accurately classify and distinguish between different image types.

Table 6. Performance comparison results.

	Accuracy	Precision	Recall	F1-Score
performance on the training set	0.989957	0.955587	0.968748	0.968128
performance on the test set	0.968812	0.917035	0.927126	0.923067

Table 7 presents the results of the comparison with other models. It has been observed that the accuracy of our model surpasses that of previous studies. While precision, recall, and f1-score were only available in certain studies, our model still exhibited superior performance compared to the previous approaches.

Table 7. Performance comparison with existing research.

Model	Accuracy	Precision	Recall	F1-Score
Proposed model	0.973134	0.930811	0.942937	0.940597
L. Nataraj et al., 2011 [22]	0.8397	-	-	-
Seok S.H. & Kim H., 2016 [23]	0.962	0.917158	0.921102	0.919126
M. Shafiq et al., 2009 [26]	0.96	-	-	-
Anderson H.S. & Roth P., 2018 [27]	0.9299	-	-	-
H. Aghakhani et al., 2020 [28]	0.92	-	-	0.92
Saxe, J. & Berlin K., 2015 [29]	0.995785	-	-	-
E. Raff et al., 2017 [30]	0.826	-	-	-

The proposed model offers several distinct advantages over simulation in a digital twin environment. First, it demonstrates the ability to effectively detect new malware that did not exist before and combat new threats. Second, the detection model is trained without separate pre-work for the detection model. Unlike previous studies, our model simplifies the overall workflow by eliminating the need for feature extraction via specific filters or APIs. Third, it simplifies and streamlines the process by eliminating the requirement to directly run the model in a separate sandbox.

Leveraging these advantages, our proposed model represents a significant advance in malware detection, providing a more streamlined and robust approach compared to existing methods in the field. Simulation results of the digital twin environment demonstrate the robustness and effectiveness of the proposed model in protecting against evolving threats.

7. Discussion

In this paper, we present an innovative intelligent detection technology that leverages generative neural networks within a digital twin-based Industrial Internet of Things (IIoT) environment. Moreover, we propose a novel method for detecting malware solely based on images of the malware using convolutional neural networks (CNN), which is a powerful technique employing deep neural networks.

Three different datasets were used to test the performance of the proposed system. The performance of the system was measured in three stages. First, in the imaging stage, the detection performance of the 64 × 64 bilinear technique was the best. Second, a new malware image was created using the image interpolation method and size determined in the previous experiment, and similarity with the original images was measured. As a result of the measurement, the FID index was 5.4403 and the correlation coefficient was 0.8698, confirming high similarity. Third, the malware detection performance using the generated malware image was measured. The performance measurement result was detected with an accuracy of 0.97, and was measured with a precision of 0.93, a recall of 0.94, and an f1-score of 0.94. This confirmed a higher level of performance than previous studies.

The proposed system does not analyze the data collected in the IIoT environment on the user's system. Instead, it utilizes the digital twin to analyze malware in the digital space. As a result, it does not adversely affect the actual system, providing a safer and more secure analysis environment. Additionally, the system offers the advantage of quick initialization if any problem occurs in the digital space. By converting malware into images that reflect their characteristics, the proposed system eliminates the need to execute or directly analyze the code, minimizing potential risks. Leveraging generative adversarial networks allows for the generation of synthetic malware, which enhances the efficiency of the analysis process. The unpredictable nature of when and in what form malware will emerge poses a challenge for traditional detection methods. However, in the digital twin environment, the proposed system can quickly respond to new forms of malware by generating and analyzing them based on existing malware. This adaptability and responsiveness make the system well-suited for addressing emerging threats in the digital space.

The table provided below (Table 8) illustrates the datasets that will be contrasted with the research findings presented in Table 7. In this paper, the Mal60 dataset, the Malimg dataset and the VXHeaven dataset were employed. To enable an objective comparison of studies, we utilized datasets from prior research. The datasets from earlier studies predominantly comprise data made available from 2010 to 2015.

Table 8. Dataset status of research for performance comparison.

Model	Dataset
L. Nataraj et al., 2011 [22]	Host-Rx reference dataset Malhuer dataset VX Heavens virus collection
Seok S.H. & Kim H., 2016 [23]	Microsoft Malware Classification Challenge VX Heavens virus collection
M. Shafiq et al., 2009 [26]	VX Heavens virus collection Malfease dataset
Anderson H.S. & Roth P., 2018 [27]	VX Heavens virus collection Malfease dataset
H. Aghakhani et al., 2020 [28]	A commercial anti-malware vendor provided executables EMBER dataset
Saxe, J. & Berlin K., 2015 [29]	Invincea's own computer systems and customers networks
E. Raff et al., 2017 [30]	Provided by an anti-virus industry partner

This study has certain limitations. During the image downsizing process aimed at reducing computational complexity, some intricate characteristics of the malicious code may diminish. Detection can be influenced by various factors such as the chosen image interpolation method, image size, and the aspect ratio of image width and height. Additionally, accuracy can be compromised when the flows and patterns of malicious codes exhibit similarity only in minute sections. Moreover, augmenting the model's reliability necessitates training it with a contemporary and comprehensive dataset.

8. Conclusions

Research efforts persist in the realm of malicious code detection. Prior investigations have involved feature extraction through methods like image conversion or direct analysis of PE files for malicious code detection. Conversely, the exploration of techniques for real-time detection within virtual machines has also been pursued. Notably, there has been a surge in research centered on detection through machine learning technologies. Nonetheless, the extraction of feature points constitutes an additional requisite step. Furthermore, the analysis conducted within a user's system might potentially trigger system-related issues, thus presenting a drawback. Machine learning, by its nature, necessitates substantial volumes of data.

Therefore, in this study, an intelligent detection technology was introduced within a digital twin environment that replicates the real-world scenario digitally. By operating within a separate space from the actual system, this approach ensures no impact on the genuine system. Moreover, even in the event of an issue, swift reinitialization is feasible. Among the techniques harnessing deep neural networks for detection, convolutional neural networks (CNN) are utilized to identify malicious code solely using images of such code. Additionally, generative neural networks were employed to augment insufficient malicious code data or generate new instances of malicious code data. A comparison was conducted against seven models from previous studies. The results indicated that the proposed system achieved an f1-score of 0.94, showcasing the effectiveness of the proposed approach. The system exhibited a slightly favorable performance outcome.

In the future, our research will focus on developing technologies capable of surmounting and rectifying the aforementioned limitations. In pursuit of this objective, we will analyze the correspondence between the technique of categorizing malicious code by its function and representing it through intricate images, and the disassembly code associated with each function. Furthermore, we will explore methods to avert the loss of intricate features during image correction. Additionally, we plan to undertake supplementary investigations that encompass other factors, including the horizontal and vertical ratios of images.

Author Contributions: Conceptualization and methodology, H.-J.C., H.-K.Y., Y.-J.S. and A.R.K.; validation, Y.-J.S. and A.R.K.; formal analysis, H.-J.C. and H.-K.Y.; resources, H.-K.Y.; supervision, Y.-J.S. and A.R.K.; writing—original draft preparation, H.-J.C.; writing—review and editing, H.-K.Y., Y.-J.S. and A.R.K. All authors have read and agreed to the published version of the manuscript.

Funding: This research was supported by the MSIT (Ministry of Science and ICT), Korea, under the Innovative Human Resource Development for Local Intellectualization support program (IITP-2023-RS-2022-00156334) supervised by the IITP (Institute for Information & communications Technology Planning & Evaluation).

Institutional Review Board Statement: Not applicable.

Informed Consent Statement: Not applicable.

Data Availability Statement: Not applicable.

Conflicts of Interest: The authors declare no conflict of interest.

References

1. Peter, O.; Pradhan, A.; Mbohwa, C. Industrial internet of things (IIoT): Opportunities, challenges, and requirements in manufacturing businesses in emerging economies. *Procedia Comput. Sci.* **2023**, *217*, 856–865. [CrossRef]
2. Sobb, T.; Turnbull, B.; Moustafa, N.; Sobb, T.; Turnbull, B.; Moustafa, N. Supply chain 4.0: A survey of cyber security challenges, solutions and future directions. *Electronics* **2020**, *9*, 1864. [CrossRef]
3. Vaza, R.N.; Prajapati, R.; Rathod, D.; Vaghela, D. Developing a novel methodology for virtual machine introspection to classify unknown malware functions. *Peer-to-Peer Netw. Appl.* **2022**, *15*, 793–810. [CrossRef]
4. Vasan, D.; Alazab, M.; Wassan, S.; Safaei, B.; Zheng, Q. Image-Based malware classification using ensemble of CNN architectures (IMCEC). *Comput. Secur.* **2020**, *92*, 101748. [CrossRef]
5. Shaukat, K.; Luo, S.; Varadharajan, V. A novel deep learning-based approach for malware detection. *Eng. Appl. Artif. Intell.* **2023**, *122*, 106030. [CrossRef]
6. Shorten, C.; Khoshgoftaar, T.M. A survey on image data augmentation for deep learning. *J. Big Data* **2019**, *6*, 60. [CrossRef]
7. Berg, S.; Kutra, D.; Kroeger, T.; Straehle, C.N.; Kausler, B.X.; Haubold, C.; Schiegg, M.; Ales, J.; Beier, T.; Rudy, M. Ilastik: Interactive machine learning for (bio) image analysis. *Nat. Methods* **2019**, *16*, 1226–1232. [CrossRef]
8. Grieves, M. Digital Twin Certified: Employing Virtual Testing of Digital Twins in Manufacturing to Ensure Quality Products. *Machines* **2023**, *11*, 808. [CrossRef]

9. Wu, J.; Yang, Y.; Cheng, X.; Zuo, H.; Cheng, Z. The development of digital twin technology review. In Proceedings of the 2020 Chinese Automation Congress (CAC), Shanghai, China, 6–8 November 2020; pp. 4901–4906. [CrossRef]
10. Lo, C.; Chen, C.; Zhong, R.Y. A review of digital twin in product design and development. *Adv. Eng. Inform.* **2021**, *48*, 101297. [CrossRef]
11. Rasheed, A.; San, O.; Kvamsdal, T. Digital twin: Values, challenges and enablers from a modeling perspective. *IEEE Access* **2020**, *8*, 21980–22012. [CrossRef]
12. Aboaoja, F.A.; Zainal, A.; Ghaleb, F.A.; Al-rimy, B.A.S.; Eisa, T.A.E.; Elnour, A.A.H. Malware detection issues, challenges, and future directions: A survey. *Appl. Sci.* **2022**, *12*, 8482. [CrossRef]
13. Bayazit, E.C.; Sahingoz, O.K.; Dogan, B. Neural network based Android malware detection with different IP coding methods. In Proceedings of the 2021 3rd International Congress on Human-Computer Interaction, Optimization and Robotic Applications (HORA), Ankara, Turkey, 11–13 June 2021; pp. 1–6. [CrossRef]
14. Bansal, M.; Goyal, A.; Choudhary, A. A comparative analysis of K-nearest neighbor, genetic, support vector machine, decision tree, and long short term memory algorithms in machine learning. *Decis. Anal. J.* **2022**, 3, 100071. [CrossRef]
15. Zheng, H.; Fu, J.; Zha, Z.-J.; Luo, J. Learning deep bilinear transformation for fine-grained image representation. In Proceedings of the 33rd International Conference on Neural Information Processing Systems, Vancouver, BC, Canada, 8–14 December 2019.
16. Khaledyan, D.; Amirany, A.; Jafari, K.; Moaiyeri, M.H.; Khuzani, A.Z.; Mashhadi, N. Low-cost implementation of bilinear and bicubic image interpolation for real-time image super-resolution. In Proceedings of the 2020 IEEE Global Humanitarian Technology Conference (GHTC), Seattle, WA, USA, 29 October–1 November 2020; pp. 1–5. [CrossRef]
17. Creswell, A.; White, T.; Dumoulin, V.; Arulkumaran, K.; Sengupta, B.; Bharath, A.A. Generative adversarial networks: An overview. *IEEE Signal Process. Mag.* **2018**, *35*, 53–65. [CrossRef]
18. Goodfellow, I.; Pouget-Abadie, J.; Mirza, M.; Xu, B.; Warde-Farley, D.; Ozair, S.; Courville, A.; Bengio, Y. Generative adversarial networks. *Commun. ACM* **2020**, *63*, 139–144. [CrossRef]
19. Goodfellow, I. Nips 2016 tutorial: Generative adversarial networks. *arXiv* **2016**. [CrossRef]
20. Pokhrel, A.; Katta, V.; Colomo-Palacios, R. Digital twin for cybersecurity incident prediction: A multivocal literature review. In Proceedings of the IEEE/ACM 42nd International Conference on Software Engineering Workshops, Seoul, Republic of Korea, 27 June–19 July 2020; pp. 671–678. [CrossRef]
21. Eckhart, M.; Ekelhart, A. Digital twins for cyber-physical systems security: State of the art and outlook. In *Security and Quality in Cyber-Physical Systems Engineering: With Forewords by Robert M. Lee and Tom Gilb*; Springer: Cham, Switzerland, 2019; pp. 383–412. [CrossRef]
22. Nataraj, L.; Yegneswaran, V.; Porras, P.; Zhang, J. A comparative assessment of malware classification using binary texture analysis and dynamic analysis. In Proceedings of the 4th ACM Workshop on Security and Artificial Intelligence, Chicago, IL, USA, 21 October 2011; pp. 21–30. [CrossRef]
23. Seok, S.; Kim, H. Visualized Malware Classification Based-on Convolutional Neural Network. *J. Korea Inst. Inf. Secur. Cryptol.* **2016**, *26*, 197–208. [CrossRef]
24. Atitallah, S.B.; Driss, M.; Almomani, I. A novel detection and multi-classification approach for IoT-malware using random forest voting of fine-tuning convolutional neural networks. *Sensors* **2022**, *22*, 4302. [CrossRef] [PubMed]
25. Gibert, D.; Mateu, C.; Planes, J.; Vicens, R. Using convolutional neural networks for classification of malware represented as images. *J. Comput. Virol. Hacking Tech.* **2019**, *15*, 15–28. [CrossRef]
26. Shafiq, M.Z.; Tabish, S.M.; Mirza, F.; Farooq, M. Pe-miner: Mining structural information to detect malicious executables in realtime. In *Recent Advances in Intrusion Detection: 12th International Symposium, RAID 2009, Saint-Malo, France, September 23–25, 2009, Proceedings*; Springer: Berlin/Heidelberg, Germany, 2009; pp. 121–141. [CrossRef]
27. Anderson, H.S.; Roth, P. Ember: An open dataset for training static pe malware machine learning models. *arXiv* **2018**. [CrossRef]
28. Aghakhani, H.; Gritti, F.; Mecca, F.; Lindorfer, M.; Ortolani, S.; Balzarotti, D.; Vigna, G.; Kruegel, C. When malware is packin'heat; limits of machine learning classifiers based on static analysis features. In Proceedings of the Network and Distributed Systems Security (NDSS) Symposium 2020, San Diego, CA, USA, 23–26 February 2020. [CrossRef]
29. Saxe, J.; Berlin, K. Deep neural network based malware detection using two dimensional binary program features. In Proceedings of the 2015 10th International Conference on Malicious and Unwanted Software (MALWARE), Fajardo, PR, USA, 20–22 October 2015; pp. 11–20. [CrossRef]
30. Raff, E.; Barker, J.; Sylvester, J.; Brandon, R.; Catanzaro, B.; Nicholas, C. Malware detection by eating a whole exe. *arXiv* **2017**. [CrossRef]
31. Kalash, M.; Rochan, M.; Mohammed, N.; Bruce, N.D.; Wang, Y.; Iqbal, F. Malware classification with deep convolutional neural networks. In Proceedings of the 2018 9th IFIP International Conference on New Technologies, Mobility and Security (NTMS), Paris, France, 26–28 February 2018; pp. 1–5. [CrossRef]
32. Singh, J.; Thakur, D.; Gera, T.; Shah, B.; Abuhmed, T.; Ali, F. Classification and analysis of android malware images using feature fusion technique. *IEEE Access* **2021**, *9*, 90102–90117. [CrossRef]
33. Github. Malimg Dataset. Available online: https://github.com/danielgibert/mlw_classification_cnn_img (accessed on 19 April 2022).
34. Kamundala, E.K.; Kim, C.H. CNN Model to Classify Malware Using Image Feature. *IISE Trans. Comput. Pract.* **2018**, *24*, 256–261. [CrossRef]

35. AlGarni, M.D.; AlRoobaea, R.; Almotiri, J.; Ullah, S.S.; Hussain, S.; Umar, F. An efficient convolutional neural network with transfer learning for malware classification. *Wirel. Commun. Mob. Comput.* **2022**, *2022*, 4841741. [CrossRef]
36. Go, J.H.; Jan, T.; Mohanty, M.; Patel, O.P.; Puthal, D.; Prasad, M. Visualization approach for malware classification with ResNeXt. In Proceedings of the 2020 IEEE Congress on Evolutionary Computation (CEC), Glasgow, UK, 19–24 July 2020; pp. 1–7. [CrossRef]
37. Bhodia, N.; Prajapati, P.; Di Troia, F.; Stamp, M. Transfer learning for image-based malware classification. *arXiv* **2019**. [CrossRef]
38. Github. Mal60 Dataset. Available online: https://github.com/pukekaka/mal60 (accessed on 30 April 2022).
39. Kang, M.C.; Kim, H.K. Rare Malware Classification Using Memory Augmented Neural Networks. *J. Korea Inst. Inf. Secur. Cryptol.* **2018**, *28*, 847–857. [CrossRef]
40. VX Heaven. Vx Heaven Virus Collection 2010-05-18. Available online: http://vxheaven.org/ (accessed on 18 May 2022).
41. VirusTotal. Virus Total. Available online: https://virustotal.com (accessed on 22 April 2022).

applied sciences

MDPI

Article

A Cross-Modal Dynamic Attention Neural Architecture to Detect Anomalies in Data Streams from Smart Communication Environments

Konstantinos Demertzis [1,*], Konstantinos Rantos [1], Lykourgos Magafas [2] and Lazaros Iliadis [3]

1 Department of Computer Science, School of Science, International Hellenic University, 65404 Kavala, Greece
2 Department of Physics, School of Science, Kavala Campus, International Hellenic University, 65404 Kavala, Greece
3 Department of Civil Engineering, School of Engineering, Democritus University of Thrace, 67100 Xanthi, Greece; liliadis@civil.duth.gr
* Correspondence: kdemertzis@emt.ihu.gr

Abstract: Detecting anomalies in data streams from smart communication environments is a challenging problem that can benefit from novel learning techniques. The Attention Mechanism is a very promising architecture for addressing this problem. It allows the model to focus on specific parts of the input data when processing it, improving its ability to understand the meaning of specific parts in context and make more accurate predictions. This paper presents a Cross-Modal Dynamic Attention Neural Architecture (CM-DANA) by expanding on state-of-the-art techniques. It is a novel dynamic attention mechanism that can be trained end-to-end along with the rest of the model using multimodal data streams. The attention mechanism calculates attention weights for each position in the input data based on the model's current state by a hybrid method called Cross-Modal Attention. Specifically, the proposed model uses multimodal learning tasks where the input data comes from different cyber modalities. It combines the relevant input data using these weights to produce an attention vector in order to detect suspicious abnormal behavior. We demonstrate the effectiveness of our approach on a cyber security anomalies detection task using multiple data streams from smart communication environments.

Keywords: cross-modal learning tasks; dynamic attention mechanism; neural architecture; anomaly detection; data streams; smart communication environments

Citation: Demertzis, K.; Rantos, K.; Magafas, L.; Iliadis, L. A Cross-Modal Dynamic Attention Neural Architecture to Detect Anomalies in Data Streams from Smart Communication Environments. *Appl. Sci.* **2023**, *13*, 9648. https://doi.org/10.3390/app13179648

Academic Editors: Peter R.J. Trim and Yang-Im Lee

Received: 1 August 2023
Revised: 24 August 2023
Accepted: 24 August 2023
Published: 25 August 2023

1. Introduction

Detecting anomalies in data streams [1] from smart communication environments is a critical problem that has significant implications for various applications, including cyber security [2], monitoring cyber-physical systems [3], and controlling the industrial ecosystem [4]. The vast amount of data generated in these environments makes it difficult to detect abnormal behavior in real-time, which can lead to significant damages and security breaches [5]. Anomaly detection in these data streams is challenging due to the volume and complexity of the data and the need for real-time detection to prevent potential damages or security breaches [6,7]. Traditional methods for anomaly detection in data streams rely on statistical techniques or rule-based systems, which may not be effective in identifying subtle or unknown anomalies [8]. Machine learning approaches, particularly deep learning methods, have shown promise in addressing this challenge by enabling automated and accurate detection of anomalies in complex data streams [9].

One of the key advantages of deep learning methods for anomaly detection is the ability to learn relevant features from the input data without relying on pre-defined rules or assumptions. Attention mechanisms, in particular, have emerged as a powerful tool for capturing relevant input data features and improving neural network performance in various applications [2]. Recent research has focused on developing novel deep-learning architectures that effectively leverage attention mechanisms to detect anomalies in data streams from smart communication environments. These architectures often use simple attention mechanisms that can adapt to changes in the input data over time and can be trained end-to-end using data streams to capture the complex interactions between sophisticated processes [10,11].

Simple attention involves computing a fixed set of attention weights for the input data learned during training based on the task-specific objective function. The network then uses these fixed attention weights to weigh the input features in subsequent neural network layers. These simple attention mechanisms have become a powerful tool for capturing relevant input data features and improving neural network performance in various applications [12].

On the other hand, dynamic attention allows the network to adjust the attention weights at each time step to give more or less importance to different parts of the input sequence depending on their relevance to the task. Dynamic attention mechanisms can be useful in applications where the types and frequencies of anomalies may change over time, allowing the model to adapt to changes in the input data [13].

Both simple and dynamic attention mechanisms have strengths and weaknesses depending on the specific application and data. Simple attention is more straightforward and can be effective in many cases. In contrast, dynamic attention can improve the model's ability to adapt to changes in the input data over time. The appropriate attention mechanism type depends on the input data's nature and task [14,15].

This paper presents a novel and holistic neural architecture called CM-DANA for detecting anomalies in data streams from smart communication environments. The model is based on a hybrid approach that combines attention mechanisms and multimodal learning techniques to capture the complex interactions between different modalities of data effectively. The CM-DANA model uses a dynamic attention mechanism that calculates attention weights for each position in the input data based on the model's current state. This attention mechanism is a location-based attention mechanism that uses the position of the input features in the sequence of real-time data streams to calculate the attention weights. The more sophisticated character of the proposed model is that it is trained end-to-end using multimodal data streams. This allows the model to attend to different features in different modalities based on the model's current state and detect suspicious abnormal behavior by combining the relevant input data from different modalities using adaptive attention weights.

The motivation for the CM-DANA model is to improve the accuracy and efficiency of anomaly detection in data streams from smart communication environments by effectively capturing relevant features and suppressing noisy or irrelevant features. The use of dynamic attention and multimodal learning techniques allows the model to attend to different features in different modalities based on the model's current state, which can improve its ability to detect suspicious abnormal behavior in real-time. Overall, the motivation for the paper is to develop a novel deep-learning architecture that can effectively detect anomalies in data streams from smart communication environments. By leveraging attention mechanisms and multimodal learning techniques, the CM-DANA model, presented for the first time in the literature, aims to be a promising approach to improving the accuracy and efficiency of anomaly detection in various applications.

2. Literature Review

Anomaly detection in data streams has been an active research area due to the increasing volume and complexity of data generated by IoT devices and smart environments [2]. Traditional anomaly detection methods, such as statistical techniques [5], clustering [16], and classification [8], have been applied to data streams [6], with varying degrees of success. However, they often struggle to adapt to the dynamic nature of data streams, which may have changing distributions and evolving patterns [5]. For example, during a timed event, the traffic pattern can change dramatically, potentially causing statistical methods that rely on historical data to label the surge in traffic as an anomaly due to the shift in statistical properties like mean and variance [17]. In addition, the traditional clustering methods might not recognize the sudden appearance of a new cluster as an anomaly, leading to delayed detection, or traditional classifiers might struggle to identify novel patterns that were not present in the training data [6]. In summary, traditional anomaly detection methods have limitations that become more pronounced in dynamic data streams with changing distributions and evolving patterns. The technical challenges of concept drift [17], high-dimensional data [7], computational efficiency [18], and feature engineering [19] contribute to their struggles in adapting to these scenarios. This has prompted the exploration of more advanced techniques, including deep learning-based approaches, which have shown better adaptability and scalability in handling the dynamic nature of data streams.

Recently, deep learning-based techniques [2] have been proposed for data stream anomaly detection, including autoencoders [20], recurrent neural networks (RNNs) [21], and convolutional neural networks (CNNs) [22]. These methods have demonstrated better adaptability and scalability compared to traditional methods, but they still face challenges in dealing with heterogeneous data types and efficiently focusing on relevant features. Specifically, deep learning techniques face significant challenges in dealing with heterogeneous data types and efficiently focusing on relevant features [2]. These challenges include handling diverse data types, ensuring feature relevance and selection, addressing data imbalance, and interpreting deep models [23]. Heterogeneous data types, such as numerical, categorical, text, image, and time series data, can be challenging to integrate and process effectively [7,24]. Researchers are exploring techniques to handle multiple data types [25], such as specialized network architectures [26] or converting different data types into a common feature space [27]. Feature engineering and selection techniques aim to identify the most informative features, while data imbalance can lead to models favoring the majority class and performing poorly in anomaly detection [28]. Interpretable models are crucial to understanding the underlying patterns learned by deep learning models, such as in manufacturing processes where engineers need to know which factors contributed to anomaly detection [29]. Researchers are developing techniques to explain deep model decisions, such as attention mechanisms, feature attribution methods, and gradient-based visualizations, to provide insights into which features were influential in making anomaly predictions [30].

Cross-modal learning [31] refers to the process of learning shared representations from multiple data modalities, such as images, text, and audio. It has shown great potential in various applications, including multimedia retrieval [32], recommendation systems [33], and multimodal sentiment analysis [25]. Several methods have been proposed for cross-modal learning, including deep neural networks [34], matrix factorization [35], and probabilistic graphical models [36]. Recently, cross-modal learning has been integrated with attention mechanisms to improve the interpretability and performance of the learned representations [37–39]. However, the application of cross-modal learning to anomaly detection in data streams from smart communication environments is still relatively unexplored. This approach offers several benefits, but also presents challenges, such as developing effective fusion strategies, addressing domain-specific issues, dealing with varying data modalities, and managing computational complexity [36]. Additionally, data privacy and ethics are critical concerns in smart communication environments, and researchers must address these concerns when designing cross-modal anomaly detection systems [25].

Attention mechanisms have been introduced in neural networks to help the model focus on the most relevant parts of the input data for a specific task [12]. The concept of attention was initially proposed in the context of Natural Language Processing (NLP) [15] and has since been extended to various domains, such as computer vision [14] and speech recognition [40]. Different types of attention mechanisms have been proposed, including self-attention [41], local attention [42], and global attention [43]. Attention mechanisms have also been combined with other neural network architectures, such as RNNs [44],

CNNs [45], and Transformer models [46], to improve their performance and interpretability. The application of attention mechanisms in anomaly detection has shown promising results, particularly in terms of handling large-scale and high-dimensional data [27]. However, incorporating dynamic attention mechanisms into cross-modal learning for anomaly detection in data streams remains a challenge. Specifically, incorporating dynamic attention mechanisms into cross-modal learning for anomaly detection in data streams requires a careful balance between adaptability, efficiency, interpretability, and performance [37]. Researchers need to devise novel approaches that address these challenges and tailor dynamic attention mechanisms to the specific requirements of dynamic data streams and multi-modal data fusion [14]. Despite the challenges, successfully implementing dynamic attention can significantly enhance the accuracy and robustness of anomaly detection systems in complex and rapidly evolving environments [12].

In summary, research gaps from the literature review in anomaly detection in dynamic environments include adapting traditional methods to handle changing distributions and patterns, integrating heterogeneous data types, improving the interpretability of deep models, exploring cross-modal anomaly detection, incorporating dynamic attention mechanisms, and addressing privacy and ethics concerns. These areas highlight opportunities for innovation and exploration in anomaly detection in smart communication environments, particularly in integrating heterogeneous data types, enhancing interpretability, and effectively utilizing dynamic attention mechanisms and cross-modal learning techniques.

By addressing these gaps, the proposed approach proposes a more effective anomaly detection method that can handle diverse data types, improve interpretability, and maintain privacy and ethics in cross-modal anomaly detection systems. Specifically, this paper presents a novel CM-DANA for detecting anomalies in data streams generated from smart communication environments. The proposed architecture leverages the advantages of cross-modal learning and dynamic attention mechanisms to effectively analyze heterogeneous data streams from different cyber modalities and identify anomalous patterns in real-time. Recent advancements inspire this approach in cross-modal learning and attention mechanisms in neural networks. Cross-modal learning has shown its potential in various applications where data comes from multiple sources or modalities, while attention mechanisms have been successful in helping models focus on relevant parts of input data for specific tasks. By combining these two concepts, our proposed approach not only improves the overall performance of anomaly detection but also enhances the interpretability and adaptability of the model in handling diverse and evolving data patterns.

The proposed method addresses research gaps in anomaly detection in dynamic data streams from smart communication environments by enhancing traditional methods, integrating heterogeneous data types, enhancing interpretable deep models, incorporating cross-modal learning, and incorporating dynamic attention mechanisms. These contributions can help develop more accurate, adaptive, and interpretable anomaly detection systems that can effectively operate in complex and rapidly evolving scenarios. By incorporating concepts from both the dynamic attention and anomaly detection domain, the proposed CM-DANA technique ensures that data from different modalities are integrated in an accurate way. By focusing on these contributions, the proposed approach makes significant strides in advancing the field of anomaly detection in dynamic data streams from smart communication environments.

3. Materials and Methods

The proposed CM-DANA consists of 4 main modules: the Feature Extraction Module, Cross-modal Learning Module, the Dynamic Attention Module, and the Anomaly Detection Module. The architecture is designed to process and analyze heterogeneous data streams from different cyber modalities, such as network traffic, log files, and user behavior patterns. The Feature Extraction Module extracts features from each modality; the Cross-modal Learning Module learns shared representations. The Dynamic Attention Module then computes attention weights to emphasize the most relevant features, forming an attention vector. Finally, the Anomaly Detection Module uses the attention vector to identify anomalous patterns.

An efficient and novel combination of intelligent algorithms is used in the CM-DANA method. Specifically, it is a combination of Convolutional Neural Networks (CNNs) for feature extraction, Transformers for cross-modal learning, Gated Recurrent Units (GRUs) for dynamic attention, and Theil-Sen Regressor as an anomaly detector. This combination leverages the strengths of each algorithm to enhance predictability performance. A high-level representation of the CM-DANA methodology is presented in the following Algorithm 1:

Algorithm 1 Pseudocode of CM-DANA methodology

```
# Feature Extraction Module
def feature_extraction(input_data):
   # Input Data Preparation
   preprocessed_data = preprocess(input_data)
   # Convolutional Layers
   convolution_output = apply_convolutional_layers(preprocessed_data)
   # Activation Functions
   activated_output = apply_activation_functions(convolution_output)
   # Pooling Layers
   pooled_output = apply_pooling_layers(activated_output)
   # Flattening
   flattened_output = flatten(pooled_output)
   # Fully Connected Layers
   features = apply_fully_connected_layers(flattened_output)
   return features
# Cross-modal Learning Module
def cross_modal_learning(modalities):
   shared_representations = []
   for modality in modalities:
      features = feature_extraction(modality)
      shared_representations.append(features)
   # Process shared representations using Transformers
   processed_representations = process_with_transformers(shared_representations)
   return processed_representations
# Dynamic Attention Module
def dynamic_attention(shared_representations):
   attention_vector = []
   for representation in shared_representations:
      attention_weights = compute_attention_weights(representation)
      attention_vector.append(weighted_sum(representation, attention_weights))
   return attention_vector
# Anomaly Detection Module
def anomaly_detection(attention_vector):
   # Use TheilSenRegressor for linear regression
   model = TheilSenRegressor()
   model.fit(attention_vector)
   # Calculate residuals
   predicted_values = model.predict(attention_vector)
   residuals = calculate_residuals(attention_vector, predicted_values)
   # Set dynamic threshold
   threshold = set_dynamic_threshold(residuals)
   # Identify anomalies
   anomalies = identify_anomalies(residuals, threshold)
   return anomalies
# CM-DANA Methodology
def CM_DANA(input_modalities):
```

Algorithm 1 *Cont.*

```
# Feature Extraction Module
extracted_features = feature_extraction(input_modalities)
# Cross-modal Learning Module
shared_representations = cross_modal_learning(extracted_features)
# Dynamic Attention Module
attention_vector = dynamic_attention(shared_representations)
# Anomaly Detection Module
anomalies = anomaly_detection(attention_vector)
return anomalies
```

The end-to-end training approach of the CM-DANA model ensures that the model learns to identify and capture the complex interactions between different modalities of data. This leads to more accurate anomaly detection in smart communication environments where data streams from multiple sources can provide valuable information about anomalies and potential threats.

It must be noted that the 4 modules of the proposed methodology introduce significant innovative aspects that collectively enhance the accuracy and efficiency of anomaly detection in the proposed CM-DANA model. Specifically, the use of CNNs for feature extraction is an innovation that tailors the architecture to the nuances of cybersecurity data. While CNNs are commonly used for image analysis, adapting them to cybersecurity data highlights a key innovation. By processing diverse modalities like network traffic, log files, and behavior patterns with CNNs, the architecture acknowledges the spatial features that hold significance in cybersecurity contexts. This customized feature extraction enhances anomaly detection's precision in identifying spatial irregularities hidden within complex data patterns.

In addition, the integration of Transformers in the Cross-modal Learning Module is an innovative approach to capturing cross-modal interactions and dependencies. Transformers were originally designed for sequence-to-sequence tasks but adapting them for cross-modal learning is a novel application. By processing different modalities with dedicated subnetworks and then aggregating shared representations using Transformers, the architecture harnesses the strength of Transformers in capturing contextual and long-range relationships within different types of data. This integration contributes to the architecture's ability to learn complex patterns across modalities.

Also, the Dynamic Attention Module introduces innovation by employing GRUs to compute attention weights. While attention mechanisms are common in machine learning, using GRUs for dynamic attention reflects an innovative application. GRUs, being recurrent neural network components, adaptively adjust attention weights based on the current state and input sequence. This dynamic attention mechanism helps the model focus on the most relevant features at each time step, allowing it to adapt to changing data patterns and improving anomaly detection accuracy.

Moreover, the application of the Theil-Sen Regressor for anomaly detection is an innovative choice. While the Theil-Sen Regressor is primarily used for linear regression, adapting it as an anomaly detection algorithm shows innovation. By fitting a linear model to the attention vector and calculating residuals, the architecture detects anomalies in a manner that accounts for potential outliers and noise, contributing to robust and accurate anomaly identification.

Collectively, the innovation of the CM-DANA methodology lies in its thoughtful combination of these components and algorithms to address the challenges of detecting anomalies across multiple cyber modalities. The details about the specific components are presented in the following subsections.

3.1. Feature Extraction Module

The feature extraction module is responsible for processing the input data from different modalities and extracting relevant features that capture the characteristics of the data. It plays a crucial role in representing the data in a format that the subsequent modules can effectively analyze. The features extracted from each subnetwork are then passed through a fusion layer, which learns to combine the multimodal features into a single shared representation. This representation is used as the input for the subsequent cross-modal learning module.

It must be noted that the input processing layer of the features extraction module takes in data streams from multiple modalities, such as data acquisition systems, sensors, or web services. CNNs are particularly effective at extracting spatial features from input data, making them suitable for processing certain modalities. Specifically, CNNs architecture (Figure 1) have shown excellent performance in extracting spatial features from data, making them suitable for processing data streams from multiple cybersecurity modalities.

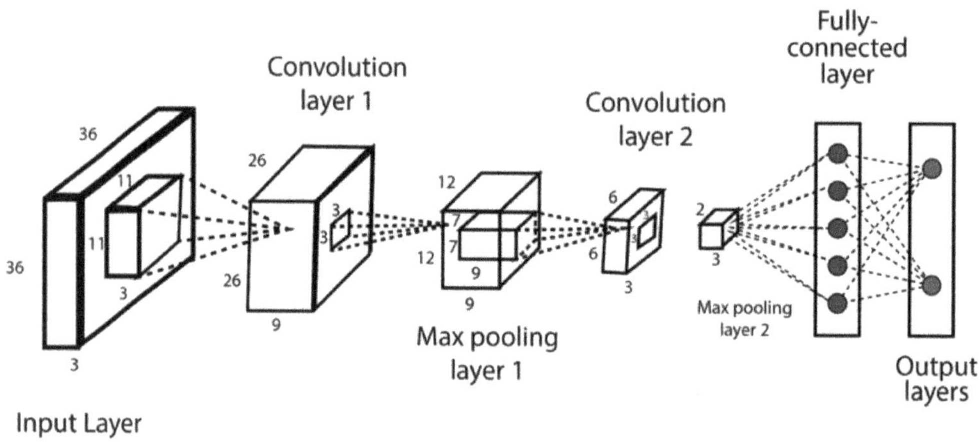

Figure 1. A Convolutional Neural Network (CNN).

The integration of Convolutional Neural Networks (CNNs) for the purpose of feature extraction within the CM-DANA architecture involves a series of sequential procedures. Specifically, commencing with Step 1, the preparation of input data is undertaken. Data originating from diverse modalities is subjected to preprocessing procedures to conform to formats conducive to CNN-compatible representations. In Step 2, the architecture employs a succession of convolutional layers to process the input data. Within these layers, convolutions are executed using adaptable filters, which effectively capture spatial features across varying levels of abstraction. The parameter adaptability, encompassing filter depth and size, assumes significance in ensuring proficient feature extraction that corresponds to the intricacy inherent in the data.

Following each convolutional layer, as elucidated in Step 3, non-linear activation functions, such as the Rectified Linear Unit (ReLU), are introduced. This introduction of non-linearity serves the purpose of capturing intricate patterns present within the data. Strategic insertion of pooling layers, as delineated in Step 4, contributes to the overall architecture. These pooling layers, which encompass MaxPooling and AveragePooling, serve the dual role of diminishing computational complexity and preserving pertinent features. The outcome of these layers is a downsampling of feature maps, thereby fostering spatial invariance.

Step 5 entails the flattening of output feature maps that are generated by the convolutional layers. This flattening operation transforms the feature maps into one-dimensional vectors, thereby preparing them for subsequent stages of processing. Transitioning to Step 6, the flattened features are directed into fully connected layers. The role of these layers is to enhance the extracted features by capturing more complex relationships and representations that exist at higher levels of abstraction. Finally, Step 7 culminates in the generation of a distinct output. The output stems from the fully connected layers and serves as a unique representation of features. This representation, in essence, encapsulates crucial spatial information inherent within the input data.

The innovation of CM-DANA becomes evident in its incorporation of CNNs tailored for anomaly detection across smart communication environments. Specifically, the proposed approach introduces a pioneering innovation that lies in the thoughtful integration of CNNs module for feature extraction, specifically designed to address the challenges of cybersecurity modalities by extracting spatial features that hold particular significance in cybersecurity contexts. Unlike conventional anomaly detection approaches, which often employ generic feature extractors, CM-DANA tailors its feature extraction to the nuances of the data, enhancing its anomaly detection prowess.

CNNs are particularly adept at capturing spatial patterns within data, while the proposed architecture leverages the inherent ability to learn hierarchies of features, enabling them to uncover intricate relationships within the data streams. This feature amplifies the model's potential to detect anomalies hidden within complex data patterns in real-time.

The CM-DANA architecture's uniqueness further emerges in its fusion of features across modalities. Extracted features from distinct subnetworks are merged through a fusion layer, creating a unified representation that encodes the combined knowledge of different data streams. By integrating CNNs for feature extraction, CM-DANA elevates this fusion process, as it now incorporates spatial insights that other architectures might overlook. This enables the architecture to capture cross-modal interactions and dependencies more effectively.

In addition, the incorporation of CNNs amplifies the architecture's ability to capture localized and global spatial features. As anomalies within smart communication environments often manifest as intricate spatial irregularities, the proposed model's innovative CNN-based feature extraction enhances its precision in pinpointing subtle anomalies that might be missed by traditional methods. This leads to more accurate and efficient anomaly detection in complex, evolving data streams.

Finally, it must be noted that the CNNs within the CM-DANA model do not operate in isolation. They serve as integral components within the cross-modal learning module, collaborating with other components to decipher complex interactions between data modalities. By enriching the feature extraction step with CNNs, the model contributes to more informative feature representations that empower subsequent modules in making more accurate anomaly detection decisions.

By leveraging CNNs to extract features, the CM-DANA architecture stands out as a promising method for capturing complex interactions between data modalities and advancing anomaly detection capabilities.

3.2. Cross-Modal Learning Module

The cross-modal learning module is responsible for processing the input data from multiple modalities and learning shared representations. Each modality is processed by a dedicated subnetwork tailored to the specific data type. Transformers have proven to be highly effective in modeling long-range dependencies and capturing contextual information. In the cross-modal learning module, transformers are used to process data patterns.

An illustration of the transformer model's core components where layers were normalized after multiheaded attention is depicted in Figure 2 [47].

Transformers excel at learning representations from sequential data and can capture the temporal relationships within cybersecurity modalities like log files, network traffic, and behavior patterns.

The foundational constituents of a transformer architecture have been extensively delineated in prior literature [12,47,48]. Firstly, the architecture inherently encompasses an Encoder–Decoder Structure, manifesting as two distinctive modules: an encoder tasked with assimilating the input sequence, and a decoder orchestrating the generation of the corresponding output sequence.

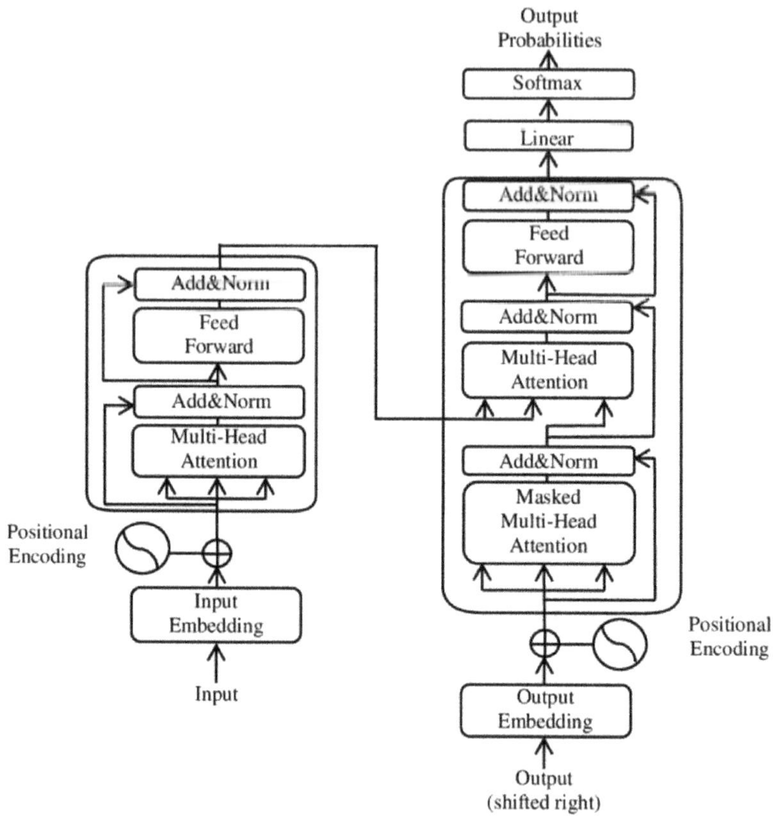

Figure 2. Transformer model's core components.

Secondly, a pivotal mechanism operative within this framework is the Self-Attention Mechanism. This mechanism engenders the capacity for individual elements within the input sequence to selectively attend to other constituent elements within the same sequence. In effect, attention weights are computed, thereby endowing the model with the faculty to emphasize pertinent informational elements during the input processing phase.

In tandem with this, the paradigm incorporates the Multi-Head Attention mechanism, which entails the integration of multiple attention layers, colloquially referred to as "heads". This arrangement facilitates the discernment of disparate forms of interrelationships existing amongst the elements comprising the input sequence. Concatenation or amalgamation of the outputs stemming from these distinct heads affords a more exhaustive and holistic representation.

Subsequently, following the application of the self-attention mechanism, the architecture integrates Feed-Forward Neural Networks. These neural networks serve to further process the representations that have been subjected to the self-attention mechanism, augmenting the model's ability to capture intricate patterns within the data.

Furthermore, an intrinsic challenge pertaining to the transformer architecture pertains to its inability to inherently fathom sequential information. To circumvent this, the framework incorporates Positional Encoding. By integrating positional encoding into the input embeddings, the model gains access to crucial positional information. This augmentation equips the transformer with the proficiency to effectively manage and interpret sequential data.

The first sublayer obtains the decoder stack's previous output, augments it with positional information, then applies multi-head self-attention to it. While the encoder is meant to attend to all words in the input sequence regardless of their position, the decoder is adjusted to only attend to the words that come before them. As a result, the prediction for a word at position i can only be based on the known outputs for the words preceding it in the sequence. This is accomplished in the multi-head attention mechanism (which implements numerous, single attention functions simultaneously) by applying a mask to the values obtained by the scaled multiplication of matrices Q and K.

Masking is accomplished by suppressing matrix values that would otherwise correspond to illegal connections [49]:

$$\text{mask}\left(QK^T\right) = \text{mask}\left(\begin{bmatrix} e_{11} & e_{12} & \cdots & e_{1n} \\ e_{21} & e_{22} & \cdots & e_{2n} \\ \vdots & \vdots & \ddots & \vdots \\ e_{m1} & e_{m2} & \cdots & e_{mn} \end{bmatrix}\right) = \begin{bmatrix} e_{11} & -\infty & \cdots & -\infty \\ e_{21} & e_{22} & \cdots & -\infty \\ \vdots & \vdots & \ddots & \vdots \\ e_{m1} & e_{m2} & \cdots & e_{mn} \end{bmatrix}$$

The second layer utilizes a multi-head self-attention technique identical to the one used in the encoder's first sublayer. On the decoder side, this multi-head mechanism takes queries from the preceding decoder sublayer as well as keys and values from the encoder output. This enables the decoder to process all of the words in the input sequence. Finally, the third layer implements a fully linked feed-forward network, similar to the one used in the encoder's second sublayer.

3.3. Dynamic Attention

The dynamic attention module computes attention weights for the shared representation generated by the cross-modal learning module. It employs a self-attention mechanism to assess the importance of each feature in the shared representation. The self-attention mechanism calculates the relevance of each feature by measuring its interaction with other features in the representation. These attention weights are then used to produce an attention vector, which is a weighted sum of the shared representation features. The attention vector captures the most relevant information across all modalities, emphasizing the features that contribute the most to the anomaly detection task.

GRUs are employed in the dynamic attention module to compute attention weights and generate the attention vector. GRUs are a type of recurrent neural network that can capture temporal dependencies and adapt to changes over time. By using GRUs, the model can dynamically adjust attention weights based on the current state and input sequence, improving the model's ability to focus on relevant features and to adapt to changes in the input data over time. It calculates attention weights for each position in the input data based on the model's current state.

The attention mechanism is a location-based attention mechanism that uses the position of the input features in the sequence of real-time data streams to calculate the attention weights. The attention mechanism is a hybrid approach that combines content-based and location-based attention. Content-based attention uses the input features to calculate the attention weights. In contrast, location-based attention uses the position of the input features in the sequence to calculate the attention weights. The attention weights are adaptive and can be adjusted at each time step to give more or less importance to different parts of the input sequence depending on their relevance to the task.

3.4. Anomaly Detection Module

The anomaly detection module of the CM-DANA model combines the relevant input data from different modalities using the adaptive attention weights to detect suspicious abnormal behavior. The Theil-Sen Regressor was used as an anomaly detection module in the CM-DANA architecture. The Theil-Sen Regressor is a robust linear regression algorithm that estimates the slope and intercept of a linear relationship between input features and target variables. While it is primarily used for regression tasks, it can also be adapted for anomaly detection by setting a threshold on the residuals used for outlier detection.

Specifically, after the dynamic attention module obtains the attention vector, it serves as the input to the anomaly detection module. The Theil-Sen Regressor fits a linear regression model to the attention vector and estimates the slope and intercept of the linear relationship. During the anomaly detection phase, it calculates the residuals by comparing the predicted values from the Theil-Sen Regressor with the actual values of the attention vector. Finally, a dynamic threshold on the residuals identifies instances where the deviation from the predicted values is significant. Data instances with residuals above the threshold are considered anomalous. Figure 3 is an example of how to fit a line through almost linear data. The orange Theil-Sen Regressor outperforms the blue linear regressor.

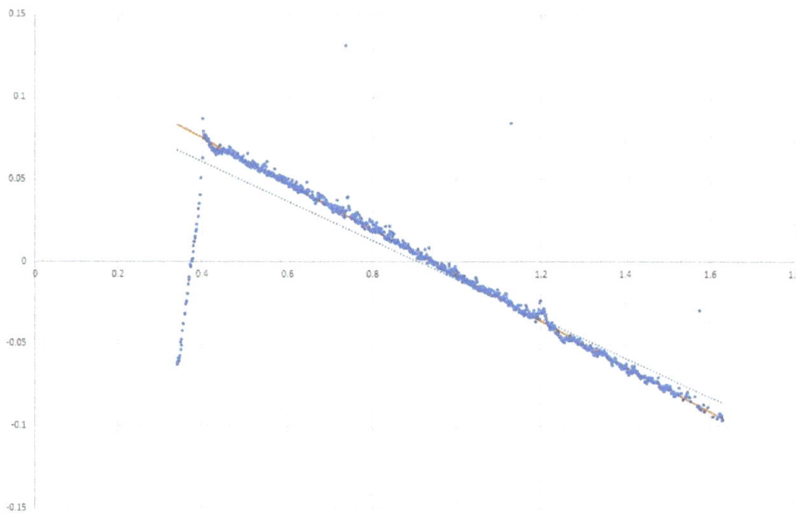

Figure 3. Theil-Sen Regressor.

4. Case Study: Application in Cybersecurity Anomaly Detection

To demonstrate how CM-DANA can identify advanced cybersecurity anomalies, we present a case study in a Smart Communication Environment. This environment generates data streams encompassing multiple modalities that can be utilized to detect security breaches, including infiltration attempts, DDoS attacks, and malicious software proliferation. The case study involves structured, semi-structured, and unstructured data streams that require sophisticated preprocessing and feature extraction techniques for accurate analysis. Intelligent models must handle temporal interdependencies and high-dimensional data streams while processing large volumes of data in near real-time. Furthermore, anomaly detection models must be adaptable to evolving data patterns for consistent performance over time.

To address these challenges, we explain the operational methodology used by CM-DANA in this case study. Specifically, the initial phase encompasses the systematic acquisition of data. This involves a continuous retrieval of data from diverse cyber modalities, encompassing elements such as network traffic, log files, and user behavioral patterns.

Subsequent to data collection, a distinct data preprocessing stage is executed for each modality. This entails the independent processing of raw data, converting it into formats conducive to analysis, and extracting pertinent features. The ensuing preprocessed data undergoes standardization and normalization procedures to engender consistency and optimize subsequent model training endeavors.

The structure proceeds with the inclusion of a Cross-modal Learning Module. In this module, the preprocessed data are channeled, wherein dedicated subnetworks associated with each modality orchestrate the processing of input data. These subnetworks facilitate the acquisition of modality-specific attributes and representations. The products of these distinct subnetworks are subsequently aggregated through a fusion technique, for instance, concatenation or summation. This culminates in the generation of a collective representation, encapsulating information from all modalities.

Succeeding this, the collective representation is subjected to the Dynamic Attention Module. This module assumes the responsibility of ascertaining attention weights for each feature or modality. Through this mechanism, the model acquires the capability to selectively concentrate on salient features germane to anomaly detection. Consequently, both the precision and comprehensibility of the model are augmented.

The ensuing step entails the Anomaly Detection Module. Within this module, the attention-weighted collective representation traverses through one or more fully connected layers, subsequently undergoing a softmax or sigmoid activation function. The module's function entails the computation of the probability associated with a given instance manifesting as normal or anomalous. Decisive outcomes are generated based on a predetermined threshold.

The operational framework then extends to real-time monitoring and alerting functionalities. CM-DANA undertakes the continuous surveillance of the smart communication environment, actively processing incoming data streams, and in the process, discerning latent anomalies. Upon anomaly identification, the system promptly generates alerts. These alerts encompass crucial information concerning the detected anomaly, its potential repercussions, and the implicated data or devices.

Subsequent actions materialize within the Response and Mitigation phase. Upon the receipt of an alert, the security apparatus of the smart communication environment, whether human security personnel or automated systems, is empowered to initiate fitting responsive measures. Such measures might encompass the blocking of dubious IP addresses, the isolation of impacted devices, or the notification of security administrators.

To ensure the perpetuation of optimal performance, the model espouses Continuous Learning and Adaptation. Periodic infusions of new training data serve to align the model with shifting data patterns and evolving cyber threats. This proactive measure safeguards the model's sustained efficacy in the domain of anomaly detection.

5. Experiments and Evaluation

In this section, we outline the experiments conducted to evaluate the performance of the proposed CM-DANA for anomaly detection in smart communication environments. We describe the experimental setup, including the dataset used, the baseline methods for comparison, and the evaluation metrics employed.

5.1. Experimental Setup

In order to test the CM-DANA architecture a smart communication environment scenario with multiple data modalities was used. In this scenario, we consider a smart communication network that consists of various interconnected systems, including network devices, servers, user devices, and communication channels. Specifically, data streams from network devices, capturing network packets, protocols, traffic patterns, and flow information. Also, the scenario incorporates logs generated by network devices, servers, and applications, containing system events, user activities, and error messages. Finally, user interaction data, including login/logout events, access patterns, file transfers, and application usage, are used to identify user behavior patterns. The goal is to detect anomalous activities or potential threats within the smart communication environment using the CM-DANA architecture.

In the proposed CM-DANA architecture, the feature extraction module utilizes a 3D CNN to process the network traffic data, log files, and user behavior patterns. The feature extraction process includes the following steps:

1. Data Preparation. Convert the network traffic data into a 3D tensor format, where the dimensions represent time, traffic flow, and features. Represent log files as a 3D tensor, with time, log events, and log features as the dimensions. Structure user

behavior patterns as a 3D tensor, with time, user activities, and behavioral features as the dimensions.

2. Input Data. Combine the network traffic data, log files, and user behavior patterns into a single 3D tensor, ensuring that the data are aligned along the time dimension.
3. Convolutional Layers. Apply two 3D convolutional layers to capture spatiotemporal features from the combined data with the following configuration:
   a. Convolutional Layer 1: Number of filters: 32, filter size: (3, 3, 3), stride: (1, 1, 1), padding: 'same'
   b. Convolutional Layer 2: Number of filters: 64, filter size: (3, 3, 3), stride: (1, 1, 1), padding: 'same'
4. Activation Function. Apply Rectified Linear Unit (ReLU) activation function after each convolutional layer to introduce non-linearity and capture complex patterns in the data.
5. Pooling Layers. Insert two 3D pooling layers. Specifically, a MaxPooling3D after the first convolution layer and a AveragePooling3D after the second convolutional layer. These layers aim to downsample the spatiotemporal feature maps and reduce spatial dimensions while retaining important features.
6. Flattening. Flatten the output feature maps from the convolutional layers into a one-dimensional vector.
7. Fully Connected Layers. Connect the flattened features to one or more fully connected layers. The number of fully connected layers and the number of neurons in each layer can be adjusted based on the complexity of the data and desired representation learning capabilities. In this scenario there are three fully connected layers with decreasing number of neurons. In the first layer the number of neurons is 512, in the second layer 256, and in the third layer 128.
8. Output. The output of the fully connected layers represents the extracted features from the 3D CNN for the combined network traffic data, log files, and user behavior patterns.

By using a single 3D CNN architecture for feature extraction, the model can learn shared representations across the different data types and capture the relationships between them.

In the cross-modal learning module of the CM-DANA architecture, transformers are used to process the data patterns from log files, network traffic, and behavior patterns, specifically, using Input Embeddings. Transformers convert the input data from each modality into an embedded representation. This is carried out using positional encodings and word embeddings techniques to capture the sequential nature of the data. Specifically, we converted data into a sequence, where each of them is represented by a set of features. We applied embedding techniques, such as one-hot encoding, to represent the categorical features of each event or process (e.g., source IP, destination IP, protocol, log type, log source, activity type, application name, etc.). Numerical features (e.g., packet size, timestamp) were scaled and normalized to a fixed range. Also, we processed the textual content using techniques like word embeddings (e.g., Word2Vec, GloVe) to capture semantic information. Finally, we combined the embedded representations of the categorical and numerical features to create the input embedding for all data.

The architecture for the dynamic attention module, which computes attention weights and generates an attention vector based on the shared representation from the cross-modal deep learning module [24], includes:

1. Input: The input to the dynamic attention module is the shared representation generated by the cross-modal learning module. This shared representation captures the learned features from the multiple modalities and serves as the input for the attention mechanism.
2. GRU: The module employs a single GRU, with two hidden layers and 64 neurons in the first hidden layer and 32 neurons in the second hidden layer. The GRU as a

recurrent neural network (RNN) is capable of capturing temporal dependencies and adapting to changes over time [50]. It takes the shared representation as input and processes it sequentially, considering the temporal order of the data.

3. Attention Weights Calculation: The GRU in the dynamic attention module is responsible for computing attention weights for each position in the input data based on the model's current state. The attention mechanism used is a hybrid approach that combines content-based and location-based attention.
   (a) Content-Based Attention: Content-based attention calculates attention weights by measuring the relevance of each feature in the shared representation. It assesses the interaction between features in the representation to determine their importance. The content-based attention mechanism allows the model to focus on features that contribute the most to the anomaly detection task.
   (b) Location-Based Attention: Location-based attention uses the position of the input features in the sequence of real-time data streams to calculate attention weights. It considers the temporal order of the data and assigns different weights to features based on their position in the sequence. Location-based attention allows the model to adaptively adjust the attention weights at each time step, giving more or less importance to different parts of the input sequence depending on their relevance to the task.
4. Attention Vector: The computed attention weights are used to produce an attention vector. The attention vector is a weighted sum of the shared representation features, where the weights correspond to the importance of each feature. The attention vector captures the most relevant information across all modalities, emphasizing the features that contribute the most to the anomaly detection task.
5. Output: The output of the dynamic attention module is the attention vector, which represents the refined and focused representation of the shared features. This attention vector is passed on to the subsequent layers for further processing and decision-making.

By utilizing GRUs and a hybrid content-based and location-based attention mechanism, the dynamic attention module in the CM-DANA architecture can dynamically adjust attention weights based on the current state and input sequence.

Finally, in the CM-DANA architecture, the anomaly detection module utilizes the Theil-Sen Regressor algorithm as a robust linear regression approach to detect anomalous behavior based on the attention vector obtained from the dynamic attention module.

Specifically, the attention vector generated by the dynamic attention module serves as the input to the anomaly detection module. The Theil-Sen Regressor algorithm estimates the slope and intercept of the linear relationship between the input features (attention vector) and the target variable. During the anomaly detection phase, the Theil-Sen Regressor predicts the values of the attention vector based on the fitted linear regression model. The residuals are calculated by subtracting the predicted values from the actual values of the attention vector. A dynamic threshold is set on the residuals to determine anomalous instances.

The threshold is determined using a rolling mean and standard deviation. Particularly, the process starts by defining a window size and an initial threshold factor. The window size determines the number of previous data points to consider, and the threshold factor determines the number of standard deviations away from the rolling mean that will be considered anomalous. Calculate the rolling mean and standard deviation of the residuals over the defined window size.

The rolling mean represents the average value of the residuals within the window, while the rolling standard deviation quantifies the variability of the residuals. Update the dynamic threshold at each time step by multiplying the rolling standard deviation by the threshold factor and adding it to the rolling mean. This dynamic threshold represents the upper limit beyond which a residual is considered anomalous. Compare the absolute value of each residual to the dynamic threshold. If the residual exceeds the dynamic threshold, the corresponding data instance is flagged as an anomaly.

Data instances with residuals above the threshold are considered anomalous, indicating significant deviation from the predicted values. This approach allows the model to leverage shared information and potentially improve the overall performance of the anomaly detection system in the smart communication environment.

5.2. Dataset

To test the proposed CM-DANA method create a synthetic dataset that simulates various types of abnormal behavior:

1. Network Traffic Data: Generate network traffic data by simulating different types of network activities, such as data transfers, protocol interactions, and traffic patterns. Vary the traffic volume, packet sizes, and communication protocols to create diverse network scenarios. Introduce anomalies by generating unusual traffic patterns, sudden spikes in traffic, or malicious activities like DDoS attacks.
2. Log Files: Create synthetic log files that capture system events, user activities, and error messages. Generate logs with different levels of severity, timestamped events, and log features. Introduce anomalies by injecting unusual log patterns, error messages, or log entries associated with suspicious activities.
3. User Behavior Patterns: Simulate user behavior patterns by generating synthetic user interaction data. Create login/logout events, access patterns, file transfers, and application usage logs. Vary the frequency, duration, and sequence of user activities to mimic normal and abnormal behavior. Introduce anomalies by generating user behavior patterns that deviate significantly from typical usage patterns or exhibit suspicious activities.
4. Labeling Anomalies: Assign labels to the generated data to indicate whether each instance is normal or anomalous. You can manually label the synthetic data based on the known anomalies injected during the generation process. Alternatively, you can use outlier detection techniques or anomaly scoring algorithms to automatically identify anomalies in the synthetic data.
5. Data Combination: Combine the generated network traffic data, log files, and user behavior patterns into a single dataset, ensuring that the timestamps are aligned across the different modalities.

Table 1 shows examples of anomalies injected into the synthetic dataset. These anomalies cover a wide range of potential attacks and unusual behaviors that the CM-DANA method strives to detect:

5.3. Results and Discussion

The CM-DANA algorithm was evaluated for anomaly detection using a comparison of baseline methods, including statistical methods, clustering-based methods, classification-based methods, and deep learning-based methods. Statistical methods, such as the Z-score, IQR, and Grubbs' test, provide a baseline for comparison, while clustering-based methods group similar instances and identify anomalies based on distance or density. Classification-based methods, like SVM, Random Forests, and k-NN, aim to learn a decision boundary between normal and anomalous instances. Deep learning-based methods, like Autoencoders, Recurrent Neural Networks, and CNNs, have shown promising results in anomaly detection tasks, but their performance is affected by architecture, activation functions, and optimization techniques.

Table 1. Anomalies injected into the synthetic dataset.

Anomaly Type	Modality	Description
DDoS Attack	Network Traffic	Introduce sudden, high-volume traffic from multiple sources, overwhelming the network.
Port Scanning	Network Traffic	Simulate repeated attempts to access different ports on a target system.
Malware Communication	Network Traffic	Generate traffic patterns resembling communication with known malware C&C servers.
Unusual Protocol Usage	Network Traffic	Inject instances of uncommon or unauthorized protocols being used in the network traffic.
Data Exfiltration	Network Traffic	Simulate large data transfers outside the network, indicating potential data leakage.
Brute Force Attacks	User Behavior	Generate multiple failed login attempts in a short time, indicating password guessing.
Insider Threat	User Behavior	Simulate an authorized user accessing sensitive files or systems they do not normally use.
Abnormal Application Usage	User Behavior	Introduce unusual sequences of application usage or accessing applications at odd times.
Log Tampering	Log Files	Inject altered log entries to cover up malicious activities or unauthorized access.
Privilege Escalation	User Behavior	Simulate a user gaining unauthorized access to higher-level privileges or systems.
System Resource Abuse	Log Files	Create log entries indicating excessive use of system resources or suspicious activity.
Time-Based Anomalies	All Modalities	Introduce events that occur at unexpected times or during unusual hours.

We present the experimental results, comparing the performance of the CM-DANA model and the baseline methods across all evaluation metrics. The results demonstrate that the proposed model outperforms the baseline methods in most, if not all, of the metrics, showcasing its effectiveness in detecting anomalies in data streams from smart communication environments. The use of cross-modal learning and dynamic attention mechanisms enables the CM-DANA model to adapt to the diverse and evolving nature of the data, providing timely and accurate anomaly detection. Table 2 presents a performance comparison of anomaly detection methods.

Table 2. Performance Comparison of Anomaly Detection Methods.

Method	Accuracy	Precision	Recall	F1 Score	AUC-ROC	AUC-PR	Time (s)
Z-score	0.76	0.62	0.78	0.69	0.78	0.65	10.5
IQR	0.80	0.65	0.80	0.71	0.82	0.67	11.2
Grubbs' test	0.74	0.58	0.76	0.66	0.75	0.62	12.8
k-means	0.82	0.69	0.82	0.74	0.83	0.70	45.6
DBSCAN	0.78	0.63	0.78	0.70	0.79	0.68	62.3
LOF	0.79	0.65	0.79	0.71	0.81	0.69	53.9
SVM	0.85	0.76	0.85	0.80	0.86	0.75	132.4
Random Forest	0.86	0.78	0.86	0.82	0.87	0.76	243.7
k-NN	0.81	0.71	0.81	0.75	0.80	0.70	76.2
Autoencoder	0.88	0.82	0.88	0.85	0.88	0.78	180.6
RNN	0.89	0.85	0.89	0.87	0.89	0.80	215.3
CNN	0.87	0.80	0.87	0.83	0.86	0.76	198.9
CM-DANA	0.92	0.88	0.92	0.90	0.92	0.85	315.2

Here are the bar plots comparing all evaluation metrics (Figure 4):

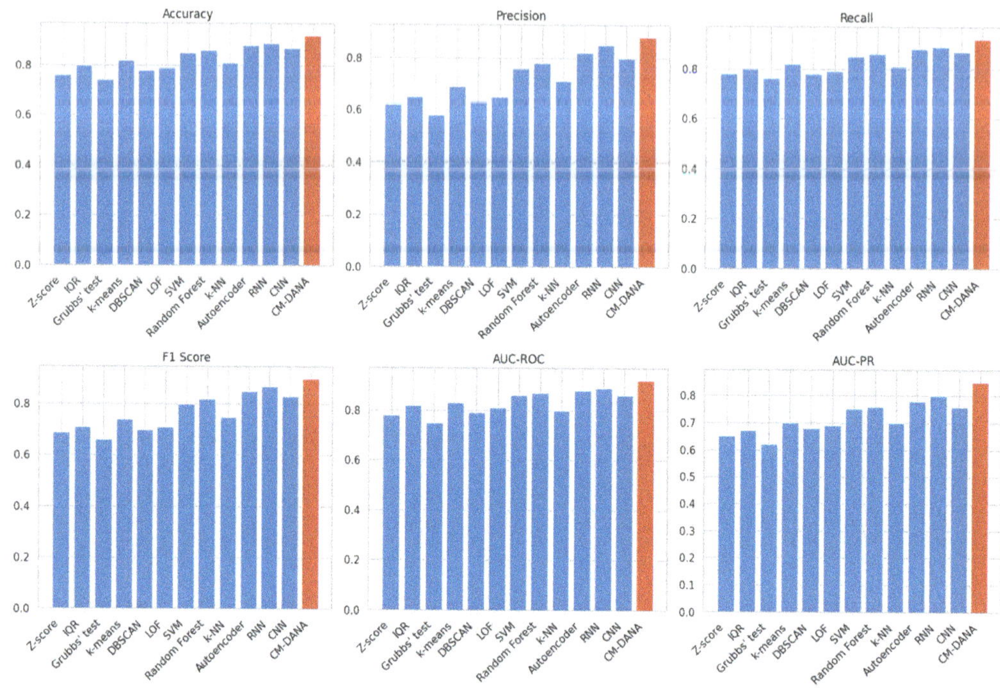

Figure 4. Bar plots comparing all evaluation metrics.

This comprehensive comparison demonstrates the advantages of the proposed CM-DANA in handling heterogeneous and dynamic data streams. Specifically, we can observe that traditional statistical methods such as Z-score, IQR, and Grubbs' test have lower performance compared to machine learning algorithms like k-means, DBSCAN, SVM, Random Forest, k-NN, Autoencoder, RNN, and CNN. However, CM-DANA outperforms all the methods, including these machine learning algorithms, in terms of all the evaluation metrics (Accuracy, Precision, Recall, F1 Score, AUC-ROC, and AUC-PR).

The CM-DANA model is trained end-to-end using multimodal data streams. This allows the model to attend to different features in different modalities based on the model's current state and detect suspicious abnormal behavior by combining the relevant input data from different modalities using adaptive attention weights.

To handle the input data as a stream of data in sliding windows, we apply a mask to the attention scores to ignore encoder outputs that are outside of the current window. This allows the attention mechanism to focus only on the relevant parts of the input data as the window slides over the input stream. Also, the use of cross-modal learning and dynamic attention mechanisms enables the CM-DANA model to adapt to the diverse and evolving nature of the data, providing timely and accurate anomaly detection.

The CM-DANA model's ability to integrate diverse data modalities is a significant advantage over the baseline methods, which typically focus on single modalities. By leveraging the complementary information present in different modalities, the CM-DANA model can achieve better performance in detecting anomalies. Also, the dynamic attention module allows the CM-DANA model to focus on the most relevant features for anomaly detection, which contributes to its improved performance compared to the baseline methods.

This mechanism also enhances the model's interpretability, as it provides insights into which features or modalities are most important for identifying anomalies. The experimental results, in addition, indicate that the CM-DANA model can effectively handle real-time data processing, making it a suitable choice for real-world applications. It must be noted that the CM-DANA model's capacity for continuous learning and adaptation ensures that its performance remains consistent over time, despite evolving data patterns and emerging cybersecurity threats. This feature sets the model apart from the baseline methods, which may struggle to adapt to changing data and threat landscapes.

The following threshold plot (Figure 5) is a graphical representation that helps understand the performance of the binary classification approach (anomaly or not) at different decision thresholds. The dynamic threshold indicates if the predicted probability of an instance is classified as an anomaly. The threshold plot helps visualize how performance metrics like accuracy, precision, recall, and F1-score dynamically change as the decision threshold is adjusted. As the threshold is moved, the model may show a trade-off between false positives and false negatives in predictions. Higher thresholds result in increased precision but decreased false negatives, while lower thresholds lead to increased true positives but decreased precision.

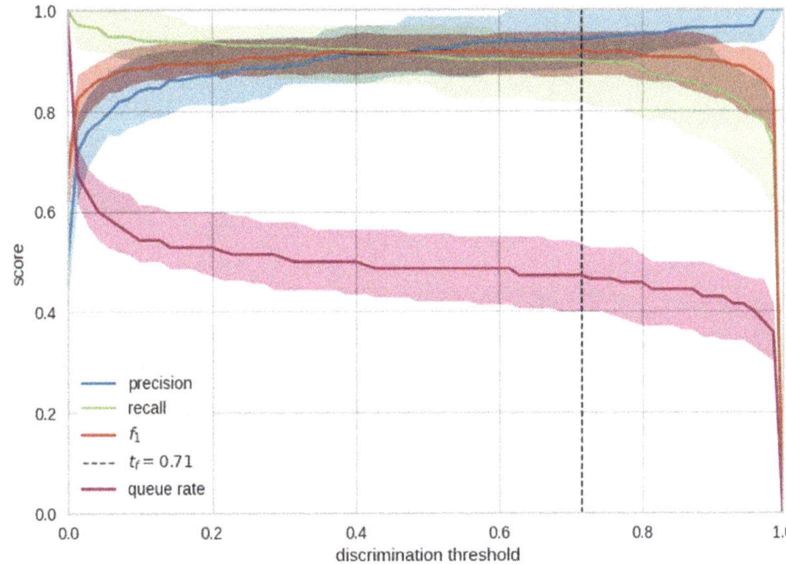

Figure 5. Threshold Plot of CM-DANA.

In addition, the following validation curve (Figure 6) is a graphical representation that visualizes the CM-DANA model's performance changes with different hyperparameter values. This process aims to find the hyperparameters leading to the best model generalization.

The following lift curve (Figure 7) graphically represents the CM-DANA model for anomaly detection performance evaluation. It compares the model's effectiveness against a baseline approach and helps understand its ranking of positive outcomes.

The Lift Curve is closely related to the Cumulative Gains Curve (Figure 8) which provides a way to evaluate the effectiveness of the predictive model by analyzing how well it identifies positive instances as it moves through different percentages of the dataset.

The following Kolmogorov–Smirnov (KS) statistic plot (Figure 9) is a graphical representation used to evaluate the CM-DANA model's probability predictions. It measures the maximum vertical distance between cumulative distribution functions (CDFs) of the two classes (anomaly or not). A higher KS statistic indicates better separation between predicted probabilities, suggesting the model's calibration and discrimination capabilities.

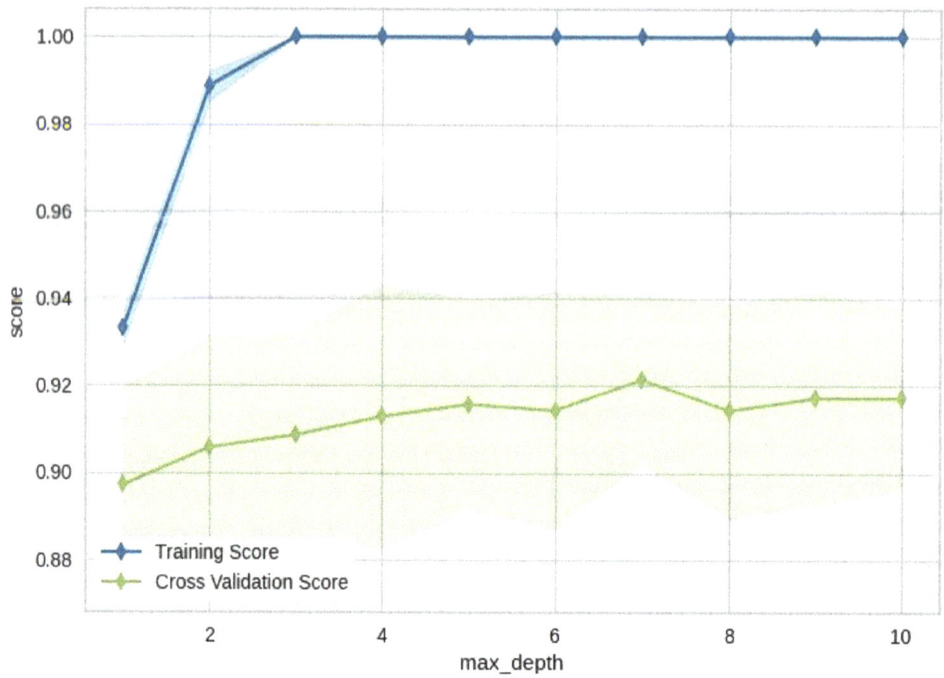

Figure 6. Validation Curve Plot of CM-DANA.

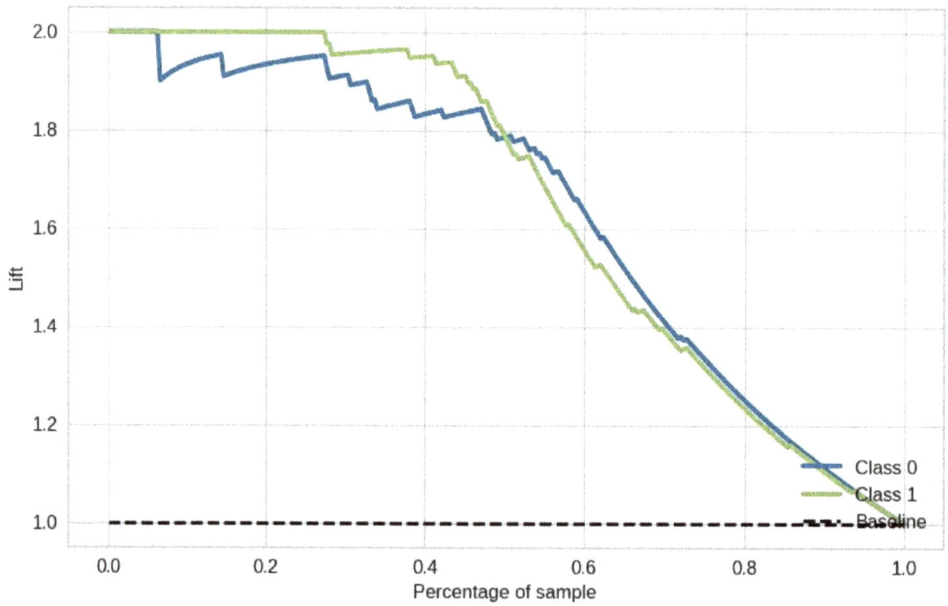

Figure 7. Lift Curve Plot of CM-DANA.

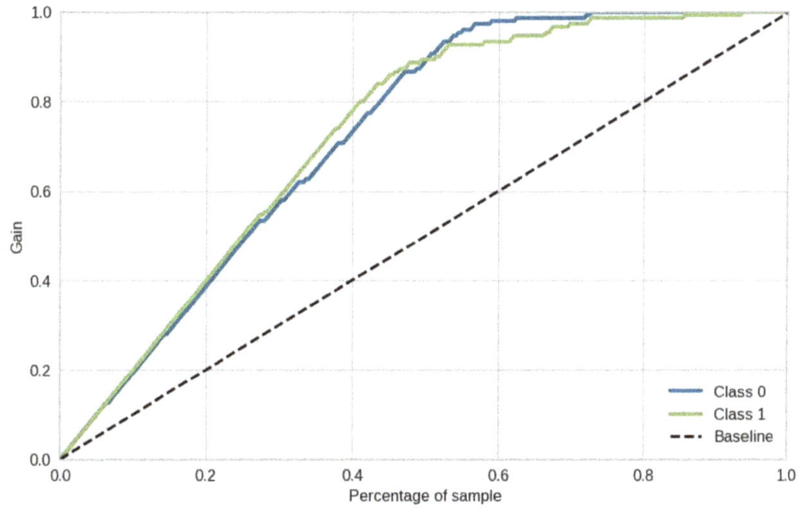

Figure 8. Cumulative Gains Curve Plot of CM-DANA.

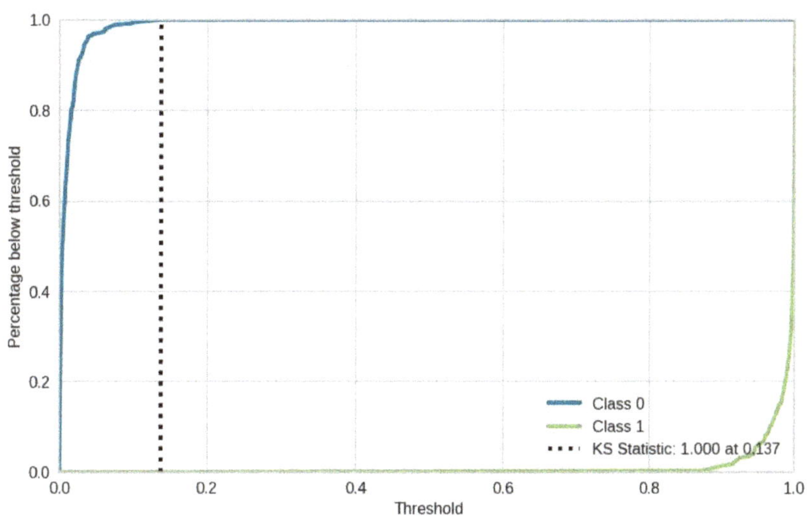

Figure 9. Kolmogorov–Smirnov (KS) Statistic Plot of CM-DANA.

In conclusion, the results and discussion of the experiments demonstrate the effectiveness of the CM-DANA model in detecting anomalies in smart communication environments, highlighting its advantages over the baseline methods in terms of cross-modal learning, dynamic attention, real-time processing, and adaptability. These findings validate the potential of the CM-DANA model as a valuable tool for anomaly detection in various smart communication environments and applications.

6. Conclusions and Future Work

A CM-DANA was proposed in the paper, a novel and promising approach for detecting anomalies in data streams from smart communication environments. The model extends the state-of-the-art attention mechanism by using a hybrid method called cross-modal attention, which combines attention weights for different modalities to capture complex interactions between them better.

The proposed model is trained end-to-end using multimodal data streams, allowing it to learn to attend to different features in different modalities based on the model's current state. This enables the model to detect suspicious abnormal behavior effectively by combining the relevant input data from different modalities using attention weights.

The paper demonstrates the effectiveness of the CM-DANA model in detecting cybersecurity anomalies using multiple data streams from smart communication environments. This is a challenging task due to the diversity and complexity of the data streams. Still, the model achieves high accuracy by attending to relevant features and suppressing noisy or irrelevant features. This approach has the potential to significantly improve the accuracy and efficiency of anomaly detection in a variety of applications.

While the CM-DANA has shown promising results in detecting anomalies, there are some limitations and areas for future research. Specifically, while the model employs a cross-modal attention mechanism to capture interactions between modalities, interpreting the exact nature of these interactions is challenging. Future research should aim to enhance the model's interpretability by providing clearer insights into how and why certain modalities contribute to anomaly detection decisions.

Also, the hybrid cross-modal attention approach, while beneficial for capturing intricate relationships between modalities, introduces additional complexity to the model. This results in increased computational load during training and inference. Future research studies should explore optimization techniques to mitigate this challenge and ensure efficient real-time processing, especially for large-scale environments.

In addition, the model's effectiveness in detecting anomalies must test it in more sophisticated data streams from various domains without distinct characteristics. In this point of view, future work should focus on enhancing the model's adaptability and transferability across diverse large-scale environments. Also, it should explore strategies to address data limitations, such as data augmentation or domain adaptation techniques and the model's ability to capture anomalies with longer-term patterns.

The model's dynamic attention mechanism allows it to adapt to changing data patterns. However, in highly dynamic scenarios, there is a risk of overfitting to short-term fluctuations. Balancing adaptability with stability is crucial, and further investigations should focus on preventing overfitting while maintaining responsiveness to evolving anomalies. Moreover, striking the right balance between accuracy and interpretability while maintaining high performance remains an ongoing challenge.

Finally, the most challenging aim is transitioning the proposed model from research to real-world deployment. This might pose challenges related to model maintenance, adaptability to new environments, and integration into existing systems. Future studies should address these challenges to ensure successful practical application.

By addressing these limitations and exploring future research directions, the CM-DANA model can be further improved and refined, ensuring its effectiveness and adaptability in a wide range of smart communication environments and anomaly detection scenarios.

Author Contributions: Conceptualization, K.D. and K.R.; methodology, K.D.; software, K.D.; validation, K.D., K.R., L.M. and L.I.; formal analysis, K.D. and K.R.; investigation, K.D. and K.R.; resources, L.M.; data curation, K.D., K.R., L.M. and L.I.; writing—original draft preparation, K.D.; writing—review and editing, K.D., K.R., L.M. and L.I.; visualization, K.D.; supervision, L.I.; project administration, L.M.; funding acquisition, K.R. All authors have read and agreed to the published version of the manuscript.

Funding: This research received no external funding.

Institutional Review Board Statement: Not applicable.

Informed Consent Statement: Not applicable.

Data Availability Statement: The dataset used to support the findings of this study is available from the corresponding author upon request.

Conflicts of Interest: The authors declare no conflict of interest.

References

1. Golab, L.; Ozsu, M.T.; Data Stream Management. Morgan & Claypool. 2010. Available online: https://books.google.gr/books/about/Data_Stream_Management.html?id=IMyogd_LF1cC&redir_esc=y (accessed on 22 July 2020).
2. Dawoud, A.; Shahristani, S.; Raun, C. Deep Learning for Network Anomalies Detection. In Proceedings of the 2018 International Conference on Machine Learning and Data Engineering (iCMLDE), Sydney, Australia, 3–7 December 2018; pp. 149–153. [CrossRef]
3. Jara, A.J.; Genoud, D.; Bocchi, Y. Big Data for Cyber Physical Systems: An Analysis of Challenges, Solutions and Opportunities. In Proceedings of the Eighth International Conference on Innovative Mobile and Internet Services in Ubiquitous Computing, Birmingham, UK, 2–4 July 2014; pp. 376–380. [CrossRef]
4. Ali, R.F.; Muneer, A.; Dominic, P.D.D.; Ghaleb, E.A.A.; Al-Ashmori, A. Survey on Cyber Security for Industrial Control Systems. In Proceedings of the 2021 International Conference on Data Analytics for Business and Industry (ICDABI), Online, 25–26 October 2021; pp. 630–634. [CrossRef]
5. Vafaie, B.; Shamsi, M.; Javan, M.S.; El-Khatib, K. A New Statistical Method for Anomaly Detection in Distributed Systems. In Proceedings of the 2020 IEEE Canadian Conference on Electrical and Computer Engineering (CCECE), London, ON, Canada, 30 August–2 September 2020; pp. 1–4. [CrossRef]
6. Jirsik, T. Stream4Flow: Real-time IP flow host monitoring using Apache Spark. In Proceedings of the NOMS 2018—2018 IEEE/IFIP Network Operations and Management Symposium, Taipei, Taiwan, 23–27 April 2018; pp. 1–2. [CrossRef]
7. Benjelloun, F.-Z.; Lahcen, A.A.; Belfkih, S. An overview of big data opportunities, applications and tools. In Proceedings of the 2015 Intelligent Systems and Computer Vision (ISCV), Fez, Morocco, 25–26 March 2015; pp. 1–6. [CrossRef]
8. Guo, S.; Liu, Y.; Su, Y. Comparison of Classification-based Methods for Network Traffic Anomaly Detection. In Proceedings of the 2021 IEEE 4th Advanced Information Management, Communicates, Electronic and Automation Control Conference (IMCEC), Chongqing, China, 18–20 June 2021; pp. 360–364. [CrossRef]
9. Dai, J.J.; Wang, Y.; Qiu, X.; Ding, D.; Zhang, Y.; Wang, Y.; Jia, X.; Zhang, C.L.; Wan, Y.; Li, Z.; et al. BigDL: A Distributed Deep Learning Framework for Big Data. In Proceedings of the ACM Symposium on Cloud Computing, Santa Cruz, CA, USA, 20–23 November 2019. [CrossRef]
10. Gallicchio, C.; Micheli, A. Deep Echo State Network (DeepESN): A Brief Survey. *arXiv* **2019**, arXiv:1712.04323.
11. Tan, C.; Sun, F.; Kong, T.; Zhang, W.; Yang, C.; Liu, C. A Survey on Deep Transfer Learning. *arXiv* **2018**, arXiv:1808.01974.
12. He, W.; Wu, Y.; Li, X. Attention Mechanism for Neural Machine Translation: A survey. In Proceedings of the 2021 IEEE 5th Information Technology, Networking, Electronic and Automation Control Conference (ITNEC), Xi'an, China, 15–17 October 2021; pp. 1485–1489. [CrossRef]
13. Tao, A.; Sapra, K.; Catanzaro, B. Hierarchical Multi-Scale Attention for Semantic Segmentation. *arXiv* **2020**, arXiv:2005.10821.
14. Sun, J.; Jiang, J.; Liu, Y. An Introductory Survey on Attention Mechanisms in Computer Vision Problems. In Proceedings of the 2020 6th International Conference on Big Data and Information Analytics (BigDIA), Shenzhen, China, 4–6 December 2020; pp. 295–300. [CrossRef]
15. Zhang, N.; Kim, J. A Survey on Attention mechanism in NLP. In Proceedings of the 2023 International Conference on Electronics, Information, and Communication (ICEIC), Singapore, 5–8 February 2023; pp. 1–4. [CrossRef]
16. Deng, D. Research on Anomaly Detection Method Based on DBSCAN Clustering Algorithm. In Proceedings of the 2020 5th International Conference on Information Science, Computer Technology and Transportation (ISCTT), Shenyang, China, 13–15 November 2020; pp. 439–442. [CrossRef]
17. Lu, J.; Liu, A.; Dong, F.; Gu, F.; Gama, J.; Zhang, G. Learning under Concept Drift: A Review. *IEEE Trans. Knowl. Data Eng.* **2018**, *31*, 2346–2363. [CrossRef]
18. Cao, K.; Liu, Y.; Meng, G.; Sun, Q. An Overview on Edge Computing Research. *IEEE Access* **2020**, *8*, 85714–85728. [CrossRef]
19. Wang, J.; Chen, J.; Lin, J.; Sigal, L.; de Silva, C.W. Discriminative feature alignment: Improving transferability of unsupervised domain adaptation by Gaussian-guided latent alignment. *Pattern Recognit.* **2021**, *116*, 107943. [CrossRef]
20. Qin, K.; Zhou, Y.; Tian, B.; Wang, R. AttentionAE: Autoencoder for Anomaly Detection in Attributed Networks. In Proceedings of the 2021 International Conference on Networking and Network Applications (NaNA), Lijiang City, China, 29 October–1 November 2021; pp. 480–484. [CrossRef]
21. Sokolov, A.N.; Alabugin, S.K.; Pyatnitsky, I.A. Traffic Modeling by Recurrent Neural Networks for Intrusion Detection in Industrial Control Systems. In Proceedings of the 2019 International Conference on Industrial Engineering, Applications and Manufacturing (ICIEAM), Sochi, Russia, 25–29 March 2019; pp. 1–5.
22. Liu, S.; Jiang, H.; Li, S.; Yang, Y.; Shen, L. A Feature Compression Technique for Anomaly Detection Using Convolutional Neural Networks. In Proceedings of the 2020 IEEE 14th International Conference on Anti-Counterfeiting, Security, and Identification (ASID), Xiamen, China, 30 October–1 November 2020; pp. 39–42. [CrossRef]
23. Sarker, I.H. Deep Learning: A Comprehensive Overview on Techniques, Taxonomy, Applications and Research Directions. *SN Comput. Sci.* **2021**, *2*, 420. [CrossRef] [PubMed]

24. Tsimenidis, S.; Lagkas, T.; Rantos, K. Deep Learning in IoT Intrusion Detection. *J. Netw. Syst. Manag.* **2021**, *30*, 8. [CrossRef]
25. Peng, C.; Zhang, C.; Xue, X.; Gao, J.; Liang, H.; Niu, Z. Cross-modal complementary network with hierarchical fusion for multimodal sentiment classification. *Tsinghua Sci. Technol.* **2022**, *27*, 664–679. [CrossRef]
26. Sanla, A.; Numnonda, T. A Comparative Performance of Real-time Big Data Analytic Architectures. In Proceedings of the 2019 IEEE 9th International Conference on Electronics Information and Emergency Communication (ICEIEC), Beijing, China, 12–14 July 2019; pp. 1–5. [CrossRef]
27. Liu, F.; Zhou, X.; Cao, J.; Wang, Z.; Wang, T.; Wang, H.; Zhang, Y. Anomaly Detection in Quasi-Periodic Time Series Based on Automatic Data Segmentation and Attentional LSTM-CNN. *IEEE Trans. Knowl. Data Eng.* **2022**, *34*, 2626–2640. [CrossRef]
28. Sani, Y.; Mohamedou, A.; Ali, K.; Farjamfar, A.; Azman, M.; Shamsuddin, S. An overview of neural networks use in anomaly Intrusion Detection Systems. In Proceedings of the 2009 IEEE Student Conference on Research and Development (SCOReD), Seri Kembangan, Malaysia, 16–18 November 2009; pp. 89–92. [CrossRef]
29. Embarak, O. Decoding the Black Box: A Comprehensive Review of Explainable Artificial Intelligence. In Proceedings of the 2023 9th International Conference on Information Technology Trends (ITT), Dubai, United Arab Emirates, 24–25 May 2023; pp. 108–113. [CrossRef]
30. Sasaki, H.; Hidaka, Y.; Igarashi, H. Explainable Deep Neural Network for Design of Electric Motors. *IEEE Trans. Magn.* **2021**, *57*, 1–4. [CrossRef]
31. Xu, X.; Lin, K.; Gao, L.; Lu, H.; Shen, H.T.; Li, X. Learning Cross-Modal Common Representations by Private–Shared Subspaces Separation. *IEEE Trans. Cybern.* **2022**, *52*, 3261–3275. [CrossRef] [PubMed]
32. Hua, Y.; Du, J. Deep Semantic Correlation with Adversarial Learning for Cross-Modal Retrieval. In Proceedings of the 2019 IEEE 9th International Conference on Electronics Information and Emergency Communication (ICEIEC), Beijing, China, 12–14 July 2019; pp. 256–259. [CrossRef]
33. Tie, Y.; Li, X.; Zhang, T.; Jin, C.; Zhao, X.; Tie, J. Deep learning based audio and video cross-modal recommendation. In Proceedings of the 2022 IEEE International Conference on Systems, Man, and Cybernetics (SMC), Prague, Czech Republic, 9–12 October 2022; pp. 2366–2371. [CrossRef]
34. Ma, M.; Liu, W.; Feng, W. Deep-Learning-based Cross-Modal Luxury Microblogs Retrieval. In Proceedings of the 2021 International Conference on Asian Language Processing (IALP), Yantai, China, 23–25 October 2021; pp. 90–94. [CrossRef]
35. Liu, X.; Hu, Z.; Ling, H.; Cheung, Y.-M. MTFH: A Matrix Tri-Factorization Hashing Framework for Efficient Cross-Modal Retrieval. *IEEE Trans. Pattern Anal. Mach. Intell.* **2021**, *43*, 964–981. [CrossRef] [PubMed]
36. Chun, S.; Oh, S.J.; de Rezende, R.S.; Kalantidis, Y.; Larlus, D. Probabilistic Embeddings for Cross-Modal Retrieval. In Proceedings of the 2021 IEEE/CVF Conference on Computer Vision and Pattern Recognition (CVPR), Nashville, TN, USA, 20–25 June 2021; pp. 8411–8420. [CrossRef]
37. Wang, X.; Liang, M.; Cao, X.; Du, J. Dual-pathway Attention based Supervised Adversarial Hashing for Cross-modal Retrieval. In Proceedings of the 2021 IEEE International Conference on Big Data and Smart Computing (BigComp), Jeju-si, Republic of Korea, 17–20 January 2021; pp. 168–171. [CrossRef]
38. Fang, Z.; Li, L.; Xie, Z.; Yuan, J. Cross-Modal Attention Networks with Modality Disentanglement for Scene-Text VQA. In Proceedings of the 2022 IEEE International Conference on Multimedia and Expo (ICME), Taipei, Taiwan, 18–22 July 2022; pp. 1–6. [CrossRef]
39. Guan, W.; Wu, Z.; Ping, W. Question-oriented cross-modal co-attention networks for visual question answering. In Proceedings of the 2022 2nd International Conference on Consumer Electronics and Computer Engineering (ICCECE), Guangzhou, China, 14–16 January 2022; pp. 401–407. [CrossRef]
40. Zhang, S.; Loweimi, E.; Bell, P.; Renals, S. Windowed Attention Mechanisms for Speech Recognition. In Proceedings of the ICASSP 2019—2019 IEEE International Conference on Acoustics, Speech and Signal Processing (ICASSP), Brighton, UK, 12–17 May 2019; pp. 7100–7104. [CrossRef]
41. Kim, M.; Kim, T.; Kim, D. Spatio-Temporal Slowfast Self-Attention Network for Action Recognition. In Proceedings of the 2020 IEEE International Conference on Image Processing (ICIP), Abu Dhabi, United Arab Emirates, 25–28 October 2020; pp. 2206–2210. [CrossRef]
42. Yan, H.; Zhang, E.; Wang, J.; Leng, C.; Liang, H.; Peng, J. Coarse-Refined Local Attention Network for Hyperspectral Image Classification. In Proceedings of the 2022 International Conference on Image Processing and Media Computing (ICIPMC), Xi'an, China, 27–29 May 2022; pp. 102–107. [CrossRef]
43. Deng, S.; Dong, Q. GA-NET: Global Attention Network for Point Cloud Semantic Segmentation. *IEEE Signal Process. Lett.* **2021**, *28*, 1300–1304. [CrossRef]
44. Shu, X.; Zhang, L.; Qi, G.-J.; Liu, W.; Tang, J. Spatiotemporal Co-Attention Recurrent Neural Networks for Human-Skeleton Motion Prediction. *IEEE Trans. Pattern Anal. Mach. Intell.* **2022**, *44*, 3300–3315. [CrossRef] [PubMed]
45. Zhang, Z.; Jiang, T.; Liu, C.; Ji, Y. Coupling Attention and Convolution for Heuristic Network in Visual Dialog. In Proceedings of the 2022 IEEE International Conference on Image Processing (ICIP), Bordeaux, France, 16–19 October 2022; pp. 2896–2900. [CrossRef]
46. Jiang, Y.; Wang, J.; Huang, T. Prediction of Typhoon Intensity Based on Gated Attention Transformer. In Proceedings of the 2022 International Conference on High Performance Big Data and Intelligent Systems (HDIS), Tianjin, China, 10–11 December 2022; pp. 141–146. [CrossRef]

47. Jia, Y. Attention Mechanism in Machine Translation. *J. Physics Conf. Ser.* **2019**, *1314*, 012186. [CrossRef]
48. Xiong, R.; Yang, Y.; He, D.; Zheng, K.; Zheng, S.; Xing, C.; Zhang, H.; Lan, Y.; Wang, L.; Liu, T. On Layer Normalization in the Transformer Architecture. *arXiv* **2020**, arXiv:2002.04745. [CrossRef]
49. Schlag, I.; Irie, K.; Schmidhuber, J. Linear Transformers Are Secretly Fast Weight Programmers. *arXiv* **2021**, arXiv:2102.11174. [CrossRef]
50. Antoniades, I.; Brandi, G.; Magafas, L.; Di Matteo, T. The use of scaling properties to detect relevant changes in financial time series: A new visual warning tool. *Phys. A Stat. Mech. Its Appl.* **2021**, *565*, 125561. [CrossRef]

applied sciences

MDPI

Article

PDF Malware Detection Based on Fuzzy Unordered Rule Induction Algorithm (FURIA)

Sobhi Mejjaouli and Sghaier Guizani *

College of Engineering, Alfaisal University, Riyadh 11533, Saudi Arabia
* Correspondence: sguizani@alfaisal.edu

Abstract: The number of cyber-attacks is increasing daily, and attackers are coming up with new ways to harm their target by disseminating viruses and other malware. With new inventions and technologies appearing daily, there is a chance that a system might be attacked and its weaknesses taken advantage of. Malware is distributed through Portable Document Format (PDF) files, among other methods. These files' adaptability makes them a prime target for attackers who can quickly insert malware into PDF files. This study proposes a model based on the Fuzzy Unordered Rule Induction Algorithm (FURIA) to detect PDF malware. The proposed model outperforms currently used methods in terms of reducing error rates and increasing accuracy. Other models, such as Naïve Bayes (NB), Decision Tree (J48), Hoeffding Tree (HT), and Quadratic Discriminant Analysis (QDA), were compared to the proposed model. The accuracy achieved by the proposed model is 99.81%, with an error rate of 0.0022.

Keywords: Portable Document Format; cyber-attack; malware detection

Citation: Mejjaouli, S.; Guizani, S. PDF Malware Detection Based on Fuzzy Unordered Rule Induction Algorithm (FURIA). *Appl. Sci.* **2023**, *13*, 3980. https://doi.org/10.3390/app13063980

Academic Editors: Peter R.J. Trim and Yang-Im Lee

Received: 7 February 2023
Revised: 6 March 2023
Accepted: 9 March 2023
Published: 21 March 2023

1. Introduction

Intelligent attacks utilizing documents with malicious codes have been increasing rapidly in recent years as file transfers expand. The majority of Internet users are aware of the risk posed by execution files that are attached to emails or web pages. However, because users are unaware of the documents, they serve as an effective means of spreading malware. Because Portable Document Format (PDF)s are more flexible than other document formats, they are one of the main attack vectors among the malware that has been identified. The majority of malicious PDF documents include JavaScript or binary scripts that exploit certain security flaws and carry out destructive deeds, as explained in [1]. On the internet, there are countless billions of PDF files. Not all of them are as benign as one might expect. In actuality, PDF files may include a variety of objects, including binary or JavaScript codes. These items might occasionally be harmful. Malware software could try to infect a computer by finding a reading weakness [2]. In 2017, Adobe Acrobat Reader was found to have sixty-eight vulnerabilities. There are about fifty of them that may be used to run arbitrary codes. Each reader has certain weaknesses, and a malicious PDF file may discover a method to exploit them [3].

1.1. Reason for the Selection of PDF Files

The PDF format is one of the most widely used file types for sharing digital documents between different platforms and applications. Refs. [4,5] contain a full description of the PDF standard. A few elements make PDFs one of the preferred file types for malware authors to disseminate dangerous content. (a) PDF is extensively utilized by people in both professional and social settings. Academic papers, technical reports, design documents, and electronic receipts are a few examples of typical instances; (b) PDF is independent of platforms and operating systems (OS). A standalone PDF reader or a modern web browser can be used to access a PDF file on a Windows PC, a Linux system, or a mobile device (with a PDF viewer plug-in); (c) it is a very versatile file format. In addition to text, PDF also allows other sorts of data, such as video files, interactive forms, links to other files, JavaScript, Flash, and unified resource locators (URLs). Additionally, different encoding and compression techniques can be utilized to reduce file size, conceal important material, or both; and (d) it is stealthy and sophisticated. In general, executable files are thought to be more dangerous than PDF files. Setting a policy to prohibit staff members from downloading executable files from the internet or including them in email attachments is a common security measure, but it is uncommon to do the same with PDF documents. The enormous ubiquity and adaptability of the PDF file format also provide attackers with several opportunities to spread malware through PDF documents.

1.2. PDF-Based Malware

Phishing and exploits are the two main types of PDF-based attacks. Phishing attempts frequently appear in emails. A typical instance is a PDF delivery or purchase confirmation receipt attached to an email that seems to be from a trustworthy online store or logistics company. Apart from the social engineering techniques used to persuade recipients to open phishing PDF attachments, the text content of such emails is largely meaningless. These PDF documents are typically one page long and include social engineering elements as well as a phishing URL that leads to a suspicious website where malicious downloads, personal data collection, and other activities can be carried out. In contrast to plain-text-based phishing efforts, PDF documents include binary or a blend of binary and ASCII languages, making them harder to detect. This is one of the factors contributing to the rise in the popularity of phishing attempts based on PDFs. Its motivation is identical to that of standard phishing scams. The information that attackers collect from victims may be used by them or may be sold on the illicit market for use in the so-called shadow economy [6].

The PDF file format is a popular option for use in offices because of its high efficiency, dependability, and interactivity. The development of non-executable file assault technologies and attack techniques such as the advanced persistent threat has seriously jeopardized PDF's security since malicious PDF files are the most researched infection routes in adversarial scenarios [7,8]. With the development of machine learning (ML) technology [9] in recent years, researchers have developed a variety of ML-based techniques to recognize distinct attack types related to PDF files. The technology for detecting malicious PDF files may be divided into techniques based on static analysis, dynamic analysis, and techniques based on a combination of static and dynamic analyses. Modern research has demonstrated that PDF detectors based on ML may achieve excellent accuracy with a remarkably low false positive rate (FPR) [10]. However, such a study focuses on the proposed Fuzzy Unordered Rule Induction Algorithm (FURIA) for malware detection in PDF files compared with Naïve Bayes (NB), Decision Tree (J48), Hoeffding Tree (HT), and Quadratic Discriminant Analysis (QDA). These models are compared based on some of the well-known assessment measures, including accuracy (ACC), F-measure (FM), recall, precision, Matthew's correlation coefficient (MCC), and mean absolute error (MAE). This study has two primary objectives:

- To propose a malware detection model that will protect the systems from any harmful activity caused by PDF malware;
- To compare the findings from the suggested and existing models in use to discover a better and more effective solution for PDF malware detection.

The main contributions of this study are summarized as follows:

- We propose a FURIA-based model for the PDF malware detection;
- We analyze the outcomes of the proposed model with four well-known ML models: NB, J48, HT, and QDA;
- We do several tests on the dataset available at: http://205.174.165.80/CICDataset/CICEvasivePDFMal2022/Dataset/ (accessed on 5 February 2023);
- We disclose the intuition of the experiments using MAE, ACC, FM, MCC, precision, and recall metrics.

The rest of this paper is organized as follows: Section 2 summarizes the literature review. Sections 3 and 4 discuss the methodology and results analysis and discussion, respectively. Finally, Section 5 concludes this work.

2. Literature Review

Several studies have been conducted on PDF malware detection using various ML and deep learning (DL) models. The use of the Portable Document Format was explained by Reum et al. [11]. They provided a comprehensive study of the JavaScript content and structure found in the XML-embedded PDF. After that, they developed a range of features, including configuration and metadata such as file size, keywords, versions, and content features, as well as encoding strategies such as keywords, names, and JavaScript-readable strings. Due to the complexity of its features and the robustness of machine learning algorithms to small modifications, adversarial examples are challenging to construct. To reduce the possibility of adversarial assaults, they also develop a recognition model utilizing black-box-style models with structure and content properties. Utilizing observable robustness features, Chen et al. [12] described how to train robust PDF malware classifiers. For instance, a classifier must always recognize PDF malware as dangerous, no matter how many pages from benign forms are placed into the document. They show how to rigorously assess a malware classifier's worst-case behavior concerning specific robustness characteristics.

ML techniques have been used to create classifiers for PDF malware in several projects. Wepawet [13] and PJScan [14] were two earlier efforts that concentrated on the harmful JavaScript that was included in PDF malware. These tools include a JavaScript code extractor and a classifier for malicious JavaScript that can be either dynamic or static. Recent PDF malware classifiers have concentrated on structural aspects of PDF files since not all PDF malware contains embedded JavaScript and because PDF malware developers have learned several ways to conceal JavaScript codes [15]. We aim to develop cutting-edge structural feature-based classifiers in this effort. There have been studies that specifically looked at the JavaScript codes in PDFs. Features based on functions, constants, objects, methods, and keywords, as well as lexical characteristics of JavaScript scripts, were established by Khitan et al. [16]. Zhang [5] also utilized elements from the PDF structure, entity characteristics, metadata information, and content statistics, along with JavaScript features, including the number of objects, number of pages, and stream filtering information. Based on the finding that malicious JavaScript functions differ from legitimate JavaScript functions, Liu et al. [17] presented a context-aware technique. This method opens the PDF file while monitoring suspicious behavior based on JavaScript statements by passing the original code as input to the "eval" function.

In [18], Smutz and Stavrou combined the PDF parser with a random forest classifier to identify fraudulent PDF files using information gleaned from document metadata and file structure. They looked into 202 features, including /Font and /JavaScript. According to Liu et al. [17], current protections against malicious PDFs are inadequate, prone to evasion, and too computationally costly to be utilized online. They recommended leveraging static and run-time features to identify JavaScript in context. A software engineering approach was used to provide a detection method based on behavioral differences in those systems since a PDF document acts identically on different platforms [19]. A malicious document, on the other hand, will act differently depending on the platform. According to Li et al. [20], the weakness of all harmful detection methods that extract JavaScript is their dependency on 3rd-party extraction tools that precisely follow the Acrobat standard. A bigger training dataset does not always result in improved detection, according to Scofield et al. [21], and very little research has been conducted to establish the minimal size of a dataset required to achieve high detection accuracy. As a consequence, ref. [21] proposes a dynamic analysis-based detection technique.

3. Research Methodology

This study aims to develop a FURIA-based model for PDF malware detection. The overall research methodology is presented in Figure 1, which starts from data acquisition to comparison and performance analysis of each employed model. The dataset used in this study has been taken from the University of New Brunswick (UNB), Canadian Institute for Cybersecurity, http://205.174.165.80/CICDataset/CICEvasivePDFMal2022/Dataset/ (accessed on 5 February 2023). The dataset consists of 33 features, of which 32 are independent and 1 is dependent. The first 11 features are removed because they do not act in the analysis phase. These attributes are known as general features, including PDF size, metadata size, encryption, header, page number, text, image number, font objects, object number, number of embedded files, and the average size of all the embedded media. For extracting such features, we have used ClassifierAttributeEvaluator methods using a ZeroR classifier and the Ranker searching method. The selected features are ranked as follows:

Selected Features: 21,7,8,10,6,5,4,3,2,9,11,20,18,19,12,17,16,15,14,13,1:21.

These features are titled as: colors, startxref, pageno, ObjStm, trailer, xref, endstream, stream, endobj, encrypt, JS, XFA, launch, EmbeddedFile, Javascript, RichMedia, JBIG2Decode, Acroform, OpenAction, AA, and obj, respectively. Table 1 presents the description of each selected feature.

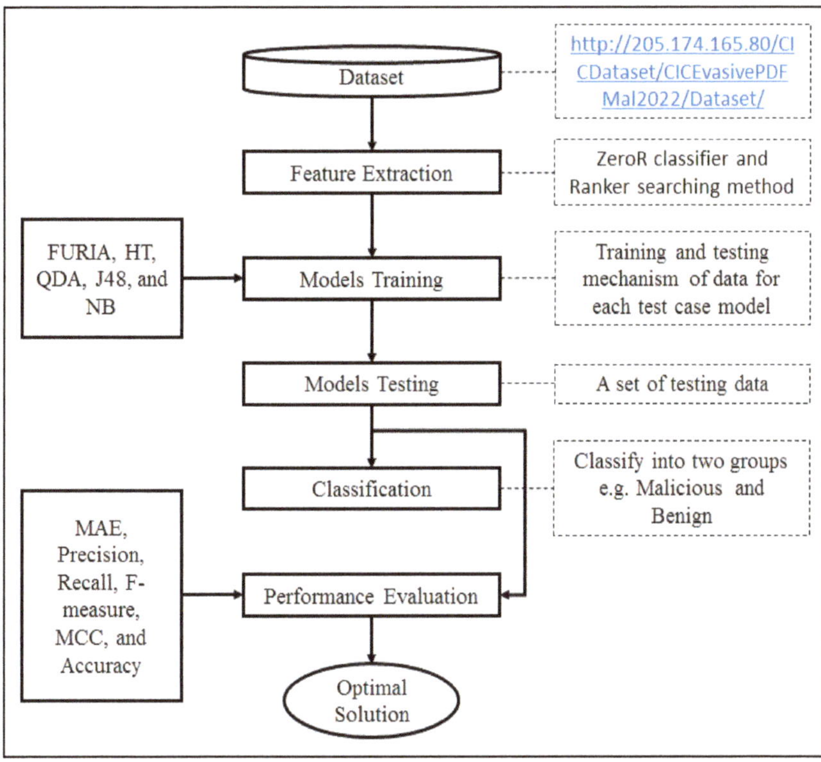

Figure 1. Methodology workflow [22].

Table 1. Selected features with their descriptions.

S No.	Feature	Description
1	Obj	This might be a sign of an attempt to obfuscate.
2	endobj	Many other forms of obfuscations are supported by PDFs, including string obfuscations in hex, octal, etc. that are typically used for evasion efforts.
3	Stream	This represents the quantity of binary data sequences in the PDF.
4	Endstream	Keywords that signify the streams' termination.
5	Xref	Size of the stream because streams may include a dangerous code.
6	Trailer	How many trailers there are in the PDF.
7	Startxref	How many keywords include "startxref," which designates the location where the Xref table is begun.
8	Pageno	Because malicious PDF files do not care how their material is presented, they often contain fewer pages—often only one blank page.
9	Encrypt	This function indicates if a PDF file is password-protected or not.
10	Objstm	streams with other items in them.
11	JS	The proportion of Javascript-containing objects.
12	Javascript	This indicates the amount of items that include a Javascript code, the most often used feature, as is clear.
13	AA	specifies a particular response to an event.
14	OpenAction	Defines a specific action to be taken when the PDF file is opened. The bulk of common malicious PDF files have been found to use this functionality in conjunction with Javascript.
15	Acroform	Form fields in Acrobat forms, which are PDF files, offer scripting technology that may be abused by hackers.
16	JBIG2Decode	A popular filter for encoding harmful stuff is JBig2Decode. How many items have nested filters? Nested filters can make decoding more challenging and may be an indicator of evasion.
17	Richmeddia	The quantity of flash files and embedded media is indicated by the number of RichMedia keywords.
18	Launch	A command or program can be run by using the term launch.
19	EmbeddedFile	PDFs can attach or embed a variety of things inside themselves that may be exploited, such as additional PDF files, Word documents, pictures, etc.
20	XFA	Certain PDF 40 files contain XFAs, which are XML Form Architectures that offer scripting technologies that can be abused by attackers.
21	Color	In the PDF, many colors are utilized.
22	Class	Classify as malicious or benign.

For model training and testing, a standard method of K-fold validation [23,24] is used. Here, the value of K is selected as 10. The performance of each employed model is evaluated using some of the standard evaluation metrics, including mean absolute error (MAE), recall, precision, Matthew's correlation coefficient (MCC), FM, and classification ACC. These measures can be calculated as follows:

$$\mathrm{MAE} = \frac{1}{2}\sum\nolimits_{j=1}^{n} |y_i - y| \quad (1)$$

$$\mathrm{Recall} = \mathrm{TP}/(\mathrm{TP} + \mathrm{FN}) \quad (2)$$

$$\mathrm{Precision} = \mathrm{TP}/(\mathrm{TP} + \mathrm{FP}) \quad (3)$$

$$\mathrm{MCC} = \frac{\mathrm{TP} * \mathrm{TN} - \mathrm{FP} * \mathrm{FN}}{\sqrt{(\mathrm{TP} + \mathrm{FP})(\mathrm{TP} + \mathrm{FN})(\mathrm{TN} + \mathrm{FP})(\mathrm{TN} + \mathrm{FN})}} \quad (4)$$

$$\mathrm{FM} = \frac{\mathrm{TP}}{\mathrm{TP} + \frac{1}{2}(\mathrm{FP} + \mathrm{FN})} \quad (5)$$

$$\mathrm{ACC} = \frac{\mathrm{TP} + \mathrm{TN}}{\mathrm{TP} + \mathrm{TN} + \mathrm{FP} + \mathrm{FN}} \tag{6}$$

Here, the important thing is to discuss the use of MAE in this study. MAE is typically used as an evaluation metric in regression problems, where the goal is to predict a continuous numerical output. However, in some cases, MAE can also be used in classification problems to evaluate the performance of the classification model. In classification, the output is a categorical variable, so using MAE as the primary evaluation metric might not be as informative as other classification-specific metrics such as accuracy, precision, recall, F1-score, or AUC-ROC. These metrics provide a more detailed understanding of how well the model performs in terms of correctly identifying positive and negative examples. However, in some cases, using MAE in classification can provide additional insights into the model's performance. In this scenario, MAE can be used to evaluate how far the predicted probabilities are from the true labels.

Fuzzy Unordered Rule Induction Algorithm (FURIA)

The FURIA is a new algorithm introduced by Huhn and Hullermeier that is responsible for generating fuzzy logic rules from a given database and classifying it using the obtained rules [22]. Fuzzy logic algorithms are well-known for their properties, such as classification rules that are simply understood by the reader, the capacity to analyze linguistic input, and the ability to enable expert judgment. They can also be used as a tool for classification purposes [25,26]. FURIA is the advanced version or derivative of the Repeated Incremental Pruning to Produce Error Reduction (RIPPER) algorithm. The RIPPER Algorithm is a classification algorithm based on rules. The training set is used to generate a set of rules. It is a famous rule induction algorithm. It generates fuzzy rules rather than traditional rules to replicate more flexible classification parameters. The fuzzy rules are constructed by substituting fuzzy intervals with a trapezoidal relevance function in association with the original RIPPER algorithm's advanced rule induction approach [27].

It has an appealing feature, which is the rule's extension. The generalization of the laws to include every option is the extension. It is a local technique that looks for information near the query. The simplest way to find the smallest generalization of a rule is to exclude those antecedents that are not met by the query [22]. The pseudo-code for a single rule, r, is shown in Algorithm 1 below, and the flowchart of FURIA is presented in Figure 2.

Algorithm 1: Generation of single ruler [22].

Let ***A*** be the set of numeric antecedents of ***r***

$While\ A \neq \varnothing\ do$
$a_{max} \leftarrow null\ (a_{max}\ denotes\ the\ antecedent\ with\ the\ highest\ purity\ )$
$pur_{max} \leftarrow 0\ (pur_{max}\ is\ the\ highest\ purity\ value,\ so\ far)$
$for\ i \leftarrow 0\ to\ size\ (A)\ do$
Compute the best fuzzification of *A[i]* in terms of purity
$pur_{A[i]}\ ] \leftarrow to\ the\ purity\ of\ this\ best\ fuzzification$
$if\ pur_{A[i]} > pur_{max}\ then$
$pur_{max} \leftarrow pur_{A[i]}$
$a_{max} \leftarrow A[i]$
end
end
$A \leftarrow A \backslash a_{max}\ Update\ r\ with\ a_{max}$
end

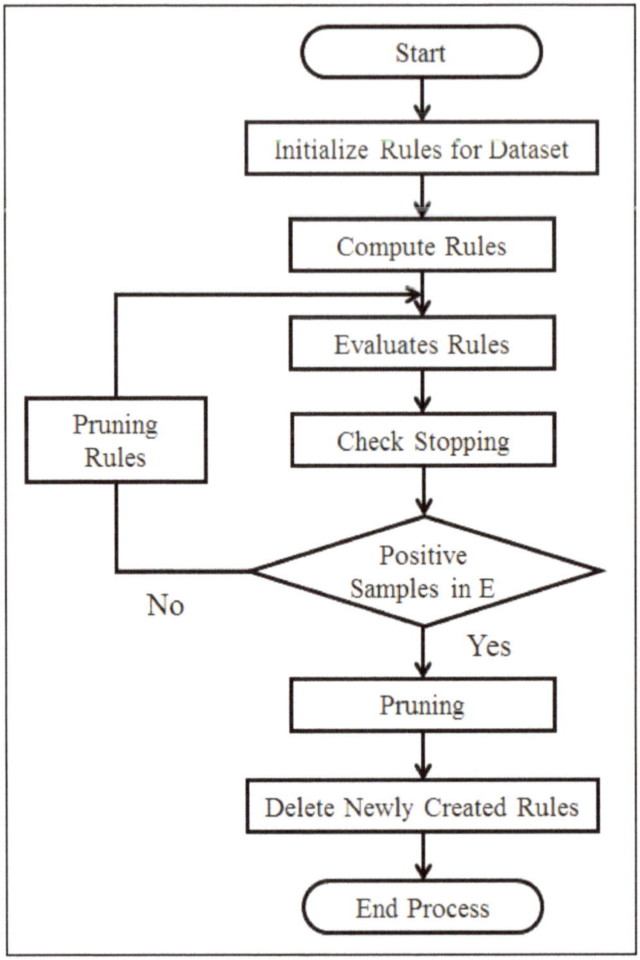

Figure 2. FURIA algorithm flowchart [22].

4. Results, Analysis, and Discussion

This section presents and discusses the study's findings. A new model, namely FURIA, is presented for PDF malware detection. FURIA and other benchmarked models are evaluated using a variety of criteria, which are ACC, FM, MCC, MAE, recall, and precision. Figure 3 illustrates the true positive rate (TPR) and false positive rate (FPR) analyses of each model compared with the proposed model. These analyses show the better performance of the FURIA, with the lowest FPR and better TPR. In both situations, it can be found that NB shows the worst outcomes. Figure 4 illustrates the outcome assessed via MAE. The MAE analysis shows the better performance of the proposed model with the lowest error rate, which is 0.0022, and the worst performance of NB with an error rate of 0.0147. All these values are achieved using a confusion matrix. Confusion matrix values achieved via each model are presented in Table 2.

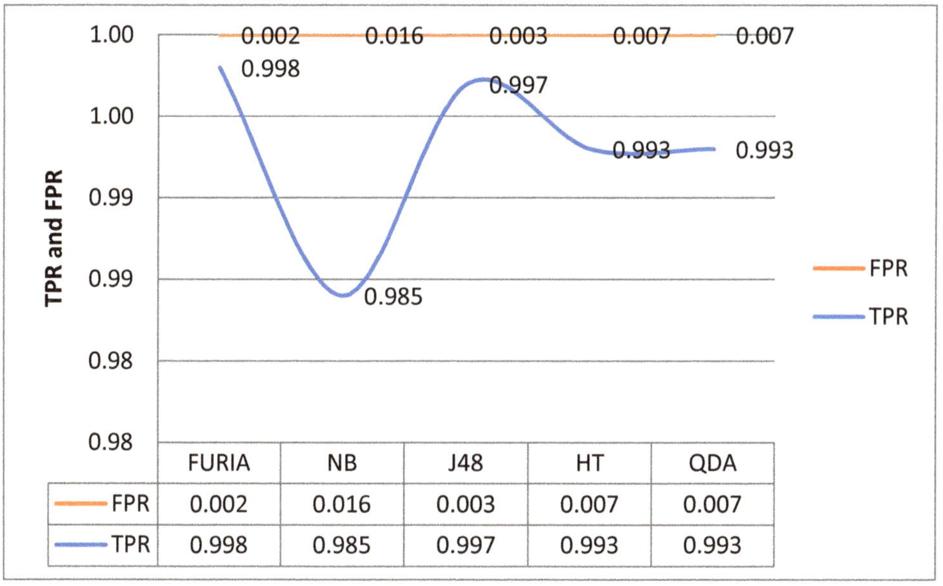

Figure 3. TPR and FPR analysis of each model.

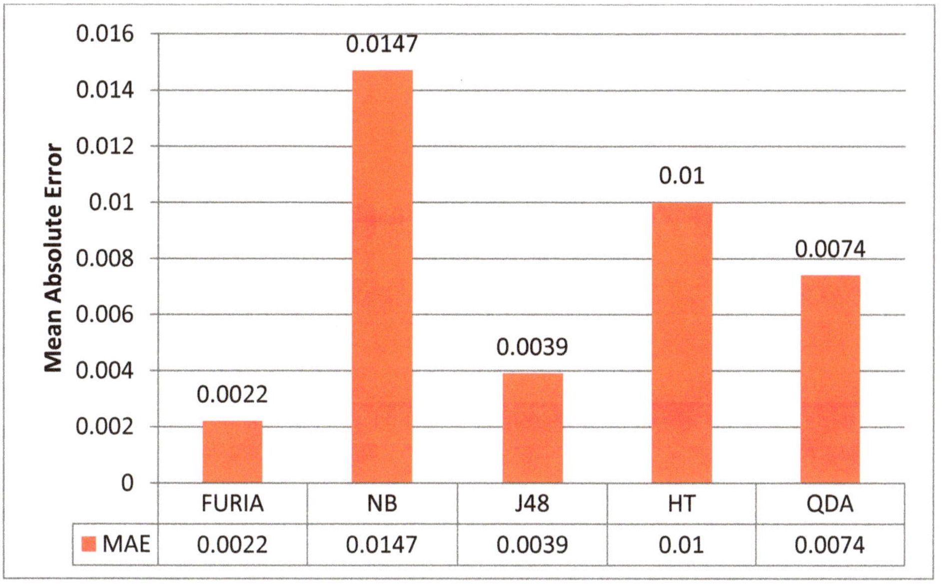

Figure 4. Proposed model comparison based on MAE.

Table 2. Confusion matrix values achieved via each employed model.

Models		No	Yes
FURIA	no	8995	11
	yes	27	10,953
NB	no	8807	199
	yes	97	10,883
J48	no	8977	29
	yes	33	10,947
HT	no	8943	63
	yes	67	10,913
QDA	no	8942	64
	yes	82	10,898

Figure 5 presents the outcomes assessed via precision, recall, and FM. These outcomes also depict the better performance of FURIA, with a value of 0.998 for precision, recall, and FM, respectively. The HT and QDA have the same outcomes of 0.993 for recall, precision, and FM, respectively, while the NB shows the poorest performance with a value of 0.985 individually for recall, precision, and FM.

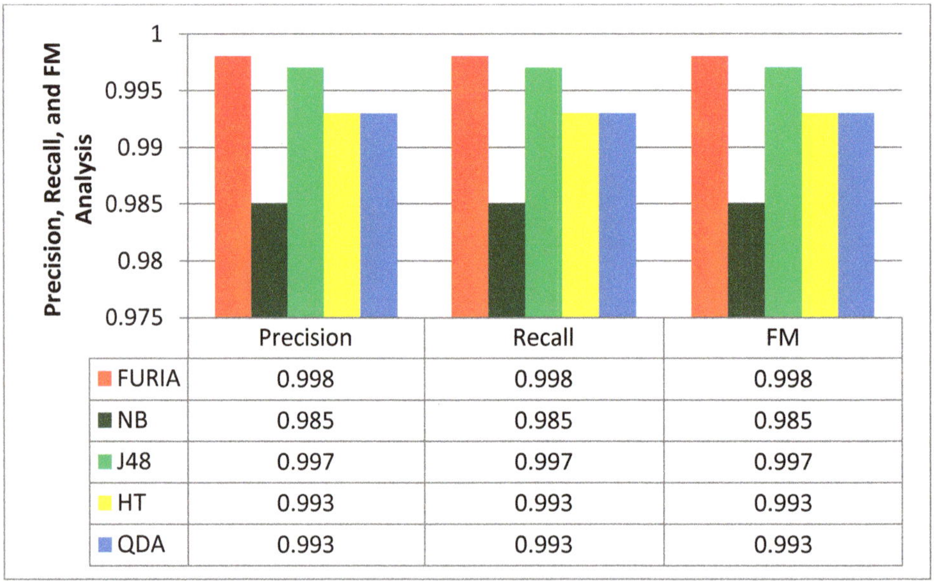

Figure 5. Models comparison based on precision, recall, and F-measure.

Figure 6 illustrates the evaluation of each model using R^2, accuracy, and a logarithmic trendline. The logarithmic trendline, which is extremely useful when the rate of change in the data is rapidly increasing or falling and then leveling out, is the best-fit curved line. The positive and negative values can both appear on a logarithmic trendline [28]. It may be obtained as follows:

$$y = a * \ln(x) + b \tag{7}$$

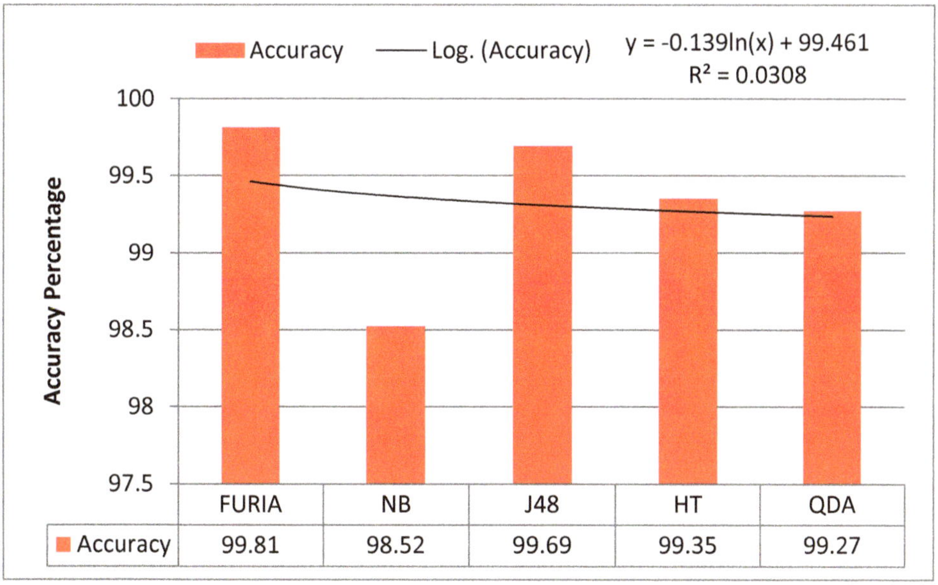

Figure 6. Accuracy analysis through each employed model.

Here, "ln" is the natural logarithmic function, and a and b are constants in the equation. The following generic equations, which differ only in the most recent input, can be used to retrieve the constants:

$$a = \text{INDEX}(\text{LINEST}(y,\ \text{LN}(x)), 1) \tag{8}$$

$$b = \text{INDEX}(\text{LINEST}(y,\ \text{LN}(x)), 1,\ 2) \tag{9}$$

R^2, also known as the coefficient of determination, is a statistical measure that represents the proportion of variance in the dependent variable (or the outcome) that can be explained by the independent variable(s) (or the predictor(s)) in a regression model [28]. It can be calculated as:

$$\text{R} - \text{squared} = \frac{\text{Explained Variation}}{\text{Total Variation}} \tag{10}$$

R^2 is consistently between 0% and 100%. The model does not take into consideration any fluctuation in the answer data around its mean, as shown by the 0%. 100% means that the model fully accounts for all the variability in the response data surrounding its meaning.

The properties of PDF encouraged hackers to take advantage of several security flaws and circumvent security measures, making the PDF format one of the most effective attack vectors for harmful malware. Therefore, it is essential for information security to accurately recognize malicious PDF files. To this end, this study proposes a model based on FURIA for PDF malware detection. According to the analysis described in the preceding section, the proposed FURIA performs better than other employed models in terms of increasing accuracy and reducing the error rate. The accuracy percentage difference (PD) between FURIA and other applied algorithms is shown in Figure 7. This analysis shows that there is very little difference between the proposed model and the J48, which is only 0.12%, while the difference between the proposed model and the NB is greater than other employed models, which show the poorest performance of the NB as compared to the proposed model. The value of PD can be obtained using Equation (11), where x1 represents the value of FURIA and x2 represents the value of other employed algorithms.

$$PD = \left(\frac{|x1 - x2|}{\frac{(x1+x2)}{2}} \right) * 100 \quad (11)$$

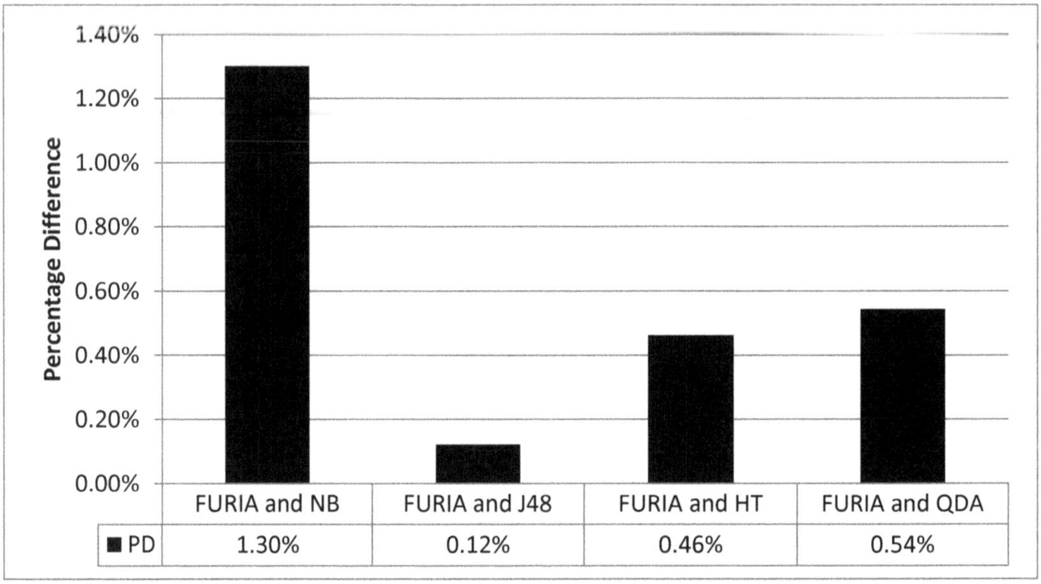

Figure 7. Accuracy percentage difference between FURIA and other employed models.

The advantage of using FURIA is that it works well on datasets with imbalanced class distributions. If a dataset has a large number of records and the majority of those records fall into one class but the remaining records fall into other classes, the dataset is said to have an unbalanced distribution of classes [22,28]. We also have an unbalanced distribution of the data in the dataset, which is why the performance of the projected model is better as compared with other employed models.

The data obtained from the UNB are used in all of the experiments. The rest of the algorithms in use are assessed using several common assessment metrics, such as MAE, precision, FM, recall, MCC, and accuracy, along with the proposed model. The models are trained and evaluated using the 10-fold cross-validation method. The threat now is that if the dataset is changed, the new results could outperform our analysis. The findings might potentially be affected by changing the criteria for data training and testing in place of the 10-fold cross-validation, for example, by using a percentage division. Another risk is that if a new algorithm is developed and it proves to be more effective than the one we now use, the results might be improved.

5. Conclusions and Future Direction

This study proposed a FURIA-based model for PDF malware detection. The proposed model is benchmarked with some of the well-known ML models, which are NB, J48, HT, and QDA. The performances of all these models are evaluated using some of the standard assessment measures that include MAE, ACC, FM, MCC, precision, and recall on the dataset taken from the UNB repository. The overall outcome presents a better performance of the proposed model, with an accuracy of 99.81% and a lowest error rate of 0.0022. FURIA outperforms other models; however, there are some limitations of FURIA. FURIA generates a large number of rules, which can make it difficult to understand and interpret the resulting model. This complexity can also lead to longer processing times and increased computational resources. Although fuzzy rules can be more interpretable than other machine learning models such as neural networks, they can still be difficult to interpret in complex data sets, which can limit their usefulness in some applications.

The FURIA-based model outperforms other well-known machine learning models for PDF malware detection; there are several potential future directions for research. There may be opportunities to refine the model further. For example, the model's parameters could be further optimized, or new features could be added to improve the model's performance. It may be beneficial to explore the potential benefits of combining the FURIA-based model with other ML models. Such hybrid models have the potential to improve the overall performance of the model.

Author Contributions: Writing—original draft, S.G.; Writing—review & editing, S.M. All authors have read and agreed to the published version of the manuscript.

Funding: This research work is supported by Alfaisal University.

Conflicts of Interest: The authors declare no conflict of interest.

References

1. Jeong, Y.S.; Woo, J.; Kang, A.R. Malware Detection on Byte Streams of PDF Files Using Convolutional Neural Networks. *Secur. Commun. Netw.* **2019**, *2019*, 8485365. [CrossRef]
2. Cuan, B.; Damien, A.; Delaplace, C.; Valois, M. Malware detection in PDF files using machine learning. In Proceedings of the ICETE 2018—The 15th International Joint Conference on e-Business and Telecommunications, Warangal, India, 18–21 December 2018; Volume 2, pp. 412–419. [CrossRef]
3. Falah, A.; Pokhrel, S.R.; Pan, L.; de Souza-Daw, A. Towards enhanced PDF maldocs detection with feature engineering: Design challenges. *Multimed. Tools Appl.* **2022**, *81*, 41103–41130. [CrossRef]
4. Docs, A.D. Adobe. Available online: https://opensource.adobe.com/dc-acrobat-sdk-docs/ (accessed on 21 November 2022).
5. Zhang, J. MLPdf: An Effective Machine Learning Based Approach for PDF Malware Detection. *arXiv* **2018**, arXiv:1808.06991.
6. Malware Analysis on PDF. Available online: https://scholarworks.sjsu.edu/etd_projects/683/ (accessed on 20 May 2019).
7. Xu, W.; Qi, Y.; Evans, D. Automatically Evading Classifiers. In Proceedings of the 23rd Annual Network and Distributed System Security Symposium—NDSS '16, San Diego, CA, USA, 21–24 February 2016; Volume 2016, pp. 21–24.
8. Chakkaravarthy, S.S.; Sangeetha, D.; Vaidehi, V. A Survey on malware analysis and mitigation techniques. *Comput. Sci. Rev.* **2019**, *32*, 1–23. [CrossRef]
9. Li, W.; Meng, W.; Tan, Z.; Xiang, Y. Design of multi-view based email classification for IoT systems via semi-supervised learning. *J. Netw. Comput. Appl.* **2019**, *128*, 56–63. [CrossRef]
10. Li, Y.; Wang, X.; Shi, Z.; Zhang, R.; Xue, J.; Wang, Z. Boosting training for PDF malware classifier via active learning. *Int. J. Intell. Syst.* **2022**, *37*, 2803–2821. [CrossRef]
11. Kang, A.R.; Jeong, Y.-S.; Kim, S.L.; Woo, J. Malicious PDF detection model against adversarial attack built from benign PDF containing javascript. *Appl. Sci.* **2019**, *9*, 4764. [CrossRef]
12. Chen, Y.; Wang, S.; She, D.; Jana, S. On training robust {PDF} malware classifiers. In Proceedings of the 29th USENIX Security Symposium (USENIX Security 20), Boston, MA, USA, 12–14 August 2020; pp. 2343–2360.
13. Cova, M.; Kruegel, C.; Vigna, G. Detection and analysis of drive-by-download attacks and malicious JavaScript code. In Proceedings of the 19th International Conference on World Wide Web, Raleigh, CA, USA, 26 April 2010; pp. 281–290.
14. Laskov, P.; Šrndić, N. Static detection of malicious JavaScript-bearing PDF documents. In Proceedings of the 27th Annual Computer Security Applications Conference, Orlando, FL, USA, 5–9 December 2011; pp. 373–382.
15. Ryan, C. *Automatic Re-Engineering of Software Using Genetic Programming*; Kluwer Academic Publishers: Dordrecht, The Netherlands, 2000.
16. Khitan, S.J.; Hadi, A.; Atoum, J. PDF forensic analysis system using YARA. *Int. J. Comput. Sci. Netw. Secur.* **2017**, *17*, 77–85.
17. Liu, D.; Wang, H.; Stavrou, A. Detecting malicious javascript in pdf through document instrumentation. In Proceedings of the 2014 44th Annual IEEE/IFIP International Conference on Dependable Systems and Networks, Atlanta, GA, USA, 23–26 June 2014; pp. 100–111.
18. Smutz, C.; Stavrou, A. Malicious PDF detection using metadata and structural features. In Proceedings of the 28th Annual Computer Security Applications Conference, Orlando, FL, USA, 7 December 2012; pp. 239–248.
19. Xu, M.; Kim, T. {PlatPal}: Detecting Malicious Documents with Platform Diversity. In Proceedings of the 26th USENIX Security Symposium (USENIX Security 17), Vancouver, BC, USA, 16–18 August 2017; pp. 271–287.
20. Li, M.; Liu, Y.; Yu, M.; Li, G.; Wang, Y.; Liu, C. FEPDF: A robust feature extractor for malicious PDF detection. In Proceedings of the 2017 IEEE Trustcom/BigDataSE/ICESS, Sydney, Australia, 1–4 August 2017; pp. 218–224.

21. Scofield, D.; Miles, C.; Kuhn, S. Fast model learning for the detection of malicious digital documents. In Proceedings of the 7th Software Security, Protection, and Reverse Engineering/Software Security and Protection Workshop, San Juan, Puerto Rico, 4–5 December 2017; pp. 1–8.
22. Hühn, J.; Hüllermeier, E. FURIA: An algorithm for unordered fuzzy rule induction. *Data Min. Knowl. Discov.* **2009**, *19*, 293–319. [CrossRef]
23. Naseem, R.; Khan, B.; Ahmad, A.; Almogren, A.; Jabeen, S.; Hayat, B.; Shah, M.A. Investigating Tree Family Machine Learning Techniques for a Predictive System to Unveil Software Defects. *Complexity* **2020**, *2020*, 6688075. [CrossRef]
24. Khan, B.; Naseem, R.; Shah, M.A.; Wakil, K.; Khan, A.; Uddin, M.I.; Mahmoud, M. Software Defect Prediction for Healthcare Big Data: An Empirical Evaluation of Machine Learning Techniques. *J. Healthc. Eng.* **2021**, *2021*, 8899263. [CrossRef] [PubMed]
25. Gasparovica, M.; Aleksejeva, L. Using Fuzzy Unordered Rule Induction Algorithm for cancer data classification. *Breast Cancer* **2011**, *13*, 1229.
26. Soares, E.; Damascena, L.; Lima, L.M.; Moraes, R.M.D. Analysis of the Fuzzy Unordered Rule Induction Algorithm as a Method for Classification. In Proceedings of the Conference: V Congresso Brasileiro de Sistemas Fuzzy, Fortaleza, Brasil, 4–6 July 2018; pp. 4–6.
27. Verma, L.; Srivastava, S.; Negi, P.C. Transactional Processing Systems A Hybrid Data Mining Model to Predict Coronary Artery Disease Cases Using Non-Invasive Clinical Data. *J. Med. Syst.* **2016**, *40*, 178. [CrossRef] [PubMed]
28. Ukanova, Z.M.; Udun, K.G.; Lemessova, Z.E.; Hamkhash, L.K.; Alchenko, E.R.; Ukasov, R.B. Detection of Paracetamol in Water and Urea in Artificial Urine with Gold Nanoparticle @Al Foil Cost-efficient SERS Substrate. *Anal. Sci.* **2018**, *34*, 183–187. [CrossRef] [PubMed]

applied sciences

MDPI

Article

Deep Learning Model with Sequential Features for Malware Classification

Xuan Wu, Yafei Song *, Xiaoyi Hou, Zexuan Ma and Chen Chen

College of Air and Missile Defense, Air Force Engineering University, Xi'an 710051, China
* Correspondence: yafei_song@163.com

Abstract: Currently, malware shows an explosive growth trend. Demand for classifying malware is also increasing. The problem is the low accuracy of both malware detection and classification. From the static features of malicious families, a new deep learning method of TCN-BiGRU was proposed in this study, which combined temporal convolutional network (TCN) and bidirectional gated recurrent unit (BiGRU). First, we extracted the features of malware assembly code sequences and byte code sequences. Second, we shortened the opcode sequences by TCN to explore the features in the data and then used the BiGRU network to capture the opcode sequences in both directions to achieve deep extraction of the features of the opcode sequences. Finally, the fully connected and softmax layers were used to output predictions of the deep features. Multiple comparisons and ablation experiments demonstrated that the accuracy of malware detection and classification were effectively improved by our method. Our overall performance was 99.72% for samples comprising nine different classes, and our overall performance was 96.54% for samples comprising two different classes.

Keywords: deep learning; malware classification; sequential feature; temporal convolutional network; bidirectional gated recurrent unit

Citation: Wu, X.; Song, Y.; Hou, X.; Ma, Z.; Chen, C. Deep Learning Model with Sequential Features for Malware Classification. *Appl. Sci.* **2022**, *12*, 9994. https://doi.org/10.3390/app12199994

Academic Editor: Peter R.J. Trim

Received: 14 September 2022
Accepted: 27 September 2022
Published: 5 October 2022

1. Introduction

With the continued development of information technology, security incidents are exponentially growing while the network is becoming increasingly sophisticated and convenient. Since the first virus, Morris worm, was discovered in the 1980s, there has been a growing international concern about cyberspace security. Currently, malware is evolving at an increasingly rapid pace, and the creators of viruses have introduced polymorphism to counteract virus detectability by constantly modifying and obfuscating malware, resulting in malware of the same type that, although having the same malicious behavior, appears to be different software. The multiplicity and amorphism of malware have made the prevention and control of cyberspace security extremely difficult. The current problem is, therefore, to quickly detect and classify malware so as to protect the network accordingly.

The problem of malware family detection is essentially a classification problem, i.e., the malicious samples to be detected are classified into different families for screening. Malware detection analysis is divided into dynamic and static analyses. The dynamic analysis approach runs in a secure and controlled environment and analyzes the behavior of malicious samples. Using a secure and controlled environment for analysis makes it easy for malicious samples to detect differences in the environment, but it is too costly for dynamic analysis to be exclusively used in the real environment. Static analysis, on the other hand, is a way to understand the logical structure of the code without executing it and make judgments accordingly. Compared with dynamic analysis, the static analysis method consumes much less time and resources; thus, this study adopted the static analysis method. This method generally extracts features through reverse engineering technology to build a model. The extractable features include string [1], opcode [2], executable file structure [3], and function call graph [4]. Opcodes are machine language instructions describing program execution operations, which are relatively more practical and reliable. The n-gram method is used to extract opcodes. The advantage of this method is that it uses great likelihood estimation and is easy to understand. After extracting the features, a model is constructed to classify the malicious families. Santos [5] et al. proposed a method to detect the maliciousness of unknown programs by calculating the frequency values of opcodes appearing in the code as features. Kang et al. [6] proposed extracting the sequence of opcodes from the disassembled files to represent the temporality of malware execution and then used the n-gram algorithm to characterize opcode sequences. Since Nataraj et al. first proposed converting malware executable files into two-dimensional grayscale maps using image texture features with a certain level of similarity in each family for training, image features have been widely used in the field of malware. In recent years, deep learning algorithms have developed rapidly in areas such as natural language processing, which has powerful learning capabilities and more advantages in mining data structures in high-dimensional data. Applying deep learning to the field of malware is a hot topic of current research. Deep learning algorithms such as the recurrent neural network (RNN) [7] and gated recurrent unit (GRU) can be used to implement malware detection. Kwon et al. [8] proposed an RNN approach using an API call function to classify malware. These authors used dynamic analysis to extract representative API call functions of nine malware families as a training set and used LSTM for classification with an average accuracy of 71%. Messay-Kebede et al. [9] proposed a detection model using both traditional machine learning methods and autoencoder-based methods. A few classes were identified by the traditional machine learning model, and others were classified with autoencoders. Gibert et al. [10] extracted byte and opcode sequences, which were fed into a classifier composed of two convolutional neural networks (CNNs). Although the structure was relatively simple, the accuracy failed to exceed that of complex classifiers. Yan et al. [11] proposed the Malnet detection model, which used CNN to learn the features of grayscale maps and LSTM to learn the opcode and then merged the classifications using a simple weighting approach. Barath et al. [12] used a CNN-LSTM approach for feature extraction and two types of machine learning for classification using support vector machines and logistic regression. Researchers Ahmadi M and Zhang Y et al. [13,14] extracted 15 and 6 features from malware, respectively, with more comprehensive information extraction, but feature extraction and selection were time-consuming and contained features that had little effect on classification.

Because a single feature has limitations, and in order to improve the ability of feature mining, the accuracy of malware classification, and reduce the interference of malware variants, packaging and obfuscation technologies, the present study proposed a multi-classification method of malware families incorporating TCN-BiGRU. The main contributions are as follows.

1. A malware detection and classification method (TCN-BiGRU) that fuses the temporal convolutional network and the bidirectional gated recurrent unit was proposed to improve the overall performance of the malware detection and classification model.
2. Opcode and bytecode sequences were fused to obtain their occurrence frequencies, reduce interference from shelling and obfuscation techniques, and improve the accuracy rate.
3. The feature extraction capability of temporal convolutional networks (TCN) for temporal data was introduced to fully learn the dependency relationship among data.
4. The output of the maximum pooling layer and the output of the average pooling layer were fused for relatively comprehensive extraction of data features.
5. The nonlinear fitting ability of a bidirectional gated recurrent unit (BiGRU) was used, and further feature extraction was conducted to learn the dependency of the before and after information in the opcode sequence, extracting the opcode features based on the time series to improve the model classification detection effect.

This paper proceeds as follows. Section 1 introduces the relevant background and related work. Section 3 presents the model. Section 4 presents the experimental results and analysis. Finally, Section 5 summarizes the experimental conclusions and discusses future research prospects.

2. Related Technology

2.1. N-Gram Method

N-gram is an important method for processing utterances in natural language processing; it uses the Markov assumption to relate the probability of occurrence of the nth word to the first n−1 words only. Based on this assumption, the probability value of the occurrence of a sentence in a text is calculated by multiplying the probability of the occurrence of each word or phrase, which is expressed in Equation (1) as follows.

$$\begin{aligned} P(T) &= P(\omega_1) \times P(\omega_2) \times \cdots \times P(\omega_n) \\ &= P(\omega_1) \times P(\omega_2|\omega_1) \times \cdots \times P(\omega_n|\omega_1\omega_2\cdots\omega_{n-1}) \end{aligned} \tag{1}$$

The n-gram in the field of malware detection refers to the n opcode or byte sequences that occur in a piece of code [15] to obtain a tighter contextual connection.

The algorithm is implemented by fixing a sliding window of size n and moving forward one opcode at a time. The value of n in the n-gram is generally an integer from 1 to 5. The computational volume of the model increases with the value of n; thus, more information is obtained, and classification accuracy is higher. At the same time, model size exponentially increases. In practical applications, the selection of n values also affects the accuracy of the model and the size of the loss value.

2.2. *Temporal Convolutional Network (TCN)*

A temporal convolutional network (TCN) is a network structure proposed by Bai, Shaojie, et al. [16] for processing time series data based on convolutional neural networks (CNNs). TCN incorporates causal convolution to make causal relationships between upper and lower layers and uses dilated convolution and skip connect to avoid the gradient disappearance problem of RNNs. The use of a temporal convolutional network model not only maintains a large receptive field for the data but also reduces computational effort to better control model memory length and improve time series classification accuracy [17].

Compared with ordinary 1D convolutional networks, TCN brings three main improvements.

(a) Causal convolution: The output value for any moment t is related to the input only before moment t and the previous layer [18]. While traditional CNN networks can see future information, causal convolution can only see past information; it is causally consequent, so causal convolution has very strict temporal constraints and is a one-way structure. When the number of convolutional kernels is 4, a single causal convolutional structure is shown in the left panel of Figure 1, and the overall structure is shown in the right panel of Figure 1. A convolution kernel of 4 means that four points are selected from the previous layer for sampling input to the next layer.

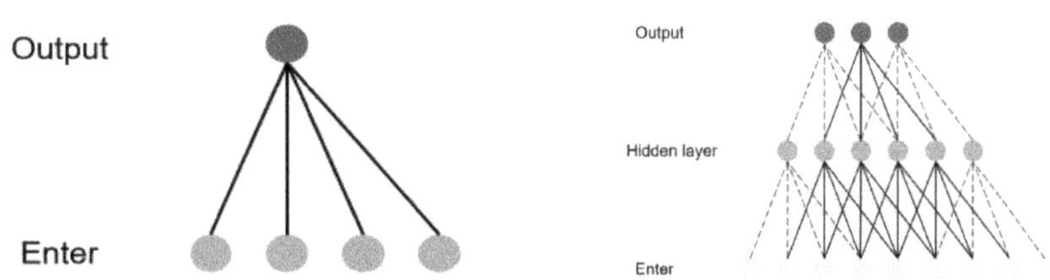

Figure 1. Causal convolution.

(b) Dilated convolution: With the gradual increase in the number of dilated convolution layers, the dilation coefficient exponentially increases, and the increase in the range of the receptive field of each layer reduces the number of convolution layers to reduce computational effort and simplify the network structure. To address the problems of traditional neural networks that require the linear stacking of multiple layers of convolution to extend the model of time series, TCN achieves a reduction in the number of convolutional layers by increasing the range of the receptive field of each layer by using dilated convolution [19], with a convolutional kernel of 4 and a dilation coefficient of 1, as shown in Figure 2. When the dilation coefficient of the input layer is 1, the samples in this model are sampled from the previous layer at an interval of 1 and input to the next layer.

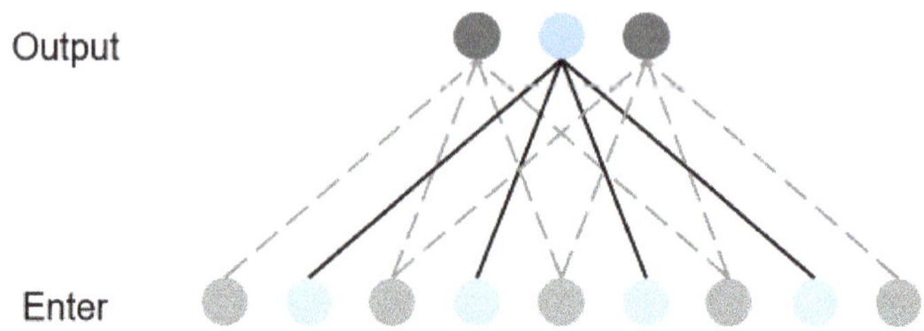

Figure 2. Dilated convolution.

The difference between dilated convolution and normal convolution is that dilated convolution allows the presence of interval sampling of the input during convolution, and the sampling rate depends on the dilation coefficient. Equation (2) of the receptive field is

$$RF = (K - 1) \times \mathrm{d} + 1 \tag{2}$$

where K is the convolution kernel size, and d is the dilation coefficient.

There are two ways for the TCN to increase the receptive field: one is to increase the size of the dilation coefficient, and the other is to choose a larger value of the convolution kernel. In the dilated convolution operation, the dilation coefficient exponentially grows with the depth of the network, so it is possible to use fewer layers to obtain a larger receptive field.

(c) Residual block: This is another important network structure in the TCN network. The residual block, shown in Figure 3, contains two layers of dilated causal convolution and nonlinear mapping. It has a constant mapping method of connection across layers, which enables the network to transfer information through a connection across layers. Through skip connect, it can not only speed up the response and convergence of the deep-level network but also solve the problem of too slow learning due to overly complex network hierarchical overlay structure. Dropout and batch normalization are also added to prevent model overfitting and speed up training [20].

The skip connect transforms the input x-value through a series of modules to output f(x); the equation for skip connect is

$$\mathrm{f}(x) = h(x) - x(1) \tag{3}$$

2.3. Bidirectional Gated Recurrent Unit (BiGRU)

As a variant of RNN, gated recurrent unit (GRU) also has a recursive structure similar to that of RNN and has the function of "memory" in processing time series data. At the same time, GRU can effectively alleviate the gradient disappearance and gradient explosion problems that may occur during RNN training, thus effectively solving the long-term memory problem. Long short-term memory (LSTM) networks are also a variant of RNN [21] and are comparable to GRU in terms of performance, but GRU is structurally simpler and can reduce computational effort and improve training efficiency [22]. The internal structure of GRU is shown in Figure 4. GRU has two inputs, the output state at the previous time and the input sequence value at the current time; the output is the state at the current time. GRU mainly updates the model state through a reset gate and an update gate. The reset gate controls the degree of forgetting historical state information so that the network can discard unimportant information; the update gate controls the weight of the past state information into the present state to help the network remember the information for a long time [23]. The internal equations of GRU are as follows:

$$\begin{cases} r_t = \sigma(W_r x_t + U_r h_{t-1}) \\ z_t = \sigma(W_z x_t + U_z h_{t-1}) \\ \widetilde{h}_t = \tanh(W_{\widetilde{h}} x_t + U_{\widetilde{h}}(r_t \odot h_{t-1})) \\ h_t = (1 - z_t) \odot h_{t-1} + z_t \odot \widetilde{h}_t \end{cases} \tag{4}$$

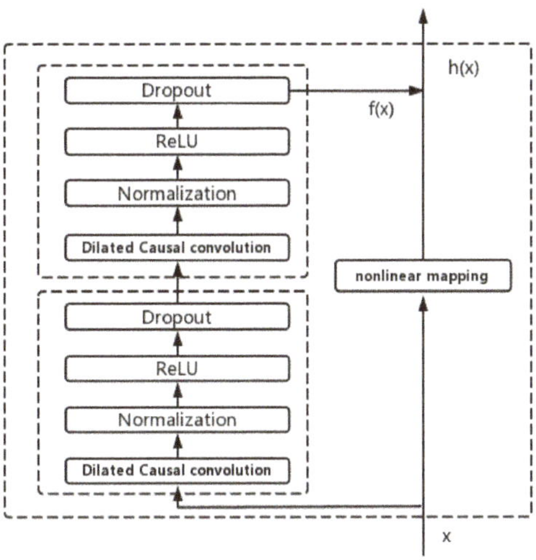

Figure 3. Residual block.

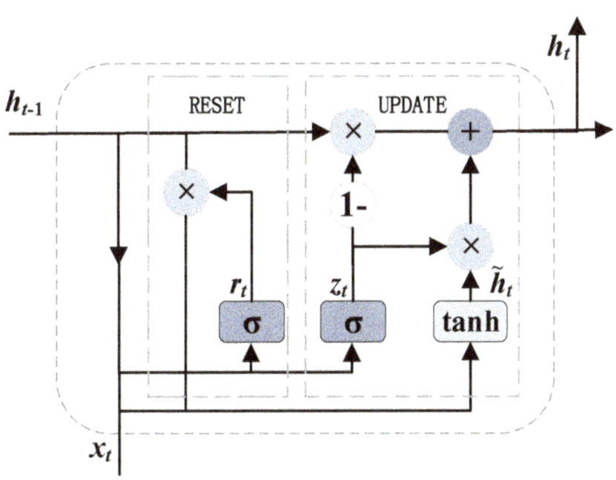

Figure 4. Gated recurrent unit.

The sigmoid activation function is shown in Equation (4) and Figure 4. It serves to convert the intermediate states to the range of 0 to 1; h_{t-1} and h_t are the output states at moments $t-1$ and t, respectively; x_t is the input sequence value at moment t (it is the candidate output state); W_r, W_z, $W_{\widetilde{h}}$, U_r and U_z are the corresponding weight coefficient matrices of each component; tanh is the hyperbolic tangent function (it is the Hadamard product of the matrix).

GRU can process the data only from forward to backward and ignores the effect of the latter moment on the data of the previous moment. To combine forward and backward data for integrated learning, BiGRU is used for further learning of the features of the malware. In the BiGRU, which consists of a forward gated recurrent unit and a backward gated recurrent unit, the network model learns the sequence from forward to backward and vice versa. The hidden layer contains two output units with the same input and is connected to the same output. The features can be better learned to increase the time series involved in training, thus providing higher accuracy for longer time series data.

3. Malware Classification Method Based on Sequence Features and Deep Learning

This section introduces the proposed TCN-BiGRU network. This network can extract past data features by one-dimensional, causal convolution with a simple structure, low memory consumption, fast operation speed, and easy superposition. The bidirectional GRU can capture a series of long-term dependencies in both directions, and the bidirectional GRU model can effectively utilize future moment information, which can compensate for the disadvantage of the one-way structure of the causal sequence in the TCN structure and the lack of comprehensive information extraction. The advantages of the two models were fully utilized and combined into a new hybrid model TCN-BiGRU, which enabled the model to conduct more comprehensive feature extraction to further improve the accuracy of malware classification and identification.

First, sample feature extraction was conducted. The originally extracted one-hot encoding and standards were normalized, after which the convolution operation was conducted using TCN to shorten the long-time sequence and extract the deep features of the network. At the same time, the maximum pooling and average pooling operations were conducted, and the extracted features were fused as the pooling output; after normalization and reconstruction, they were passed into the BiGRU network for the deep extraction of temporal features to complete malware detection classification. Finally, the most suitable hyperparameters were selected for the model to improve detection performance. The malware classification process included three stages: pre-processing, feature extraction and training, and classification. The model structure is shown in Figure 5.

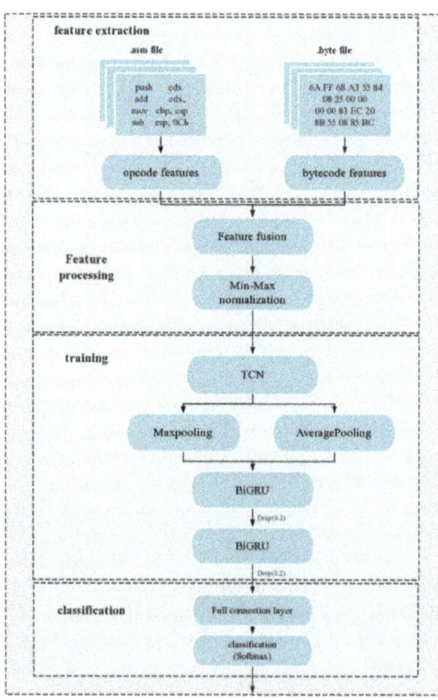

Figure 5. Malware detection model structure.

3.1. Features Extraction

(1) Malware opcode features

Programs are sequential instructions, and the underlying operation of a computer consists of the execution of instructions. Instructions generally comprise two parts: opcodes and operands. One of the static features commonly used in malware detection is the opcode feature. Batch disassembly is performed using the IDA pro tool on an executable PE file to obtain the .asm file with opcode sequence. The .asm file is generally divided into three segments: .text, .data, and .bss. The opcodes in the .text segment are shown in Figure 6.

```
.text:00401016 C2 04 00                                    retn    4
.text:00401016                                  ; ---------------------------------------------------
.text:00401019 CC CC CC    CC CC CC CC                     align 10h
.text:00401020 C7 01 08    BB 42 00             mov    dword ptr [ecx],   offset off_42BB08
.text:00401026 E9 26 1C    00 00                           jmp    sub_402C51
.text:00401026                                  ; ---------------------------------------------------
.text:0040102B CC CC CC    CC CC                           align 10h
.text:00401030 56                               push   esi
.text:00401031 8B F1                                       mov    esi, ecx
.text:00401033 C7 06 08    BB 42 00             mov    dword ptr [esi],   offset off_42BB08
.text:00401039 E8 13 1C    00 00                           call   sub_402C51
.text:0040103E F6 44 24    08 01                           test   byte ptr [esp+8], 1
.text:00401043 74 09                                       jz     short loc_40104E
.text:00401045 56                               push   esi
.text:00401046 E8 6C 1E    00 00                           call   ??3@YAXPAX@Z   ; operator delete(void *)
.text:0040104B 83 C4 04                                    add    esp, 4
.text:0040104E
.text:0040104E                                  loc_40104E:                      ; CODE XREF: .text:00401043↑j
.text:0040104E 8B C6                                       mov    eax, esi
.text:00401050 5E                               pop    esi
.text:00401051 C2 04 00                                    retn   4
.text:00401051                                  ; ---------------------------------------------------
.text:00401054 CC CC CC    CC CC CC CC CC CC CC CC   CC    align 10h
.text:00401060 8B 44 24    08                              mov    eax, [esp+8]
```

Figure 6. Original opcode sequence.

An opcode can usually be divided into four categories: data movement, arithmetic or logic, control flow types, and others, from which the more important opcodes are filtered to extract the opcode sequence text. The opcode codes category is shown in Table 1.

Table 1. Operation codes.

Category	Operation Codes
Data move	mov, movzx, push, pop, lea, xchg
Arithmetic/logic	add, sub, inc, dec, imul, or, xor, shl, shr, ror, rol
Control flow	jmp, jz, cmp, jnb, call, retf, retn
Other	nop

Figure 6 shows a sample of 0A32eTdBKayjCWhZqDOQ. The opcodes in each .asm file are sequentially extracted by regularization. The extracted opcode sequence text is shown in Figure 7.

```
mov,dword,jmp,byte,push,mov,mov,dword,call,test
,jz,xor,div,mov,pop,pop,retn,dword,push,mov,mov,
,offset,lea,push,mov,dword,mov,push,mov,mov,dw
,or,cmp,jbe,mov,jmp,mov,mov,mul,mov,shr,shr,cm
jbe,cmp,dword,jb,mov,jmp,lea,lea,lea,call,mov,add,
v,call,test,jbe,mov,push,lea,cmp,jb,mov,jmp,test,jnz,
op,pop,retn,lea,cmp,dword,lea,jb,mov,jmp,mov,mo
ov,xor,call,pop,retn,dword,dword,dword,dword,pus
,mov,jb,mov,xor,mov,pop,pop,mov,pop,pop,retn,b
```

Figure 7. Opcode sequence after pre-processing.

The opcode sequences differ. Some are extremely long, so to extract more complete information, the method based on n-gram in natural language processing is used to extract the opcode features. By treating each opcode as a word, the n-gram method takes subsequences of the opcode sequence according to the magnitude of the n value with a sliding window, and then the frequency of the corresponding subsequence is calculated. Then, a word frequency threshold is set, and the subsequence with a particular number of occurrences above the threshold is retained. The retained subsequence is a feature of the malware.

(2) Malware bytecode features

The malware itself is a file consisting of a series of bytes. One idea is to convert the binary file of malware into a grayscale image using the similarity between the values of the bytes and the range of pixel values taken in the grayscale image. The classification of malware families is achieved based on the texture similarity of grayscale images of the same family of malware and the different textures due to the different structures of different families of malware. To detect similar variants of malware, binary files can be better differentiated such that the impact of obfuscation is reduced.

Malware is converted into a sequence consisting of a binary, and the hexadecimal .byte file is read in binary, then divided by 16 bits in order, and converted into decimal values within [0, 256). The first line number of each byte file is ignored, and only the hexadecimal values after the line number are extracted. Only the values and letters in the byte file are kept, and the rest of the symbols are replaced with zeros, thus converting the malware file into a one-dimensional vector of decimal numbers.

The steps for extracting bytecode sequence features from malware are shown in Algorithm 1.

Algorithm 1: The hex file is converted to a sequence of decimal values within [0, 256).

```
Input: hexadecimal file;
Output: a one-dimensional vector-matrix representation of file byte sequence.
  function getMatrixfrom(file)
    f = open(file,"rb"); /*read the file in binary */
   hexst = binascii.hexlify(f); /*convert binary file to a hexadecimal string */
    Byte = np.array([int(hexst[i : i + 2], 16) for i in range(0, len(hexst), 2)]);
/*convert the string to an unsigned decimal number by byte division into a byte*/
     return byte;
  end function
```

Similarly, the length of the sequence of each sample varies. To extract more complete information, intercept a particular length, then use each decimal number within that length as a feature, then calculate the frequency of each decimal value.

3.2. Feature Pre-Processing

After the malware features (opcode and bytecode features) were extracted, we checked whether this data had missing values, treated the missing values as 0 uniformly, and then performed standard normalization on the malware feature data. Data normalization reduced the variance of the features to a smaller interval, reduced the impact of the difference in the size of different feature values, and improved the convergence rate of the model. Current normalization methods are commonly used to normalize the values to (0,1) and $(-1, 1)$. The normalization method used in this study was maximum-minimum normalization, which scales the values to the interval (0,1), as shown in Equation (5).

$$x' = \frac{x - M_{min}}{M_{max} - M_{min}} \tag{5}$$

where x' is the scaled value, M_{min} is the smallest value in the feature dimension, and M_{max} is the largest value in the feature dimension.

In the process of malicious code feature extraction, there are many zero values. This method can retain the zeros in the features and can handle the data values with small variances in the features.

3.3. Combine TCN and BiGRU for Feature Extraction

The advantages of the TCN model are extraction of past data by one-dimensional causal convolution to guarantee temporality, time savings via the skip connect block, extraction of temporal features by dilated convolution, and the fusion of the average pooling layer with the maximum pooling layer. The advantage of using the GRU model is its nonlinear fitting ability to efficiently extract the data features and its faster convergence speed than the LSTM model [24]. The two-way GRU model better captures the sequence features of the opcode by collecting information forward and backward, thus improving the accuracy of model classification. These two models are integrated into the TCN-BiGRU model to obtain better accuracy as well as lower loss values. The structure of the integrated model is shown in Figure 8.

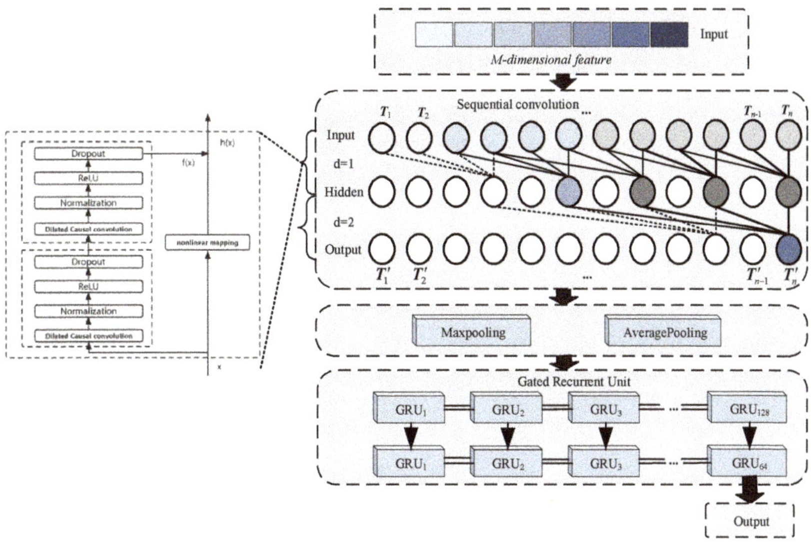

Figure 8. TCN-BiGRU model.

In Figure 8, the TCN-BiGRU model structure includes:

a. Input layer: processed malicious code opcode feature data and shape (total number of samples, time step, and feature dimension).

b. Time series convolutional network layer: the feature vectors T_j were extracted via TCN, and the residual units were set up in two layers. A residual unit consisted of two convolutional units and one nonlinear mapping, and the convolutional kernel weights were normalized. The residual unit in Figure 8 was used only as the input layer to the hidden layer; the same was true for the hidden layer to the output layer. The convolution kernel size value was 4, and the dilation coefficient was (1, 2). Dropout was added to prevent overfitting in training.

c. The different features extracted from the average pooling layer, as well as the maximum pooling layer, were fused as pooling outputs. We merged the average with the maximum pooling layer.

d. The combined pooling layer consisted of a maximum pooling and an average pooling layer, each of which was calculated as shown in Equation (6). Maximum pooling and average pooling were obtained by traversing the pooling window with the input from the previous layer of the network. The pooled maximum and average values were then summed and passed to the next layer of the model structure.

$$\begin{cases} h_{\max} = \max pool(h) \\ h_{avg} = avgpool(h) \\ h_{fuse} = h_{\max} \oplus h_{avg} \end{cases} \quad (6)$$

where h is the input from the upper layer network into the fused pooling layer; $h_{\max}$ is the maximum pooling output; h_{avg} is the average pooling output; and h_{fuse} is the output obtained by combining maximum pooling and average pooling in parallel.

e. Bidirectional gated recurrent unit layer: The figure shows the structure of the GRU unit when it had two layers. The output vector of the TCN model was first used as the input of the GRU to extract the long-term correlation in the time series. Then the data were output with the results obtained from two layers of BiGRU.

f. Output layer: Output the result of the last moment of the BiGRU to the classification layer.

3.4. Classification Output Layer

The classification output layer contained fully connected and softmax layers. The fully connected layer was used to obtain the display expression of the classification, and the softmax function was used to calculate the classification result of malicious code y. The structure of the classification output layer is shown in Figure 9.

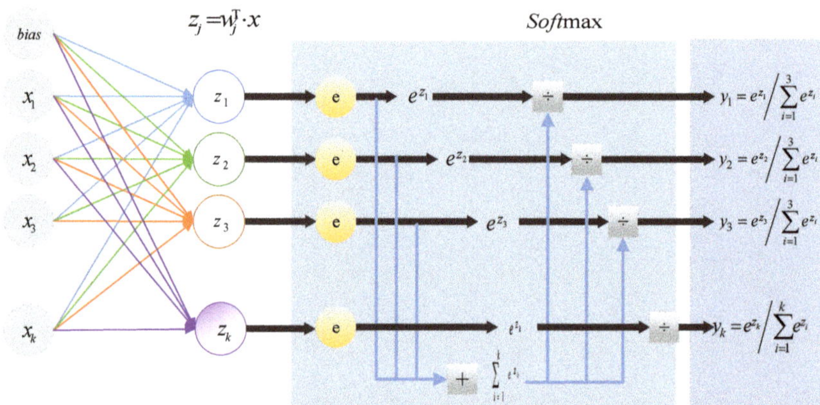

Figure 9. The classification output layer.

The fully connected layer multiplied the weight matrix by the input vector and added a bias to map n (−∞, +∞) real numbers to K (−∞, +∞) real numbers (fractions); Softmax mapped K real numbers. The real numbers of (−∞, +∞) were mapped to K (0,1) real numbers (probabilities) while ensuring that their sum was 1.

$$y_i = \text{softmax}(z) = \text{softmax}(w^T x + b_i) \quad (7)$$

where y denotes the probability of classification into malicious family type I; w denotes the weight matrix of the fully connected layer; and b is the bias vector of class i; at time t, replace x with h_{tn}.

The softmax layer superimposed the input features linearly with the weights. The number of neurons in the softmax layer was set by the number of malicious code types.

4. Experiments and Analysis of Results

4.1. Experimental Setup

To test the performance of the malicious code classification method fusing TCN and BiGRU, the following experiments were implemented:

Experiment 1: Feature selection experiment
Experiment 2: TCN-BiGRU model performance analysis experiment
Experiment 3: Comparison experiments of different pooling methods
Experiment 4: Model ablation comparison experiment
Experiment 5: Comparison experiments of different classification algorithms.

4.2. Experimental Environment and Data Set

The experimental environment was a computer configured with Win10, Intel Core (TM)-9880H CPU @ 2.30 GHz, 64 GB RAM, Quadro RTX 4000 GPU; the programming environment was PyCharm2021.2.2, using the Python 3.7 language in a CUDA 11.0 accelerated environment. The neural network model used TensorFlow 2.4.1 and Keras 2.4.3 versions of the deep learning framework.

The experimental datasets were from the open-source dataset provided by Microsoft [15], and the PE samples were from the Datacon Open Data Project provided by Qianxin (China) [25]. The malicious code families in the dataset provided by Microsoft were divided into 9 categories, with 10,868 malware samples. Each sample file had two formats: .asm and .bytes; the PE samples provided by Qianxin had two categories, containing a large amount of mining-type malicious code and non-mining samples. These are the latest real samples captured from the existing network; thus, these samples are likely to contain a large number of shelling samples and resource obfuscation samples. To prevent samples' mistaken execution from infecting the environment, the MZ and PE headers, as well as the import and export table parts, were removed. To ensure that the dataset samples have a certain level of diversity, similar samples were filtered. Therefore, in actual use, the MZ and PE headers were artificially added to extract the opcode features, and the samples were disassembled into .asm files using IDA tools. The family name, type number, number of samples, and expression number of the malware dataset used are shown in Figures 10 and 11.

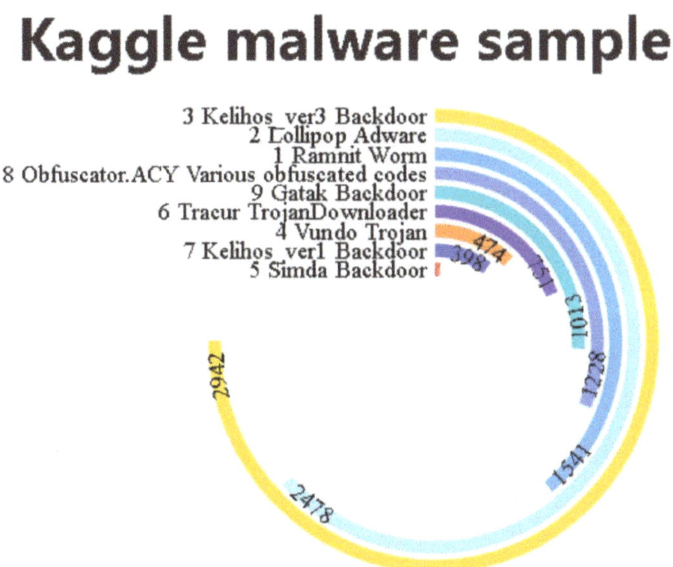

Figure 10. Kaggle malware sample.

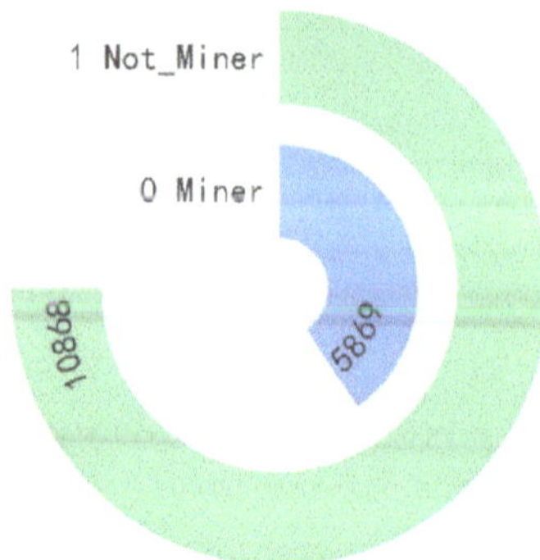

Figure 11. Datacon sample.

To fully evaluate our method, experiments were conducted on two different datasets according to different methods to fully validate the model. The first method used 9 malicious families in Kaggle malicious samples labeled 1–9, and the dataset was noted as 9-class-data. The second method used 0, 1 sub-table labeling on Datacon samples as sample labels, and the dataset was noted as 2-Class-Datacon; a five-fold cross-validation method was used to randomly divide the data into 10 parts, selecting 9 of these parts as the training set and 1 part as the test set.

4.3. Experimental Evaluation Criteria

The experiment selected common evaluation criteria in the field of malware classification detection: accuracy (Acc), precision (PR), recall (RR), and f1-score (F1) to evaluate the classification of the network. These criteria were calculated as follows:

$$\begin{cases} Acc = \frac{TP+TN}{TP+TN+FP+FN} \\ PR = \frac{TP}{TP+FN} \\ RR = \frac{TP}{TP+FP} \\ F1 = \frac{2\times PR\times RR}{PR+RR} \end{cases} \tag{8}$$

where q is the number of samples; d is the number of categories; the value type is the processed one-hot code (string consisting of 0 or 1), and $\hat{y}_{id}$ is the output value of the softmax function ($\sum_{d=1}^{d=9} \hat{y}_{id} = 1$). TP is the true class (meaning that malware was correctly classified as malware), FN is the false negative class (meaning that malware was incorrectly classified as normal software), FP is the false positive class (meaning that normal software was incorrectly classified as malware), and TN is the true negative class (meaning that normal software was correctly classified as normal software).

Model performance was presented using a visual representation of the confusion matrix, as shown in Table 2.

Table 2. Confusion Matrix.

Location	Real Label	
	For Malware	For Not Malware
Malware	TP	FP
Not malware	FN	TN

4.4. Feature Selection Experiments

After obtaining the opcode sequence, the n-gram method was used for feature extraction of the instruction file from the .asm file. The frequency f of the instruction n-gram in the .asm file was calculated as the feature and then used as input.

Normalization pre-processes the values of the frequency of the feature extracted from the n-gram of the malicious code, and the one-hot encoding method pre-processes the values of the malicious family categories. For example, the Ramnit family can be represented as 000000001.

In the experiment, the feature extraction was first performed by selecting $N = 3$, and then the instruction frequency threshold was selected as 300. Then, the test was conducted by increasing the value at 200 intervals, and the highest value selected was 1100. The experimental results showed that when instruction frequency increased, classification accuracy showed a trend of first rising and then decreasing, and the classification effect was best when the frequency was selected as 700, as shown in the lower panel of Figure 12.

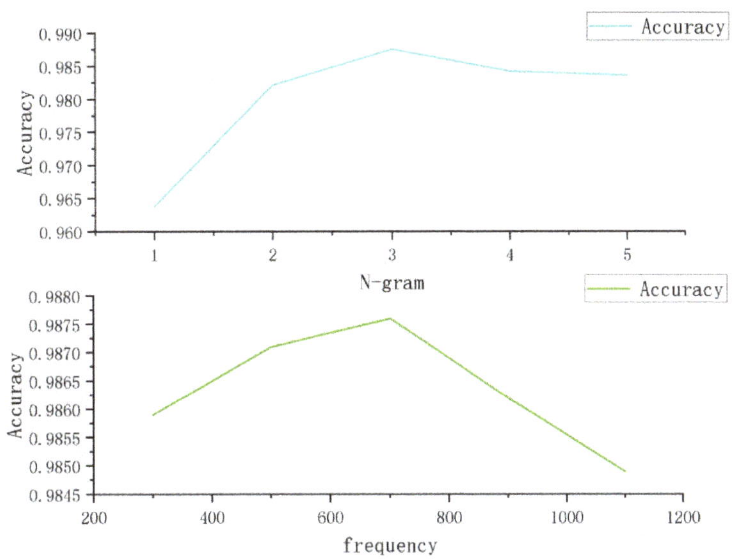

Figure 12. Experiment on the selection of N value and frequency.

In the experiments on change in the N value, the comparison experiment of N value was conducted using the frequency with the best effect in the instruction frequency experiment, i.e., a frequency of 700. As the N value increased, classification accuracy showed a trend similar to that of frequency, which also showed a trend of increasing and then decreasing. Through analysis of the experimental results, the classification effect was best at $N = 3$, and the experimental results are shown in the upper panel of Figure 13. Therefore, in subsequent experiments with the classification model, $N = 3$ was selected as the feature for input, and a frequency of 700 was selected as the input to the model.

For the byte code feature, the byte sequence with a length of fewer than 1500 bytes frequency was selected, then the experiment was conducted by increasing the frequency at an interval of 1000, and sequences with no more than 4500 in length were selected for the experiment. The experiment found that accuracy gradually decreased, so sequences within a length of 1000 were selected for training. The experiment then found that accuracy decreased compared to sequences within a length of 1500; thus, in subsequent features, we selected byte code with lengths below a 1500 bytes frequency for the fusion experiment.

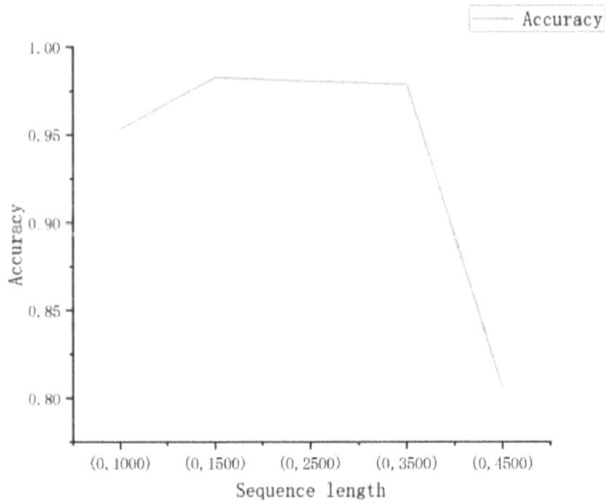

Figure 13. Sequence length selection experiment.

4.5. TCN-BiGRU Model Performance Analysis Experiments

In the TCN-BiGRU model, the choice of some hyperparameters in the model could impact the experimental results. A single feature (n-gram method) was used in tuning the model to optimize model parameters. Two hyperparameters, the number of filters and the number of convolutional kernels, were selected among the optimization class parameters, and the number of BiGRU layers and the number of neurons per layer were selected as variables from the model class parameters. The number of model iterations was set at 50, the dilation coefficient in TCN was exponentially increased by 2, the dilation coefficient was set to (1, 2), the optimization algorithm was chosen as Adamax, and the learning rate was set at 0.002. To avoid the overfitting problem, a dropout layer was added, and the value was taken as 0.2. To make the experimental data more accurate and valid, a five-fold cross-validation method was used. The prediction data obtained from the experiments regarding the classification of malicious code families when setting a different number of filters, the number of convolutional kernels, and the number of neurons are shown in Tables 3–5, respectively.

Table 3. Parameter setting of model.

Model Parameter	Real Label
Batch size setting	64
Optimizer	Adamax
Optimizer learning rate	0.002
Epoch setting	50
Number of TCN filters	7
Number of TCN convolution kernels	4
TCN dilation coefficient	(1, 2) cc
Number of BiGRU units	32\32
Dropout rate	0.2

Using the grid search algorithm, parameter search experiments were conducted for the filters (5, 7, 10, 15, 20) and the number of convolutional kernels (2, 3, 4, 5, 6) to finally determine the optimal parameter settings for the model, as shown in Table 4.

According to the values of each parameter obtained from the above experiments, the fusion of two features with $N = 3$, frequency = 700 and the first 1500 byte sequences was performed again using the TCN-BiGRU classification model.

Table 4. Parameter setting of model.

Malicious Code Family	Precision	Recall	F_1-Score
1	0.99	1.00	1.00
2	1.00	1.00	1.00
3	0.99	1.00	1.00
4	1.00	1.00	1.00
5	1.00	1.00	1.00
6	0.99	1.00	0.99
7	0.94	1.00	0.97
8	0.99	0.99	0.99
9	0.99	0.99	0.99
accuracy	-	-	0.99
Overall	99.55%	99.54%	99.54%

Table 5. Parameter setting of model.

Malicious Code Family	Precision	Recall	F_1-Score
0	0.94	0.93	0.93
1	0.96	0.97	0.96
accuracy	-	-	0.95
Overall	96.37%	96.63%	96.50%

The confusion matrix for the classification of the 9-class-data dataset is shown in Figure 14, with "Real label" on the vertical axis indicating true malicious code and "Prediction" on the horizontal axis indicating the prediction made by the model.

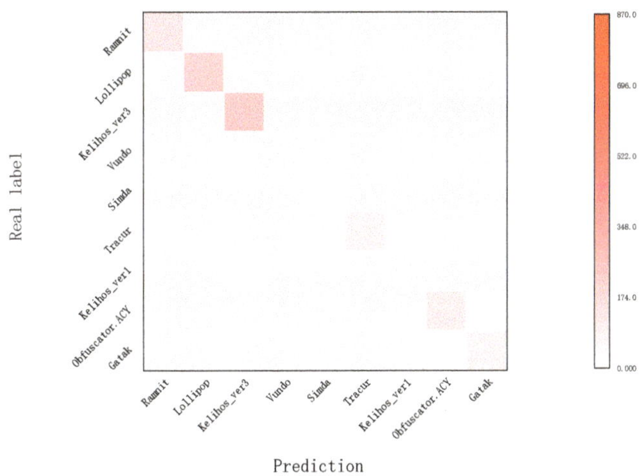

Figure 14. Confusion matrix.

Table 5 shows in more detail the precision, recall, and F_N-score (N = 1) of the predictions for each category. Note that for the Lollipop class, Vundo class, and Simda class, the classification is 100%. In the Kelihos_ver1 class, the classification is poor, with an accuracy of only 94%, while the remainder reached more than 99%. The family class Kelihos_ver1 belongs to the backdoor virus type in the broad category, while there are three families that are all backdoor viruses. Their poor classification was probably due to confusion with similar families.

Table 6 details the precision, recall, and f_N-score (N = 1) for each category of predictions on the 2-Class-Datacon dataset. The table shows that the results were better for the Not_Miner classification on the 2-Class-Datacon dataset, with an accuracy greater than

96% and recall at 97%. The overall accuracy of the 2-Class-Datacon dataset was slightly worse, probably due to the presence of many shelled samples and resource confusion as this dataset was collected from the current network. As for whether model generalization ability was good on the 2-Class-Datacon dataset, model ablation was set, and different comparison tests were performed for verification.

Table 6. Ablation experiment of model.

Model	Dataset	Accuracy	Precision	Recall	F_1-Score
TCN	9-class-data	99.36%	99.37%	99.36%	99.36%
	2-class-Datacon	94.62%	95.98%	95.81%	95.89%
GRU	9-class-data	99.36%	99.29%	99.35%	99.32%
	2-class-Datacon	95.7%	95.8%	95.72%	95.76%
TCN-GRU	9-class-data	99.54%	99.46%	99.54%	99.50%
	2-class-Datacon	95.52%	95.62%	95.63%	95.62%
TCN-BiGRU	9-class-data	99.72%	99.55%	99.54%	99.54%
	2-class-Datacon	96.54%	96.37%	96.63%	96.50%

4.6. Model Ablation Experiments

To verify the detection effect of the model proposed, model ablation experiments were performed. Under the same experimental conditions, TCN, GRU, TCN-GRU, and our model were compared on two different datasets to detect the corresponding results of each model for various indexes of the dataset. The detection results are shown in Table 6.

Observe from Table 7 that the proposed model significantly improved the classification effect of malicious samples, with accuracy up to 99.72% and 96.54% on the two datasets, respectively. The 9-Class-Datacon dataset has an accuracy improvement of 0.36%, 0.36%, and 0.18% using TCN, GRU, and TCN-GRU, respectively. The accuracy of the 2-Class-Datacon dataset was improved by 1.92%, 0.84%, and 1.02% using TCN, GRU, and TCN-GRU, respectively. Observing the results of the accuracy, completeness, and F1 values of the two datasets on the three models TCN, GRU, and TCN-GRU, it was found that the proposed TCN-BiGRU model outperformed TCN, GRU, and TCN-GRU in all indexes, thus verifying that the combination of both TCN and BiGRU in the model improved the detection effect for malicious code.

Table 7. Accuracy for different pooling methods.

Dataset	No Pooling	Average Pooling	Maximum Pooling	Pooling Fusion
9-class-data	99.45%	99.54%	99.45%	99.72%
2-class-Datacon	94.92%	95.10%	95.28%	96.54%

4.7. Comparison Experiments of Different Pooling Methods

To solve the problem of insufficient feature extraction abilities of the model, this study proposed a pooling fusion method that simultaneously averaged and maximized the pooling of data and performed parallel pooling. This section presents a comparison experiment on the effect of different pooling methods on the performance of malicious code classification. The model adopted four schemes: no pooling, average pooling, maximum pooling, and pooling fusion. The classification accuracy for both datasets is shown in Table 8.

From Table 8, observe that the method using pooling fusion has higher detection accuracy compared with schemes that perform average pooling or maximum pooling alone. By using pooling fusion to combine these two features and complement each other, we better reflected the nature of the network attack data and obtain higher identification accuracy. This experiment demonstrated that our pooling fusion method can significantly improve the ability of the model to extract features.

4.8. Comparison Experiments for Classification Algorithms

Regarding model performance, comparative experiments were conducted with reference to existing literature. Comparison experiments were done on the 9-Class-data dataset with reference to the relevant literature [9,10,26–28]. Experimental results are shown in Table 8. Among the five comparative studies, two focused on machine learning, one was related to gene sequence classification, and the remaining two concerned deep learning models. Burnaev et al. [26] used opcode features and grayscale map features, which were extracted and later detected by svm for classification. Narayanan et al. [27] processed grayscale graphs converted from malware, downscaling the features by PCA, and then classifying them using the machine learning model known as K nearest neighbor. Drew et al. [28] used a genetic detection method similar to Strand to classify text. Gibert et al. [10] extracted byte and opcode sequences, and then used a classifier composed of two CNNs for classification. Yan et al. [11] extracted features via a CNN model for grayscale maps and LSTM model for opcode features, then fused the results for classification. These methods produced good results, but there remained a gap between them and the method of this study. Under the accuracy evaluation criterion, our proposed TCN-BiGRU model integrating opcode and byte features achieved 99.72% accuracy; the accuracy values of the five comparison studies were all below 99.72%. Therefore, our proposed model incorporating both features and fusing TCN and BiGRU performed best.

Table 8. 9-class-data dataset_model comparison.

Model	No Pooling	Average Pooling
One-class SVM [26]	Opcode + Grayscale map	92%
PCA and kNN [27]	Grayscale map	96.6%
Strand Gene Sequence [28]	Asm sequence	98.59%
Orthrus [10]	Byte + Opcode	99.24%
MalNet [11]	Opcode + Grayscale map	99.36%
Model in this paper	Opcode + Byte	99.72%

For the 2-Class-Datacon dataset, we referred to the literature [29–31] to perform comparison experiments with the results shown in Table 9. Among these three comparison studies, one was on integration learning, one was on deep learning, and one was on machine learning. Guo et al. [29] extracted grayscale maps of malicious samples, extracts feature with different parameters for GIST descriptors, and then adopted the KNN and random forest algorithms to integrate classification by voting algorithm. Saadat et al. [30] also processed malicious sample images; it first pre-trained a good convolutional neural network model and then used the Xgboost algorithm for classification. Liu et al. [31] extracted the assembly instructions of malware samples. The assembly instructions were then pre-processed and downscaled using the LDA algorithm and were finally trained with the random forest algorithm for classification. These methods produced good results, but there remain gaps between them and the method proposed in this paper. Under the ACC evaluation criterion, our TCN-BiGRU model integrating opcode and byte features reached 96.54% accuracy; the ACC values of the three comparison papers were all under 96.54%. After the above comparative experiments on two datasets, it was proved that our proposed model integrating both features and fusing TCN and BiGRU performed best and had strong generalization capability.

Table 9. 2-Class-Datacon model comparison.

Model	No Pooling	Average Pooling
KNN + RandomForest [29]	Grayscale map	93.03%
CNN + Xgboost [30]	Grayscale map	93.44%
LDA + RandomForest [31]	Opcode	95.58%
Model in this paper	Opcode + Byte	96.54%

5. Conclusions

Threats to cyberspace security are increasing, and classification of the massive number of viruses has become an increasingly critical issue. This study proposed a static classification model of malicious code fused with TCN and BiGRU to extract and integrate the opcode features and byte features of malicious code. The model focusd on the potential features of the data and obtained the long-term dependencies existing in the sequences through a BiGRU network in both directions. It showed several advantages, such as high classification detection rate, anti-shelling, and obfuscation on both datasets. It also showed good generalizability and adaptability to high data volume requirements. However, the method used for feature extraction was relatively simple and did not bring out the full performance of the features. In follow-up work, we will use a natural language classification model to further process the samples.

Author Contributions: X.W.: conceptualization, methodology, writing—original draft; Y.S.: formal analysis, writing—review and editing; Z.M.: conceptualization; X.H.: formal analysis; C.C.: methodology. All authors have read and agreed to the published version of the manuscript.

Funding: This work is supported by the National Science Foundation of China (61806219, 61703426 and 61876189), the National Science Foundation of Shaanxi Provence (2021JM-226), the Young Talent fund of the University and Association for Science and Technology in Shaanxi, China (20190108, 20220106), and the Innovation Capability Support Plan of Shaanxi, China (2020KJXX-065).

Institutional Review Board Statement: Not applicable.

Informed Consent Statement: Not applicable.

Data Availability Statement: All data used in this paper can be obtained by contacting the authors of this study.

Conflicts of Interest: The authors declare no conflict of interest.

References

1. Zhao, J.; Zhang, S.; Liu, B.; Cui, B. Malware detection using machine learning based on the combination of dynamic and static features. In Proceedings of the 27th International Conference on Computer Communication and Networks (ICCCN), Hangzhou, China, 11 October 2018.
2. Guo, H.; Wu, J.T.; Huang, S.G.; Pan, Z.L.; Shi, F.; Yan, Z.H. Research on malware detection based on vector features of assembly instructions. *Inf. Secur. Res.* **2020**, 6, 113–121.
3. Raff, E.; Sylvester, J.; Nicholas, C. Learning the pe header, malware detection with minimal domain knowledge. In *Proceedings of the 10th ACM Workshop on Artificial Intelligence and Security*; Association for Computing Machinery: New York, NY, USA, 2017; pp. 121–132.
4. Zhao, S.; Ma, X.; Zou, W.; Bai, B. DeepCG: Classifying metamorphic malware through deep learning of call graphs. In *Proceedings of the International Conference on Security and Privacy in Communication Systems*; Springer: Berlin, Germany, 2019; pp. 171–190.
5. Santos, I.; Brezo, F.; Ugarte-Pedrero, X.; Bringas, P.G. Opcode sequences as representation of executables for data-mining-based unknown malware detection. *Inf. Sci.* **2013**, 227, 64–82. [CrossRef]
6. Kang, B.; Yerima, S.Y.; McLaughlin, K.; Sezer, S. N-opcode Analysis for Android Malware Classification and Categorization. In Proceedings of the 2016 International Conference on Cyber Security and Protection of Digital Services (Cyber Security), London, UK, 9 July 2016.
7. Pascanu, R.; Stokes, J.W.; Sanossian, H.; Marinescu, M.; Thomas, A. Malware classification with recurrent networks. In Proceedings of the 2015 IEEE International Conference on Acoustics, Speech and Signal Processing (ICASSP), South Brisbane, QLD, Australia, 19–24 April 2015; pp. 1916–1920.
8. Kwon, I.; Im, E.G. Extracting the Representative API Call Patterns of Malware Families Using Recurrent Neural Network. In *Proceedings of the International Conference on Research in Adaptive and Convergent Systems*; Association for Computing Machinery: New York, NY, USA, 2017; pp. 202–207.
9. Messay-Kebede, T.; Narayanan, B.N.; Djaneye-Boundjou, O. Combination of Traditional and Deep Learning based Architectures to Overcome Class Imbalance and its Application to Malware Classification. In Proceedings of the NAECON 2018-IEEE National Aerospace and Electronics Conference, Dayton, OH, USA, 23–26 July 2018; pp. 73–77.
10. Gibert, D.; Mateu, C.; Planes, J. Orthrus: A Bimodal Learning Architecture for Malware Classification. In Proceedings of the 2020 International Joint Conference on Neural Networks (IJCNN), Glasgow, UK, 19–24 July 2020; pp. 1–8.
11. Yan, J.; Qi, Y.; Rao, Q. Detecting malware with an ensemble method based on deep neural network. *Secur. Commun. Netw.* **2018**, *2018*, 7247095. [CrossRef]

12. Narayanan, B.N.; Davuluru, V.S.P. Ensemble Malware Classification System Using Deep Neural Networks. *Electronics* **2021**, *9*, 721. [CrossRef]
13. Ahmadi, M.; Ulyanov, D.; Semenov, S.; Trofimov, M.; Giacinto, G. Novel Feature Extraction, Selection and Fusion for Effective Malware Family Classification. In *Proceedings of the 6th ACM Conference on Data and Application Security and Privacy*; Association for Computing Machinery: New York, NY, USA, 2016; pp. 183–194.
14. Zhang, Y.; Huang, Q.; Ma, X.; Yang, Z.; Jiang, J. Using Multi-features and Ensemble Learning Method for Imbalanced Malware Classification. In Proceedings of the 2016 IEEE Trustcom/BigDataSE/ISPA, Tianjin, China, 23–26 August 2016; pp. 965–973.
15. Bai, J.R.; Wang, J.F. Improving malware detection using multiview ensemble learning. *Secur. Commun. Netw.* **2016**, *9*, 4227–4241. [CrossRef]
16. Bai, S.; Kolter, J.Z.; Koltun, V. An empirical evaluation of generic convolutional and recurrent networks for sequence modeling. *arXiv* **2018**, arXiv:1803.01271.
17. Fan, Y.Y.; Li, C.J.; Yi, Q.; Li, B.Q. Classification of Field Moving Targets Based on Improved TCN Network. *Comput. Eng.* **2021**, *47*, 106–112.
18. Yating, G.; Wu, W.; Qiongbin, L.; Fenghuang, C.; Qinqin, C. Fault Diagn-osis for Power Converters Based on Optimized Temporal Convolutional Network. *IEEE Trans. Instrum. Meas.* **2020**, *70*, 1–10. [CrossRef]
19. Huang, Q.; Hain, T. Improving Audio Anomalies Recognition Using Temporal Convolutional Attention Network. In Proceedings of the ICASSP 2021 IEEE International Conference on Acoustics, Speech and Signal Processing (ICASSP), Toronto, ON, Canada, 6–11 June 2021; pp. 6473–6477.
20. Zhu, R.; Liao, W.; Wang, Y. Short-term prediction for wind power based on temporal convolutional network. *Energy Rep.* **2020**, *6*, 424–429. [CrossRef]
21. Xu, Z.; Zeng, W.; Chu, X.; Cao, P. Multi-Aircraft Trajectory Collaborative Prediction Based on Social Long Short-Term Memory Network. *Aerospace* **2021**, *8*, 115. [CrossRef]
22. Liu, Y.; Ma, J.; Tao, Y.; Shi, L.; Wei, L.; Li, L. Hybrid Neural Network Text Classification Combining TCN and GRU. In Proceedings of the 2020 IEEE 23rd International Conference on Computational Science and Engineering (CSE), Guangzhou, China, 29 December–1 January 2020; pp. 30–35.
23. Sun, Y.C.; Tian, R.L.; Wang, X.F. Emitter signal recognition based on improved CLDNN. *Syst. Eng. Electron.* **2021**, *43*, 42–47.
24. Wang, Y.; Liao, W.L.; Chang, Y.Q. Gated Recurrent Unit Network-Based Short-Term Photovo-ltaic Forecasting. *Energies* **2018**, *11*, 2163. [CrossRef]
25. Qi An Xin Technology Research Institute. DataCon: Multidomain Large-Scale Competition Open Data for Security Research. Available online: https://datacon.qianxin.com/opendata (accessed on 11 November 2021). (In Chinese).
26. Burnaev, E.; Smolyakov, D. One-class SVM with privileged information and its application to malware detection. In Proceedings of the 2016 IEEE 16th International Conference on Data Mining Workshops (ICDMW), Barcelona, Spain, 12–15 December 2016; pp. 273–280.
27. Narayanan, B.N.; Djaneye-Boundjou, O.; Kebede, T.M. Performance analysis of machine learning and pattern recognition algorithms for malware classification. In Proceedings of the 2016 IEEE National Aerospace and Electronics Conference (NAECON) and Ohio Innovation Summit (OIS), Dayton, OH, USA, 25–29 July 2016; pp. 338–342.
28. Drew, J.; Hahsler, M.; Moore, T. Polymorphic malware detection using sequence classifcation methods and ensembles: BioSTAR 2016 Recommended Submission. *EURASIP J. Inf. Secur.* **2017**, *2017*, 2. [CrossRef]
29. Guo, H.; Huang, S.; Zhang, M. Classification of malware variant based on ensemble learning. In *International Conference on Machine Learning for Cyber Security*; Springer: Cham, Switzerland, 2020; pp. 125–139.
30. Saadat, S.; Joseph Raymond, V. Malware classification using CNN-Xgboost model. In *Artificial Intelligence Techniques for Advanced Computing Applications*; Springer: Singapore, 2021; pp. 191–202.
31. Liu, Y.; Wang, Z.; Hou, Y. A method for feature extraction of malicious code based on probabilistic topic models. *J. Compute. Res. Dev.* **2019**, *56*, 2339–2348.

applied sciences

MDPI

Article

Time Aware F-Score for Cybersecurity Early Detection Evaluation

Manuel López-Vizcaíno *,†, Francisco J. Nóvoa †, Diego Fernández † and Fidel Cacheda †

Center for Information and Communications Technologies Research (CITIC), Department of Computer Science and Information Technologies, University of A Coruña, 15071 A Coruña, Spain; fidel.cacheda@udc.es (F.C.)

* Correspondence: manuel.fernandezl@udc.es

† These authors contributed equally to this work.

Abstract: With the increase in the use of Internet interconnected systems, security has become of utmost importance. One key element to guarantee an adequate level of security is being able to detect the threat as soon as possible, decreasing the risk of consequences derived from those actions. In this paper, a new metric for early detection system evaluation that takes into account the delay in detection is defined. Time aware F-score (TaF) takes into account the number of items or individual elements processed to determine if an element is an anomaly or if it is not relevant to be detected. These results are validated by means of a dual approach to cybersecurity, Operative System (OS) scan attack as part of systems and network security and the detection of depression in social media networks as part of the protection of users. Also, different approaches, oriented towards studying the impact of single item selection, are applied to final decisions. This study allows to establish that nitems selection method is usually the best option for early detection systems. TaF metric provides, as well, an adequate alternative for time sensitive detection evaluation.

Keywords: early detection; machine learning; classification algorithms; network security; social networks; time-aware metrics

Citation: López-Vizcaíno, M.; Novoa, F.J.; Fernández, D.; Cacheda, F. Time Aware F-Score for Cybersecurity Early Detection Evaluation. *Appl. Sci.* **2024**, *14*, 574. https://doi.org/10.3390/app14020574

Academic Editors: Peter R. J. Trim and Yang-Im Lee

Received: 19 November 2023
Revised: 4 January 2024
Accepted: 5 January 2024
Published: 9 January 2024

1. Introduction

The relevance of the early detection problem has been explored in many fields. Although a thorough study was proposed there was limited research on its evaluation metrics and methodologies. Mostly there is little work related to non-dataset-dependent metrics and lack of interpretability and better discrimination among results. In a real world environment, where a true streaming evaluation is applied, no data from the full dataset is available, so parameters for the metric must be extracted from the problem.

Many fields could benefit from an early detection approach as stopping any kind of anomaly or problem as soon as possible would minimise the risks of unwanted results.

Particularly, in the field of network and systems security, the longer the time passes and the more phases of the attack are achieved, the higher the probabilities of significant damage.

In the early detection in social media field, the detention of fake news or rumours as well as cyberbullying would decrease their consequences. Those consequences include depression, self-esteem decrease, suicide or suicide ideations, for example.

The lack of a proper dataset independent early detection metric with good interpretability was solved with the definition of TaF. This is a non-dataset-dependent metric for the evaluation of early detection systems, presented as an alternative to the issues found in the state-of-the-art metrics. Additionally, the evaluation of multiple alternatives for detection systems functions could improve the results, thanks to the variety of options available for the different problems.

To summarise, the fundamental issue found in this topic was related to the proper time aware evaluation and thus the main contribution of this paper is the definition of a non-dataset-dependent time-aware metric. Also, another contribution is the study of different approaches for taking final decisions based on the elements that are being processed.

The remaining sections of this paper is organised as follows: After the Introduction section, a Related Works containing relevant references to analyse the state-of-the-art. Then, Section 3 present the formal representation of the methods used in the experiments, a brief analysis of the datasets used, the models applied and the metrics used for the evaluation of performance. Next, on the Results section values from the experiments are presented in terms of figures, tables and a study of the results. Lastly, Discussion and Conclusions outline the outcome of the experiments, displaying also the implications and limitations of this research.

2. Related Works

As stated in the previous section, the relevance of the early detection problem has been explored in many fields. Although, a thorough study was proposed there was limited research on its evaluation metrics and methodologies. Best efforts as evaluation metric definition for the early detection problem was presented by *Early Risk Detection Error* ($ERDE$) [1], *F-latency* [2] an also by [3] where an alternative, *Time aware Precision* (TaP) non-dataset-dependent metric was proposed.

At the 2017 CLEF (Conference an Labs for the Evaluation Forum), $ERDE$ was presented in the workshop for early prediction (*eRisk*) as the metric for the evaluation of the tasks. Then, by using the time-aware proposed metric, participants detection systems for early detection of situations such as depression or other behaviours and disorders using social media network data [1,4].

F-latency, a latency-weighted $F1$, was created to avoid heuristically defined parameters, as it was one of the problems detected by the authors in the definition of $ERDE$. Those parameters were replaced by dataset defined values for the evaluation of depression detection in social media [5].

Time aware Precision (TaP), was presented in [3] based on the study of $ERDE$ and *F-latency* as well as other non time aware metrics to overcome problems with the definition of both. Such as, interpretability, better discrimination among results and to obtain a dataset agnostic approach. In a real world environment, where a true streaming evaluation is applied, no data from the full dataset is available, so parameters for the metric must be extracted from the problem.

In terms of metrics those three, $ERDE$ [1] *F-latency* [5] and TaP [3], are the best effort presented for early detection evaluation with specific latency dependent metrics.

Also, it must be mentioned that disregarding specific latency aware metrics, some examples can be found by using traditional metrics like precision, recall or F-score, for the early detection problem. For example [6] or [7], where $F1$ is measured at specific points without time penalization.

Particularly, in the field of network and systems security, early detection of attacks could prevent an increase of the threat minimising the outcome, as presented in [8]. The longer the time passes and the more phases of the attack are achieved, the higher the probabilities of significant damage. As development in this area, several works had been published, such as [9,10] in order to detect attacks at their early stages, usually not taking into account the time in their evaluation. Even, if it is incorporated, usually is just measured by the amount of time until the detection, as shown in [11,12], where a distributed denial of service (DDOS) is targeted.

In the early detection in social media some works have been presented to target different aspects of those interactions and the problems that could derive from them. For example, in order to stop as soon as possible the diffusion of fake news and rumours, some solutions are presented as in [13,14] but relying only in the time required for the detection as latency evaluation. This, combined with the increase of cyberbullying [15] and the possible consequences of it and the diffusion and spread of fake news and rumours could lead to depression, self-esteem decrease, suicide ideations or even suicide [16].

Recently, a thorough comparison between different approaches for cyberbullying detection was presented in [17], where as part of machine learning evaluation, Logistic Regression is shown as an alternative. As part of the cyberbullying detection problem, some session based studies had been presented, using specifically Large Language Models (LLM) for the early detection task [18] . Lately, also [19] applies machine learning models as Decision Tree (DT), Random Forest (RF) and AdaBoost (AB) for cybersecurity threats in form of Networks Attacks.

Summarising, in terms of metrics, several attempts have been made towards an early detection evaluation, such as *ERDE* [1], *F-latency* [5] and *TaP* [3]. Although, some present several problems for real world applications and their interpretation.

3. Materials and Methods

In this section we present a thorough description of the elements used in the tests for obtaining the results. First, an introduction to the early detection representation and its particularities is included in the Methods section. Then, a comprehensive explanation of the datasets is shown, followed by the models used for the decision making phase. Finally, the metric used for the evaluation is presented.

3.1. Methods

The early detection problem presents certain characteristics that must be taken into account, and it must be formally described in order to define proper methods. Although a thorough description of the formal definition of the early detection problem can be found in [3], a summary is presented next for ease of interpretation. To do so, we define $E = \{e_1, e_2, \ldots, e_{|E|}\}$ as the set of entities susceptible of being classified, in this case $|E|$ describes the amount of entities. Each one of them (e) is composed from a series of items (I_e), with a label (l_e) for the class of the entity. This could represent either that the entity is anomalous or not, $l_e = true$ and $l_e = false$ respectively. It must be added that although this represents a binary classification task, early detection systems could provide a third value showing that the decision has not been taken yet (i.e., a delay).

The set of items for a particular entity is expected to change over time and is represented by $I_e = (< I_1^e, t_1^e >, < I_2^e, t_2^e >, \ldots, < I_n^e, t_n^e >)$, being $< I_k^e, t_k^e >, k \in [1, n]$ the tuple that represents, for each entity e, its k-th item I_k^e, and the timestamp associated with the particular item I_k^e is denoted as t_k^e.

Also, it is important to notice that the following statement must be true, as items must be time ordered:

$$\forall < I_k^e, t_k^e > : \; t_k^e \; before \; t_{k+1}^e$$

An item, I_k^e, is characterized by a feature vector, and, particularly, it can be inferred that all items linked to an entity, $I_k^e, k \in [1, n]$, share the same feature vector. Those attribute values will be expected to change over time.

$$I_k^e = \left[f_{k_1}^e, f_{k_2}^e, \ldots, f_{k_m}^e \right], k \in [1, n]$$

Due to the independence of entities, the sequence of items I_e for each entity $e \in E$ may exhibit varying lengths, denoted as n. It is important to emphasize that the number of features, m, remains consistent across all items.

In this kind of problem, for a given entity e, the goal is to identify any anomalous behaviour while examining the fewest number of items from I_e possible.

The objective function will be defined as $f(l_e, I_e \times [1..n]) \to \{0, 1, 2\}$. If an entity e is deemed anomalous following the analysis of i_1 to I_k items, this function will output 1 (i.e., *positive*). If an entity e is determined to be normal (i.e., non-anomalous or *negative*) after processing an amount of k items and the preceding ones, then $f(l_e, I_e, k) = 0$. Lastly, $f(l_e, I_e, k) = 2$ represents that no definitive decision can be made regarding e entity after processing an amount k of items, indicating the need for further processing (i.e., *delay*). As a result, when $f(l_e, I_e, k)$ yields outputs 0 and 1, the processing of items $I_{k+1}, \ldots, I_n$ is unnecessary. Conversely, if the output is 2, it necessitates the processing of additional items $I_{k+1}, \ldots, I_n$ until a conclusive output is obtained or the sequence of items is exhausted.

Among the different existent evaluation methods (e.g., batch, streaming, time-based, etc.) we choose streaming evaluation as it provides the best representation for each individual entity.

In this case, I_e, which represents the sequence formed by individual items, is treated individually and following the original sequence. As a result, for each entity e, the function $f(l_e, I_e, k)$, $k \in [1, n]$, must be executed until a conclusive output is achieved or until the maximum number of *items* (n) is reached. Similar to previous scenarios, when item k, I_k, is being processed, the model can incorporate all preceding items, $I_1, \ldots, I_{k-1}$, as illustrated in Figure 1.

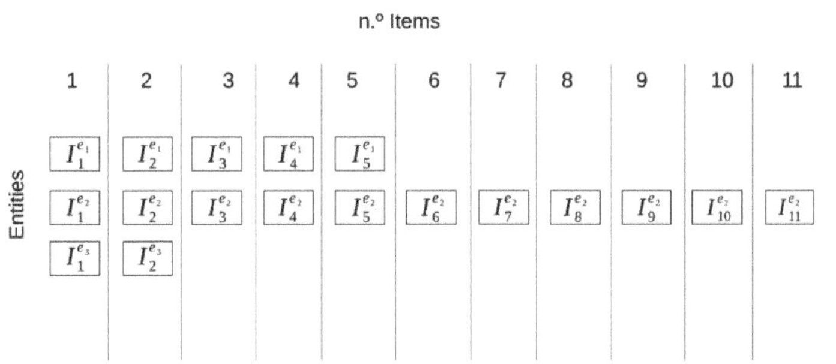

Figure 1. Items distribution for streaming evaluation.

3.2. Datasets

For the sake of variety in the experiments developed in this paper, we selected two different datasets. In order to study the outcome under particular circumstances, we choose two where both their nature and particular characteristics differ. The first one, obtained from Kitsune dataset [2], is the OS Scan Attack, which presents some particularities that will be shown later in this section. The second one was specially collected for the eRisk 2017 Workshop [1]. In this case, posts from the website Reddit were included after a selection process and being identified as written by users with depression or users without depression. As for Kitsune OS Scan Attack, an analysis is included next.

Kitsune dataset is composed by data traffic from a video monitoring network and includes different attacks performed over the network through several days. It was designed and utilized to evaluate the Intrusion Detection System—Network based (NIDS) described in [2]. In particular we use the OS Scan Attack, which examines the devices connected trying to detect hosts and which operating systems (OS) they are using, to find potential vulnerabilities, and it will be referred as "Network Attack" from now on. This dataset is composed by a set of network packets from which a group of engineered features are extracted and tagged as "Attack" or "Normal". In order to further analyse the traffic, we divided it into bidirectional flows as described in [20,21], defined as the aggregation of packets with same pair source IP address—destination IP address, source port—destination port and protocol over a defined period of time. To account for the time division, we used timestamp of the packets as a split point using 0.1 s as threshold for the inter-packet time and 1 s as threshold for the flow span as described in [22]. For the experiments performed, we used the features described in [2]. As presented in the original paper only characteristics of the packet itself and its relation in time with the rest of the traffic are used in the creation of the dataset without including specific information of the flow they belong to.

Table 1 summarizes the main statistics for the datasets, among which OS Scan Attack from Kitsune dataset can be found. The amount of individual packets reaches nearly 1.7 million, distributed in 75,700 bidirectional flows. As the dataset represents an OS scan attack, most of anomalous flows are going to be around 2 packets in size, like it can be seen in the average of packets per flow for attack class, which accounts for the request and the reply from the device under attack. This, results on getting the majority of *Entities* of Anomalous type even if the greater part of *items* or packets belong to the Normal class.

The eRisk depression dataset, referred as "Depression Dataset" from now on was expressly collected for the 2017 edition of eRisk workshop on Early Detection [1]. It is composed by a set of publicly available Reddit posts published by users in about a year of use period. Those posts were later marked as "Depressed" or "Non-depressed" guided by self-reported diagnoses of depression. In this case, subjects correspond with the *Entities* (E) and each of the *items* (I_i^e) with the subject's posts. As for the features used in the experiments we will use the ones defined in [23,24]. Although, we will not include features created by the aggregation of a sequence of posts by only including individual post characteristics.

Also, we include in Table 1, along with the ones of the previous described dataset, the main statistics computed for the subjects and posts considered. This dataset is significantly smaller both in terms of number of *Entities* and *Items* but it has a higher average ratio of *items* per *entity*. With only 887 *Entities* and slightly over 500,000 items, it achieves nearly 600 items for each entity. Being the amount of "Anomalous" cases the 15.22% of the total subjects, it is relevant to display that this ratio reaches almost half the value for "Anomalous" than for "Normal" cases.

Table 1. Summary of the main characteristics of the datasets.

Dataset	Entities & Items	Normal	Anomalous	Total
Depression	Entities	752	135	887
	Items	481,837	49,557	531,394
	Items per entity	640.7	367.1	599.1
Network attacks	Entities	10,045	65,655	75,700
	Items	1,566,602	131,249	1,697,851
	Items per entity	155.96	1.99	22.43

3.3. Models

The set of models used in the experimental evaluation of the proposed selectors and metric is composed by some well known of-the-shelf state-of-the-art machine learning algorithms. The selection of this set was made based on the work presented in [25] for detection of cyberbullying in Vine social network. This methods were also tested on the same datasets for the early detection problem as shown in [3]. Particularly, the implementation by *scikit-learn* [26] was used, and the description of the models with the parameters used is listed below:

- *LinearSVC:* Support Vector Classification implementation with a 'linear' kernel that improves parameter selection and scalability.
  - Parameters: C = 1, *class_weight* = *'balanced'*, *dual* = *False*, *max_iter* = 1000
- *ExtraTree:* Meta estimator that uses averaging of randomized decision trees for classification.
  - Parameters: *n_estimators* = 50, *bootstrap* = *False*, *class_weight* = *None*
- *AdaBoost:* Meta estimator that refits a classifier by updating weights of incorrectly classified instances.
  - Parameters: *n_estimators* = 1000, *learning_rate* = 2.0, *algorithm* = *'SAMME.R'*

- *Random Forest:* Meta estimator which uses a determined number of decision trees and applies averaging to obtain the classifier.
  - Parameters: $n\_estimators = 500$, $class\_weight = None$, $max\_features = 'sqrt'$, $max\_depth = 7$, $bootstrap = False$
- *Logistic Regression:* Conformed by a *logit MaxEnt* classifier, where the maximum entropy classifier is combined with a logistic function.
  - Parameters: $C = 0.1$, $class\_weight = 'balanced'$, $dual = False$, $penalty = 'l2'$, $solver = 'sag'$

After models for detection in individual items are trained, we apply a selector system to take the final decision based on individual items. Those decision functions are tested by using the results previously obtained with standard final deciders as baselines. In this case, four different functions are applied to this task, particularly:

- *mean*: Uses an aggregation of probabilities for each class, and decides one or the other based on those values.
- *bigger*: Uses the bigger class probability to provide the output of the classifier.
- *last*: Uses the last item processed to decide if the result is normal or anomalous
- *2last*: Uses the combination of the class probabilities of the last two values in order to provide the final output.

3.4. Metrics

For the early detection problem evaluation it is important to obtain models that not only make correct predictions but also that those are taken as soon in time as possible. In order to do so, a metric that is able to consider the time used for making the prediction is needed. Several metrics had been presented to fulfill that task, among which: Early Risk Detection Error *ERDE* [1], *F-latency* [5] and Time aware Precision *TaP* [3].

In this paper, we define the metric *Time aware F-score* or *TaF* for short, to overcome some limitations of the previous metrics and to ease the interpretation of the results. This metric presents a behaviour more similar to *F-latency*, than previous ones, but improving the configuration parameters. Reflecting the same idea as *TaP*, a penalization point and the degree of penalization is defined as problem-based instead of based on the particular values of the dataset. In this sense, it is more natural to do so that to set it based on the mean of those values. This last point is also crucial because in a real world streaming situation, it will not be possible to get the complete image or range of values before one specific item is processed.

The metric has been defined as follows:

$$TaF_{o,\lambda}(e_i,k) = \begin{cases} 1 & \textit{if } TP \wedge k \leq o \\ 1 - pf_{o,\lambda}(k) & \textit{if } TP \wedge k > o \\ 0 & \textit{if delay} \end{cases}$$

$$TaF(E,k) = \frac{\sum_{e_i \in E} TaF(e_i,k)}{|E_{TP}| + \frac{1}{2}(|E_{FP}| + |E_{FN}|)}$$

The cases defined in $TaF_{o,\lambda}$ use the same principle as in the previous metrics (*ERDE* and *TaP*) but just for the correctly detected positive cases (true positives, TP). The maximum value is obtained if the item is identified before the point of measure o, otherwise a penalization function named $pf(k)_{o,\lambda}$ is applied.

For the final $TaF(E,k)$ value an aggregation of intermediate results is performed in order to use the final value as a penalized count of true positive cases. That is to use it instead of $|E_{TP}|$, applying, then, the F-score function. The number of real true positive cases (TP) is depicted as $|E_{TP}|$, whereas negative cases are shown as $|E_{FP}|$ and $|E_{FN}|$ for false positives and false negatives respectively.

For this metric, the penalty function is defined as follows, to give values in the range [1, 0] which generates an output in the metric in the same range. This maintains the relation with the output values of F-score.

$$pf(k)_{o,\lambda} = -1 + \frac{2}{1 + e^{-\lambda(k-o)}}$$

The values of penalization shown in Figure 2 for different values of the parameter λ display how the results for individual entities e_i affects to the final value of TaF. In every instance a proper prediction is emitted although, the amount of *items* required to achieve that prediction varies as presented on X-axis. It must be observed that the output range of the function for TaF metric is [0, 1].

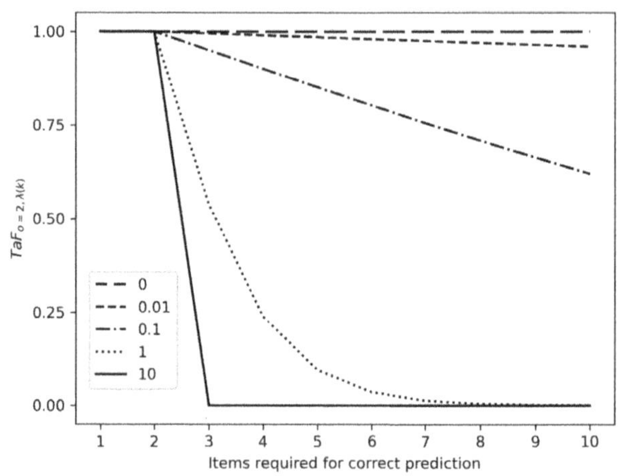

Figure 2. TaF configured with point of measure $o = 2$ and various values of λ. The number o items used to reach a correct prediction by the system are represented on the X-axis, supposing a delay was produced fro previous items. In this case, when $x = 5$ the system outputs the proper prediction on item $k = 5$ and, therefore, it generates a delay earlier on (i.e., $x < 5$).

4. Results

Results for eRisk dataset over Random Forest, Extra Tree, Ada Boost and Logistic Regression models are presented on Figures 3 and 4. Figure 3 shows the results obtained with Time aware F-score (TaF) metric, when compared with Figure 4, which shows the results when *F-latency* metric is used, some differences must be noticed.

First, TaF parameters are not dataset defined but problem defined, as presented in the previous section, in contrast to other metrics. Besides that, it can be seen that more differences between the four models, and mostly between different selectors for each model, can be spotted by using this metric. Particularly, it provides an improved representation of selectors' performances.

Also, when specific values are observed in Figures 3 and 4, some differences are shown between selectors. One particular observation can be made at 2 items for *2last*, with a decrease in performance for both metrics. This can be explained mainly because of an absence of many predictions in the previous point combined with a poor performance when the first item is included for these decisions. The results could imply that the first element of the sequence is mostly incorrectly classified and it does not improve the general output of the algorithm at that point.

Finally regarding Figure 3, when the number of processed items increase, best behaviour for the majority of the models is obtained with bigger selection function, although for Extra Tree *2last* obtains the best results. This function is also the second better for Ada Boost and Random Forest, reaching even better values than bigger function for lower amount of items processed.

Furthermore, Table 2 presents results obtained with *TaF* with the same combination of models and selectors when applied on Kitsune dataset.

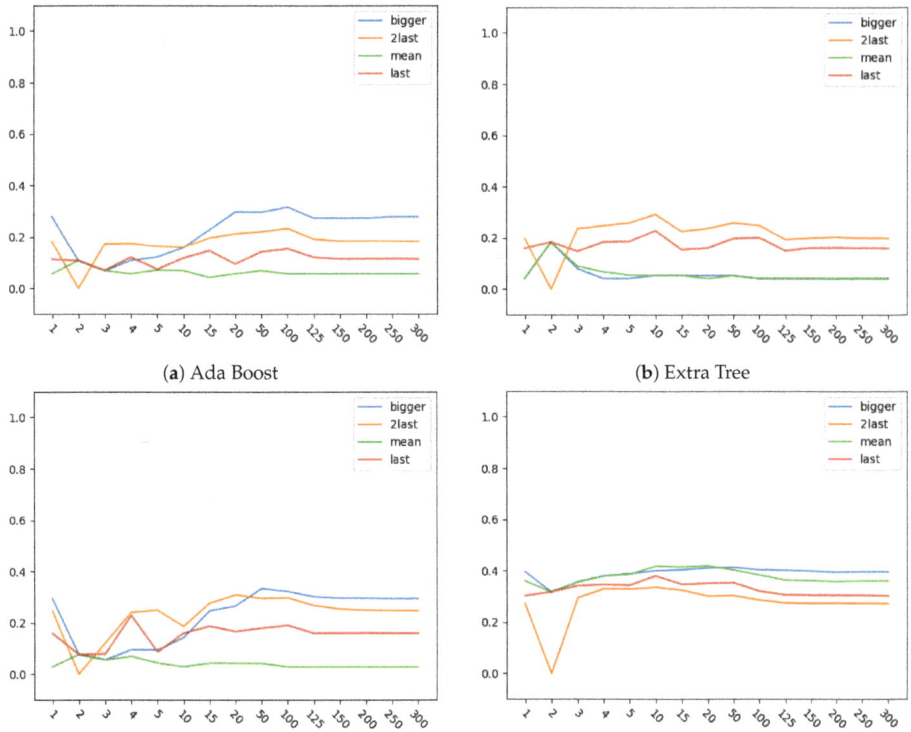

(**a**) Ada Boost (**b**) Extra Tree

(**c**) Random Forest (**d**) Logistic Regression

Figure 3. Results for combinations of selectors and models (Random Forest (RF), Extra Tree (ET), AdaBoost (AB) and Logistic Regression (LR)) for each number of items analyzed with eRisk dataset using Time aware F-score (TaF) as metric.

In this case, as the problem analysed is specific of an OS Scan Attack and defined flows present distinct characteristics, models trained display high values of TaF metric regardless the combination of models and selectors. Even though that is the case for this particular, it allows to observe how higher values of the metric seem less influenced by differences in selectors, and small differences can be observed in the model (Logistic Regression) where those values are lower.

Results on Kitsune dataset display also particularities when the amount of items processed is lower, and this situation is enhanced by the fact that this dataset presents a high number of entities with just two items. Also, it must be remembered that those small entities are in their vast majority attack entities.

Although no direct comparison with previous results referenced is possible due to the use of a new metric presented in this paper, the analysis of both *F-latency* and TaF results for eRisk dataset (Figures 3 and 4) provides a base to connect it with previous experiments. Besides, the use of one of the selection methods for the early detection systems that was already applied for the same Machine Learning models supplies the needed base for the required results interpretation.

In view of the conducted experiments and the results obtained, TaF metric is shown as an improved alternative to the early detection metrics. This means that this metric could be applied to evaluate any problem from the domain overcoming the limitations present on other time aware metrics. In terms of improvements of the early detection by machine learning models, it can be seen that a further exploration of selection methods could be applied. By expanding the ways the final decision is taken, performance might be improved.

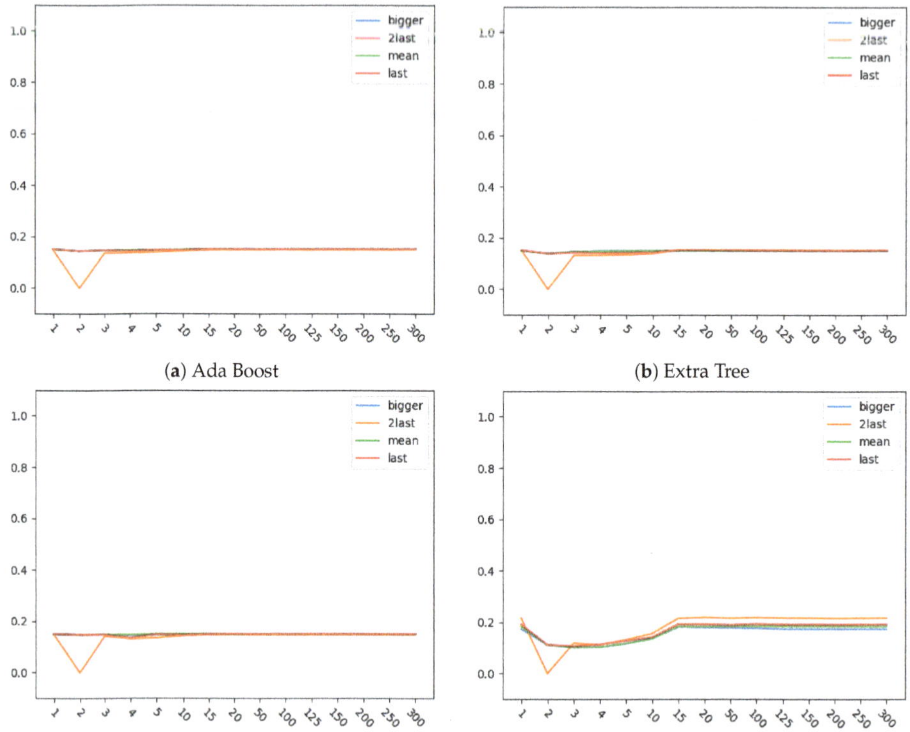

(**a**) Ada Boost (**b**) Extra Tree

(**c**) Random Forest (**d**) Logistic Regression

Figure 4. Results for combinations of selectors and models (**a**) Forest (RF), Extra Tree (ET), Ada Boost (AB) and Logistic Regression (LR)) for each number of items analyzed with eRisk dataset using *F-latency* as metric.

Table 2. Results for combinations of selectors and models (Random Forest (RF), Extra Tree (ET), Ada Boost (AB) and Logistic Regression (LR)) for each number of items analyzed with Kitsune dataset.

Model	Selector	1	2	3	4	5	25	50	75	100	125	150	175	200	225	300
RF	bigger	0.9950	1.0000	0.9950	0.9950	0.9950	0.9950	0.9950	0.9950	0.9950	0.9950	0.9950	0.9950	0.9950	0.9950	0.9950
	2last	0.9950	0.0000	0.9950	0.9950	0.9950	0.9950	0.9950	0.9950	0.9950	0.9950	0.9950	0.9950	0.9950	0.9950	0.9950
	mean	0.9950	1.0000	0.9950	0.9950	0.9950	0.9950	0.9950	0.9950	0.9950	0.9950	0.9950	0.9950	0.9950	0.9950	0.9950
	last	0.9950	1.0000	0.9950	0.9950	0.9950	0.9950	0.9950	0.9950	0.9950	0.9950	0.9950	0.9950	0.9950	0.9950	0.9950
ET	bigger	0.9950	1.0000	0.9950	0.9950	0.9950	0.9950	0.9950	0.9950	0.9950	0.9950	0.9950	0.9950	0.9950	0.9950	0.9950
	2last	0.9950	0.0000	0.9950	0.9950	0.9950	0.9950	0.9950	0.9950	0.9950	0.9950	0.9950	0.9950	0.9950	0.9950	0.9950
	mean	0.9950	1.0000	0.9950	0.9950	0.9950	0.9950	0.9950	0.9950	0.9950	0.9950	0.9950	0.9950	0.9950	0.9950	0.9950
	last	0.9950	1.0000	0.9950	0.9950	0.9950	0.9950	0.9950	0.9950	0.9950	0.9950	0.9950	0.9950	0.9950	0.9950	0.9950
AB	bigger	0.9950	1.0000	0.9950	0.9950	0.9950	0.9950	0.9950	0.9950	0.9950	0.9950	0.9950	0.9950	0.9950	0.9950	0.9950
	2last	0.9950	0.0000	0.9950	0.9950	0.9950	0.9950	0.9950	0.9950	0.9950	0.9950	0.9950	0.9950	0.9950	0.9950	0.9950
	mean	0.9950	1.0000	0.9950	0.9950	0.9950	0.9950	0.9950	0.9950	0.9950	0.9950	0.9950	0.9950	0.9950	0.9950	0.9950
	last	0.9950	1.0000	0.9950	0.9950	0.9950	0.9950	0.9950	0.9950	0.9950	0.9950	0.9950	0.9950	0.9950	0.9950	0.9950
LR	bigger	0.9896	0.9946	0.9896	0.9896	0.9896	0.9896	0.9896	0.9896	0.9896	0.9896	0.9896	0.9896	0.9896	0.9896	0.9896
	2last	0.9948	0.0000	0.9948	0.9948	0.9948	0.9948	0.9948	0.9948	0.9948	0.9948	0.9948	0.9948	0.9948	0.9948	0.9948
	mean	0.9896	0.9946	0.9896	0.9896	0.9896	0.9896	0.9896	0.9896	0.9896	0.9896	0.9896	0.9896	0.9896	0.9896	0.9896
	last	0.9896	0.9946	0.9896	0.9896	0.9896	0.9896	0.9896	0.9896	0.9896	0.9896	0.9896	0.9896	0.9896	0.9896	0.9896

5. Discussion

From the results obtained from this experiments we can extract two main conclusions. The first one is the strengths of *TaF* as metric for early detection problem evaluation, as it allows to better differentiate between outputs than *F-latency* metric. Being also problem dependent instead of datasets dependent, which was previously presented as a limitation for real world streaming environments evaluation.

Second conclusion extracted from this experiments is the ability of 2*last* as selection function to extract the best results when little information is known, being just overcome by bigger, when bigger amount of items were processed. As the objective of these systems is to detect as soon as possible this situation, the use of a *nlast* approach could improve general performance of systems. Problems detected when less than n items are being processed could be avoided by the combination of two different selection functions depending on the amount of items. Which could even be explored in future works with the analysis of the application of different selection functions to the *nlast* items processed.

6. Conclusions

Finally, from the research presented in this paper it can be extracted that due to the limitations of existent time aware metrics, the alternative provided by TaF contributes to a proper evaluation of detection systems in the early detection problem.

Also, and under the evaluation supplied by TaF metric, it is highlighted the importance of specific selection methods for early detection problem. A broader study could be performed with the inclusion of more selection methods and datasets with different particularities, but it is interesting to see how the number of items has an impact both on the penalisation and on the amount of information used by the system.

Author Contributions: Conceptualization, F.C. and M.L.-V.; Methodology, D.F. and F.J.N.; Software, M.L.-V.; Data curation, F.C.; Resources: D.F. and F.J.N.; Writing—original draft, M.L.-V.; Writing—review & editing, F.J.N., D.F. and F.C. All authors have read and agreed to the published version of the manuscript.

Funding: This work was supported in part by the Ministry of Economy and Competitiveness of Spain and Fondo Europeo de Desarrollo Regional (FEDER) Funds of the European Union under Project PID2019-111388GB-I00; and in part by the Centro de Investigación de Galicia—Centro de Investigación en Tecnologías de la Información y las Comunicaciones (CITIC) Funded by Xunta de

Galicia and the European Union (European Regional Development Fund–Galicia 2014-2020 Program), under Grant ED431G 2019/01.

Institutional Review Board Statement: Not applicable.

Informed Consent Statement: Not applicable.

Data Availability Statement: Network data were obtained from dataset creators and are available at the DOI 10.24432/C5D90Q. Depression data were obtained from eRisk organization with their permission (https://erisk.irlab.org/2017/, accessed on 20 November 2023).

Conflicts of Interest: The authors declare no conflict of interest.

Abbreviations

The following abbreviations are used in this manuscript:

AB	AdaBoost
DDOS	Distributed Denial Of Service
ERDE	Early Risk Detection Error
ET	Extra Tree
FN	False Negative
FP	False Positive
LR	Logistic Regression
MDPI	Multidisciplinary Digital Publishing Institute
NIDS	Network Intrusion Detection Sytem
OS	Operative System
TaF	Time aware F-score
TaP	Time aware Precision
TP	True Positive
TN	True Negative
RF	Random Forest

References

1. Losada, D.E.; Crestani, F. A Test Collection for Research on Depression and Language Use. In *Experimental IR Meets Multilinguality, Multimodality, and Interaction;* Springer: Cham, Switzerland, 2016; pp. 28–39. [CrossRef]
2. Mirsky, Y.; Doitshman, T.; Elovici, Y.; Shabtai, A. Kitsune: An Ensemble of Autoencoders for Online Network Intrusion Detection. In Proceedings of the Network and Distributed Systems Security (NDSS) Symposium 2018, San Diego, CA, USA, 18–21 February 2018.
3. Lopez-Vizcaino, M.F.; Novoa, F.J.; Fernandez, D.; Cacheda, F. Measuring Early Detection of Anomalies. *IEEE Access* **2022**, *10*, 127695–127707. [CrossRef]
4. Losada, D.E.; Crestani, F.; Parapar, J. eRisk 2020: Self-harm and Depression Challenges. In *Advances in Information Retrieval;* Springer International Publishing: Cham, Switzerland, 2020; pp. 557–563.
5. Sadeque, F.; Xu, D.; Bethard, S. Measuring the latency of depression detection in social media. In Proceedings of the WSDM 2018—11th ACM International Conference on Web Search and Data Mining, Marina Del Rey, CA, USA, 5–9 February 2018; pp. 495–503. [CrossRef]
6. Chinchor, N. MUC-4 Evaluation Metrics. In Proceedings of the 4th Conference on Message Understanding, McLean, VA, USA, 16–18 June 1992; pp. 22–29.
7. Samghabadi, N.S.; Monroy, A.P.L.; Solorio, T. Detecting Early Signs of Cyberbullying in Social Media. In Proceedings of the Second Workshop on Trolling, Aggression and Cyberbullying, Marseille, France, 11–16 May 2020; European Language Resources Association (ELRA): Paris, France, 2020; pp. 144–149.
8. Hutchins, E.M.; Cloppert, M.J.; Amin, R.M. Intelligence-Driven Computer Network Defense Informed by Analysis of Adversary Campaigns and Intrusion Kill Chains. *Lead. Issues Inf. Warf. Secur. Res.* **2011**, *1*, 113–125.
9. Narayanan, S.N.; Ganesan, A.; Joshi, K.; Oates, T.; Joshi, A.; Finin, T. Early detection of cybersecurity threats using collaborative cognition. In Proceedings of the 4th IEEE International Conference on Collaboration and Internet Computing, CIC 2018, Philadelphia, PA, USA, 18–20 October 2018; pp. 354–363. [CrossRef]
10. Pivarníková, M.; Sokol, P.; Bajtoš, T. Early-Stage Detection of Cyber Attacks. *Information* **2020**, *11*, 560. [CrossRef]
11. Xu, C.; Lin, H.; Wu, Y.; Guo, X.; Lin, W. An SDNFV-Based DDoS Defense Technology for Smart Cities. *IEEE Access* **2019**, *7*, 137856–137874. [CrossRef]
12. Privalov, A.; Lukicheva, V.; Kotenko, I.; Saenko, I. Method of Early Detection of Cyber-Attacks on Telecommunication Networks Based on Traffic Analysis by Extreme Filtering. *Energies* **2019**, *12*, 4768. [CrossRef]

13. Zhou, X.; Jain, A.; Phoha, V.V.; Zafarani, R. Fake News Early Detection: A Theory-driven Model. *Digit. Threat. Res. Pract.* **2020**, *1*, 12. [CrossRef]
14. Zhao, Z.; Resnick, P.; Mei, Q. Enquiring Minds: Early Detection of Rumors in Social Media from Enquiry Posts. In Proceedings of the 24th International World Wide Web Conference, Florence, Italy, 18–22 May 2015; International World Wide Web Conferences Steering Committee: Geneva, Switzerland, 2015; pp. 1395–1405. [CrossRef]
15. *Cyber Bullying: Common Types of Bullying 2019*; Statista: New York, NY, USA, 2019.
16. Royen, K.V.; Poels, K.; Daelemans, W.; Vandebosch, H. Automatic monitoring of cyberbullying on social networking sites: From technological feasibility to desirability. *Telemat. Inform.* **2015**, *32*, 89–97. [CrossRef]
17. Teng, T.H.; Varathan, K.D. Cyberbullying Detection in Social Networks: A Comparison Between Machine Learning and Transfer Learning Approaches. *IEEE Access* **2023**, *11*, 55533–55560. [CrossRef]
18. Yi, P.; Zubiaga, A. Session-based cyberbullying detection in social media: A survey. *Online Soc. Netw. Media* **2023**, *36*, 100250. [CrossRef]
19. Dhanya, K.A.; Vajipayajula, S.; Srinivasan, K.; Tibrewal, A.; Kumar, T.S.; Kumar, T.G. Detection of Network Attacks using Machine Learning and Deep Learning Models. *Procedia Comput. Sci.* **2023**, *218*, 57–66. [CrossRef]
20. Aitken, P.; Claise, B.; Trammell, B. *RFC 7011*; Specification of the IP Flow Information Export (IPFIX) Protocol for the Exchange of Flow Information; RFC Editor: Marina del Rey, CA, USA, 2013. [CrossRef]
21. Trammell, B.; Boschi, E. *RFC 5103*; Bidirectional Flow Export Using IP Flow Information Export (IPFIX); RFC Editor: Marina del Rey, CA, USA, 2008. [CrossRef]
22. Lopez-Vizcaino, M.; Novoa, F.J.; Fernandez, D.; Carneiro, V.; Cacheda, F. Early Intrusion Detection for OS Scan Attacks. In Proceedings of the 2019 IEEE 18th International Symposium on Network Computing and Applications, NCA 2019, Cambridge, MA, USA, 26–28 September 2019. [CrossRef]
23. Cacheda, F.; Fernández, D.; Novoa, F.J.; Carneiro, V. Analysis and Experiments on Early Detection of Depression. In Proceedings of the Conference and Labs of the Evaluation Forum, Avignon, France, 10–14 September 2018.
24. Cacheda, F.; Fernandez, D.; Novoa, F.J.; Carneiro, V. Early Detection of Depression: Social Network Analysis and Random Forest Techniques. *J. Med. Internet Res.* **2019**, *21*, e12554. [CrossRef] [PubMed]
25. Rafiq, R.I.; Hosseinmardi, H.; Mattson, S.A.; Han, R.; Lv, Q.; Mishra, S. Analysis and detection of labeled cyberbullying instances in Vine, a video-based social network. *Soc. Netw. Anal. Min.* **2016**, *6*, 88. [CrossRef]
26. Scikit-Learn: Machine Learning in Python—Scikit-Learn 1.1.2 Documentation. Available online: https://scikit-learn.org/stable/ (accessed on 20 December 2023).

applied sciences

MDPI

Article

Phishing Node Detection in Ethereum Transaction Network Using Graph Convolutional Networks

Zhen Zhang [1], Tao He [1], Kai Chen [1], Boshen Zhang [1], Qiuhua Wang [1,2] and Lifeng Yuan [1,2,*]

[1] School of Cyberspace, Hangzhou Dianzi University, Hangzhou 310018, China
[2] Data Security Governance Zhejiang Engineering Research Center, Hangzhou Dianzi University, Hangzhou 310018, China
* Correspondence: yuanlifeng@hdu.edu.cn

Abstract: As the use of digital currencies, such as cryptocurrencies, increases in popularity, phishing scams and other cybercriminal activities on blockchain platforms (e.g., Ethereum) have also risen. Current methods of detecting phishing in Ethereum focus mainly on the transaction features and local network structure. However, these methods fail to account for the complexity of interactions between edges and the handling of large graphs. Additionally, these methods face significant issues due to the limited number of positive labels available. Given this, we propose a scheme that we refer to as the Bagging Multiedge Graph Convolutional Network to detect phishing scams on Ethereum. First, we extract the features from transactions and transform the complex Ethereum transaction network into three simple inter-node graphs. Then, we use graph convolution to generate node embeddings that leverage the global structural information of the inter-node graphs. Further, we apply the bagging strategy to overcome the issues of data imbalance and the Positive Unlabeled (PU) problem in transaction data. Finally, to evaluate our approach's effectiveness, we conduct experiments using actual transaction data. The results demonstrate that our Bagging Multiedge Graph Convolutional Network (0.877 AUC) outperforms all of the baseline classification methods in detecting phishing scams on Ethereum.

Keywords: phishing node detection; Ethereum; graph convolutional network; node classification, transaction network

Citation: Zhang, Z.; He, T.; Chen, K.; Zhang, B.; Wang, Q.; Yuan, L. Phishing Node Detection in Ethereum Transaction Network Using Graph Convolutional Networks. *Appl. Sci.* **2023**, *13*, 6430. https://doi.org/10.3390/app13116430

Academic Editor: Yoshiyasu Takefuji

Received: 16 April 2023
Revised: 19 May 2023
Accepted: 20 May 2023
Published: 24 May 2023

1. Introduction

Ethereum, which provides Turing completeness in smart contracts, has become the largest smart contract platform. Meanwhile, Ether, i.e., the cash in Ethereum, has become one of the most popular cryptocurrencies. Hence, it is not surprising that Ethereum has been targeted extensively by cybercriminals. For example, according to the 2022 crypto crime report of Chainalysis, illicit transaction activity has reached an all-time-high value, and scams are the largest form of cryptocurrency-based crime by transaction volume, with over $7.7 billion of cryptocurrency taken from victims worldwide [1].

Among the various types of cybercriminal activity on Ethereum, phishing scams are notably prevalent and highly damaging and have garnered significant attention [2]. Currently available approaches to the detection of phishing primarily concentrate on identifying the specific characteristics of fraudulent emails and websites [3–7]. However, such methods are ineffective against scams that trick users into transferring cryptocurrency to Ethereum addresses that belong to or are controlled by scammers.

To detect phishing attempts on Ethereum, many novel methods using the transaction network were proposed [8–10]. The nodes of the transaction network represent Ethereum accounts, and the edges represent transactions between accounts. Specifically, these models transformed the phishing scam detection task into a node classification task [11]. Recently, researchers have constructed a transaction subgraph for each target node and used the features of the transaction subgraph as the features of the target node. Thus, the problem of phishing node detection in the Ethereum transaction network can be converted from a node classification task to a graph classification task [12–14]. These existing works, however, have some limitations in handling large graphs and have few positive labels.

Figure 1 is a schematic representation of the Ethereum transaction network, where multiple edges are merged into a single edge and the directions of the edges are hidden. The Ethereum transaction network is a multidigraph and contains rich information about node behavior patterns. From Figure 1, we can identify the following characteristics of the Ethereum transaction network.

1. There is a very large amount of transaction data. Although only a few dozen nodes are shown in the figure, there are hundreds of edges among the nodes. Thus, it can be concluded that the transaction network is very complex.
2. There is an intricate relationship between the nodes. Figure 1 shows that the nodes in the transaction network are connected closely with other nodes, and there are multiple edges between nodes.
3. There is an imbalance in the data. Based on the figure, phishing nodes only account for a small proportion of the data compared to normal nodes, and this indicates that a serious data imbalance problem exists in the Ethereum transaction data.

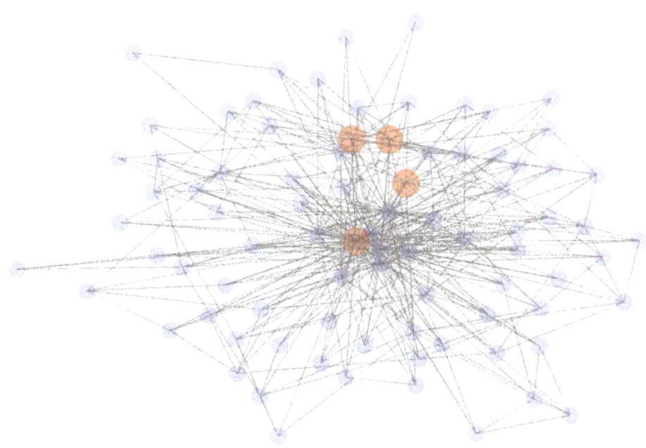

Figure 1. Part of the Ethereum transaction network in which multiedges between nodes are simplified to a single edge. In this network, scam nodes and normal nodes are marked in red and blue, respectively.

Based on the observations stated above, we can identify the reason for the limitations in the model's performance. First, a great deal of computational resources are needed to process large-scale transaction data. Second, the intricate relationships mean that naive edge handling approaches can lead to a loss of transaction information. Last but not least, the serious data imbalance problem causes models to inadequately learn phishing features. Most researchers alleviate these challenges via graph sampling (e.g., subgraph extraction) and graph filtering mechanisms. However, these methods have difficulty in obtaining global structural information. The lack of global structural information in the node embedding impacts the final phishing node detection.

Therefore, to fully benefit from the structural information of the transaction network and effectively solve the data imbalance problem, we propose a Bagging Multiedge Graph Convolutional Network (BM-GCN) model. The model simplifies the complex relationships between the nodes by breaking down the entire transaction network into three inter-node graphs. The advantage of this is that it facilitates the extraction of node features while preserving global structure information. The inter-node graph refers to the fact that each pair of nodes (a, b) in the graph has at most two edges, one from a to b and one from b to a. Specifically, in our approach, we preprocess the transaction data and generate three inter-node graphs to represent the property on the transaction graph. Then, the GCN model is utilized as the embedding generation method to make use of structural information in the graph. During the training of the GCN model, a bagging strategy is adopted to mitigate the impact of imbalanced data and unlabeled nodes. Consequently, our model can deal with large-scale data. To the best of our knowledge, this work is the first example that uses a bagging GCN for the detection of Ethereum phishing scams.

The remainder of this article is organized as follows. Section 2 addresses the research related to the subject of this article, and Section 3 presents the motivation for the research. Section 4 describes the BM-GCN model and the strategy that was used when we were training the model. In Section 5, we introduce the method of evaluating the effectiveness of the BM-GCN model and analyze the experimental results. Section 6 summarizes the contributions of this paper and presents our future research plans.

2. Related Work

2.1. Scams on Blockchain Platforms

With the development of blockchain technology and the growth of its community, the number of fraud attacks on digital currencies is increasing, and this has prompted researchers to analyze the scams. Vasek et al. [15] presented the first survey of Bitcoin-based scams; after gathering and combining the various reports of scams, they categorized the scams into four groups, i.e., Ponzi schemes, mining scams, scam wallets, and fraudulent exchanges. Since Ethereum is an extension of Bitcoin, it can also be categorized in these ways.

Bitcoin Ponzi schemes have received a great deal of attention because they are a classical form of economic deception. Vasek et al. [16] identified why Ponzi scams occur frequently in this ecosystem. Bartoletti et al. [17] analyzed this type of scheme on Ethereum and studied how Ponzi schemes are promoted on the web. To fuel the detection of Ponzi schemes on smart contracts, Chen et al. [18] provided an open dataset by gathering real-world samples, and they used a random forest model built on account features and code features to identify latent smart Ponzi schemes.

2.2. Detection of Phishing Scams on Ethereum

Phishing scams are among the most severe cybercrimes aimed at Ethereum users, and many efforts have been made to detect phishing [3]. Wu et al. [8] proposed trans2vec, which used a weighted random walk to generate the embeddings of nodes, and then employed a one-class SVM model to classify the embeddings to detect phishing nodes. Chen et al. [10] extracted graph-based cascade features from transaction records and developed a lightGBM-based dual-sampling ensemble algorithm to identify phishing accounts. Chen et al. [9] obtained statistics on the transaction information as features of nodes and then used a graph convolutional network (GCN) and autoencoder technology to extract the structural features of the subgraph. The output of the GCN and handcrafted features are concatenated to obtain the final result for classification. To detect potential phishing scammers, Zhang et al. [14] proposed a multi-channel graph classification model (MCGC) with multiple feature extraction channels for GNN to extract richer information from the input graph.

Although the approaches mentioned above have been able to complete the detection of Ethereum phishing, their methods of processing graph data are designed for simple subgraphs, thus ignoring the global structural information of the Ethereum transaction network. In addition, they do not work in multiedge graphs. To make full use of the transaction information and structural information, we propose a novel method that transforms the transaction network into some inter-node graphs for feature extraction.

2.3. Graph Embedding

Graph embedding transforms the data on the graph into a low-dimensional space while retaining the graph's structural information and properties as much as possible [19]. This operation facilitates subsequent analytical tasks in both homogeneous and heterogeneous networks. Graph embedding methods can be roughly divided into three categories, i.e., random walk, matrix decomposition, and deep learning. The basic idea of random-walk-based graph embedding is to utilize SkipGram on a path set sampled by a truncated random walk on the graph data to obtain a node embedding [20,21]. Matrix decomposition methods factorize a proximity matrix that represents node relationships to obtain the node embedding [22]. For example, ProNE [23] learns embedding both rapidly and efficiently via matrix factorization with spectral propagation. The core idea of the deep learning methodology is to obtain a graph embedding directly from the graph structure through a deep neural network. For example, Kipf et al. [24] proposed the graph convolutional network (GCN), which introduced a variant of convolutional neural networks that can use graphs directly and match neighborhoods in the spatial domain.

3. Research Motivations

The Ethereum transaction network is a multiedge graph with a large number of transactions. In such a graph, phishing nodes generally make up only a tiny percentage of the nodes. Therefore, there are several factors that can impact the classification performance when constructing a phishing node detection scheme on the graph.

3.1. Challenges

Transaction graph has complex inter-node relationships

Generally, in the Ethereum transaction network, there are multiple transactions with varying amounts occurring at different times between two nodes. In other words, there will be multiple adjacent edges between nodes. Figure 2 shows a simple transaction graph with only five nodes. The simple addition of weights leads to the unexpected fusion of the features, which limits the effective utilization of the discrete properties.

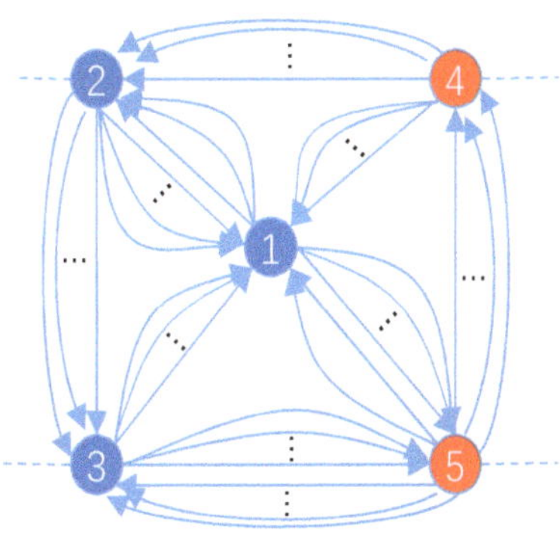

Figure 2. A simple multiedge graph example in the transaction network. It can be observed in this figure that there are many edges between nodes, and these edges have different amounts of transactions and time. Thus, it is challenging to merge them. When the number of transactions is greater than three, we use the symbol "..." to represent the remaining transactions.

Significant imbalance between phishing and normal nodes

In the Ethereum transaction network example presented in [25], there were 2,973,489 nodes and 13,551,303 transactions, but only 1165 phishing nodes. In other words, there was a significant imbalance between phishing and normal nodes, which can impact the results of the classification.

Unlabeled nodes

The labeling of phishing nodes relies on reports from users of specific websites, such as etherscamdb.info and etherscan.io. In other words, these websites can track phishing incidents only if they are reported, and significant numbers of frauds and scams are not reported [26,27]. Therefore, having these unknown/undetected phishing nodes in the "normal node" set can skew and impact the classifier's performance, a situation that is also referred to as a Positive Unlabeled (PU) learning challenge.

3.2. Potential Solutions

We posit the potential of using the following approaches to mitigate the challenges discussed in Section 3.1.

1. To address the multiedge graph problem, we extract three features from the transactions, i.e., inter-node interaction, transaction time variance, and transaction frequency. For each feature, we replace the edges between two nodes that have the same direction with a single directed edge and construct a feature graph to represent the information contained in the multiedges.
2. We use the bagging strategy [28] to deal with both data imbalances and the PU problem. In doing so, we use bootstrap aggregating techniques to leverage unlabeled data and mitigate the limitations associated with the PU problem. In addition, the sampling method used in the bagging strategy also minimizes the impact of the imbalance in the data on the classification results.

4. Proposed BM-GCN Model

4.1. Representing the Features of the Graph

Due to the complexity of Ethereum transaction networks, the use of GCN directly in the original network cannot effectively encode the topology around the nodes. Therefore, we consider extracting features from the original transaction network and transforming the complex network into three simple graphs, i.e., a node interaction graph, a time variance graph, and a transaction frequency graph. Then, we use the corresponding adjacency matrices, A_i, A_v, and A_f, to represent the three feature graphs (see also Figure 3). Note that the transactions are directed and the matrices are not symmetrical.

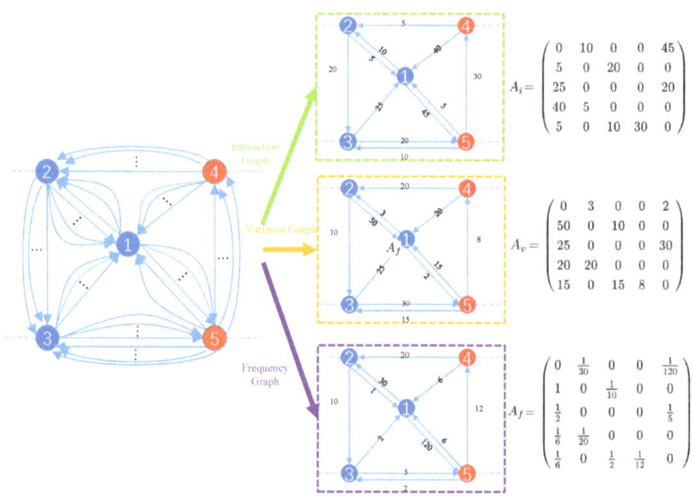

Figure 3. Feature representation: The property of the transaction graph is extracted into three inter-node graphs, and the matrices in the right part show the feature representation of each graph. The numbers in the graphs and matrices are only examples, not the actual information in the transaction network.

4.1.1. Node Interaction Graph

Transaction records provide a significant amount of information to build inter-node graphs. For example, if many transaction records exist between node i and node j, there will be a closer relationship between these nodes than between nodes with fewer interactions. With this in mind, we constructed an interaction graph to indicate whether there are frequent interactions between two nodes. We denote $I_{i,j}$ as the trade number from i to j, and we build the interaction graph as follows:

$$G_i(V,E) \quad weight = TransactionNumber \tag{1}$$

$$A_i = \begin{pmatrix} 0 & I_{0,1} & I_{0,2} & \cdots & I_{0,N-1} \\ I_{1,0} & 0 & \cdots & \cdots & I_{1,N-1} \\ I_{2,0} & I_{2,1} & 0 & \cdots & I_{2,N-1} \\ \vdots & \vdots & \vdots & \ddots & \vdots \\ I_{N-1,0} & \cdots & \cdots & \cdots & 0 \end{pmatrix} \tag{2}$$

4.1.2. Time Variance Graph

Intuitively, the interval of transaction time, which shows the changes in trading time between two nodes, can be used effectively to describe the transaction relationship between nodes. To introduce the time feature of transactions between nodes, we use the variance of the time of the transactions to construct the second inter-node graph. Let $v_{i,j}$ denote the variance in the transaction time from node i to j; the mean value of the transaction time from i to j is $\overline{t_{i,j}}$, the total number of transactions from i to j is $n_{i,j}$, and the time of k-th transaction is τ_k. The graph of the time variance is constructed as follows:

$$G_v(V,E) \quad weight = TransactionTimeVariance \tag{3}$$

$$v_{i,j} = \frac{\sum_{k=1}^{n_{i,j}} (\tau_k - \overline{t_{i,j}})^2}{n_{i,j}} \tag{4}$$

$$A_v = \begin{pmatrix} 0 & v_{0,1} & v_{0,2} & \cdots & v_{0,N-1} \\ v_{1,0} & 0 & \cdots & \cdots & v_{1,N-1} \\ v_{2,0} & v_{2,1} & 0 & \cdots & v_{2,N-1} \\ \vdots & \vdots & \vdots & \ddots & \vdots \\ v_{N-1,0} & \cdots & \cdots & \cdots & 0 \end{pmatrix} \tag{5}$$

4.1.3. Transaction Frequency Graph

We use the frequency of transactions between nodes as the weight to construct a graph; specifically, we introduce additional time information into our model, which reflects the average duration of the intervals of the transactions from node i to node j, also written as $f_{i,j}$. We denote the transaction frequency from node i to j as the reciprocal of $f_{i,j}$. This also ensures that high-frequency nodes have high weights. The frequency graph can be represented as follows:

$$G_f(V,E) \quad weight = TransactionFrequency \tag{6}$$

$$f_{i,j} = \begin{cases} 0, & n_{i,j} = 1 \\ \frac{\sum_{k=1}^{n_{i,j}-1} \tau_{k+1} - \tau_k}{n_{i,j}}, & n_{i,j} \geq 2 \end{cases} \tag{7}$$

$$A_f = \begin{pmatrix} 0 & \frac{1}{f_{0,1}} & \frac{1}{f_{0,2}} & \cdots & \frac{1}{f_{0,N-1}} \\ \frac{1}{f_{1,0}} & 0 & \cdots & \cdots & \frac{1}{f_{1,N-1}} \\ \frac{1}{f_{2,0}} & \frac{1}{f_{2,1}} & 0 & \cdots & \frac{1}{f_{2,N-1}} \\ \vdots & \vdots & \vdots & \ddots & \vdots \\ \frac{1}{f_{N-1,0}} & \cdots & \cdots & \cdots & 0 \end{pmatrix} \tag{8}$$

4.2. GCN for Inter-Node Graphs

In this section, we model the phishing detection problem as a binary classification. The inputs of this model are the three feature graphs discussed earlier and the outputs of this model are the prediction labels of Ethereum nodes.

In our model, we use the layer-wise propagation rule of Kipf et al. [24] to build a multilayer GCN. The rule is as follows:

$$H^{(l+1)} = \sigma\left(\tilde{M}^{-\frac{1}{2}}\tilde{A}\tilde{M}^{-\frac{1}{2}}H^{(l)}W^{(l)}\right) \tag{9}$$

In the above equation, $\tilde{A} = A + I_N$ denotes the adjacency matrix of the graph G with self-connections added, I_N is the identity matrix, $\tilde{M}_{ii} = \sum_j \tilde{A}_{ij}$ and $W^{(l)}$ are the trainable weight matrices, $\sigma(\cdot)$ is the activation function, and $H^{(l)} \in \mathbb{R}^{N\times D}$ is the matrix of activations in the l-th layer, $H^{(0)} = X$.

Then, we use the propagation rule mentioned above to build a GCN. For each feature graph, we use graph convolution to generate the embedding of the feature, which is shown on the left side of Figure 4. The input graph G is denoted by $G = \{n_1, n_2, ..., n_{|V|}\}$, where n_i is the i-th node, and x_i is the representation of n_i. For the three feature graphs $\{G_i, G_v, G_f\}$ in our model, we denote the vector of the i-th node as $\{x_i^i, x_i^v, x_i^f\}$.

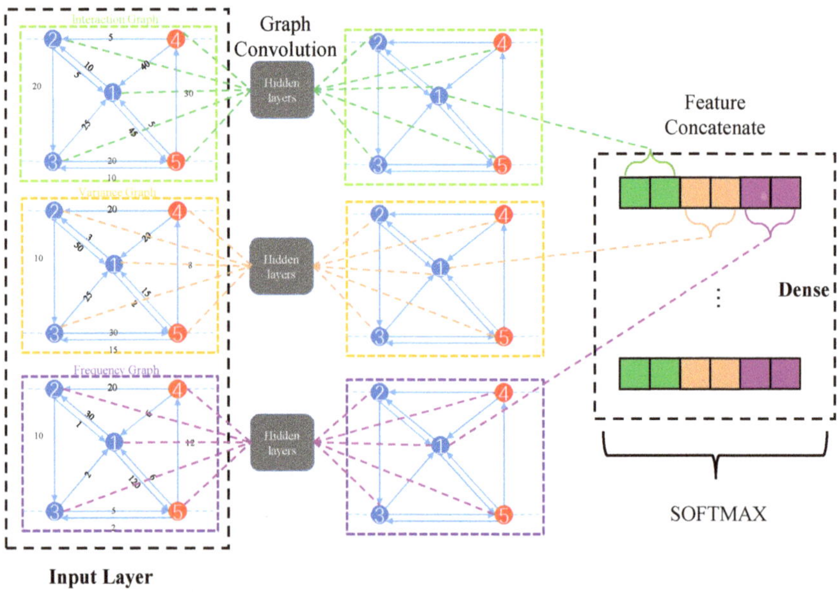

Figure 4. GCN classification model. On the left is the GCN used to learn the different structures of the three feature graphs. Although the three graphs have the same topology, the weights of their edges are different. On the right is the concatenation of the output of the previous GCNs with the dense layer and softmax layer for classification.

In order to predict the labels of nodes, we concatenate the outputs of three GCN models as $X_i = (x_i^i : x_i^v : x_i^f)$, and use a dense layer $y = f(w \cdot X + b)$ and a softmax layer to obtain the predictions of node labels. The softmax function is as follows:

$$p_i = \frac{\exp(y_i)}{\sum_{k=1}^{n}\exp(y_k)} \tag{10}$$

4.3. Bagging

Considering that there are many unlabeled phishing nodes in the transaction data and the distribution of positive and negative examples in the data is very asymmetrical, we use the transductive bagging strategy [28] to construct a bagging learning approach dealing with both data imbalances and the PU problem in the transaction graph.

The method that we propose for PU learning in the transaction data is presented in Algorithm 1. It creates a training set, S, by combining all positive nodes and sampled unlabeled nodes randomly and using S to train a classifier. Then, labeled and unlabeled samples are treated as positive and negative, respectively. For each S, the algorithm uses the Adam optimizer to update the w parameter of the model.

Algorithm 1 Bagging learning

Input: $\mathcal{P}, \mathcal{U}, K$ = size of bootstrap samples, T = number of bootstraps
Output: a function $f : \mathcal{X} \to \mathbb{R}$
  for $t = 1$ to T **do**
    Draw a bagging sample U_t of size K in U.
    Make a bootstrap set S from P and U_t with corresponding labels.
    Use bootstrap set S to train the classifier f to discriminate P against U_t.
    while stopping criterion not met **do**
      Update w with Adam optimizer
    end while
  end for
  return f

5. Evaluation

In this section, we demonstrate that our approach can deal with both data imbalances and the PU problem while fully utilizing the graph structure information for the detection of phishing.

First, we introduce the dataset and metrics used in the evaluation. Next, we evaluate the performance of Wu et al.'s approach [8] over different network scales and different negative–positive ratios (NP ratios). To verify the effectiveness of our approach, we conduct the following evaluations: (1) we evaluate the effects of feature numbers to determine their performance with varying NP ratios; (2) we evaluate the effectiveness of our approach in dealing with data imbalances; (3) we evaluate the effectiveness of our approach in dealing with the PU problem. Finally, we present a comparative summary of the performance of our approach with several graph-embedding-based methods.

5.1. Dataset and Evaluation Metrics

We evaluated our model using the dataset of Chen et al. [25] which is available at https://xblock.pro/#/dataset/13. The dataset contains 2,973,489 Ethereum accounts, 13,551,303 transactions, and 1165 labeled accounts. The transaction time in the dataset starts on 7 August 2015 and ends on 19 January 2019. We constructed a transaction graph using accounts as nodes and transactions as edges, and we transformed it into three inter-node graphs as the input to our classification model.

We used the Area Under the Curve (AUC) of the Receiver Operating Characteristic (ROC) curve as an evaluation metric. In the testing phase, we calculated both the True Positive Rate (TPR) and the False Positive Rate (FPR) of the classification result, with T as the varying parameter, where T is the threshold of probability X that the node is classified as "positive" if $X > T$ and "negative" otherwise. Then, the ROC curve was defined by FPR and TPR as the x and y axes, respectively. To evaluate the performance of each baseline model, we used different ratios of both positive and negative instances.

Since the performance of schemes given different positive and negative proportions varies dramatically, we evaluated the classification results of several models using different NP ratio numbers.

5.2. Baseline Methods

We empirically compared the performance of our proposed approach with the performance of the Support Vector Machine (SVM), Logistic Regression (LR), and Random Forest (RF).

1. SVM represents the examples as vectors in space, and it chooses a hyperplane that represents the largest separation between examples in order to classify them.
2. As a statistical classification method, LR models a binary dependent variable using a logistic function and obtains the corresponding probability of the class of examples.
3. RF is an ensemble learning method that constructs a large number of decision trees at training time and outputs the modes of the classes as the classification result.

Since the above classification approaches require vectors as input, we utilized Deepwalk [20], trans2vec [8], ProNE [23], and NETSMF [22] to obtain node embeddings for the baseline methods.

DeepWalk is the first Word2vec-based node vectorization model. It uses the random walking paths of nodes on the network to imitate the process of generating text, and then treats the paths of the nodes as the equivalents of sentences and applies the language model to vectorize each node. Trans2vec introduces biased random walks to determine whether each walk is affected by transaction time bias or amount bias, and then it concatenates these two biases to balance their effects.

Different from DeepWalk and Trans2vec, ProNE and NETS-MF use matrix factorization directly to embed graphs. We introduce them into the baseline system to evaluate our scheme from another perspective. The embedding vector generated by ProNE contains both localized smoothing information and global clustering information, making it able to utilize the graph information more effectively. NETSMF is proposed to provide an efficient way to obtain embeddings from large graphs.

The baseline models were DeepWalk-SVM, DeepWalk-LR, DeepWalk-RF, Trans2vec-SVM, Trans2vec-LR, Trans2vec-RF, ProNE-SVM, ProNE-LR, ProNE-RF, NETSMF-SVM, NETS-MF-LR, and NETSMF-RF. We ran these baseline models on the entire transaction network and obtained the corresponding embeddings for all nodes. To ensure that the comparison was relatively fair, we used the publicly released source codes in the DeepWalk and ProNE papers and their default parameters. For Trans2vec, we added random walking weights in the source code of DeepWalk, following the parameters proposed in [8] to build this baseline model. Moreover, for NETSMF, we also used their source code. However, we reduced the number of training rounds to 80 due to the high usage of memory during training. (After only 80 rounds of training on the Ethereum transaction data, NETSMF had used more than 200 GB of memory.)

5.3. Findings

In our evaluations, we followed the guideline in [28] to set our bagging parameters, which were $T = 100$ and $K = 1165$. The parameters of our GCN were as follows: the number of hidden layers was 3, and the units of hidden layers 1, 2, and 3 were 16, 16, and 8, respectively. The maximum epoch number per bag was 20, the learning rate was 0.01, and the dropout rate was set to 0.5. We selected the value of the NP ratio among the following: 5, 10, 20, 50, 100, 200, 500, and "All", where "All" means that we used all nodes in the experiment (the NP ratio was 2,972,324:1165).

Evaluation of Wu et al.'s method: Figure 5 shows the performance of Wu et al.'s approach [8] for various NP ratios and graph scales. For each scale, we constructed three test graphs following the approach, and we used their average AUC when evaluating the performance. The average node and edge numbers are provided in Table 1. The following two limitations can be observed in their scheme.

1. As the NP ratio increases, the performance of their scheme decreases consistently. Specifically, the average classification AUC value of Trans2vec decreased from 0.886

to 0.732 when the NP ratio increased from 1 to 25. In other words, Trans2vec is not capable of dealing with data imbalances.

2. As the network scales, the performance of their model decreases gradually. In other words, the scale of the network impacts the node representation capabilities of their scheme and degrades the classification performance (i.e., Trans2vec is not inadequate for large-scale transaction graphs).

Table 1. Average scale of different test graphs

Scale	Node Number	Edge Number
1	32,582	70,082
3	39,606	95,154
5	45,397	134,552
10	59,250	188,875
15	76,344	241,639
20	90,388	283,260
30	119,368	375,159
50	162,388	535,819
All	2,973,489	13,551,303

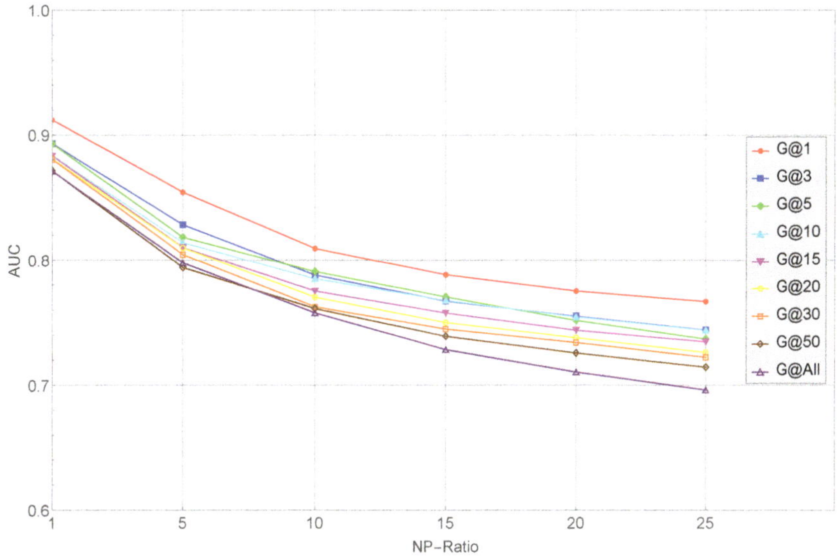

Figure 5. Curves of average AUC of Wu et al.'s model [8], with varying NP ratios. Each curve represents a network scale, and it refers to the number after "G" in the legend. It indicates the proportion of positive and negative examples when the network is initialized.

Effect of different features: We evaluated the effect of different feature combinations on the proposed BM-GCN model using two NP ratios, i.e., 50 and "All". Table 2 displays the aggregate AUC of the GCN with different feature combinations. We observe that the classification performance is most significantly improved by the feature G_i out of the three analyzed. For multiple feature combinations, we found that the combination of G_i, G_f, and G_v worked best. G_i is an indicator that describes the total number of transactions between nodes, which evidently reflects the closeness of the relationships between nodes. It outperforms the other two. G_f and G_v describe the time properties of transactions from the perspective of transaction frequency and changes in transaction time. This combination effectively improves the AUC value as they complement each other. The combination of all three features allows us to achieve the best classification performance. This implies that the features reflect the topological characteristics of the nodes to a certain extent, and our transaction feature extraction scheme is effective.

Table 2. AUC with different features

Features	NP-ratio@50	NP-ratio@All
G_f	0.852591	0.851575
G_v	0.863451	0.848624
G_i	0.872630	0.867106
$G_f + G_v$	0.861069	0.867045
$G_f + G_i$	0.867851	0.855019
$G_v + G_i$	0.869612	0.875120
$G_i + G_f + G_v$	**0.883325**	**0.875443**

Bagging vs. no bagging: In this section, we maintain the NP ratio to evaluate the impact of removing the bagging strategy on the classification performance of the model. As shown in Table 3, in the case of no bagging, the classification performance of the model decreases rapidly as the NP ratio increases. Even when the NP ratio is equal to 50, the AUC of the model drops below 0.5. The findings show that if the bagging strategy is not used in the model's training process, the original GCN solution will not be able to cope with extreme data imbalances in the Ethereum transaction network and detect phishing nodes effectively.

Table 3. AUC without bagging strategy

NP Ratio	No Bagging	Bagging
5	0.855752	0.870561
10	0.798114	0.880036
20	0.745765	0.877302
50	0.495823	0.883325

Evaluation of the PU problem: Next, we utilized the spy technique [29] to set false negative examples to evaluate the robustness of our model with respect to the PU problem. In the evaluation, we selected 15% positive examples and set them as negative examples, and we placed them in the training set to simulate the PU problem in the training set. Then, we checked whether these examples that were intentionally marked as negative examples could be detected by the model. Specifically, we evaluated the classification performance of the BM-GCN model with 173 spy nodes at NP ratios of 5, 20, 50, and "All".

Table 4 shows the model's capability of recovering spy nodes' labels. It also indicates that the AUC value of classification increases as the NP ratio increases, which intuitively reflects the negative impact of unlabeled data. For small datasets, such as when the NP ratio equals 5, there are only 4141 negative examples. Thus, the introduction of 173 unlabeled nodes confuses the model significantly, resulting in the degradation of its performance. However, even in the worst case, 97.6 of the 173 spy nodes are restored successfully by the model. Thus, our model effectively avoids the adverse impact of the unlabeled nodes on the results of the classification. In addition, it illustrates that our model can deal with the PU problem in the Ethereum transaction data.

Table 4. Results of the spy test

NP Ratio	Restored Nodes	AUC
5	97.6	0.819079
20	115.8	0.852759
50	123.0	0.858245
All	133.6	0.870821

Baseline evaluation: Figure 6 shows the aggregate AUC of our approach and all of the baselines with varying NP ratios. We can see that the model (GCN with $G_i + G_f + G_v$) outperforms all of the baseline systems. First, on the entire range of NP ratios, our model achieves a higher AUC than all of the baselines. Second, the BM-GCN model achieves an average AUC of 0.877, whereas the Deepwalk-LR only achieves an average AUC of 0.661. Thus, we conclude that our BM-GCN model uses more transaction information than other models. Moreover, our model is more robust than all of the baseline models. For example, Figure 6 shows that the performance of all the baselines decreased rapidly as the NP ratio increased, but our scheme remained stable. This implies the potential in using our BM-GCN model for larger datasets.

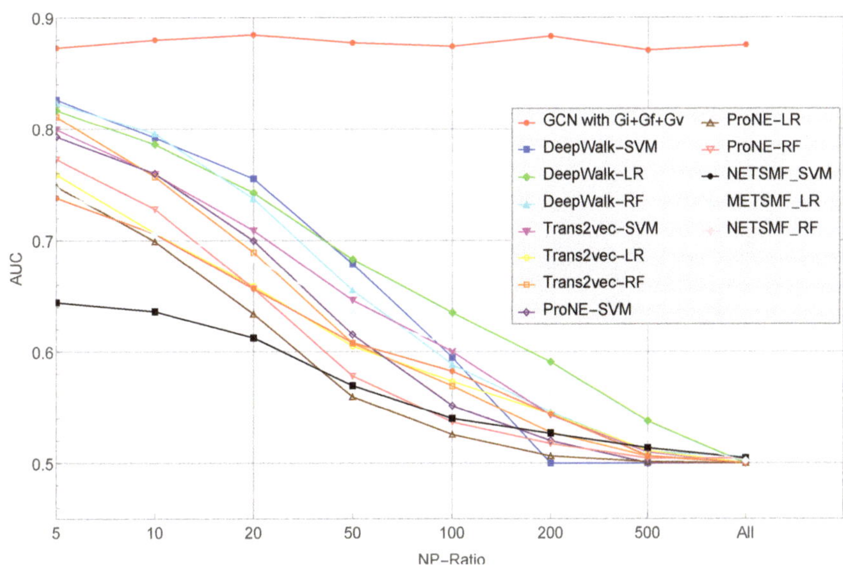

Figure 6. Curves showing the average AUC values of our model and the baseline models as the NP ratio is increased from 5 to "All".

Comparison with other methods: Table 5 shows the comparison between the proposed method and other methods in terms of the AUC metric. All the methods use transaction data for phishing node detection. However, compared to other methods that directly use raw transaction data or relevant statistical features, BM-GCN extracts the global structural features and preprocesses the raw transaction information into three types of interactive information, i.e., node interaction, time variance, and transaction frequency. As shown in Table 5, our method achieves the best results.

Table 5. Comparison with other methods

Methods	Features	AUC
Chen et al. [9]	Handcrafted features + local structural features	0.5866
Chen et al. [10]	Handcrafted features	0.8071
Zhang et al. [14]	Hierarchical structural features	0.8274
BM-GCN	Global structural features	0.8771

6. Conclusions

In this work, we introduce a BM-GCN model to detect phishing scams targeting Ethereum. This model extracts features of transactions by converting the multiedge transaction graph into several simple graphs. A bagging strategy is introduced during the training of the BM-GCN model to deal with the PU problem and the data imbalance problem in the transaction data. Compared with the baselines, BM-GCN is more effective in three respects: (1) it fully uses complex relations in multiedges; (2) it is able to cope with the problems of data imbalance and unlabeled nodes in the Ethereum transaction network; and (3) the model performs well on both small- and large-scale graphs.

Future research will include conducting systematic statistical tests to make the experimental results more convincing and extending this work to evaluate Ethereum-related transactions in real time. These tasks will require collaboration with the relevant stakeholders.

Author Contributions: Conceptualization, Z.Z. and T.H.; Data curation, K.C. and B.Z.; Formal analysis, K.C. and B.Z.; Funding acquisition, Z.Z., Q.W. and L.Y.; Investigation, Z.Z., K.C. and B.Z.; Methodology, Z.Z. and T.H.; Project administration, Z.Z.; Resources, K.C. and B.Z.; Software, T.H., K.C. and B.Z.; Supervision, Z.Z., Q.W. and L.Y.; Validation, T.H., K.C. and B.Z.; Visualization, T.H.; Writing—original draft, Z.Z. and T.H.; Writing—review and editing, Z.Z., T.H., Q.W. and L.Y. All authors have read and agreed to the published version of the manuscript.

Funding: This research was funded by the "Pioneer" and "Leading Goose" R&D Program of Zhejiang (Grant No. 2023C03203, 2023C03180, 2022C03174).

Institutional Review Board Statement: Not applicable.

Informed Consent Statement: Not applicable.

Data Availability Statement: All data included in this study are available upon request by contacting the corresponding author.

Conflicts of Interest: The authors declare no conflict of interest.

References

1. Chainalysis Team. The 2022 Crypto Crime Report. 2022. Available online: https://blog.chainalysis.com/reports/2022-crypto-crime-report-introduction/ (accessed on 8 April 2022).
2. Onyema, E.; Dinar, A.; Ghouali, S.; Merabet, B.; Merzougui, R.; Feham, M. Cyber Threats, Attack Strategy, and Ethical Hacking in Telecommunications Systems. In *Security and Privacy in Cyberspace*; Springer: Berlin/Heidelberg, Germany, 2022; pp. 25–45.
3. Varshney, G.; Misra, M.; Atrey, P.K. A survey and classification of web phishing detection schemes: Phishing is a fraudulent act that is used to deceive users. *Secur. Commun. Networks* **2016**, *9*, 6266–6284. [CrossRef]
4. Xiang, G.; Hong, J.; Rose, C.P.; Cranor, L. CANTINA+: A Feature-Rich Machine Learning Framework for Detecting Phishing Web Sites. *ACM Trans. Inf. Syst. Secur.* **2011**, *14*, 1–28. [CrossRef]
5. Kausar, F.; Al-Otaibi, B.; Al-Qadi, A.; Al-Dossari, N. Hybrid client side phishing websites detection approach. *Int. J. Adv. Comput. Sci. Appl.* **2014**, *5*, 132–140. [CrossRef]
6. Ramesh, G.; Krishnamurthi, I.; Kumar, K.S.S. An efficacious method for detecting phishing webpages through target domain identification. *Decis. Support Syst.* **2014**, *61*, 12–22. [CrossRef]
7. Chen, T.C.; Stepan, T.; Dick, S.; Miller, J. An Anti-Phishing System Employing Diffused Information. *ACM Trans. Inf. Syst. Secur.* **2014**, *16*, 1–31. [CrossRef]
8. Wu, J.; Yuan, Q.; Lin, D.; You, W.; Chen, W.; Chen, C.; Zheng, Z. Who Are the Phishers? Phishing Scam Detection on Ethereum via Network Embedding. *arXiv* **2019**, arXiv:1911.09259.
9. Chen, L.; Peng, J.; Liu, Y.; Li, J.; Xie, F.; Zheng, Z. Phishing scams detection in ethereum transaction network. *ACM Trans. Internet Technol. (TOIT)* **2020**, *21*, 1–16. [CrossRef]

10. Chen, W.; Guo, X.; Chen, Z.; Zheng, Z.; Lu, Y. Phishing Scam Detection on Ethereum: Towards Financial Security for Blockchain Ecosystem. In Proceedings of the 29th International Joint Conference on Artificial Intelligence, Yokohama, Japan, 7–15 January 2020; pp. 4506–4512.
11. Wu, J.; Liu, J.; Zhao, Y.; Zheng, Z. Analysis of cryptocurrency transactions from a network perspective: An overview. *J. Netw. Comput. Appl.* **2021**, *190*, 103139. [CrossRef]
12. Yuan, Z.; Yuan, Q.; Wu, J. Phishing Detection on Ethereum via Learning Representation of Transaction Subgraphs. *Blockchain Trust. Syst.* **2020**, *1267*, 178–191.
13. Wang, J.; Chen, P.; Yu, S.; Xuan, Q. Tsgn: Transaction subgraph networks for identifying ethereum phishing accounts. In Proceedings of the International Conference on Blockchain and Trustworthy Systems, Guangzhou, China, 5–6 August 2021; Springer: Berlin/Heidelberg, Germany, 2021; pp. 187–200.
14. Zhang, D.; Chen, J.; Lu, X. Blockchain Phishing Scam Detection via Multi-channel Graph Classification. In Proceedings of the International Conference on Blockchain and Trustworthy Systems, Guangzhou, China, 5–6 August 2021; Springer: Berlin/Heidelberg, Germany, 2021; pp. 241–256.
15. Vasek, M.; Moore, T. There's No Free Lunch, Even Using Bitcoin: Tracking the Popularity and Profits of Virtual Currency Scams. In Proceedings of the International Conference on Financial Cryptography and Data Security, San Juan, Puerto Rico, 26–30 January 2015; pp. 44–61.
16. Vasek, M.; Moore, T. Analyzing the Bitcoin Ponzi Scheme Ecosystem. In *International Conference on Financial Cryptography and Data Security*; Springer: St. Kitts, Saint Kitts and Nevis, 2019; pp. 101–112.
17. Bartoletti, M.; Carta, S.; Cimoli, T.; Saia, R. Dissecting Ponzi schemes on Ethereum: Identification, analysis, and impact. *Future Gener. Comput. Syst.* **2020**, *102*, 259–277. [CrossRef]
18. Chen, W.; Zheng, Z.; Ngai, E.C.; Zheng, P.; Zhou, Y. Exploiting Blockchain Data to Detect Smart Ponzi Schemes on Ethereum. *IEEE Access* **2019**, *7*, 37575–37586. [CrossRef]
19. Cai, H.; Zheng, V.W.; Chang, K.C. A Comprehensive Survey of Graph Embedding: Problems, Techniques and Applications. *arXiv* **2018**, arXiv:1709.07604.
20. Perozzi, B.; Al-Rfou, R.; Skiena, S. DeepWalk: Online learning of social representations. In Proceedings of the 20th ACM SIGKDD International Conference on Knowledge Discovery and Data Mining, New York, NY, USA, 24–27 August 2014; pp. 701–710.
21. Grover, A.; Leskovec, J. Node2vec: Scalable Feature Learning for Networks. In Proceedings of the 22nd ACM SIGKDD International Conference on Knowledge Discovery and Data Mining, New York, NY, USA, 13–17 August 2016; pp. 855–864.
22. Qiu, J.; Dong, Y.; Ma, H.; Li, J.; Wang, C.; Wang, K.; Tang, J. NetSMF: Large-Scale Network Embedding as Sparse Matrix Factorization. In Proceedings of the World Wide Web Conference, San Francisco, CA, USA, 13–17 May 2019; pp. 1509–1520.
23. Zhang, J.; Dong, Y.; Wang, Y.; Tang, J.; Ding, M. ProNE: Fast and Scalable Network Representation Learning. In Proceedings of the 28th International Joint Conference on Artificial Intelligence, Macao, China, 10–16 August 2019; pp. 4278–4284.
24. Kipf, T.N.; Welling, M. Semi-supervised classification with graph convolutional networks. *arXiv* **2016**, arXiv:1609.02907.
25. Chen, L.; Peng, J.; Liu, Y.; Li, J.; Xie, F.; Zheng, Z. XBLOCK Blockchain Datasets: InPlusLab Ethereum Phishing Detection Datasets. 2019. Available online: http://xblock.pro/ethereum/ (accessed on 8 April 2020).
26. Team, C. Crypto Crime Series: Decoding Ethereum Scams. 2019. Available online: https://blog.chainalysis.com/reports/ethereum-scams (accessed on 8 April 2020).
27. Redman, J. Data Shows Ethereum is the 'Cryptocurrency of Choice for Scams'. 2019. Available online: https://news.bitcoin.com/data-shows-ethereum-is-the-cryptocurrency-of-choice-for-scams/ (accessed on 8 April 2020).
28. Mordelet, F.; Vert, J.P. A bagging SVM to learn from positive and unlabeled examples. *Pattern Recognit. Lett.* **2014**, *37*, 201–209. [CrossRef]
29. Liu, B.; Dai, Y.; Li, X.; Lee, W.S.; Yu, P.S. Building text classifiers using positive and unlabeled examples. In Proceedings of the 3rd IEEE International Conference on Data Mining, Melbourne, FL, USA, 9–12 November 2003; pp. 179–188.

applied sciences

MDPI

Article

Machine Learning Algorithms for Raw and Unbalanced Intrusion Detection Data in a Multi-Class Classification Problem

Mantas Bacevicius and Agne Paulauskaite-Taraseviciene *

Faculty of Informatics, Kaunas University of Technology, Studentu 50, 51368 Kaunas, Lithuania; mantas.bacevicius@ktu.edu
* Correspondence: agne.paulauskaite-taraseviciene@ktu.lt

Abstract: Various machine learning algorithms have been applied to network intrusion classification problems, including both binary and multi-class classifications. Despite the existence of numerous studies involving unbalanced network intrusion datasets, such as CIC-IDS2017, a prevalent approach is to address the issue by either merging the classes to optimize their numbers or retaining only the most dominant ones. However, there is no consistent trend showing that accuracy always decreases as the number of classes increases. Furthermore, it is essential for cybersecurity practitioners to recognize the specific type of attack and comprehend the causal factors that contribute to the resulting outcomes. This study focuses on tackling the challenges associated with evaluating the performance of multi-class classification for network intrusions using highly imbalanced raw data that encompasses the CIC-IDS2017 and CSE-CIC-IDS2018 datasets. The research concentrates on investigating diverse machine learning (ML) models, including Logistic Regression, Random Forest, Decision Trees, CNNs, and Artificial Neural Networks. Additionally, it explores the utilization of explainable AI (XAI) methods to interpret the obtained results. The results obtained indicated that decision trees using the CART algorithm performed best on the 28-class classification task, with an average macro *F1-score* of 0.96878.

Keywords: intrusion; machine learning; XAI; imbalanced dataset; multi-class classification

Citation: Bacevicius, M.; Paulauskaite-Taraseviciene, A. Machine Learning Algorithms for Raw and Unbalanced Intrusion Detection Data in a Multi-Class Classification Problem. *Appl. Sci.* **2023**, *13*, 7328. https://doi.org/10.3390/app13127328

Academic Editors: Peter R. J. Trim and Yang-Im Lee

Received: 29 May 2023
Revised: 15 June 2023
Accepted: 19 June 2023
Published: 20 June 2023

1. Introduction

Intrusion Detection Systems (IDS) play a key role in strengthening organizational security by providing an additional layer of defense to detect and respond in time to potential threats that may have bypassed preventive measures. These systems can detect a wide range of threats, including malware, phishing attacks, and other types of cyber-attacks, that may be missed by more traditional security solutions [1]. As digital networks become more complex and interconnected, the potential for vulnerabilities and weaknesses in the system increases, making it more difficult to identify and prevent all potential threats. IDSs can help to address this problem by providing continuous monitoring and analysis of network traffic, helping to identify potential threats before they can cause significant damage, but the development of a reliable and effective system may still be a challenging task.

IDSs must be designed to identify and respond to a wide range of cyber threats, including new and emerging attack techniques and to process massive amounts of network traffic data in real time, which can be challenging in terms of both computational resources and data processing speed [2]. This requires careful optimization of IDS algorithms and systems to ensure that they can process large volumes of data in a timely and efficient manner.

IDSs typically rely on a variety of techniques, such as signature-based detection [3,4], anomaly detection [5,6] and machine learning algorithms [7,8], to identify potential threats. Nevertheless, addressing class imbalance is a frequent challenge in machine learning and data analysis, especially when working with datasets in which the occurrence of one class is significantly less frequent compared to the others. In the context of network traffic analysis, this can be a problem because most of the traffic is typically benign, while malicious traffic is relatively rare. Many papers propose to deal with this problem by leaving only two types of attack—benign and malignant [9–11]—to reduce the number of classes in the grouping [12], but knowing which type of attack is also important.

Class imbalances can pose a number of challenges for researchers and data analysts, such as reducing the effectiveness of certain classification models and causing interpretation and generalization problems that may lead to biased or inaccurate results [13]. However, there are other factors that can also contribute to this issue. One of these factors is the overlap of classes due to a lack of feature separation, which can make it difficult for the algorithm to distinguish between the different classes. In addition, a lack of attributes that are specific to a certain decision boundary can make it difficult for the algorithm to accurately classify data. In recent years, deep learning techniques, including convolutional neural networks (CNNs) [14,15], recurrent neural networks (RNNs)) [16,17], and deep belief networks (DBNs)) [18], have gained significant attention for their promising capabilities in various classification tasks. Particularly in the field of cybersecurity attack classification, deep learning approaches have been increasingly utilized to exploit their potential.

While accuracy is very important, there is a growing demand for algorithmic transparency and the "white box" principle, particularly in the field of intrusion detection. This demand is driven by the need to understand and explain the decisions made by an IDS and to ensure that these decisions are fair, unbiased and explainable [19,20].

In recent years, the emergence of the XAI (explainable Artificial Intelligence) paradigm has played an important role in making algorithms more transparent and understandable [21]. XAI aims to create transparent, interpretable and explainable artificial intelligence models to make it easier to understand how these models make decisions. In the context of intrusion detection, XAI can help identify the features that are most influential in the decision-making process and why a certain output is obtained [22–24]. Similarly, in intrusion detection, an XAI system can explain to the intelligent system user why a particular network activity has been flagged as suspicious or anomalous and provide insights into the underlying patterns or trends that led to the detection. By providing interpretable insights, XAI can help security analysts to understand how the system is identifying and classifying network traffic and to make more informed decisions on how to respond to potential threats [25,26].

This study addresses the challenges of understanding network intrusion multi-class classification results on a highly imbalanced dataset (CIC-IDS2017 and CSECIC-IDS2018) using different machine learning (ML) models, such as Logistic Regression, Random Forest, Decision trees, Multilayer Perceptron and dense-layer network. The paper conducted a comparative analysis of the classification results of machine learning models using various accuracy metrics. Moreover, specific explainable AI (XAI) methods, including local and global explanation models, have been employed to evaluate the capabilities and robustness of XAI in identifying the crucial features within the dataset that significantly influence the classification results.

2. Related Works

The CIC-IDS2017 dataset is a well-known benchmark dataset for Intrusion Detection Systems (IDS). It contains network traffic data collected from real-world environments with different types of attacks and normal traffic. The initial dataset consisted of 79 features and 15 classes (one benign, 14 malicious). To enhance the classification performance of this dataset, numerous studies have been conducted, incorporating detailed analyses such as feature selection, class grouping, data cleaning, and processing. For this task, a wide range of classifiers have been employed, including classical methods, such as Random Forest RF [27,28] and MLP (multi-layer perceptron) [13,28] as well as deep learning architectures [29–33]. Table 1 presents the results of specific multi-class classification studies, displaying the *F1-score* and the corresponding number of classification classes. The majority of the results demonstrated an accuracy exceeding 90%, regardless of whether the classification involved 15 or fewer classes.

Table 1. Multi-classification accuracy results for the CIC-IDS2017 dataset.

Method	Classifier	*F1-Score*, (%)	Classes
Zachariah Pelletier [27]	ANN RF	96.53 96.24	13
Amer A.A. Alsameraee et al. [28]	RF MLP Naive Bayes (NB)	98.36 80.63 90.82	6
Mariama Mbow et al. [29]	LSTM 4 layers CNN	98.65 98.98	15
Hongpo Zhang [30]	SGM-CNN	96.36	15
Razan Abdulhammed et al. [31]	PCA+RF	99.60	15
Petros Toupas et al. [32]	8 hidden layers DenseNet	94.10	13
Yong Zhang [33]	Parallel cross convolutional neural network (PCCN)	99.68	12

Due to class imbalance, which is prevalent in the CIC-IDS2017 dataset and numerous other cybersecurity datasets, one class, specifically Benign, dominates the others, accounting for 80.3% of the raw data. To tackle this issue, various approaches have been employed, resulting in variations in the number of classes across different studies. Methods such as resampling [12], ensemble learning [34] or employing simple binary classification [9,34] have been frequently utilized in addressing class imbalance. One of the main issues encountered in the CIC-IDS2017 dataset is the incorrect aggregation of packets into streams, primarily caused by inadequate protocol detection. This problem has an impact on the number of classes or attributes within the dataset [35,36]. As a result, flows in the dataset may be too short or too long or may consist of packets from different conversations, resulting in inaccurate flow attributes. In addition, there are problems with TCP session termination in the dataset, so similar flows may have different labels. The presence of inaccurate flow attributes resulting from incorrect packet aggregation can introduce complexities that may confuse machine learning algorithms, consequently rendering the detection of network intrusions more challenging.

The CSE-CIC-IDS-2018 dataset contains a total of 28 classes of network traffic, each representing a different type of network activity. However, in classification tasks, these classes are often aggregated into 7 groups, namely "Benign", "DDoS", "DoS", "Brute Force", "Bot", "Infiltration", and "Web." Occasionally, the classes are collapsed into 10 groups, and the maximum number of classes observed for classification is 14, as depicted in Table 2.

Table 2. Multi-classification accuracy results for the CSECIC-IDS2018 dataset.

Method	Classifier	*F1-Score*, (%)	Classes
Jumabek, Alikhanov et al. [37]	DT(CART) RF CatBoost	87.36 88.13 88.84	13
L. Liu et al. [38]	DSSTE + miniVGGNet DSSTE + AlexNet DSSTE + LSTM	97.04 96.49 96.50	14
Jofrey L. Leevy [39]	CatBoost LightGBM DT LR NB RF XGBoost	91.65 94.69 88.58 49.47 24.31 92.94 93.64	7

Table 2. *Cont.*

Method	Classifier	*F1-Score*, (%)	Classes
Farhan, Baraa Ismael et al. [40]	LSTM	99.00	10
Ilhan Firat Kilincer et al. [41]	LGBM (Light GBM) XGBoost	99.94 99.92	7
Saud Alzughaibi et al. [42]	MLP-BP (Multi-layer perceptron + backpropagation) MLP-PSO (multi-layer perceptron + particle swarm optimization)	99.20 97.60	7

It has been observed that hybrid deep learning architectures tend to outperform other machine learning algorithms in terms of accuracy results. All these studies have shown that the majority of *F1-scores* are above 95% for classifications of up to 14 classes. In addition, such common metrics as recall, precision, ACC (Accuracy) and *F1-score* are provided, but the weighted average *F1-score* is a more useful metric in this case, as it takes into account the relative importance of each class in terms of its support, which is particularly important for unbalanced datasets. Macro averaging is another commonly used technique for evaluating performance metrics in highly imbalanced datasets; however, none of these metrics were provided in the studies.

Many studies have reported promising results using deep learning architectures on this dataset, classifying network traffic as either benign or malicious (as a binary classification task) [9,11,43]. However, this does not mean that increasing the number of classes will always result in decreased accuracy, and such trends have not been observed.

Observations have indicated that the impact of model architecture or hyperparameters on accuracy is minimal [9,44]. Instead, factors such as training and testing data, data processing and preparation, the number of sampled features and other similar considerations have a more significant influence. Nevertheless, there remains uncertainty regarding whether more complex hybrid models actually yield more accurate results. Additionally, questions persist regarding the rationale and benefits of merging classes, conducting extensive data cleaning, or reducing features. Thus, in this study, experiments were carried out separately on the raw CIC-IDS2017 and CSE-CIC-IDS-2018 datasets, i.e., keeping the initial number of attack types (the number of classes) at 15 and 28, respectively. For further investigation, these datasets were merged into a single dataset, which resulted in 19,063,686 instances, 79 features (initially 85) and 28 classes.

3. Dataset

The CIC-IDS2017 and CSE-CIC-IDS-2018 (an extension to the CIC-IDS-2017 dataset with more network traffic data) datasets have become popular in the research community, particularly in the field of IDSs. The CIC-IDS2017 dataset was generated from real network recordings.

The data collection period started on Monday 3 July 2017 at 9.00 a.m. and ended on Friday 7 July 2017 at 5.00 p.m. for a total of 5 days. Each dataset record contains 79 parameters, the last of which is an output with 15 different names (classes) indicating to which type of malicious activity, if any, the network packet described in the dataset record belongs. All possible output values and their proportions in the overall dataset are shown in Figure 1.

The CSE-CIC-IDS-2018 dataset is larger than the CIC-IDS2017 dataset in terms of the number of flows it contains, with more than 80 million flows compared to about 3 million flows in the CIC-IDS2017 dataset. This dataset includes not only the attack types identified in the CIC-IDS2017 dataset, but also several new attack types commonly found in web applications, such as SQL injection, cross-site scripting (XSS) and command injection. These attacks can be particularly damaging because they target vulnerabilities in the application itself, allowing attackers to gain unauthorized access to sensitive data or to take control of the system. SQL injection attacks and XSS attacks are two common types of web application attacks that can be very harmful. SQL injection attacks occur when an attacker is able to inject malicious SQL code into a web application's input fields. If the application does not properly validate or sanitize the user input, the malicious code can be executed by the database, allowing the attacker to access, modify or delete sensitive data. XSS attacks, on the other hand, involve injecting malicious scripts into a web application's pages. This can allow attackers to steal sensitive information, such as cookies or login credentials, or to gain control of the user's browser, redirecting them to other malicious sites or launching further attacks.

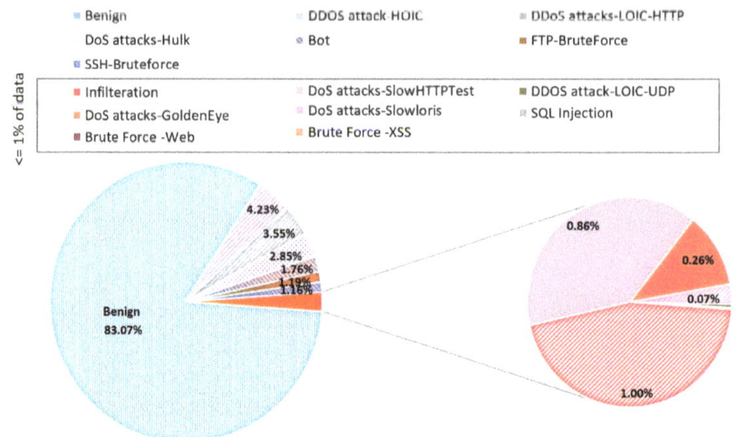

Figure 1. Distribution of output class in the CIC-IDS2017 dataset.

Regarding the output imbalance problem, we can see that the situation is very similar to the 2017 dataset (the dominant class accounts for 82.66% of the dataset), so it would make sense to merge them (see Figure 2). By merging the datasets, the total number of examples in the minority class would increase. Additionally, by keeping only the entries that have a certain number of entries, it would be possible to remove any outliers or rare classes that may not have enough examples to train the classification model effectively.

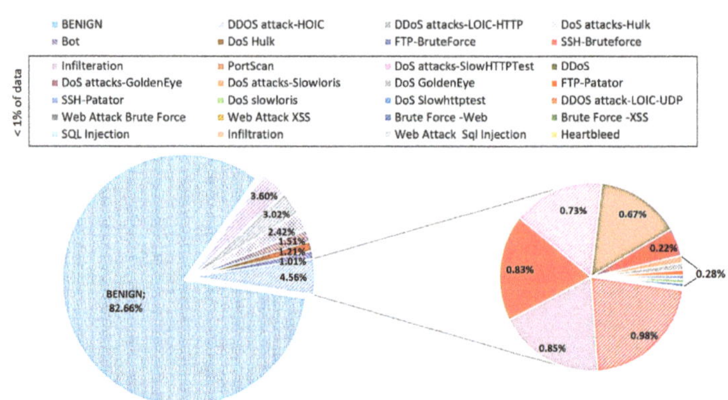

Figure 2. Distribution of output types in the CSE-CIC-IDS-2018 dataset.

4. Materials and Methods

4.1. Logistic Regression

Logistic regression is a binary classification algorithm used to predict a binary outcome (i.e., 0 or 1). For multi-class classification tasks, it is better to use an extension of logistic regression, such as Softmax regression, One-vs-Rest (OvR) or One-vs-One (OvO). In this study, we used Softmax regression, also known as multinomial logistic regression. This approach produces a single model that directly models the probability distribution of all

K classes. The model learns a set of K linear functions, one for each class, and uses a softmax function to normalize the output to the probability distribution of all K classes. The class with the highest probability was chosen as the output during the prediction. The softmax function for class k is defined as follows:

$$(y = k|x) = \frac{e^{f_k(x)}}{\sum_{j=1}^{K} e^{f_j(x)}} \text{ for } k = 1, 2, \ldots, K \quad (1)$$

where $f_k(x)$ is a score of class k for input x. The score can be defined as a linear function.

During training, the parameters of the model (i.e., the weights and biases of the linear functions) are learned using an optimization algorithm that minimizes a loss function. The most commonly used loss function for softmax regression is cross-entropy loss. The cross-entropy loss for a single example (x, y) is defined as follows:

$$L(x, y) = -\sum_{k=1}^{K} y_k \cdot logP(y = k|x) \quad (2)$$

where y_k is the indicator function, taking the value 1 if the true label is k and 0 otherwise. The class with the highest probability was chosen as the output for the prediction:

$$\hat{y} = \underset{k}{\text{arg}max} \ P(y = k|x) \quad (3)$$

where $\hat{y}$ is the predicted class label.

4.2. Decision Trees and Random Forest

Decision trees (DTs) can be used to solve both classification and regression problems and are particularly useful when the data have complex non-linear relationships. DT can be a powerful tool for building intrusion detection technologies, such as firewalls and IDS [45–47]. However, DT can be prone to overfitting and can have high variance, which can be addressed by techniques such as pruning or the ensemble method. In this research, we have implemented two DT algorithms—ID3 and CART—and a random forest composed from a set of CART-types trees.

The ID3 (Iterative Dichotomiser 3) algorithm uses Information Gain (IG) to find the best feature to split the data at each node of the decision tree. Information Gain is a measure of the reduction in entropy that can be achieved by splitting the data on a particular feature. The feature with the highest IG was chosen as the split feature at each node, as it is expected to provide the most useful discrimination between the different classes.

$$Entropy(S) = -\sum_{i=1}^{n} p_i \cdot log_2(p_i), \quad (4)$$

where S represents the dataset that entropy is calculated, i represents the classes in set, and p_i represents the proportion of data points that belong to class i to the number of total data points in set S. The Information Gain formula is provided below:

$$IG(S, A) = Entropy(S) - \sum_{v \in Values(A)} \frac{|S_v|}{|S|} Entropy(S_v), \quad (5)$$

where S_v is the set of rows in S for which the feature column A has value v, $|S_v|$ is the number of rows in S_v and likewise $|S|$ is the number of rows in S.

CART (Classification and Regression Trees) is another DT algorithm that can handle both categorical and continuous input variables and is less prone to overfitting than ID3. CART is used for binary splitting, which involves splitting the data into two groups based on the value of a single input variable. To perform binary splitting, the algorithm first evaluates the *Gini* impurity index:

$$Gini = 1 - \sum_{i=1}^{C} (p_i)^2, \quad (6)$$

where C is the total number of classes, and p_i probability of selecting a data point with class i. The *Gini* impurity of a pure node (the same class) is equal to zero.

A random forest (RF) is a decision tree-based algorithm that uses an ensemble of decision trees to make predictions [48]. Moreover, it is a very popular technique for network intrusion detection [46,49]. In our case, employing RF, several CART-type decision trees were created using randomly selected subsets of training data and features. Each tree in the forest made a prediction based on the values of the features and provided a class label. The final prediction of the random forest was the dominant class.

$$RF = f(DT_1, DT_2, \ldots, DT_n) \tag{7}$$

where $f = mode$, n-number of tress DT in random forest. The *mode* function returns the most frequently occurring predicted class across all trees.

Theoretically, the *RF* algorithm has several advantages over a single decision tree in that it can handle high-dimensional, less balanced datasets with many attributes and can reduce overfitting. For this purpose, regularization, bagging and boosting techniques have been applied, as well as pre-pruning methods, which involve setting conditions for the growth of decision trees during the construction of the forest.

4.3. ANN-Type Models

In this study, we implemented two ANN-type models: (1) a multi-layer perceptron (MLP) and (2) a deep learning architecture—the dense-layer network.

Multilayer perceptron (MLP) is a type of ANN composed of neurons that are interconnected and function as individual information processing units, which allows it to learn complex functional relationships between input features and the output [50]. We have implemented MLP with 4 hidden layers, with 64 nodes in the first layer, 128 nodes in the second and third layers, and 64 nodes in the fourth layer. The strength of the L2 regularization term is set to 0.001, which means that the model provides a moderate amount of regularization to the weights of the model. The Rectified Linear Unit (ReLU) function is used as the activation function in the hidden layers. To prevent overfitting, early stopping is enabled and the strength of the L2 regularization term is set to 0.01.

A dense-layer network, also known as a fully connected network, is a type of neural network architecture commonly used in deep learning. In a dense network, each neuron in a given layer is connected to all the neurons in the previous layer, resulting in a dense, fully connected graph of nodes [51]. A neural network with 23 dense layers was used as a deep learning model, incorporating batch normalization, a ReLu activation function and an Add operation with a total of 2,714,780 parameters. The structure of the model architecture is provided in Figure 3.

4.4. Explainable Artificial Intelligence Methods

Machine learning algorithms are increasingly being used in various applications of cybersecurity, such as malware detection, intrusion detection and vulnerability assessment. However, the black-box nature of many machine learning models can make it difficult to understand why they make certain predictions or decisions. XAI technologies are very relevant in the field of cybersecurity, where the consequences of errors or biases in machine learning models can be severe.

Before applying XAI methods, it is important to understand their scope, the stage of system development where XAI can be used and what it can explain [52]. Two different XAI models can be distinguished according to the scope of the explanation: (1) Local; (2) Global models. At the local level, explainability focuses on providing explanations for individual predictions or decisions made by an AI model. Local explainability methods help understand the factors or features that influence a specific outcome. Meanwhile, global explainability aims to explain the overall behavior and decision-making process of an AI model across its entire input space. It focuses on providing insights into the model's general characteristics, biases and patterns.

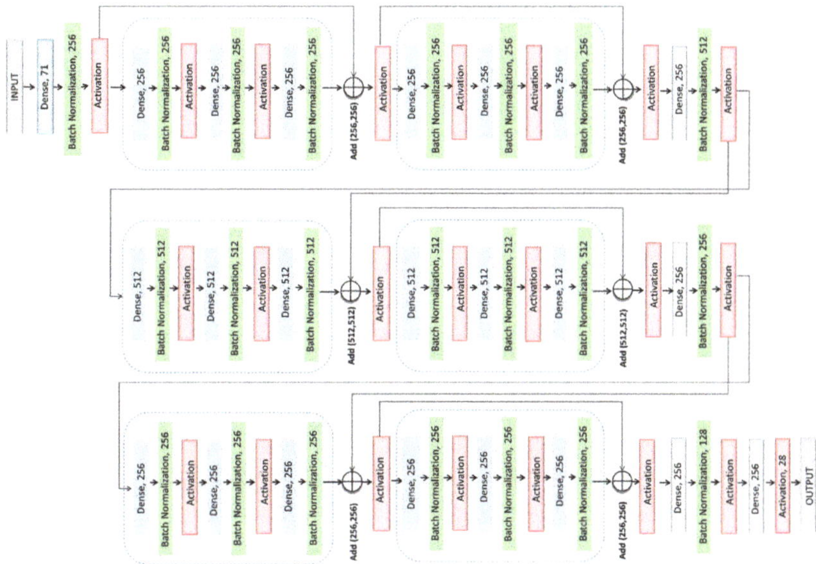

Figure 3. The architecture of dense-layer deep neural networks used for intrusion classification tasks.

Next, it is important to assess at which stage an explanation of the ML model is relevant. Pre-model explainability is the process of explaining an AI model before it is implemented or trained. It focuses on the development and selection of models or algorithms that are intrinsically explainable, such as rule-based algorithms, decision trees or linear models that are transparent from the outset. Post-model explainability involves explaining an already trained or deployed AI model. These methods aim to provide insights into the decision-making process of black-box models, such as deep neural networks or complex machine learning algorithms. Such methods often rely on techniques such as feature importance analysis, model-agnostic explanations or surrogate models.

Feature importance analysis assesses the importance or relevance of individual features or variables in the model's decision-making process. They aim to identify the features that have the most significant impact on the model's predictions. Techniques such as permutation importance, partial dependence plots or SHAP (Shapley Additive Explanations) values can be employed for feature importance analysis. Model-agnostic approaches, such as LIME (Local Interpretable Model-Agnostic Explanations), provide explanations for black-box models by approximating their decision boundaries using surrogate models. These surrogate models are more interpretable, allowing for insights into the behavior of the black-box model.

In this study, we implemented two different models to assess the explainability of our dataset and its benefits and drawbacks, including the feasibility of using the XAI method and the robustness of the interpretation:

- LIME is an example of a Local Explanation Model (model-agnostic approach), which provides simplified, interpretable models that explain the behavior of complex models for individual instances, enabling local explanations that shed light on the factors influencing specific predictions.
- SHAP is a method that provides both global and local explanations for machine learning models. SHAP can provide global explanations, summarizing the overall importance of the attributes, while local explanations reveal the contribution of each attribute to individual predictions in the context of model behavior.

4.5. Accuracy Evaluation Metrics

Different accuracy measures have been calculated to evaluate experimental results, such as Root Mean Square Error (*RMSE*), Mean Absolute Percentage Error (*MAPE*) and *F*1-*score*.

RMSE is simply the square root of the mean square error, with the only difference being that *MSE* measures the variance of the residuals, while *RMSE* measures the standard deviation of the residuals:

$$RMSE = \sqrt{MSE}, \text{ where } MSE = \frac{1}{n}\sum_{t=1}^{n}|y_t - \hat{y}_t|^2 \quad (8)$$

where n–the number of time point, y_t–is the actual value at a given observation in a dataset t, and $\hat{y}_t$– is the predicted value.

MAPE is the most commonly used metric for evaluating the accuracy of a prediction model. It calculates the average percentage difference between the actual and predicted values of a variable:

$$MAPE = \frac{100\%}{n}\sum_{t=1}^{n}\left|\frac{y_t - \hat{y}_t}{y_t}\right| \quad (9)$$

The *F*1-*score* is a widely used metric for evaluating the performance of a classification model, especially in scenarios in which we want to balance both precision and recall.

$$Precision = \frac{TP}{TP + FP} \quad (10)$$

$$Recall = \frac{TP}{TP + FN} \quad (11)$$

$$F1\ score = \frac{2 \times Precision \times Recall}{Precision + Recall} \quad (12)$$

For multi-class classification, the *F*1-*score* for each class is calculated using the one-against-one (OvR) method. In this approach, the metrics for each class are determined separately, as if a separate classifier were used for each class. However, instead of assigning several *F*1-*scores* to each class, it is more appropriate to derive an average and obtain a single value to describe the overall performance. There are three types of averaging methods commonly used for *F*1-*score* calculation in multi-class classification, but weighted averaging is most relevant for such unbalanced data. Weighted averaging calculates the *F*1-*score* for each class separately and then takes the weighted average of these scores, where the weight for each class is proportional to the number of samples in that class. In this case, the *F*1-*score* result is biased toward the larger classes.

$$Weighted_{avg}F1\ Score = \sum_{i=1}^{n} w_i \times F1\ Score_i \quad (13)$$

$$w_i = \frac{k_i}{N} \quad (14)$$

where N–total number of samples, number of samples k_i in class i.

The macro average calculates the *F*1-*score* for each class separately and derives an unweighted average of these scores. This means that each class is treated equally, regardless of the number of samples it contains (support value):

$$Macro_{avg}F1\ Score = \frac{\sum_{i=1}^{n} F1\ Score_i}{n} \quad (15)$$

where n–number of classes.

5. Results

The results in Figure 4 show the weighted average *F1-scores* of six different prediction algorithms trained on three datasets. The results showed that the decision tree algorithms (including RF) were the most accurate and their results were quite similar to all datasets. The CART decision tree algorithm was the most accurate, achieving 99.22% of the weighted average *F1-score* over the 3 datasets. However, this accuracy result was only ~0.6% better than with RF or another decision tree model-ID3. The worst classification results were shown by the dense-layer net and LR models. In terms of datasets, the CSE-CIC-IDS-2018 dataset (88.81) gave the worst results and the CIC-IDC2017 dataset the best (91.48%). Meanwhile, the merged dataset achieved an average accuracy of 90.04%.

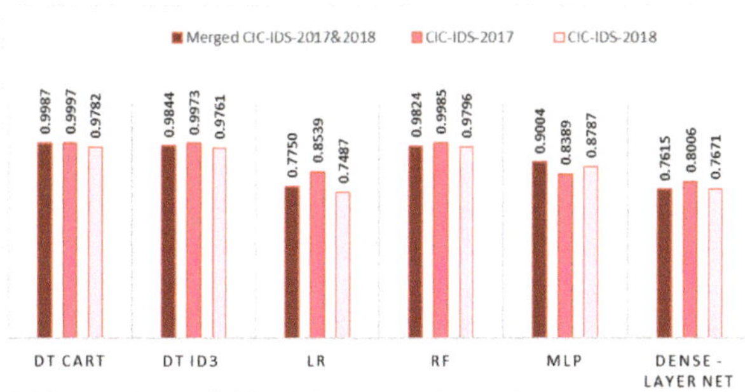

Figure 4. Weighted average *F1-score* values.

To understand these performance results, the experimental results for each algorithm (including different accuracy metrics) are provided below.

Table 3 shows classification results of six different ML models providing precision, recall and *F1-score* values, including two performance metrics. Taking into account all 3 performance metrics, it was evident that the DT (CART)-based model demonstrated the highest level of accuracy, with *F1-score* values ranging from 96.87% to 99.87%.

Table 3. Classification results of ML algorithms on a merged dataset.

Model	Performance Metrics	*Precision*	*Recall*	*F1-Score*
LR	Accuracy	0.83741	0.83741	0.83741
	Weighted average	0.72454	0.83741	0.77504
	Macro average	0.14042	0.14438	0.13566
DT(CART)	Accuracy	0.99874	0.99874	0.99874
	Weighted average	0.99874	0.99874	0.99874
	Macro average	0.97305	0.96603	0.96878
DR(ID3)	Accuracy	0.98551	0.98552	0.98552
	Weighted average	0.98327	0.98552	0.98436
	Macro average	0.74171	0.70811	0.71516
RF	Accuracy	0.98489	0.98489	0.98489
	Weighted average	0.98106	0.98489	0.98237
	Macro average	0.88152	0.81499	0.83882
MLP	Accuracy	0.93029	0.93029	0.93029
	Weighted average	0.87388	0.93018	0.90038
	Macro average	0.27089	0.27662	0.27127
Dense-layer net	Accuracy	0.83276	0.83276	0.83276
	Weighted average	0.72457	0.83276	0.76153
	Macro average	0.15026	0.05313	0.06269

The *F1-score* accuracy results for each algorithm for each class of the dataset are presented below (see Figure 5). The figure shows that the LR model failed to detect such intrusion classes as "Bot", "Heartbleed", "Infiltration", "PortScan", "SSH-Patator" and "FTP -Patator". *F1-score* values greater than 0 were obtained for such classes as "Dos slowloris" (0.18%), "DoS Attack-slowloris" (40.73%) and "DDoS" (54.25%), although these scores are not high. Higher values were obtained for only two classes—"Benign" (91.62%) and "DDOS attack-LOIC-UDP" (70.53%) (see Figure 5a). The DT-based algorithms achieved quite high scores, but the most accurate was the DT(CART) algorithm, which reached high *F1-score* results (>90%) for the majority of classes (see Figure 5b). Only three attack classes had lower results: "Infiltration" (68.96%), "SQL Injection" (77.08%) and "Web Attack Sql Injection" (85.71%). The ID3 algorithm did not identify classes, such as "Web Attack Sql Injection", "Heartbleed" and "SQL Injection", but attacks, such as "SSH-Bruteforce", "FTP-BruteForce" and "DoS attacks-SlowHTTPTest", were identified with the highest accuracy >99.9% (see Figure 5c).

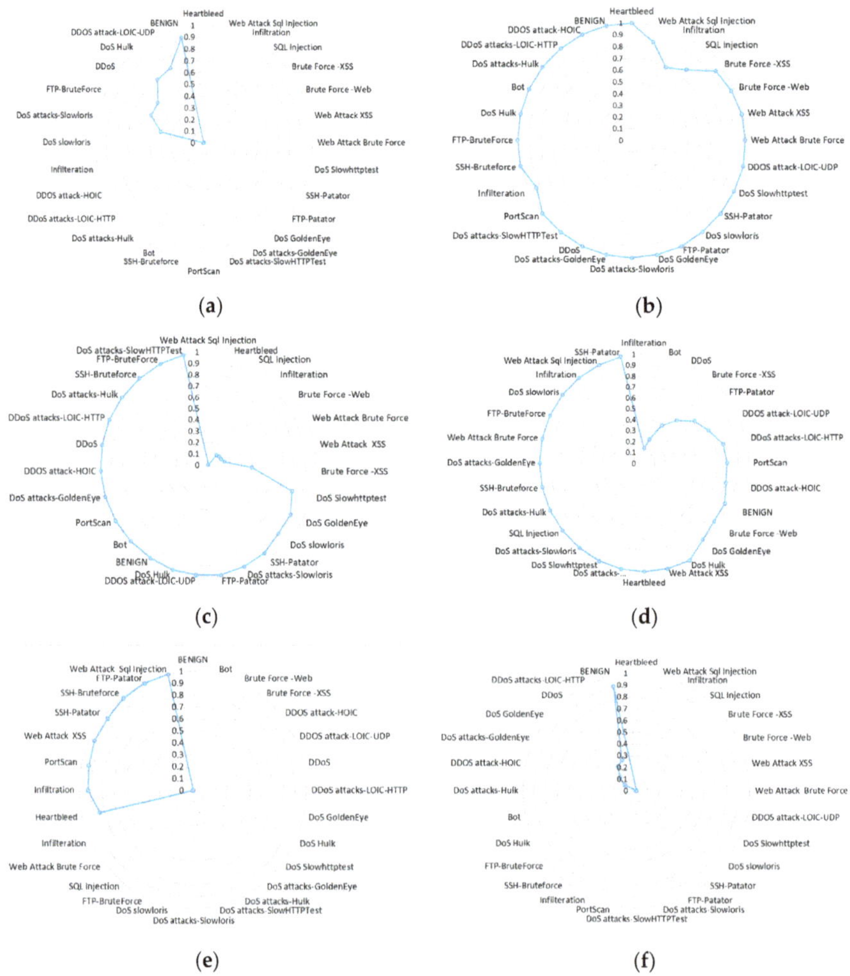

Figure 5. Radar charts for comparing classification models in terms of *F1-score* accuracy for all intrusion classes. (**a**) LR; (**b**) DT(CART); (**c**) DT (ID3); (**d**) RF; (**e**) MLP; (**f**) Dense-layer net.

Concerning the results obtained from RF, the majority of classes were successfully classified with an accuracy rate exceeding 99%. Meanwhile, prediction success rates of "Web Attack-Brute Force", "Web Attack-Sql Injection" and "Web Attack-XSS" data classes were significantly lower, ranging from 33.3% to 77.3%. Among the different data classes, the RF algorithm achieved the lowest score for "Infiltration" attacks (13.45%) and the highest score (99.99%) for "SSH-Patator" attacks, as illustrated in Figure 5d.

The classification results with MLP showed (Figure 5e) that the algorithm identified certain classes of attacks very well, i.e., "SSH-Bruteforce", "FTP-Patator" and "Web Attack Sql Injection", but did not identify others at all (the *F*1-*score* value was 0.). The worst results were obtained with the dense-layer net model, which only classified the "Benign" class correctly with 90.81% accuracy. Most intrusion attack classes had a classification accuracy of 0% value, and only a few had an accuracy in the range of 10–20% ("DDoS" type attacks).

As the dataset was heavily imbalanced, it was important to verify how accuracy correlated with sample size. The testing sample size for each of the classes is given in Figure 6.

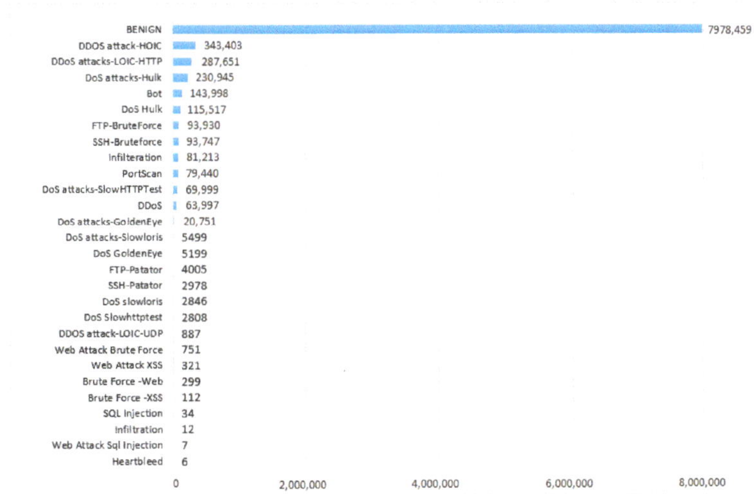

Figure 6. Average support values (number of testing samples) for testing.

Support value dependencies on *F*1-*score* accuracy are provided in Figure 7. As the difference in support values was very significant, it was appropriate to consider a logarithmic scale for the support values. Near-linear dependencies were observed in the DT CART and ID3 models. However, the DT ID3 and RF models had more outliers. For example, the "Infilteration" class with a high support value only achieved 13.45% of *F*1-*scores* for the RF model and 11.09% for the DT ID3 model (see Figure 7c,d). Despite the low overall accuracy of the MLP model, the results presented in Figure 7e show that higher *F*1-*scores* were only obtained at higher values of support.

Figure 8 shows misclassification results for all six models, where it was important to evaluate the misclassification of malicious attacks as benign because this can lead to potential security risks and threats going undetected. Identifying and correctly classifying malicious attacks is important, but confusing certain malicious classes may be less critical than classifying them as benign. Figure 8 shows only malicious attacks, which are shown in red if the attack was correctly classified, in green if it was identified as benign and in gray if it was classified as malicious but the attack class was incorrectly assigned. The percentage values represented all of these cases for each attack class.

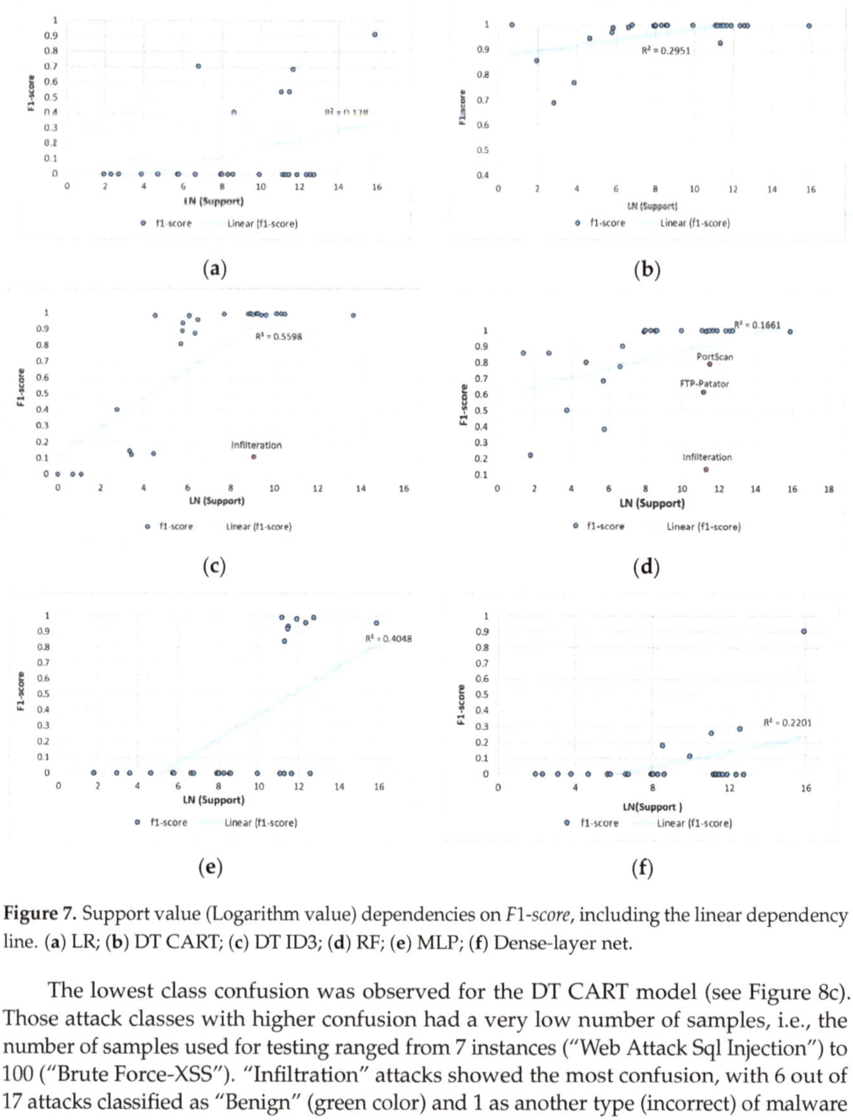

Figure 7. Support value (Logarithm value) dependencies on *F1-score*, including the linear dependency line. (**a**) LR; (**b**) DT CART; (**c**) DT ID3; (**d**) RF; (**e**) MLP; (**f**) Dense-layer net.

The lowest class confusion was observed for the DT CART model (see Figure 8c). Those attack classes with higher confusion had a very low number of samples, i.e., the number of samples used for testing ranged from 7 instances ("Web Attack Sql Injection") to 100 ("Brute Force-XSS"). "Infiltration" attacks showed the most confusion, with 6 out of 17 attacks classified as "Benign" (green color) and 1 as another type (incorrect) of malware attack (gray color).

Comparing the DT ID3 algorithm and RF, we can see that the results were more satisfactory with RF because they classified less malicious attacks as benign. The worst situation was for "Infilteration" type attacks, as most of them were classified as benign (almost 90%) (see Figure 8d). The situation with regard to the classification of "Infilteration" attacks was the same as for the DT ID3 model (Figure 8b). In addition, more than 60% of the sample for the other seven types of attacks were also classified as benign. For the LR model (Figure 8a), we can see that the majority of malicious attacks were classified as benign ("Bot", "Brute Force-Web", "Heartbleed" "DoS GoldenEye", etc.). The most inappropriate model was the dense-layer net, which classified almost all malicious attacks as benign. The MLP model identifyied at least seven types of malicious attacks with high accuracy, but the rest were classified as benign attacks (see Figure 8e).

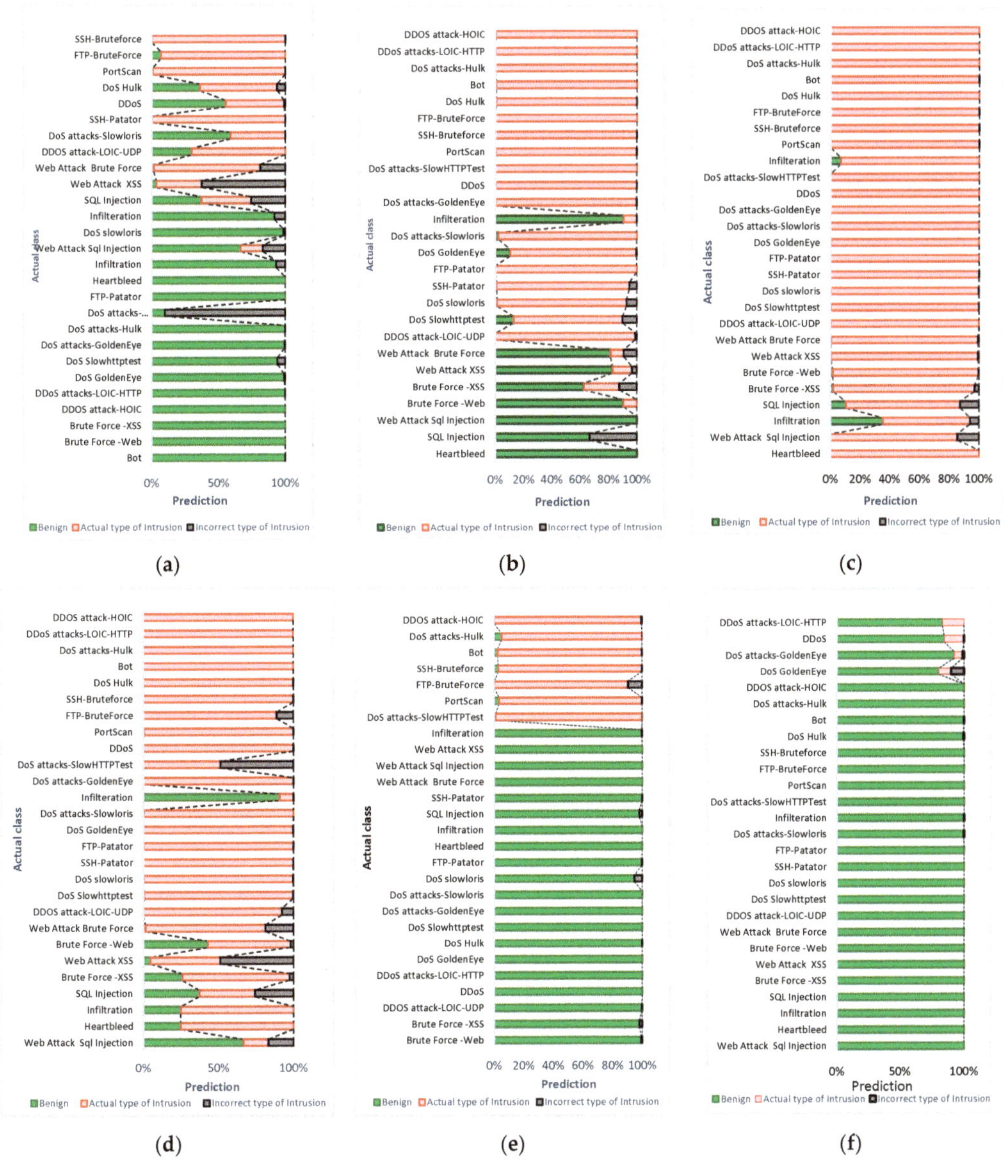

Figure 8. Misclassification results of different types of malicious attacks. (**a**) LR; (**b**) DT CART; (**c**) DT ID3; (**d**) RF; (**e**) MLP; (**f**) Dense-layer net.

6. XAI-Based Explanations

Experiments with two XAI methods—LIME and SHAP—were carried out to investigate their ability to explain classification results with more complex models, such as MLP. First, two multilayer perceptron models were created and trained on different datasets. The first model was trained with our joined CIC-IDS2017/-2018 dataset. The second model was trained with the same dataset, only with an addition of "unrelated_column" values, which were set randomly.

Both models' explanations were first generated with LIME. For explanations, input values were selected that generated model predictions, indicating a network anomaly. In this case, inputs passed to both models indicated a network anomaly—"DDoS" attack.

Figure 9 shows which attributes had the greatest impact on the prediction, where the green columns indicate that they have a positive effect, i.e., they increase the model's score, while the red columns decrease the score. From the data, it was observed that not only were the features that had the biggest impact on the model output different, but in "unrelated_column", they had the most importance, although its values were completely random. However, in both cases, this instance was classified as a benign attack for the original dataset (see Figure 9a) with 83% probability and for the modified dataset (see Figure 9b) with 90% probability). Further explanations of both models were developed using the SHAP approach.

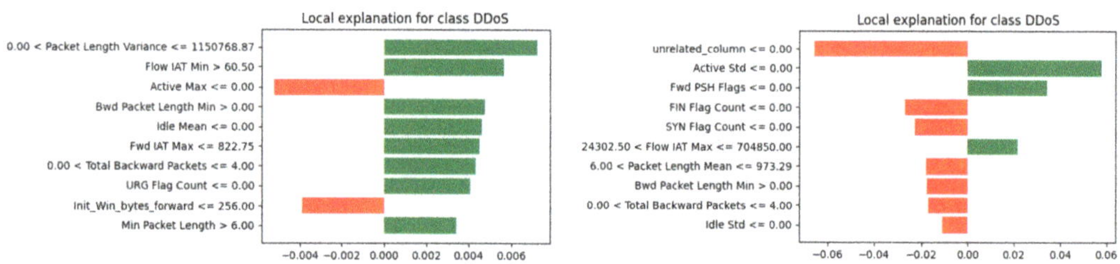

(a) (b)

Figure 9. LIME local explanations. (**a**) Explanation with original dataset; (**b**) Explanation with added "unrelated_column".

In this scenario, a sample instance of data had been submitted, the output of which was a "Benign" class. The data presented in Figure 10b shows that the results are similar to those of the LIME models, considering that "unrelated_column" had the largest impact on the decision.

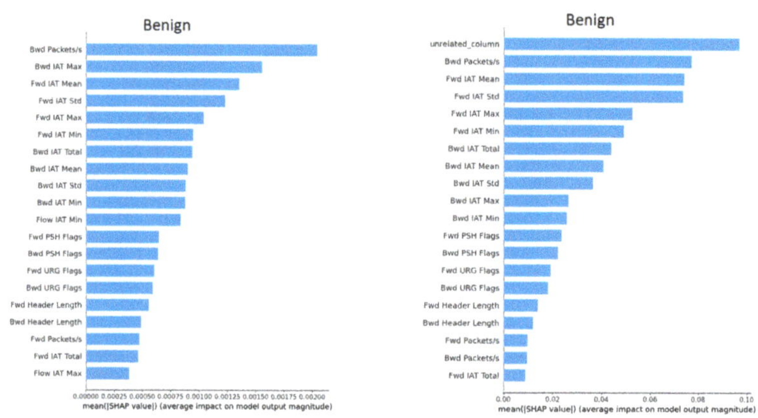

(a) (b)

Figure 10. SHAP local explanations. (**a**) Model explanation with original dataset; (**b**) Model explanation with added "unrelated_column" random values.

When comparing the LIME and SHAP local explanation cases, almost every time the LIME explanations were generated, a different result was obtained, whereas SHAP was more stable in assessing the influence of attributes.

While performing the aforementioned experiments, certain drawbacks were observed of both LIME and SHAP explanation methods. For instance, LIME implementation is limited to providing explanations for individual instances within the dataset and lacks the ability to generate global explanations like SHAP. This means that in order to obtain a view of dataset-wide scope, the user has to generate explanations of every single instance of the dataset. Due to this issue, the process of analyzing and interpreting results can become complex, especially when working with datasets that contain over 19 million data entries, as is the case in our situation. Even in the case when a user is interested in only one data instance of the dataset, additional interpretability problems arise when a multiclass model is used. In this research, an MLP model was used to generate the LIME explanations, which resulted in separate explanations for all 28 classes, taking into account the influence of 79 features. The most interesting results for the eight classes are shown in Figure 11.

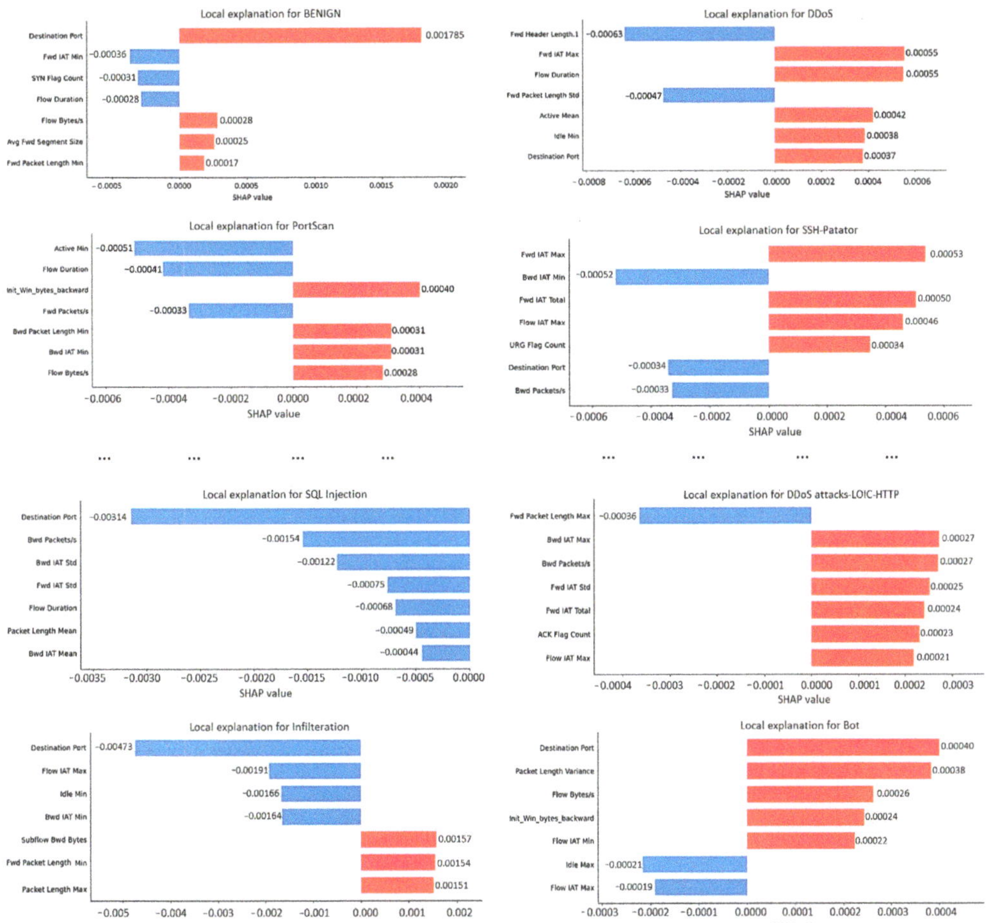

Figure 11. LIME local explanations for all eight separate classes in the dataset: red for positive impacts, blue for negative impacts.

Meanwhile, SHAP can provide global explanations, and while this enables us to see explanations for the whole dataset, explanations can still be hard to interpret when using multiclass output models. Figure 12, provides global explanations for 28 classes, providing an importance score for the most dominant features. Although this allows for a detailed visualization of the explanation, the amount of data might be too confusing for the non-specialist end user.

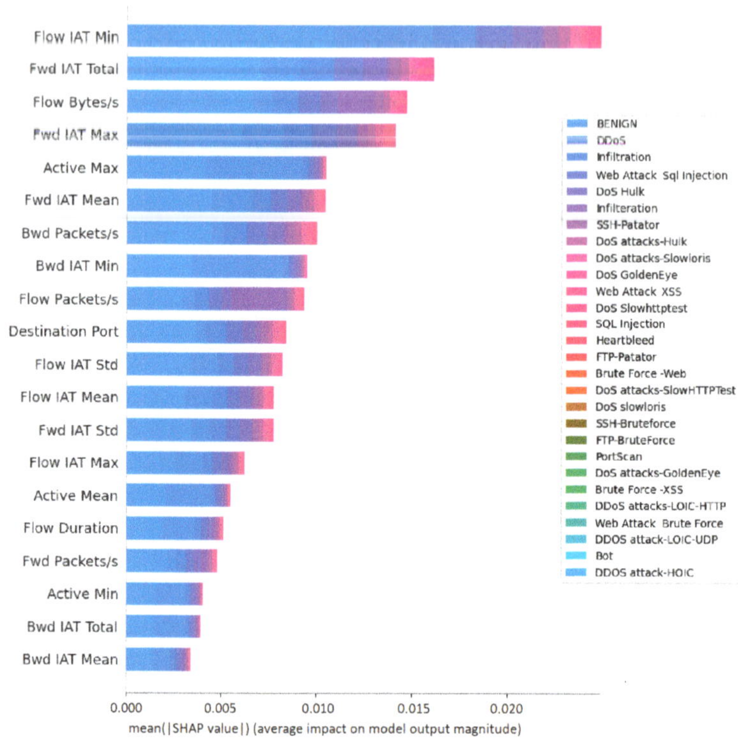

Figure 12. SHAP global explanations for all 28 classes in the dataset.

7. Discussion

We conducted several experiments, varying the hyperparameters, to improve the accuracy of the best models in our investigation study. Regarding the CART tree, in the initial phase, the whole tree was generated to estimate its maximum size and performance. Subsequently, we carried out experiments by manipulating parameters, such as the maximum depth of the tree, the minimum number of samples required for a split, and the criteria (Gini impurity or entropy) used for splitting. During the initial stage, the CART tree achieved a maximum depth of 51 while utilizing a minimum of 2 samples as the criterion for node splitting and employing the "gini" criterion. To optimize the model's performance, we applied pre-pruning techniques and limited the depth to 27. The decision was made to stop the growth of the tree at level 27 due to the observation that beyond this depth, there was confusion between the "BENIGN" class and the "Infiltration" class. Figure 8c provides further evidence of the significant confusion observed between these classes.

The ID3 algorithm, another DT algorithm employed in this study, initially generated a maximum depth of 48. Through experimentation, we progressively reduced the depth to 32, 25 and ultimately 19. Notably, a depth of 19 yielded the most accurate results. This stopping of the tree growth was appropriate because only the "BENIGN" and "Infilteration" classes were continuously mixed in the lower layers of the tree, which resulted in lower accuracy of the latter class (confusion with "BENIGN" reached 89.57%). Stopping the growth of the tree at a depth of 19 led to a minor enhancement in the accuracy, specifically for the "BENIGN" class, which already achieved high accuracy. However, the overall increase in the model's accuracy remained relatively modest (1.04%). In addition, it was observed that splitting leaves with fewer than 6 elements was not very appropriate. However, considering the limited number of instances within certain specific classes of the dataset, such as "Heartbleed", "SQL Injection" and "Web Attack SQL Injection", the minimum number of instances required for splitting was adjusted to 3.

For the RF algorithm, we employed entropy criteria, which provided better results than "gini" in our case. We set the number of features to consider when searching for the optimal split as the square root of the total number of features (sqrt) provided in the dataset. The bootstrap method was utilized in our RF model; hence, out-of-bag samples were employed for estimating the generalization score. The number of trees in the forest was set to 100.

The observation reveals that conducting more comprehensive studies encompassing a wider range of hyperparameters would yield substantial benefits. Hence, we have intentions to persist with such studies in the future, as the results and insights will be crucial for the development of new XAI models.

Moreover, in this study, we carried out experiments using additional machine learning models, but only those models with the most promising practical results were selected for deeper analysis. As our future research goal is to develop robust and stable XAI models, it was necessary to test models with different levels of explainability (or interpretability) (see Figure 13). The accuracy results of the KNN model and LSTM are provided in Table 4. Throughout our experimentation, we explored different deep learning models, including VGG-19, AlexNet and several others. However, it was the LSTM model that gave the most accurate results. The main reason why these models were not included in the more detailed ones is the low average macro-level *F1-score* (<0.7).

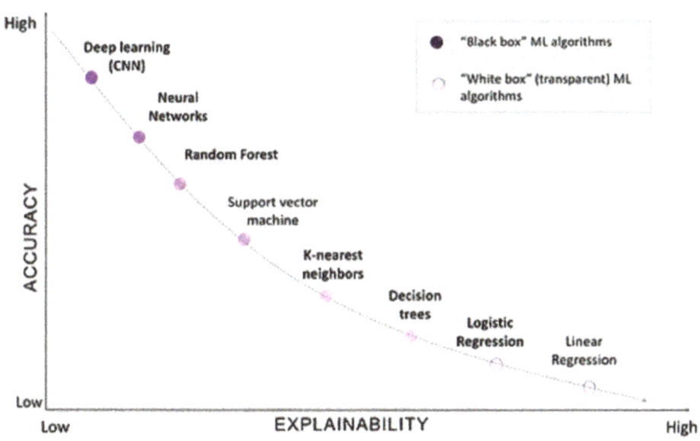

Figure 13. Graphical representation of the trade-off between the accuracy and interpretability of ML algorithms (groups of ML algorithms used in the study are presented in bold).

Table 4. Classification results of additional models.

Model	Performance Metrics	*Precision*	*Recall*	*F1-Score*
KNN	Accuracy	0.99399	0.99399	0.99399
	Weighted average	0.99367	0.99399	0.99342
	Macro average	0.81778	0.78565	0.79845
LSTM	Accuracy	0.89023	0.89023	0.89023
	Weighted average	0.83573	0.89029	0.85358
	Macro average	0.21564	0.19212	0.18313

KNN has fast learning capabilities; however, when it comes to making a decision, such as predicting a class based on new data, it takes much longer compared to the other machine learning algorithms implemented in our study. Therefore, we eliminated KNN as a real-time and practical algorithm, even though the accuracy rates obtained were actually very high (see Table 4), and it was the second best algorithm after CART.

8. Conclusions

This study addressed the challenges of understanding the results of multi-class classification of network intrusions in highly imbalanced data, including the CIC-IDS2017 and CSE-CIC-IDS-2018 datasets. In the research, machine learning models of different complexity and explainability were included in order to evaluate and understand whether more complex ML models were indeed capable of providing higher classification results for the classification of 28 classes of intrusions. The study compared the classification performance of six machine learning models using various measures of classification accuracy. The results revealed that the DT model utilizing the CART algorithm achieved the highest performance in the multi-class classification task, achieving an *F1-score* of 0.998 and an average macro *F1-score* of 0.969. The lowest results were obtained with the dense-layer network, with an *F1-score* of 0.833 and an average macro *F1-score* of 0.063. Another decision tree algorithm, namely ID3, along with the RF model, resulted in slightly lower but still significant results, achieving an *F1-score* of 0.985.

In addition, experiments were carried out with the XAI methods, LIME and SHAP, to assess the potential and reliability of identifying the most important features of the dataset. Although these methods provide a list of the most influential features, it has been observed that the local explanation is often unstable, and each regeneration may lead to a completely different result. Even a specially added column that has no relevance to the problem may have the greatest influence on the decision. Another observation is that a global interpretation of all classes is not very explicit, and it is quite difficult to understand the visualization. Therefore, it is likely that when we have more than multi-class and multi-feature datasets, it would be more useful to use numerical or aggregated results of explanations.

Author Contributions: Conceptualization, A.P.-T.; data curation: M.B. and A.P.-T.; investigation, M.B. and A.P.-T.; methodology, M.B. and A.P.-T.; software, M.B. and A.P.-T.; resources M.B. and A.P.-T.; validation M.B.; writing—original draft, M.B. and A.P.-T.; writing—review and editing, M.B. and A.P.-T.; supervision, A.P.-T. All authors have read and agreed to the published version of the manuscript.

Funding: This research received no external funding.

Data Availability Statement: Data is available upon reasonable request to the corresponding author.

Conflicts of Interest: The authors declare no conflict of interest.

References

1. Ozkan-Okay, M.; Samet, R.; Aslan, O.; Gupta, D. A Comprehensive Systematic Literature Review on Intrusion Detection Systems. *IEEE Access* **2021**, *9*, 157727–157760. [CrossRef]
2. Li, Y.; Liu, Q. A comprehensive review study of cyber-attacks and cyber security; Emerging trends and recent developments. *Energy Rep.* **2021**, *7*, 8176–8186. [CrossRef]
3. Jin, S.; Chung, J.-G.; Xu, Y. Signature-Based Intrusion Detection System (IDS) for In-Vehicle CAN Bus Network. In Proceedings of the IEEE International Symposium on Circuits and Systems (ISCAS), Daegu, Republic of Korea, 22–28 May 2021; pp. 1–5.
4. Erlacher, F.; Dressler, F. FIXIDS: A high-speed signature-based flow intrusion detection system. In Proceedings of the IEEE/IFIP Network Operations and Management Symposium, Taipei, Taiwan, 23–27 April 2018; pp. 1–8.
5. Preuveneers, D.; Rimmer, V.; Tsingenopoulos, I.; Spooren, J.; Joosen, W.; Ilie-Zudor, E. Chained Anomaly Detection Models for Federated Learning: An Intrusion Detection Case Study. *Appl. Sci.* **2018**, *8*, 2663. [CrossRef]
6. Yang, Z.; Liu, X.; Li, T.; Wu, D.; Wang, J.; Zhao, Y.; Han, H. A systematic literature review of methods and datasets for anomaly-based network intrusion detection. *Comput. Secur.* **2022**, *116*, 102675. [CrossRef]
7. Lan, Y.; Truong-Huu, T.; Wu, J.; Teo, S.G. Cascaded Multi-Class Network Intrusion Detection with Decision Tree and Self-attentive Model. In Proceedings of the IEEE International Conference on Data Mining Workshops (ICDMW), Orlando, FL, USA, 28 November–1 December 2022; pp. 1–7. [CrossRef]
8. Saranya, T.; Sridevi, S.; Deisy, C.; Chung, T.D.; Khan, M. Performance Analysis of Machine Learning Algorithms in Intrusion Detection System: A Review. *Procedia Comput. Sci.* **2020**, *171*, 1251–1260. [CrossRef]
9. Alsyaibani, O.M.A.; Utami, E.; Hartanto, A.D. An Intrusion Detection System Model Based on Bidirectional LSTM. In Proceedings of the 3rd International Conference on Cybernetics and Intelligent System (ICORIS), Makasar, Indonesia, 25–26 October 2021; pp. 1–6.

10. Iwendi, C.; Khan, S.; Anajemba, J.H.; Mittal, M.; Alenezi, M.; Alazab, M. The Use of Ensemble Models for Multiple Class and Binary Class Classification for Improving Intrusion Detection Systems. *Sensors* **2020**, *20*, 2559. [CrossRef]
11. Khan, M.A. HCRNNIDS: Hybrid Convolutional Recurrent Neural Network-Based Network Intrusion Detection System. *Processes* **2021**, *9*, 834. [CrossRef]
12. Ho, Y.-B.; Yap, W.-S.; Khor, K.-C. The Effect of Sampling Methods on the CICIDS2017 Network Intrusion Data Set. IT Con-vergence and Security. *Lect. Notes Electr. Eng.* **2021**, *782*, 33–41.
13. Bulavas, B.; Marcinkevicius, V.; Rumiński, J. Study of Multi-Class Classification Algorithms' Performance on Highly Imbalanced Network Intrusion Datasets. *Informatica* **2021**, *32*, 441–475. [CrossRef]
14. Tran, T.P.; Nguyen, V.C.; Vu, L.; Nguyen, Q.U. DeepInsight-Convolutional Neural Network for Intrusion Detection Systems. In Proceedings of the 8th NAFOSTED Conference on Information and Computer Science (NICS), Hanoi, Vietnam, 21–22 December 2021; pp. 120–125.
15. Atefinia, R.; Ahmadi, M. Network intrusion detection using multi-architectural modular deep neural network. *J. Supercomput.* **2021**, *77*, 3571–3593. [CrossRef]
16. Yin, C.; Zhu, Y.; Fei, J.; He, X. A Deep Learning Approach for Intrusion Detection Using Recurrent Neural Networks. *IEEE Access* **2017**, *5*, 21954–21961. [CrossRef]
17. Ravi, V.; Kp, S.; Poornachandran, P. Evaluation of Recurrent Neural Network and its Variants for Intrusion Detection System (IDS). *Int. J. Inf. Syst. Model. Des.* **2017**, *8*, 43–63.
18. Sohn, I. Deep belief network based intrusion detection techniques: A survey. *Expert Syst. Appl.* **2021**, *167*, 114170. [CrossRef]
19. Lundberg, H.; Mowla, N.-I.; Thar, K.; Mahmood, A.; Gidlund, M.; Raza, S. Experimental Analysis of Trustworthy In-Vehicle Intrusion Detection System Using eXplainable Artificial Intelligence (XAI). *IEEE Access* **2022**, *10*, 102831–102841. [CrossRef]
20. Patil, S.; Varadarajan, V.; Mazhar, S.-M.; Sahibzada, A.; Ahmed, N.; Sinha, O.; Kumar, S.; Shaw, K.; Kotecha, K. Explainable Artificial Intelligence for Intrusion Detection System. *Electronics* **2022**, *11*, 3079. [CrossRef]
21. Arrieta, A.-B.; Díaz-Rodríguez, N.; Ser, J.-D.; Bennetot, A.; Tabik, S.; Barbado, A.; Garcia, S.; Gil-Lopez, S.; Molina, D.; Benjamins, R.; et al. Explainable Artificial Intelligence (XAI): Concepts, taxonomies, opportunities and challenges toward responsible AI. *Inf. Fusion* **2020**, *58*, 82–115. [CrossRef]
22. Hartl, A.; Bachl, M.; Fabini, J.; Zseby, T. Explainability and Adversarial Robustness for RNNs. In Proceedings of the IEEE Sixth International Conference on Big Data Computing Service and Applications (BigDataService), Oxford, UK, 3–6 August 2020; pp. 148–156.
23. Hariharan, S.; Robinson, R.R.R.; Prasad, R.R.; Thomas, C.; Balakrishnan, N. XAI for intrusion detection system: Comparing explanations based on global and local scope. *J. Comput. Virol. Hacking Tech.* **2022**, *19*, 217–239. [CrossRef]
24. Mahbooba, B.; Timilsina, M.; Sahal, R.; Serrano, M. Explainable Artificial Intelligence (XAI) to Enhance Trust Management in Intrusion Detection Systems Using Decision Tree Model. *Complexity* **2021**, *2021*, 6634811. [CrossRef]
25. Kuppa, A.; Le-Khac, N.-A. Black Box Attacks on Explainable Artificial Intelligence (XAI) methods in Cyber Security. In Proceedings of the International Joint Conference on Neural Networks (IJCNN), Glasgow, UK, 19–24 July 2020; pp. 1–8.
26. Kuppa, A.; Le-Khac, N.-A. Adversarial XAI Methods in Cybersecurity. *IEEE Trans. Inf. Forensics Secur.* **2021**, *16*, 4924–4938. [CrossRef]
27. Pelletier, Z.; Abualkibash, M. Evaluating the CIC IDS-2017 Dataset Using Machine Learning Methods and Creating Multiple Predictive Models in the Statistical Computing Language R. *Int. Res. J. Adv. Eng. Sci.* **2020**, *5*, 187–191. Available online: http://irjaes.com/wp-content/uploads/2020/10/IRJAES-V5N2P184Y20.pdf (accessed on 28 May 2023).
28. Alsameraee, A.A.A.; Ibrahem, M.K. Toward Constructing a Balanced Intrusion Detection Dataset. *Samarra J. Pure Appl. Sci.* **2021**, *2*, 132–142. [CrossRef]
29. Mbow, M.; Koide, H.; Sakurai, K. An Intrusion Detection System for Imbalanced Dataset Based on Deep Learning. In Proceedings of the Ninth International Symposium on Computing and Networking (CANDAR), Matsue, Japan, 22–26 November 2021; pp. 38–47.
30. Zhang, H.; Huang, L.; Wu, C.Q.; Li, Z. An effective convolutional neural network based on SMOTE and Gaussian mixture model for intrusion detection in imbalanced dataset. *Comput. Netw.* **2020**, *177*, 107315. [CrossRef]
31. Abdulhammed, R.; Musafer, H.; Alessa, A.; Faezipour, M.; Abuzneid, A. Features Dimensionality Reduction Approaches for Machine Learning Based Network Intrusion Detection. *Electronics* **2019**, *8*, 322. [CrossRef]
32. Toupas, P.; Chamou, D.; Giannoutakis, K.M.; Drosou, A.; Tzovaras, D. An Intrusion Detection System for Multi-class Classification Based on Deep Neural Networks. In Proceedings of the 18th IEEE International Conference on Machine Learning and Applications (ICMLA), Boca Raton, FL, USA, 16–19 December 2019; pp. 1253–1258.
33. Zhang, Y.; Chen, X.; Guo, D.; Song, M.; Teng, Y.; Wang, X. PCCN: Parallel Cross Convolutional Neural Network for Abnormal Network Traffic Flows Detection in Multi-Class Imbalanced Network Traffic Flows. *IEEE Access* **2019**, *7*, 119904–119916. [CrossRef]
34. Mhawi, D.N.; Aldallal, A.; Hassan, S. Advanced Feature-Selection-Based Hybrid Ensemble Learning Algorithms for Network Intrusion Detection Systems. *Symmetry* **2022**, *14*, 1461. [CrossRef]
35. Rosay, R.; Cheval, E.; Carlier, F.; Leroux, P. Network Intrusion Detection: A Comprehensive Analysis of CIC-IDS2017. In Proceedings of the 8th International Conference on Information Systems Security and Privacy, Online, 9–11 February 2022; pp. 25–36.

36. Lanvin, M.; Gimenez, P.-F.; Han, Y.; Majorczyk, F.; Me, L.; Totel, E. Errors in the CICIDS2017 dataset and the significant differences in detection performances it makes. In Proceedings of the 17th International Conference Risks and Security of Internet and Systems, Sousse, Tunisia, 7–9 December 2022; pp. 18–33.
37. Alikhanov, J.; Jang, R.; Abuhamad, M.; Mohaisen, D.; Nyang, D.; Noh, Y. CatBoost-Based Network Intrusion Detection on Imbalanced CIC-IDS-2018 Dataset. *J. Korean Inst. Commun. Inf. Sci.* **2021**, *46*, 2191–2197.
38. Liu, L.; Wang, P.; Lin, J.; Liu, L. Intrusion Detection of Imbalanced Network Traffic Based on Machine Learning and Deep Learning. *IEEE Access* **2021**, *9*, 7550–7563. [CrossRef]
39. Leevy, J.L.; Hancock, J.; Zuech, R.; Khoshgoftaar, T.M. Detecting cybersecurity attacks across different network features and learners. *J. Big Data* **2021**, *8*, 38. [CrossRef]
40. Farhan, B.I.; Jasim, A.D. Performance analysis of intrusion detection for deep learning model based on CSE-CIC-IDS2018 dataset. *Indones. J. Electr. Eng. Comput. Sci.* **2022**, *26*, 1165–1172. [CrossRef]
41. Kilincer, I.F.; Ertam, F.; Sengur, A. A comprehensive intrusion detection framework using boosting algorithms. *Comput. Electr. Eng.* **2022**, *100*, 107869. [CrossRef]
42. Alzughaibi, S.; El Khediri, S. A Cloud Intrusion Detection Systems Based on DNN Using Backpropagation and PSO on the CSE-CIC-IDS2018 Dataset. *Appl. Sci.* **2023**, *13*, 2276. [CrossRef]
43. Jinsi, J.; Jose, D.V. Deep Learning Algorithms for Intrusion Detection Systems in Internet of Things Using CIC-IDS 2017 Dataset. *Int. J. Electr. Comput. Eng. (IJECE)* **2023**, *13*, 1134–1141.
44. Wang, Y.-C.; Houng, Y.-C.; Chen, H.-X.; Tseng, S.-M. Network Anomaly Intrusion Detection Based on Deep Learning Ap-proach. *Sensors* **2023**, *23*, 2171. [CrossRef] [PubMed]
45. Ingre, B.; Yadav, A.; Soni, A.K. Decision Tree Based Intrusion Detection System for NSL-KDD Dataset. In Proceedings of the Information and Communication Technology for Intelligent Systems (ICTIS 2017), Ahmedabad, India, 25–26 March 2017; Volume 2, pp. 207–218.
46. Brabec, J.; Machlica, L. Decision-Forest Voting Scheme for Classification of Rare Classes in Network Intrusion Detection. In Proceedings of the IEEE International Conference on Systems, Man and Cybernetics, Miyazaki, Japan, 7–10 October 2018; pp. 3325–3330. [CrossRef]
47. Sahani, R.; Shatabdinalini; Rout, C.; Badajena, J.C.; Jena, A.K.; Das, H. Classification of Intrusion Detection Using Data Mining Techniques. In *Progress in Computing, Analytics and Networking*; Springer: Singapore, 2018; pp. 753–764. [CrossRef]
48. Ren, Q.; Cheng, H.; Han, H. Research on machine learning framework based on random forest algorithm. *AIP Conf. Proc.* **2017**, *1820*, 080020. [CrossRef]
49. Alshamy, R.; Ghurab, M.; Othman, S.; Alshami, F. Intrusion Detection Model for Imbalanced Dataset Using SMOTE and Random Forest Algorithm. *Commun. Comput. Inf. Sci.* **2021**, *1487*, 361–378. [CrossRef]
50. Vang-Mata, R. *Multilayer Perceptrons: Theory and Applications*; Nova Science Publishers: Hauppauge, NY, USA, 2020; p. 153, ISBN 978-1-53617-364-2.
51. Huang, G.; Liu, Z.; Maaten, L.; Weinberger, K. Densely Connected Convolutional Networks. In Proceedings of the IEEE Conference on Computer Vision and Pattern Recognition (CVPR), Honolulu, HI, USA, 21–26 July 2017; pp. 2261–2269.
52. Kamath, U.; Liu, J. *Explainable Artificial Intelligence: An Introduction to Interpretable Machine Learning*; Springer: Berlin/Heidelberg, Germany, 2021; p. 310. [CrossRef]

applied sciences

MDPI

Article

A Multi-Model Proposal for Classification and Detection of DDoS Attacks on SCADA Systems

Esra Söğüt * and O. Ayhan Erdem

Department of Computer Engineering, Faculty of Technology, Gazi University, Ankara 06560, Turkey; ayerdem@gazi.edu.tr
* Correspondence: esrasogut@gazi.edu.tr

Abstract: Industrial automation and control systems have gained increasing attention in the literature recently. Their integration with various systems has triggered considerable developments in critical infrastructure systems. With different network structures, these systems need to communicate with each other, work in an integrated manner, be controlled, and intervene effectively when necessary. Supervision Control and Data Acquisition (SCADA) systems are mostly utilized to achieve these aims. SCADA systems, which control and monitor the connected systems, have been the target of cyber attackers. These systems are subject to cyberattacks due to the openness to external networks, remote controllability, and SCADA-architecture-specific cyber vulnerabilities. Protecting SCADA systems on critical infrastructure systems against cyberattacks is an important issue that concerns governments in many aspects such as economics, politics, transport, communication, health, security, and reliability. In this study, we physically demonstrated a scaled-down version of a real water plant via a Testbed environment created including a SCADA system. In order to disrupt the functioning of the SCADA system in this environment, five attack scenarios were designed by performing various DDoS attacks, i.e., TCP, UDP, SYN, spoofing IP, and ICMP Flooding. Additionally, we evaluated a scenario with the baseline behavior of the SCADA system that contains no attack. During the implementation of the scenarios, the SCADA system network was monitored, and network data flow was collected and recorded. CNN models, LSTM models, hybrid deep learning models that amalgamate CNN and LSTM, and traditional machine learning models were applied to the obtained data. The test results of various DDoS attacks demonstrated that the hybrid model and the decision tree model are the most suitable for such environments, reaching the highest test accuracy of 95% and 99%, respectively. Moreover, we tested the hybrid model on a dataset that is used commonly in the literature which resulted in 98% accuracy. Thus, it is suggested that the security of the SCADA system can be effectively improved, and we demonstrated that the proposed models have a potential to work in harmony on real field systems.

Keywords: critical infrastructure; SCADA; cybersecurity; DDoS; deep learning; testbed

Citation: Söğüt, E.; Erdem, O.A. A Multi-Model Proposal for Classification and Detection of DDoS Attacks on SCADA Systems. *Appl. Sci.* **2023**, *13*, 5993. https://doi.org/10.3390/app13105993

Academic Editor: Luis Javier Garcia Villalba

Received: 31 March 2023
Revised: 7 May 2023
Accepted: 11 May 2023
Published: 13 May 2023

1. Introduction

Facilities that produce, store, and transmit natural resources, such as water, oil, and natural gas, or energy sources, such as hydroelectric, solar, and nuclear, constitute critical infrastructures. Space, satellite, air, sea, or train transportation systems are also in these groups. These systems spread and work over small or large areas. Some systems monitor, control, and, when necessary, intervene in processes and events in critical infrastructures from a central point. One of them is the Supervisory Control and Data Acquisition (SCADA) system. For example, municipalities use SCADA systems to monitor water levels, pipe pressure, and the temperature in tanks located in utility water distribution facilities.

The reports and research published every year in the field of cybersecurity suggest one should always be ready for attacks that may occur from the inside or the outside [1]. Ensuring the cybersecurity of SCADA systems in the cyber world is a crucial issue and has become mandatory. Since cyberattacks against SCADA systems are dangerous for critical infrastructure systems, these attacks should be investigated [2]. According to a special report by the National Institute of Standards and Technology, cyberattacks to control systems can disrupt the reliable operation of industrial processes. Therefore, providing cybersecurity is imperative [3]. Defence Research and Development Canada published a report aimed to increase the cyber resilience of Canada's critical infrastructure. According to the report, changes to the standard network configurations of SCADA networks can greatly improve the protection of control system fields [4]. In the study conducted by the U.S. Department of Energy Office of Electricity Delivery and Energy Reliability, security vulnerabilities found to be common in control systems such as SCADA were discussed and grouped by severity. In addition, security recommendations were provided for asset owners and system vendors [5]. Due to the architectural structure of SCADA systems, their integration with advanced technology has not been fully solved. On the other hand, internet usage, access to external networks, and remote control are increasing worldwide. These developments enhance the functionality of traditional SCADA systems, but they also bring many security vulnerabilities.

Critical infrastructures are designed to enable citizens to maintain their lives in better conditions. Problems experienced in the functioning of these structures may affect not only the relevant area but also the whole country. For example, the failure of electricity generation, storage, and transmission facilities can cause massive chaos in a country and directly affect other electrical systems. Countries experiencing power outages have realized how crucial such blackouts are. Attacks on critical infrastructures can destructively impact the economy, security, or health. The consequences of cyberattacks against SCADA systems may be far beyond estimates. As a result, necessary measures should be taken for the cybersecurity of SCADA systems. Security system developments, such as attack detection and prevention, will considerably contribute to the continuity of a country's critical infrastructures.

Models that include machine learning, deep learning, or artificial intelligence algorithms used in attack detection studies may also serve in SCADA systems. Studies to determine the "attacks" and "attack types" can contribute to the cybersecurity of SCADA systems. There are different types of cyberattacks, and distributed denial-of-service (DDoS) attacks are more common than other attacks. In particular, handling DDoS attacks for SCADA systems is essential in cybersecurity. Since the attack detection models in the algorithms have different structures, the analyses also give different results. For example, an attack detection model providing high performance on one dataset can deliver poor performance on another, or different models on a dataset may not yield the same highly successful results. For these reasons, developing an attack detection model that provides high performance for a particular dataset is essential.

The current study aimed to detect DDoS attacks that may occur against a SCADA system used in critical infrastructures and to determine the type of DDoS attack. For this purpose, a testbed was prepared that enables the processing of cyber and physical processes. Various DDoS attacks and tests were implemented on the testbed to damage the processes of the SCADA system and measure the system's reaction against attacks. Attack detection is essential to ensure the cybersecurity of the system. For this purpose, the network traffics in the baseline situation without any attack and the situations in which DDoS attacks were applied and recorded. Deep learning and machine learning algorithms were used to analyze the recorded network traffic packets and to determine whether there is an attack or not. In addition to intrusion detection, these algorithms have also been studied to determine the type of attack. The deep learning-based convolution neural network (CNN) model, long short-term memory (LSTM) model, and hybrid model using LSTM-CNN algorithms together were evaluated. Machine learning-based 13 algorithms such as K-Nearest Neighbors (KNN), LogitBoost, Naive Bayes, PART, decision tree, and random forest were used. High success rates were obtained with deep learning-based LSTM-CNN hybrid model and machine learning-based decision tree model. It is aimed to provide different perspectives for ensuring the cybersecurity of SCADA systems and to prepare suitable models for determining the type of attack.

The main contributions of the present study are as follows:

- A testbed environment containing a SCADA system was prepared and different components, software and hardware were used from the studies in the literature.
- Various DDoS attacks (five different) and the baseline situation were evaluated together to add diversity to the literature.
- A new dataset was prepared to contribute to the literature by including various DDoS attacks and a baseline situation, enabling detection and identification of attack types.
- CNN and LSTM algorithms were used as separate models for attack detection and attack type determination. In addition, LSTM and CNN algorithms were evaluated together and used as a hybrid model. In the studies in the literature we examined, there are no such separate and hybrid uses in this way. By using a hybrid model, a higher success rate was obtained than using separate models. In addition to deep learning-based models, machine learning-based models were also prepared and evaluated in the study. Analyses were performed with 13 different machine learning algorithms and the highest success rate was obtained with the decision tree model.
- A commonly used dataset in the literature was selected and tested to evaluate the adequacy of the hybrid model. According to the results obtained, a high accuracy rate was achieved.

This study comprises six chapters. The first part provides an overview of SCADA systems, shows the security vulnerabilities, and explains the importance of ensuring the cybersecurity of SCADA systems. The second part examines the studies that detect attacks against SCADA systems using their own datasets, ready-made datasets, or their own testbeds. The third chapter discusses SCADA systems and cyberattacks against these systems. The fourth section covers the prepared testbed environment, DDoS attacks against this environment, the obtained dataset, the success metrics, and the proposed models. The fifth section presents the analyses for DDoS attack detection for SCADA system, experimental results of the proposed models, and the results of other studies in the literature. The study results and recommendations for future studies are summarized in the sixth section.

2. SCADA Systems and Cybersecurity

This section gave information about what SCADA systems are, what components they consist of, the cybersecurity of these systems, and possible attacks.

2.1. Scada System

SCADA systems perform control and monitoring tasks in critical infrastructure or facilities. Critical infrastructures, such as power generation plants, wind energy turbines, and natural gas distribution facilities are vital structures that produce and (or) transmit natural gas, oil, water, and similar resources to another place. To give more examples, many systems such as municipal water distribution facilities, airlines, and ship systems are also critical infrastructure systems and have a significant place nationally and internationally. SCADA systems are also used in production facilities, factories, or public institutions apart from these infrastructures.

SCADA systems consist of a master terminal unit (MTU), remote terminal units (RTUs), and a communication network. The MTU controls the processes in the system using a human–machine interface (HMI). There is data exchange and command transmission between RTUs and MTU. RTUs transmit the data collected from the field sensors to the MTU, and RTUs carry out the commands from MTU. Modbus, DNP3, and Profibus communication protocols—specific to SCADA systems—are used for communication between basic units. The sensors and actuators on the RTUs abide by the commands. Elements such as pumps and relays serve as actuators. The HMI also demonstrates the data obtained from the sensors [6,7].

2.2. *Cybersecurity and Attacks in Scada Systems*

Most structures where SCADA systems are used have not direct connections to the internet and work independently from external networks. Developing technologies expand the area of internet usage, and this situation also affects SCADA systems. Innovations such as the co-usage of different technologies and remote accessing the system via the internet create new cybersecurity problems for SCADA systems. SCADA systems, which cannot keep up with the developing technology, have many architecture-related security problems. For example, the frequently used Modbus protocol has many vulnerabilities that can be attacked, such as by a man-in-the-middle, command injection, and denial-of-service (DoS) [8–10]. SCADA systems in different sectors become attractive targets for malicious people who are aware of these situations and work in the national or international arena. Figure 1 shows the sectors where SCADA systems serve. Each can contain various threats that malicious applications can attack.

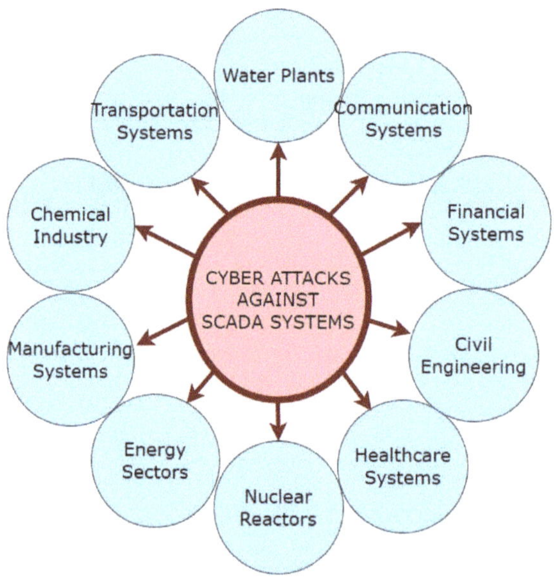

Figure 1. Sectors where SCADA systems are used.

Today, many attack scenarios may occur in SCADA systems, such as emerging new vulnerabilities, existing old security gaps, vulnerability exploitation, damaging systems, or rendering the system inoperable. Possible scenarios may cover malicious remote control of the system and power cut threats. For example, cyberattacks can occur by targeting electricity generation or distribution facilities. As a result of these attacks, there may be power cuts; cities may suddenly go dark. A nuclear power plant's centrifuges were remotely disrupted through the Stuxnet, which is one of the most dangerous attacks. In this attack, while the system was physically damaged, the field operators noticed the problem much later [11]. Another example of an attack is the remote poisoning of the Florida City Water Supply. The attackers seized the water facility and tried to increase the sodium hydroxide level in the city water. Once the authorities realized the situation, they quickly intervened and prevented the attack [12]. As can be understood from these examples, cyberattacks can also affect some or all of the SCADA systems. In addition, the experienced problems may adversely trigger other systems associated with SCADA systems. Today, actions to disrupt public peace, complicate their daily life, or harm their health have become possible using the vulnerabilities in SCADA systems. For these reasons, cybersecurity in SCADA systems is necessary today.

SCADA systems are vulnerable to numerous attacks due to their tasks, traditional architectural structure, and built-in communication technologies. Specially developed attack techniques make SCADA systems targets for aggressive attempts, and the security risk of these systems is increasing day by day. Various attacks are made against SCADA systems, such as man-in-the-middle, data injection, command injection, DoS, and DDoS [10,13,14]. Among these, DDoS attacks are common and dangerous attacks that can affect any SCADA system. These attacks aim to disrupt control and process operations and render the system out of use [15,16]. DDoS attacks against SCADA systems used in critical infrastructures may cause devastating harm to these infrastructures.

3. Literature Studies on SCADA Security

In the literature, studies carried out on the detection of DDoS attacks against SCADA systems are considerably popular. These studies have frequently used machine learning- and deep learning-based methods for attack detection. Some of these works are summarized below.

Marcio Andrey Teixeira et al. performed a study to detect cyberattacks on SCADA systems. The authors created a dataset using a test environment. They employed random forest, decision tree, logistic regression, Naive Bayes, and KNN algorithms in their study for attack detection [17].

Thomas Morris and colleagues worked on potential cyberattacks at Mississippi State University's SCADA Security Lab and investigated the security vulnerabilities of the most widely used communication protocols in SCADA systems. They aimed to detect attacks and minimize their effects with the security mechanisms developed with neural network methods [18].

Nader et al. carried out a study on the security of industrial control systems and critical infrastructures. They emphasized that traditional attack detection systems could not detect attacks newly developed and unregistered in databases. They used data from a water distribution system in France in the study and proposed machine learning algorithms for attack detection [19].

Focusing on the developments in information and communication technologies, Y. Yang et al. have emphasized that the complexity and security vulnerabilities in SCADA procedures are gradually increasing. They stated that new security measures were necessary for new-generation SCADA designs integrated into the internet and different systems. Therefore, they proposed an attack detection system with a behavior-based and multilayer framework [20].

Almalawi et al. proposed two approaches to detect attacks against SCADA systems. The first was to determine whether the data in the system were consistent or inconsistent. The second approach was to obtain proximity detection rules from specified situations. They stated that the KNN-based attack detection system showed significant accuracy [21].

Meir Kalech proposed techniques based on temporal pattern recognition for cyberattack detections in SCADA systems. The study proposed two algorithms based on Hidden Markov models (HMM) and artificial neural network-based self-organizing maps (ANN-based SOM). According to the results obtained, they stated that it was easier to detect cyberattacks [22].

Jun Gao et al. discussed temporally uncorrelated and correlated attacks against SCADA systems. They detected attacks using the feedforward neural network (FNN) and LSTM algorithms based on deep learning. The FNN-LSTM model, on the other hand, succeeded in detecting both types of cyberattacks, regardless of their temporal correlations [23].

While intrusions into SCADA systems will continue, defense mechanisms against different attack vectors remain insufficient. Therefore, Maglaras et al. conducted a study to ensure the cybersecurity of SCADA systems. Accordingly, they proposed an integrated attack detection mechanism against cyberattacks that captures network traffic, divides traffic by source, and creates a set of one-class support vector machine (OCSVM) models [24].

Gao et al. proposed two models aimed at detecting attacks against SCADA systems. These models have many-to-many (MTM) and many-to-one (MTO) architectures. The models used the LSTM algorithm. Both detection systems performed well in detecting temporally uncorrelated attacks [25].

There are many studies aimed at detecting unauthorized access to SCADA systems. Shitharth and Winston developed an intrusion detection system that classifies attacks based on optimization. They proposed intrusion weighted particle-based cuckoo search optimization (IWP-CSO) and hierarchical neuron architecture-based neural network (HNA-NN) techniques [26,27].

This study focused on DDoS attacks in SCADA architecture and presented models that work efficiently to detect attacks. In the literature, studies that detect attacks on SCADA systems have been examined and it has been seen that machine learning algorithms such as random forest, decision tree, logistic regression, Naive Bayes, KNN, and SVM are used more frequently than other algorithms for detection. In addition to these, there are models in which neural networks and deep learning algorithms such as LSTM are used. In this study, deep learning-based models (CNN, LSTM, LSTM-CNN hybrid) and machine learning-based models (13 models such as KNN, LogitBoost, Naive Bayes and decision tree) were proposed. The attack detection accuracy rates of the examined studies and this study are placed in the table in the Section 5.2. Thus, a general review and comparison is provided for the studies.

4. Materials and Method

This section elaborated on the prepared testbed, fictionalized the cyberattacks using scenarios, and gave information about the dataset's features obtained from the testbed. This section also covered the metrics to analyze the dataset as well as their explanations. Information was given about the proposed models and their architectural structures. The topics in this section are summarized in Figure 2.

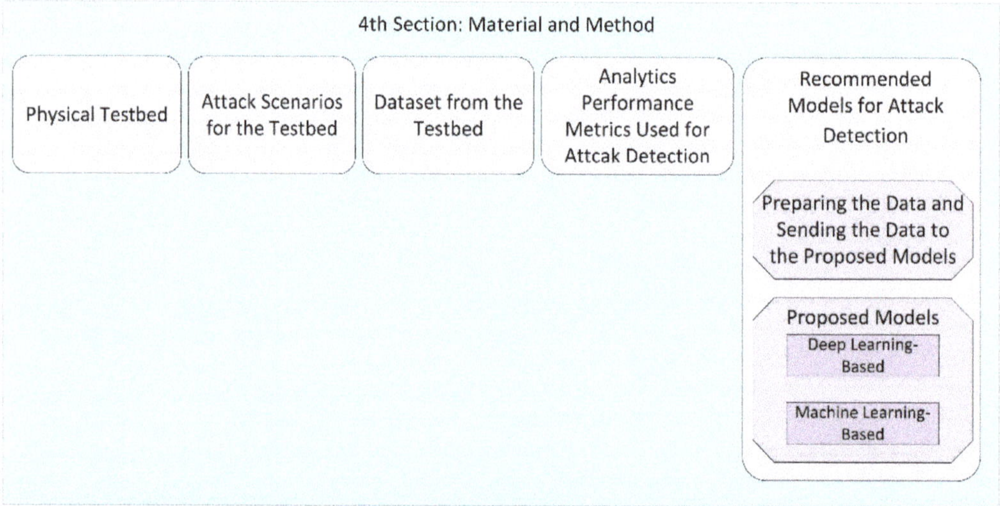

Figure 2. Organizational chart of the Materials and Method section.

4.1. Physical Testbed

The test environment aimed to simulate the industrial control systems of a plant as approximately as possible without completely copying them [28]. In addition, it aimed to contribute to the performance of national and international industrial control system standards and directives. The preparation and use of a testbed provides a suitable environment for performing real cyberattacks and even observing the results of the attack.

In order to contribute to cybersecurity research, a testbed environment including a SCADA system was prepared in the study. In this environment there are storage tanks, specific processes are operated, and Modbus TCP/IP communication is used. A SCADA system is usually realized by integrating Modbus communication protocol [29]. A simplified version of a real water plant was shown in this testbed. The SCADA system controls and monitors the water circulation processes and the status of the storage tanks. This section explained the configuration and architectural structure of the prepared SCADA system test environment. The equipment used in the test environment was selected from the components frequently used in real SCADA systems. The architectural structure of the test environment is shown in Figure 3.

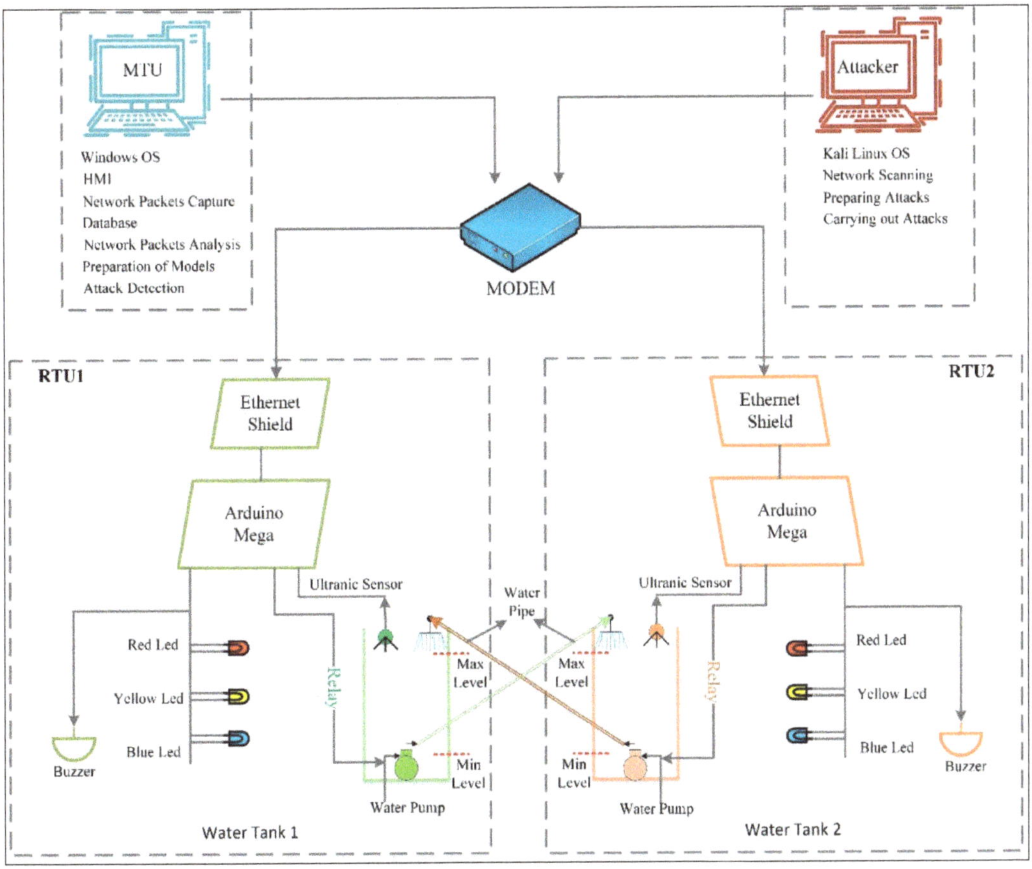

Figure 3. The architectural structure of the testbed.

As shown in Figure 3, there are two water circulation and storage tanks in RTUs. There were sensors and actuators connected to RTUs. The sensors monitor the water levels in the tanks, and the water pumps operate according to the levels. In order to prevent problems such as the overflowing of the tanks and running out of water in the tanks, the water level is continuously controlled. In addition, an alarm was generated according to the state of the water level and, thus, attracting the attention of the operator who is interested in the system. LEDs and buzzers are designed for alarm events. Modbus TCP/IP wired and wireless communication protocols were used for communication in the environment. The data received from the RTUs were transmitted to the MTU and the processes were followed through the HMI simulators on the MTU. Incoming data were checked and stored, and new commands were sent to RTUs.

Attackers scanned the network and attacked the appropriate RTU. Whether there is an attack or not is checked on MTU. When 5 different DDoS attacks were applied to the RTU, network traffic packets were listened to and recorded separately for each attack. In addition, the same listening and recording operations were performed for baseline operation without attack. Google Colab, an environment offered by Google Research, enables Python coding for machine learning, data analysis, and training. The data were preprocessed in this environment to make the packets suitable for analysis. Pandas' libraries were added to this environment and different models were generated for attack detection using deep learning and machine learning algorithms.

4.2. Attack Scenarios for the Testbed

This section elaborates on the baseline situation of the testbed and the attacks against the testbed. DDoS attacks, one of the most common attacks on SCADA systems, were discussed and attacks against an RTU selected by the attacker were performed. Different types of DDoS attack scenarios were implemented and aimed to affect the operation of the system. These scenarios were:

1. Baseline (normal or no-attack) situation;
2. TCP flooding attack scenario;
3. UDP flooding attack scenario;
4. SYN flooding attack scenario;
5. Spoofing IP flooding attack scenario;
6. ICMP flooding attack scenario.

In the baseline situation scenario (when the SCADA system was not under attack), the obtained network traffic was listened to and recorded. In this scenario, water circulated continuously between the water tanks and the necessary operations were performed automatically according to the change in the water level. The pinging method was used to establish communication between RTU and MTU.

Specific coding was made by the attacker for each attack type and 5 different DDoS attacks were performed against the target RTU. Each of TCP, UDP, SYN, Spoofing IP and ICMP flooding attacks were carried out at different times and separately. Each of these attack scenarios were executed for approximately 2 min. During the attacks, the target RTU system processes were interrupted for a short period of time. Processes such as water recirculation and alarm generation were disrupted. These adverse conditions also affected the other RTU system and the operation of the entire testbed system was interrupted for short periods of time. Abnormal situations such as incorrect measurement of the tank water level or buzzer alarming at the wrong time were observed. When the execution of the attack scenarios ended, the system operation slowly recovered and, after a while, the system returned to its former state. If the time taken to restore the system operation is too long to be tolerated, irreversible major problems may occur for SCADA systems. For this reason, it is important to attack SCADA systems and monitor and analyze the attack responses. In this study, this issue is emphasized.

4.3. Dataset from the Testbed

This section provides information about the total dataset obtained as a result of the scenarios performed separately on the testbed. Network traffic packets of each scenario were collected with Wireshark network listening and analysis tool. Then, the packets of these 6 scenarios were collected in a single file and the total dataset was created. The features frequently used in the literature and specific to the Modbus TCP/IP protocol were determined for this dataset [23,25]. Table 1 shows the features used in this research.

Table 1. Features used in the dataset and their descriptions.

No	Features	Descriptions
1	No	Data number
2	Time	Time
3	SourceIP	Source Internet Protocol
4	DestinationIP	Destination Internet Protocol
5	SourcePort	Source port
6	DestinationPort	Destination port
7	Protocol	Protocol
8	Length	Data packet length
9	Info	Information about packet
10	Modbus_ByteCount	Modbus protocol data area (in bytes) size
11	Modbus_ResponseTime	Modbus protocol response time
12	Modbus_ReqFrame	Modbus protocol message format
13	DeltaTime	Duration between the start and end of an operation
14	ModbusEventCount	Number of Modbus device transactions
15	TimeSince_FirstFrameInThisTCPStream	Time elapsed since the first frame in this TCP stream
16	TimeSince_PreviousFrameInThisTCPStream	Time elapsed since the previous frame in this TCP stream
17	TimeDeltaFromPrevious_CapturedFrame	Time difference from the previous captured frame
18	TimeDeltaFromPrevious_DisplayedFrame	Time difference from the previous displayed frame
19	TimeSince_ReferenceOrFirstFrame	Time elapsed since the reference or first frame
20	FrameLength_OnTheWire	Frame length on the wire
21	FrameLength_StoredIntoTheCaptureFile	Frame length stored into the capture file
22	TimeToLive	Time to live
23	TotalLength	Total length
24	FrameLengthStoredIntoTheCaptureFile	Frame length stored into the capture file
25	ModbusTCPLength	Modbus TCP packet length
26	ModbusByteCount	Modbus packet byte count
27	ModbusTimeFromRequest	Modbus packet time from request
28	TCPHeaderLength	TCP header length
29	ModbusRegNum	Modbus register number
30	Register Value (UINT16)	Modbus register value
31	Class	Classification column

A new and comprehensive dataset consisting of 30 attributes, 1 deterministic class, and a total of 22.768 samples was obtained. While preparing the dataset, the attacks were observed on the SCADA system and abnormal situations were noticed clearly by the operator. It is detected whether there is a DDoS attack and if there is an attack, which of the 5 different types is determined. This dataset is suitable for training and testing deep learning and machine learning models. Due to these properties, a new perspective and contribution to the literature is presented.

4.4. The Performance Analysis Metrics in Attack Detection

Performance metrics serve for the evaluation and comparison of the deep learning, and the machine learning algorithms for the model. While working on a problem, using these metrics makes it easier to propose more solutions and apply the proposed methods. To determine the most effective method in problem solving, the performance information of each method is obtained one by one. Then the method producing the highest success rate is selected. Table 2 shows the confusion matrix containing the values for performance metrics.

Table 2. Confusion Matrix.

		Actual Values	
		Positive	**Negative**
Predictive Values	**Positive**	TP	FP
	Negative	FN	TN

The values in the confusion matrix show the actual values and the estimation values [30]. The fact that the value with a positive label in reality also has a positive label in the prediction part makes it a true positive (*TP*). The fact that the value with a negative label is positively labeled in the prediction part makes it a false positive (*FP*). The fact that the value with a positive label is negatively labeled in the prediction portion makes it a false negative (*FN*). The fact that the value with a negative label also has a negative label in the prediction part makes it a true negative (*TN*). The success metrics calculated with the values on the confusion matrix are below.

Accuracy is the ratio of the correctly predicted values to the total values. Equation (1) shows this situation:

$$\text{Accuracy} = \frac{TP + TN}{TP + FP + TN + FN} \tag{1}$$

Precision is the ratio of the correctly predicted positive values to the predicted values with a positive label. Equation (2) shows this ratio:

$$\text{Precision} = \frac{TP}{TP + FP} \tag{2}$$

The recall is the ratio of correctly predicted positive values to the values with a positive label. Equation (3) shows this ratio [31]:

$$\text{Recall} = \frac{TP}{TP + FN} \tag{3}$$

F-1 Score—ranging from 0 to 1—is the harmonic mean of precision and recall values. Equation (4) calculates this average:

$$\text{F1 Score} = 2 \times \frac{\text{Precision} \times \text{Recall}}{\text{Precision} + \text{Recall}} \tag{4}$$

The current study used accuracy, precision, recall, and f1-score among traditional performance metrics. While comparing the literature studies, this research preferred the frequently used "accuracy success metric".

4.5. Recommended Models for Attack Detection

In this section, attacks against the testbed containing the SCADA system were detected. For this purpose, a deep learning-based model and a machine learning-based model were applied to the previously prepared dataset. The analysis results obtained were compared with each other according to particular metrics.

In order to achieve successful results in proposed models, data were preprocessed, and experiments were performed. Innovative and different approaches were proposed for the cybersecurity of a physical testbed containing a SCADA system.

4.5.1. Preparing the Data and Transmitting Them to the Proposed Models

This section concerns turning the dataset into an analyzable state and designing the appropriate models, which Figure 4 summarizes. Several data pre-processes were determined to make the dataset analyzable. The primary operations were deleting attributes and instances deemed unnecessary or containing too many null values. The next step was completing the attributes containing missing data using the mean method. Another process is to convert data types to the same type using categorization processes. In the end, a dataset with 25 features was obtained.

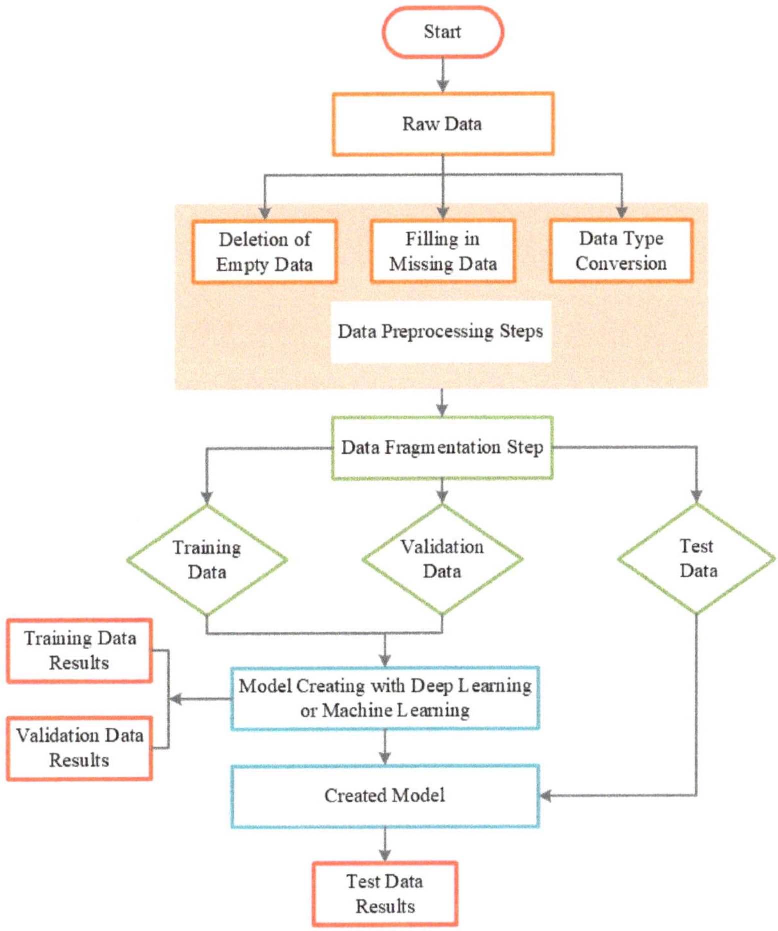

Figure 4. Processing the data and delivering them to the proposed models.

After preprocessing, the dataset was divided into parts for training, validation, and testing in the data fragmentation stage. The split ratios here were kept constant in the models used. The obtained training and validation data were combined and sent to the proposed models. Tests were carried out on the model using the test data, and analysis results were obtained for training, validation, and test data. According to the results, the attack-detection success of the proposed model was evaluated. These stages were essential for obtaining the most suitable model which achieved the highest success rate in detecting the attack.

4.5.2. Recommended Models

LSTM and CNN, which are important deep learning algorithms, were used alone and in combination with different algorithms in the literature. In this study, CNN and LSTM algorithms were evaluated and tested separately in order to contribute to the literature. Then, a hybrid model was created by considering these two algorithms together and tests were performed. Tests with different properties were applied to the LSTM model and CNN model. The parameter values in the LSTM-CNN hybrid model architecture were changed and different test procedures were performed. These models were analyzed separately and their attack detection success rates were discussed.

In addition to these, machine learning algorithms that are frequently used in the literature were determined. By using these algorithms, suitable models for attack detection were obtained. All prepared models were compared according to the determined success metrics and the results are presented in the Section 5.1. Information about the models used, their architectural structures, and parameter values were given in this section.

A 70% randomly selected dataset—that is, 15,937 rows of data—was used in the models' training. The remaining 30% was split into two to evaluate the testing and validation of the proposed model. Accordingly, 3415 rows were used for data validation and 3416 rows (including the class column) for testing. There were 22,768 rows of data (samples) in total.

Deep Learning-Based Models

The deep learning-based models used in the study are explained in this section. Analyses were made on the LSTM models, the CNN models, and the hybrid models in which LSTM-CNN were used together.

Since categorical data were included for all three proposed models, categorical_crossentropy was chosen as the loss function. Adaptive moment estimation (ADAM) was used as the optimization algorithm because it works efficiently on datasets containing many parameters [32]. In order to ensure stability, the rectified linear units' (ReLU) activation function was preferred. ReLU has a simple computational form and determines the output by evaluating the input [33]. The batch size was left by default. The softmax function was used to finish the classification.

LSTM-Based Models

In this model, analyses were performed on the LSTM algorithm. The LSTM algorithm is an iterative neural network and has been used frequently recently. Due to its structure, it is very effective in catching long-term addictions. It can store information for a long time with its special memory cell architecture. LSTM consists of repetitive sequential blocks known as memory blocks.

In this algorithm, there are input, output, and forget gates that enter and exit between cells and regulate the flow of information. For the iteration process, the input is generated, the predicted output value is obtained according to the current situation, and the next output vector is generated. Figure 5 shows the architecture of the LSTM-based deep learning models.

The first proposed model was based on deep learning using the LSTM algorithm. LSTM networks contain a sequential input layer. In the proposed LSTM network architecture, the LSTM layer was placed after the input layer. Next came a smoothing layer and, finally, the fully connected classification and output layers. In the study, two LSTM models with 200 epoch and 300 epoch parameters were prepared (LSTM1a and LSTM1b). Other parameters selected for the models were mentioned at the beginning of the chapter.

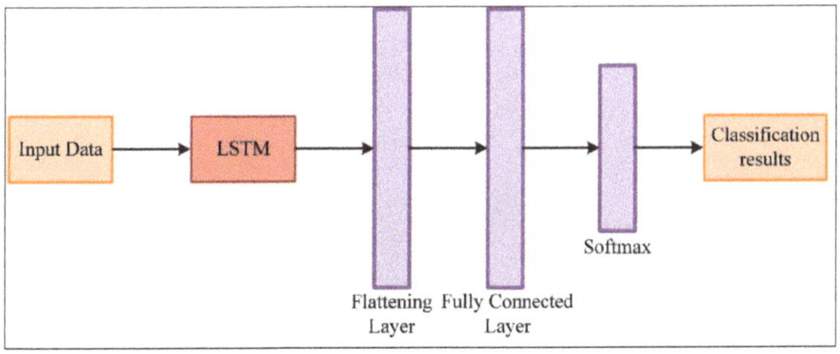

Figure 5. The architecture of the LSTM-based deep learning models.

CNN-Based Models

The CNN algorithm was studied in this model. The CNN algorithm is a variant of feed forward neural network. The architecture of the CNN algorithm is similar to the multilayer perceptron and consists of three layers. These are the convolution layer, pooling layer, and fully connected layer [34]. Multiple filters are included in this algorithm to extract or retrieve hidden features from the dataset. Figure 6 shows the architecture of the CNN-based deep learning models.

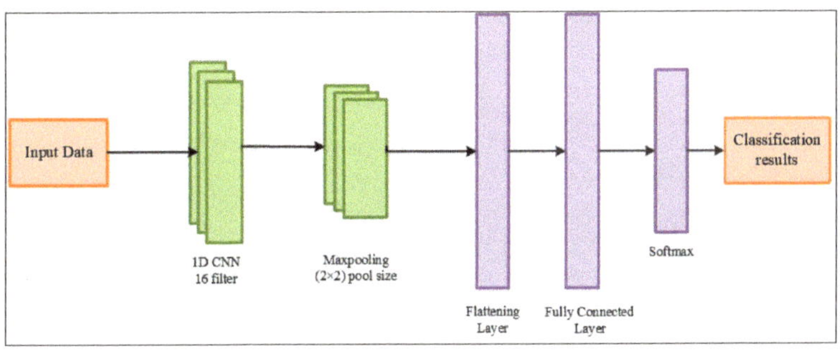

Figure 6. The architecture of the CNN-based deep learning models.

Deep learning was performed with the CNN1a model using 200 epochs and the CNN1b model using 300 epochs. Both models employed a 1-D CNN layer and pooling layer followed by normalization and flattening. Finally, connected, classification, and output layers were used. Other parameters selected for these models were explained at the beginning of the chapter.

Hybrid-Based Models

In the other model proposed in the study, LSTM and CNN algorithms were used as a hybrid and analyses were carried out. Three different models (HYBRID1, HYBRID2, and HYBRID3) were prepared using hybrid deep learning. The architecture of the first model (HYBRID1) is shown in Figure 7 as the others were prepared with reference to the first model.

In this hybrid model (HYBRID1), deep learning was performed using LSTM and CNN algorithms. Pooling layers and 1-D CNN layers were used. Normalization processes were done and, after the last pooling layer, the LSTM layer was placed in the model. Smoothing, fully connected, classification, and output layers were used. The HYBRID1a model with 200 epochs and the HYBRID1b model with 300 epochs were obtained.

In the second hybrid model (the HYBRID2), unlike the HYBRID1, normalization and activation processes were applied twice. Then, the HYBRID2a model was obtained by applying 200 epochs to the model and the HYBRID2b model was obtained by applying 300 epochs to the model.

The kernel size in the 1-dimensional CNN layers in the HYBRID1 model was increased and the number of filtering operations was reduced. In this way, the HYBRID3 model was obtained. The HYBRID3a model for 200 epochs and the HYBRID3b model for 300 epochs were prepared.

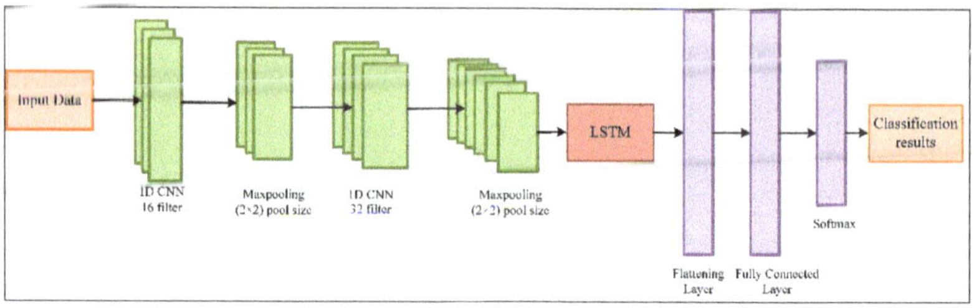

Figure 7. The architecture of the hybrid deep learning model (HYBRID1).

Machine Learning Based Models

When the literature was examined, it was seen that machine learning methods are also used in the detection of attacks on SCADA systems. Algorithms such as random forest, decision tree, logistic regression, Naive Bayes, and KNN were frequently used in the literature. In this study, in addition to deep learning algorithms, machine learning algorithms were also evaluated. Machine learning models were prepared for the detection of DDoS attacks and DDoS attack types for the testbed environment using the SCADA system. The results obtained were given in Table 3.

Table 3. Performance values of the proposed models.

Models		Accuracy (%)	Precision (%)	Recall (%)	F1-Score (%)
Deep Learning Based Models					
LSTM	LSTM1a	84.60	86.03	84.60	83.73
	LSTM1b	84.28	84.63	84.28	83.63
CNN	CNN1a	93.53	94.01	93.53	93.57
	CNN1b	94.26	94.79	94.26	94.35
LSTM-CNN HYBRID	HYBRID1a	94.09	94.23	94.09	94.12
	HYBRID1b	93.97	93.99	93.97	93.97
	HYBRID2a	93.91	94.05	93.91	93.93
	HYBRID2b	91.92	92.33	91.92	91.93
	HYBRID3a	92.77	92.99	92.77	92.82
	HYBRID3b	**94.73**	**94.90**	**94.73**	**94.74**
Machine Learning Based Models					
Lazy	KStar	79.93	81.93	79.95	79.03
	LWL	66.00	59.62	66.02	58.53
	KNN	86.15	86.08	86.15	86.11
Meta	LogitBoost	83.91	88.33	83.93	83.13
	AdaBoost	42.96	-	43.01	-
Bayes	NaiveBayes	84.03	85.43	84.00	83.54
	BayesNet	85.24	86.44	85.20	84.82
Rules	ZeroR	22.55	-	22.51	-
	PART	79.24	91.32	79.23	77.14
	DecisionTable	59.39	-	59.40	-
Trees	DecisionTree	**98.77**	**98.77**	**98.77**	**98.77**
	RandomForest	95.84	97.21	95.84	96.51
	RandomTree	83.07	85.71	83.14	82.44

KStar, locally weighted learning (LWL) and KNN algorithms from lazy learning methods were preferred. LogitBoost and AdaBoost algorithms from Meta Learning Methods and Naive Bayes and Bayes Net algorithms from Bayesian methods were used. ZeroR, PART, and decision Table algorithms based on rules and the decision tree, random forest, and random tree algorithms based on trees were analyzed.

5. Experimental Results

This section discussed the analysis results of the proposed deep learning-based, and machine learning-based models. In addition, the discussion section covered the comparison between previous studies and the current study for attack detection success. The analysis results of the proposed hybrid model on a different dataset were also placed in the table in the Discussion section.

5.1. Results

In the study, the steps mentioned in Title 4 were carried out on the dataset. Table 4 shows statistical information about network traffic captured while applying attack scenarios and the baseline situation. Captured packets in network traffic represent samples in datasets.

Table 4. Statistical information about network packets of attack scenarios.

	Attack Scenarios Values					
Measurement	**Normal**	**TCP Flooding**	**UDP Flooding**	**SYN Flooding**	**Spoofing IP Flooding**	**ICMP Flooding**
Total number of packets	3391	5253	3118	3238	3217	4551
Average packet size (bytes)	109	60	89	60	143	60
Total size of packet (bytes)	370,724	315,180	277,679	194,280	461,615	273,084
Duration of capture (ms)	530	294	253	163	286	271

As shown in Table 4, while there were 3391 packages in the normal situation scenario, there were 19,377 packages in the attack scenarios. The distribution of the number of packages was in a balanced state in all scenarios. The average sizes of packets (in bytes) were the same for TCP, SYN, and ICMP flooding attack scenarios. The spoofing IP flooding attack scenario had the maximum value. When the attack scenarios were analyzed separately, the total packet sizes (in bytes) took different values. In attack scenarios, when the packet capture times were examined, the most listening was done for the baseline situation. The least time was spent on the SYN Flooding attack scenario.

Analyses were made to reveal the attacks and DDoS attack types on the system. Suggestions were made for the attack detection system. Table 3 presents the analysis of the proposed models results.

When the performance results were examined, it was seen that the HYBRID3b model was more successful in analyzing and classifying DDoS attack data among deep learning algorithms. Among the machine learning algorithms, the highest success rate was obtained with the decision tree model. Considering the accuracy, precision, recall, and f1-score success metrics, these two models were found to be the most suitable models for attack detection. LSTM models from deep learning algorithms and the ZeroR model from machine learning algorithms performed the attack detection with the lowest success rate. The confusion matrix values obtained with the HYBRID3b model were placed in Table 5 and are shown below.

Table 5. Confusion matrix values of the proposed HYBRID3b model.

		Predicted Class							
		Normal (%)	TCP Flooding (%)	UDP Flooding (%)	SYN Flooding (%)	Spoofing IP Flooding (%)	ICMP Flooding (%)	TP Rate (%)	FN Rate (%)
Actual Class	**Baseline Situation**	88.27	0.00	8.65	0.00	3.08	0.00	88.30	11.70
	TCP Flooding	0.00	100	0.00	0.00	0.00	0.00	100	0.00
	UDP Flooding	3.67	0.00	92.01	0.00	4.32	0.00	92.10	7.90
	SYN Flooding	0.00	0.00	0.00	100	0.00	0.00	100	0.00
	Spoofing IP Flooding	6.34	0.00	9.90	0.00	83.76	0.00	83.80	16.20
	ICMP Flooding	0.00	0.00	0.00	0.00	0.00	100	100	0.00

The confusion matrix values in Table 5 were evaluated according to the accuracy metric frequently used in the literature [35]. Accordingly, among the attack types, TCP, SYN, and ICMP Flooding attacks were correctly detected with 100%. The worst detection performance was achieved in the spoofing IP flooding attack with a rate of 84%. For this attack, 423 of 505 samples were correctly detected. In the baseline situation, 459 of 520 samples were determined as not attacked and a high detection rate of 88% was obtained. The UDP flooding attack detection also had a rate close to the non-attack detection rate. The confusion matrix values obtained with the decision tree model were placed in Table 6 and shown below.

Table 6. Confusion matrix values of the proposed the decision tree model.

		Predicted Class							
		Normal (%)	TCP Flooding (%)	UDP Flooding (%)	SYN Flooding (%)	Spoofing IP Flooding (%)	ICMP Flooding (%)	TP Rate (%)	FN Rate (%)
Actual Class	**Baseline Situation**	98.08	0.00	0.58	0.00	1.34	0.00	98.10	1.90
	TCP Flooding	0.00	100	0.00	0.00	0.00	0.00	100	0.00
	UDP Flooding	0.86	0.00	95.25	0.00	3.89	0.00	95.30	4.70
	SYN Flooding	0.00	0.00	0.00	100	0.00	0.00	100	0.00
	Spoofing IP Flooding	0.59	0.00	2.18	0.00	97.23	0.00	97.30	2.70
	ICMP Flooding	0.00	0.00	0.00	0.00	0.00	100	100	0.00

The values in Table 6 were evaluated according to the accuracy metric. As in the HYBRID3b model, all TCP, SYN, and ICMP flooding attacks were correctly detected with 100% in the decision tree model. The UDP flooding attack was the worst-detected attack with 95%. For this attack, 441 of 463 samples were correctly detected. In the baseline situation, 510 of 520 samples were determined as non-attack and a high detection rate of 98% was obtained. Spoofing IP flooding attack detection also had a rate close to the non-attack detection rate.

5.2. Discussion

The analysis results of the two models with the highest success rates among the models proposed in the study (deep learning-based and machine learning-based) were compared with the analysis results of the studies in the literature. The results obtained are given in Table 7.

Table 7. Comparison of studies in the literature.

References	Datasets	Algorithms	Detection Rate (%)
[17]	Their own dataset	Random Forest	99.89
		Decision Tree	99.89
		Logistic Regression	99.59
		Naive Bayes	99.60
		KNN	72.29
[18]	Mississippi State University SCADA Laboratory	Neural Network	Average 83.00
[19]	Water distribution system real dataset	SVDD	Average 84.00
		Robust SVM	Average 76.00
		Slab SVM	Average 82.00
		Proposed Method	Average 91.00
[20]	Their own dataset	Hybrid SCADA-IDS	100
[21]	DUWWTP Dataset	KNN	92.86
[22]	CyberGym SCADA Lab dataset Ben-Gurion University of the Negev SCADA Lab Dataset	ANN-based SOM	Average 85.00
		HMM	Average 88
[23]	Their own dataset	FNN	Average 99.00
		LSTM	Average 99.00
		FNN-LSTM	Average 99.00
[24]	Their own dataset	OCSVM	96.30
[25]	Their own dataset	MTO-based LSTM	Average 99.00
		MTM-based LSTM	Average 98.00
[26]	ADFA-LD Dataset	IWP-CSO + SVM	91.50
		HNA-NN	83.20
		IWP-CSO + HNA-NN	93.10
		SVM	74.90
Our Study	Our Dataset	HYBRID3b Model	94.73
		Decision Tree Model	98.77
	Mississippi State University SCADA Laboratory	HYBRID3b Model	98.09

The study addresses the detection of DDoS attacks against the physical testbed using SCADA systems. For this, approaches based on deep learning and machine learning were used. The number of previous studies that detected attacks using ready-made datasets was very high. Fewer studies created a testbed for attack detection, prepared their own dataset, and performed analyses using the dataset. Both types of studies were equally included and reviewed.

Machine learning-based classifier methods such as KNN, Naive Bayes, and random forest were generally used in attack detection. There were also studies based on deep learning approaches such as LSTM and neural networks. As can be seen in Table 7, different algorithms were used for various datasets in the detection of attacks on SCADA systems. Each dataset had different characteristics and should be evaluated on its own.

In the studies examined in the literature, machine learning and deep learning approaches had achieved an average of over 90% success in attack detection on SCADA systems. As a result of the analyses performed in this study, two models based on deep learning and machine learning were proposed. With the hybrid model using LSTM and

CNN algorithms together, 95% success was achieved. A higher success rate of 99% was achieved with the decision tree-based model.

When we consider the existing research on the subject, we obtained promising models by creating an appropriate testbed, utilizing relevant technologies, and preparing dataset features. It is difficult to directly compare the performances of different models with the results obtained from studies using different datasets. Therefore, a different technique was used to demonstrate the performance of our proposed deep learning-based hybrid model. A dataset [5], which is frequently used in the literature and included in the benchmark table, was selected and evaluated for analysis. This dataset prepared by Morris et al. was analyzed with our proposed HYBRID3b model and a high success rate was obtained for attack detection. This result was shown in the last row of Table 7.

It has been observed that our proposed models have higher or very close performances compared to other models in the literature. Due to the diversification and development of attacks, it is important to carry out new analyses on different environments, and this has been achieved in this study. As a result, it is important that attack detection studies for SCADA systems are frequently updated and diversified.

6. Conclusions

The continuous functionality of a SCADA system enables smooth operation of cri-tical infrastructure systems. DDoS attacks against SCADA systems may interrupt the whole system causing functionality lost. Interruption in the operation of the SCADA system can be costly from both financial and time aspects. The methods proposed in the study will reinforce SCADA systems against cyberattacks. Thus, early DDoS attack detection on the system will be possible, and it will be easier to prevent disaster scenarios.

In this study, DDoS attacks were performed against the prepared testbed using the SCADA system. The obtained data both under the attacks and without attacks were recorded. LSMT, CNN, LSTM-CNN hybrid, and machine learning-based models were tested on the preprocessed dataset. After modifying the parameters of the models, various versions were obtained and used. The deep learning-based LSTM-CNN hybrid model achieved a classification accuracy of 95.00%, and the machine learning-based decision tree model achieved a classification accuracy of 99%. For a further evaluation of the success of the hybrid model, tests were conducted on a commonly used dataset in the literature which resulted in a high success rate of 98%. A higher success rate was achieved compared to the study in the literature using this dataset.

In addition to DDoS attack detection, DDoS attack type detection was also performed. With deep learning-based and machine learning-based models, all TCP, SYN, and ICMP flooding attacks were correctly detected. These models will provide high success and efficiency in the detection of such attacks. In this respect, it is aimed to contribute to the literature and provide guidance for future studies.

More detection studies should be carried out to reduce the effects of DDoS attacks on SCADA systems. Since SCADA systems are used in many different sectors, studies should be diversified by using different and new technologies and environments. In future studies, it should be an aim to prepare the SCADA system testbed environment more comprehensively and effectively. It should be an aim to apply different type of attacks other than DDoS attacks to this environment and to diversify the models used for their detection.

Author Contributions: The first author (E.S.) conducted experiments, wrote the manuscript, and executed software process. The second author (O.A.E.) was responsible for supervision and correcting the direction of the work. All authors have read and agreed to the published version of the manuscript.

Funding: The authors received no financial support for the research, authorship, or publication of this paper.

Institutional Review Board Statement: Not applicable.

Informed Consent Statement: Not applicable.

Data Availability Statement: Not applicable.

Conflicts of Interest: The authors declare no conflict of interest.

References

1. Fanuscu, M.C.; Kocak, A.; Alkan, M. Detection of Counter-Forensic Incidents Using Security Information and Incident Management (SIEM) Systems. In Proceedings of the 2022 15th International Conference on Information Security and Cryptography (ISCTURKEY), Ankara, Turkey, 19–20 October 2022; pp. 74–79. [CrossRef]
2. Domínguez, M.; Prada, M.A.; Reguera, P.; Fuertes, J.J.; Alonso, S.; Morán, A. Cybersecurity training in control systems using real equipment. *IFAC-PapersOnLine* **2017**, *50*, 12179–12184. [CrossRef]
3. Stouffer, K.; Pillitteri, V.; Lightman, S.; Abrams, M.; Hahn, A. Guide to Industrial Control Systems (ICS) Security. *NIST Spec. Publ.* **2015**, *800*, 16. [CrossRef]
4. Fabro, M. *Study on Cyber Security and Threat Evaluation in SCADA Systems*; Lofty Perch Inc Markham, Defence Research and Development Canada: Markham, ON, Canada, 2012; pp. 13–16.
5. Fink, K.R.; Spencer, D.F.; Wells, R.A. *Lessons Learned from Cyber Security Assessments of Scada and Energy Management Systems*; United States Department of Energy Office of Electricity Delivery and Energy Reliability: SW Washington, DC, USA, 2006.
6. Dominguez, M.; Fuertes, J.J.; Prada, M.A.; Alonso, S.; Morán, A.; Perez, D. Design of Platforms for Experimentation in Industrial Cybersecurity. *Appl. Sci.* **2022**, *12*, 6520. [CrossRef]
7. Söğüt, E.; Erdem, O.A. Endüstriyel Kontrol Sistemlerine (SCADA) Yönelik Siber Terör Saldırı Analizi. *J. Polytech.* **2019**, *23*, 557–566. [CrossRef]
8. Zhang, L. An Implementation of SCADA Network Security Testbed. Master's Thesis, University of Victoria, Victoria, BC, Canada, 2015.
9. Gao, W.; Morris, T.H. On Cyber Attacks and Signature Based Intrusion Detection for Modbus Based Industrial Control Systems. *J. Digit. Forensics Secur. Law* **2014**, *9*, 3. [CrossRef]
10. Queiroz, C.; Mahmood, A.; Tari, Z. SCADASim—A Framework for Building SCADA Simulations. *IEEE Trans. Smart Grid* **2011**, *2*, 589–597. [CrossRef]
11. Farwell, J.P.; Rohozinski, R. Stuxnet and the Future of Cyber War. *Survival* **2011**, *53*, 23–40. [CrossRef]
12. Available online: https://www.securityweek.com/remote-hacker-caught-poisoning-florida-city-water-supply/ (accessed on 5 March 2023).
13. Tesfahun, A.; Bhaskari, D.L. A SCADA testbed for investigating cyber security vulnerabilities in critical infrastructures. *Autom. Control. Comput. Sci.* **2016**, *50*, 54–62. [CrossRef]
14. de Brito, I.B.; de Sousa, R.T., Jr. Development of an open-source testbed based on the modbus protocol for cyber-security analysis of nuclear power plants. *Appl. Sci.* **2022**, *12*, 7942. [CrossRef]
15. Khan, A.A.Z. Misuse intrusion detection using machine learning for gas pipeline SCADA networks. In Proceedings of the International Conference on Security and Management (SAM), Las Vegas, NV, USA, 29 July–1 August 2019; pp. 84–90.
16. Polat, H.; Türkoğlu, M.; Polat, O.; Şengür, A. A novel approach for accurate detection of the DDoS attacks in SDN-based SCADA systems based on deep recurrent neural networks. *Expert Syst. Appl.* **2022**, *197*, 116748. [CrossRef]
17. Teixeira, M.A.; Salman, T.; Zolanvari, M.; Jain, R.; Meskin, N.; Samaka, M. SCADA System Testbed for Cybersecurity Research Using Machine Learning Approach. *Futur. Internet* **2018**, *10*, 76. [CrossRef]
18. Morris, T.; Srivastava, A.; Reaves, B.; Gao, W.; Pavurapu, K.; Reddi, R. A control system testbed to validate critical infrastructure protection concepts. *Int. J. Crit. Infrastruct. Prot.* **2011**, *4*, 88–103. [CrossRef]
19. Nader, P.; Honeine, P.; Beauseroy, P. Detection of cyberattacks in a water distribution system using machine learning techniques. In Proceedings of the 2016 Sixth International Conference on Digital Information Processing and Communications (ICDIPC), Beirut, Lebanon, 21–23 April 2016; pp. 25–30. [CrossRef]
20. Yang, Y.; McLaughlin, K.; Sezer, S.; Littler, T.; Im, E.G.; Pranggono, B.; Wang, H.F. Multiattribute SCADA-Specific Intrusion Detection System for Power Networks. *IEEE Trans. Power Deliv.* **2014**, *29*, 1092–1102. [CrossRef]
21. Almalawi, A.; Yu, X.; Tari, Z.; Fahad, A.; Khalil, I. An unsupervised anomaly-based detection approach for integrity attacks on SCADA systems. *Comput. Secur.* **2014**, *46*, 94–110. [CrossRef]
22. Kalech, M. Cyber-attack detection in SCADA systems using temporal pattern recognition techniques. *Comput. Secur.* **2019**, *84*, 225–238. [CrossRef]
23. Gao, J.; Gan, L.; Buschendorf, F.; Zhang, L.; Liu, H.; Li, P.; Dong, X.; Lu, T. Omni SCADA Intrusion Detection Using Deep Learning Algorithms. *IEEE Internet Things J.* **2020**, *8*, 951–961. [CrossRef]
24. Maglaras, L.A.; Jiang, J.; Cruz, T. Integrated OCSVM mechanism for intrusion detection in SCADA systems. *Electron. Lett.* **2014**, *50*, 1935–1936. [CrossRef]
25. Gao, J.; Gan, L.; Buschendorf, F.; Zhang, L.; Liu, H.; Li, P.; Dong, X.; Lu, T. LSTM for SCADA Intrusion Detection. In Proceedings of the 2019 IEEE Pacific Rim Conference on Communications, Computers and Signal Processing (PACRIM), Victoria, BC, Canada, 21–23 August 2019; pp. 1–5. [CrossRef]
26. Shitharth, S.; Prince Winston, D. An enhanced optimization based algorithm for intrusion detection in SCADA network. *Comput. Secur.* **2017**, *70*, 16–26. [CrossRef]

27. ADFA. Intrusion Detection Datasets. 2013. Available online: https://research.unsw.edu.au/projects/adfa-ids-datasets (accessed on 1 January 2023).
28. An Industrial Control System Cybersecurity Performance Testbed. 2015. Available online: http://nvlpubs.nist.gov/nistpubs/ir/2015/NIST.IR.8089.pdf (accessed on 25 December 2022).
29. Yang, Y.-S.; Lee, S.-H.; Chen, W.-C.; Yang, C.-S.; Huang, Y.-M.; Hou, T.-W. Securing SCADA Energy Management System under DDos Attacks Using Token Verification Approach. *Appl. Sci.* **2022**, *12*, 530. [CrossRef]
30. Güllü, M.; Akcayol, M.A.; Barışçı, N. Machine Learning-Based Comparative Study for Heart Disease Prediction. *Adv. Artif. Intell. Res.* **2022**, *2*, 51–58. [CrossRef]
31. Duman, E. Implementation of XGBoost Method for Healthcare Fraud Detection. *Sci. J. Mehmet Akif Ersoy Univ.* **2022**, *5*, 69–75.
32. Hernandez-Suarez, A.; Sanchez-Perez, G.; Toscano-Medina, L.K.; Olivares-Mercado, J.; Portillo-Portilo, J.; Avalos, J.-G.; Villalba, L.J.G. Detecting Cryptojacking Web Threats: An Approach with Autoencoders and Deep Dense Neural Networks. *Appl. Sci.* **2022**, *12*, 3234. [CrossRef]
33. Oyucu, S. A Novel End-to-End Turkish Text-to-Speech (TTS) System via Deep Learning. *Electronics.* **2023**, *12*, 1900. [CrossRef]
34. Krithivasan, K.; Pravinraj, S.; Shankar Sriram, V.S. Detection of Cyberattacks in Industrial Control Systems Using Enhanced Principal Component Analysis and Hypergraph-Based Convolution Neural Network (EPCA-HG-CNN). *IEEE Trans. Ind. Appl.* **2020**, *56*, 4394–4404. [CrossRef]
35. Demirtas, M.; Koc, K. Parameter Extraction of Photovoltaic Cells and Modules by INFO Algorithm. *IEEE Access* **2022**, *10*, 87022–87052. [CrossRef]

applied sciences

MDPI

Article

Location-Aware Measurement for Cyber Mimic Defense: You Cannot Improve What You Cannot Measure

Zhe Huang [1], Yali Yuan [1,2,*], Jiale Fu [3], Jiajun He [3], Hongyu Zhu [1] and Guang Cheng [1,4]

1 School of Cyber Science and Engineering, Southeast University, Nanjing 211189, China; chengguang@seu.edu.cn (G.C.)
2 State Key Laboratory for Novel Software Technology, Nanjing University, Nanjing 210093, China
3 School of Mathematics, Southeast University, Nanjing 211189, China
4 Jiangsu Province Engineering Research Center of Security for Ubiquitous Network, Nanjing 211189, China
* Correspondence: yaliyuan@seu.edu.cn

Abstract: Cyber mimic defense is designed to ensure endogenous security, effectively countering unknown vulnerabilities and backdoors, thereby addressing a significant challenge in cyberspace. However, the immense scale of real-world networks and their intricate topology pose challenges for measuring the efficacy of cyber mimic defense. To capture and quantify defense performance within specific segments of these expansive networks, we embrace a partitioning approach that subdivides large networks into smaller regions. Metrics are then established within an objective space constructed on these smaller regions. This approach enables the establishment of several fine-grained metrics that offer a more nuanced measurement of cyber mimic defense deployed in complex networks. For example, the common-mode index is introduced to highlight shared vulnerabilities among diverse nodes, the transfer probability computes the likelihood of risk propagation among nodes, and the failure risk assesses the likelihood of cyber mimic defense technology failure within individual nodes or entire communities. Furthermore, we provide proof of the convergence of the transfer probability. A multitude of simulations are conducted to validate the reliability and applicability of the proposed metrics.

Keywords: cyber mimic defense; complex network measurement; metrics

Citation: Huang, Z.; Yuan, Y.; Fu, J.; He, J.; Zhu, H.; Cheng, G. Location-Aware Measurement for Cyber Mimic Defense: You Cannot Improve What You Cannot Measure. *Appl. Sci.* **2023**, *13*, 9213. https://doi.org/10.3390/app13169213

Academic Editors: Peter R.J. Trim and Yang-Im Lee

Received: 30 June 2023
Revised: 9 August 2023
Accepted: 9 August 2023
Published: 13 August 2023

1. Introduction

The increasingly widespread application of the Internet in various social and economic sectors is leading to an increasingly severe challenge for cyberspace. Network security threats are becoming more diverse, complex, frequent, and widespread. In the current online environment, there exists a significant asymmetry between network attacks and defenses [1], often favoring the attackers. From the defensive perspective, it is generally difficult to anticipate when and how attacks will occur, making it challenging to deploy targeted defense strategies.

Traditional defense techniques, such as firewalls and intrusion detection techniques [2,3], typically rely on known attack signatures to identify and match target behaviors, leaving them at a disadvantage against unknown vulnerabilities and backdoors in cyber warfare. A series of novel proactive defense technologies are proposed to address this issue, such as honeypots [4] and Moving Target Defense (MTD) [5–7]. These methods effectively improved the situation and significantly increased the difficulty and cost for attackers to launch their attacks. However, they still have limitations: honeypot technology requires a significant amount of prior knowledge from attackers [8], and MTD possesses time sensitivity and uncontrollability, and the high-frequency variability, particularly, leads to a decline in system performance [7].

In fact, there is no defense strategy that can achieve absolute security. Due to the stage-specific nature of technological development and the level of awareness, vulnerabilities or backdoor issues in software and hardware design cannot be completely avoided. In the absence of the ability to eliminate inherent flaws and lack of prior knowledge, addressing the threat of unknown vulnerabilities and backdoors remains a significant challenge in cybersecurity. The emergence of the Cyber Mimic Defense (CMD) theory [9] provided a new idea and paradigm to tackle this problem and demonstrated effective defense capabilities in areas such as Software-Defined Networking (SDN) [10], cloud computing [11], distributed systems [12], etc.

The core framework of Cyber Mimic Defense (CMD) is "Dynamic Heterogeneous Redundancy" (DHR) [9], which is characterized by the following: (1) Dynamic: selecting a set of functional executors based on scheduling policies at the current moment and continuously changing this set to conceal the internal structure. (2) Heterogeneous: utilizing multiple heterogeneous executors with significantly different implementation methods to achieve the same functionality. (3) Redundancy: employing multiple executors and using an adjudication mechanism to determine the final system output.

Through its structural effects, CMD achieves endogenous defense effects that are independent of attack characteristics, effectively countering various attacks and unknown threats. However, there is currently a lack of universal metrics to directly measure the effectiveness of cyber mimic defense technology when applied to modern networks. We cannot improve what we cannot measure [13], and this principle applies to cyber mimic defense technology as well. The vast scale of real-world networks and the complexity of their topology pose challenges for evaluating the effectiveness of the cyber mimic defense. Therefore, there is an urgent need to develop general quantitative evaluation metrics for cyber mimic defense systems.

It is widely recognized that no single metric is powerful enough to fully reflect the impact of all relevant behaviors and defense strategies on the network. Therefore, we establish multi-dimensional evaluation metrics to assess the effectiveness of cyber mimic defense technology from various perspectives. We also found that most existing security strategies are typically evaluated based on the entire network. However, in many cases, even the best defense strategy may not necessarily extend security uniformly across the entire network, especially in large networks with hundreds or thousands of nodes. If we use security metrics based on the overall network assessment and observe improved security, it could be misleading as the security improvement may be limited to certain parts of the network. As mentioned earlier, asymmetry in network attacks and defenses exists, particularly in large-scale networks. The location from which attackers launch their attacks is difficult to predict, and the scope of protection provided by defense strategies is often limited. In such cases, global metrics fail to clearly reflect the defensive performance. In other words, metrics can reflect the overall defensive performance but cannot pinpoint the exact location of changes in defensive information. Therefore, it is necessary to adopt a location-aware method.

The main contributions of our work are summarized as follows. Firstly, to capture variations in attack-defense performance within specific local networks, we utilize a network partitioning method (i.e., Louvain algorithm) to segment the extensive network into smaller segments and perform correlation metrics on these segments to achieve a more fine-grained assessment. Subsequently, we establish relevant security metrics within the partitioned objective space for cyber mimic defense. These metrics encompass various quantitative measures such as the common-mode index and failure risk, which are tailored to accurately reflect the effectiveness of cyber mimic defense systems. This results in the creation of an innovative approach for effectively assessing cyber mimic defense deployed in complex networks. Finally, simulated attacks are carried out to verify the applicability and effectiveness of the proposed metrics.

This paper is organized in the following way. In Section 2, we introduce the preliminaries; in Section 3, we give the network partitioning method; in Section 4, we define metrics in the constructed objective space; in Section 5, we perform simulation experiments and give results; in Section 6, we list related work; and in Section 7, we make a conclusion of our work and propose a vision for future development.

2. Preliminaries

In this section, we provide a concise overview of the CMD framework, including its key concepts, and illustrate the actual topology of modern networks.

2.1. CMD Framework

As depicted in Figure 1, the cyber mimic defense system [9] primarily comprises six components: input agent, online set of executors, back-up executor pool, arbiter, scheduler, and output agent. During runtime, the input agent acquires input data and duplicates them to distribute among the heterogeneous executors in the online executor set. Each executor independently processes the data and produces an output. The arbiter then adjudicates the outputs of each executor based on a predefined algorithm to determine the final output. Additionally, the arbiter provides feedback on the adjudication to the scheduler, which uses this information to dynamically update the online executor set using specific strategies. The functional details of each component are described as follows.

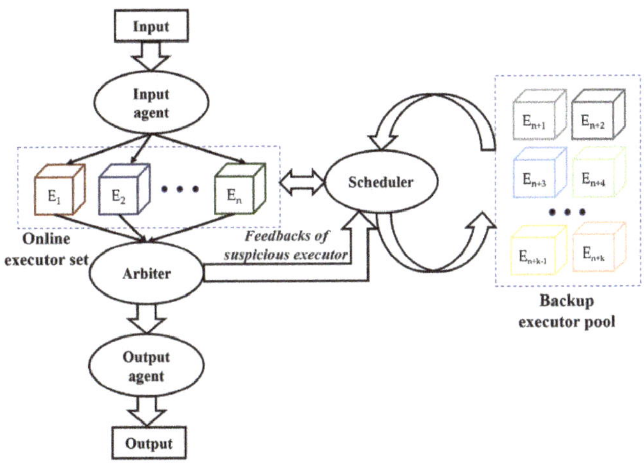

Figure 1. The Framework for Cyber Mimic Defense.

Input Agent: The input agent obtains the input, and copies and distributes it to each online executor.

Online executor set: Each online executor possesses an equivalent function and operates independently to process input. It is crucial to ensure a high degree of heterogeneity among the executors to mitigate common-mode vulnerabilities effectively. Once the calculations are completed, the results are transmitted to the arbiter for further processing.

Backup executor pool: The backup of online executors. The scheduler periodically or selectively chooses an instance to replace the currently active online executor set.

Arbiter: The arbiter adjudicates the output results of each executor based on a predefined algorithm and provides feedback regarding any suspicious executor to the scheduler.

Scheduler: The scheduler dynamically dispatches executors based on the operational status of online executors, handling tasks such as offline cleaning of suspicious executors and periodic replacement of executors.

Output Agent: The output agent obtains the voting results from the arbiter, formats it if necessary, and then outputs the final result.

With its endogenous security mechanism rooted in dynamic, heterogeneous, and redundancy strategies, cyber mimic defense establishes a spatiotemporal inconsistency scenario, preventing attackers from replicating past successes. It enhances the concealability and camouflage of the target defense scenario and behavior. Even in the event of an attack, the attacker cannot simultaneously breach all the actuators (exponential difficulty) [9], ensuring that the functions protected by the imitation system remain undisturbed and achievable. As a result, cyber mimic defense gains a more robust advantage when dealing with persistent, stealthy, and high-intensity offensive and defensive scenarios, especially in the presence of uncertain threats including unknown vulnerabilities, backdoors, and viruses.

2.2. Network Topology

Contrary to popular belief, most real-world networks do not exhibit random structures. Instead, they often follow a scale-free network concept [14], where a small number of nodes have a large number of connections, while the majority have only a few connections. Scale-free networks are complex networks characterized by a degree distribution that closely follows a power-law distribution. In such networks, the probability of a node having k connections (i.e., degree k) follows a power-law distribution, denoted as $P(k) \sim k^{-\gamma}$. The power exponent γ represents the structural properties of the network. The scale-free network exhibits significant heterogeneity, and the distribution of connections among its nodes is remarkably uneven, effectively simulating real-world network conditions.

In the era of the Internet of Everything, network-connected devices are diverse and encompass not only computers but also switches, sensors, smart home devices, and more. These devices can be regarded as nodes within the network. Let N represent the graph that illustrates the physical topology of the network, and we use a binary group of nodes and their connections to represent the network, denoted as $N = \langle V, E \rangle$. Here, V refers to the devices in the network, collectively known as hosts, and E represents the undirected edge connections between them. Subsequent studies are based on the proposed network topology as described above.

3. Network Partitioning

In this section, we discuss the necessity of performing network partitioning on complex networks for evaluating CMD and present how to utilize the Louvain algorithm for network partitioning.

Due to the immense scale of modern networks, conducting an evaluation of CMD technology from a global perspective in complex network environments is often imprecise and challenging. The location of attacker intrusions is random and unpredictable, and the effective coverage of CMD technology typically cannot encompass the entire large-scale network. This means that attacks conducted outside the effective range of CMD may remain largely unaffected. If an overall improvement in security is observed from global evaluation metrics, it could lead to misjudgments about the effectiveness of CMD technology. Therefore, it is necessary to adopt a divide-and-conquer approach, evaluating the effectiveness of CMD technology within smaller regions to enhance the accuracy and applicability of the metrics.

As mentioned in Section 2.2, there is significant heterogeneity in the connections between nodes in real-world networks. This often leads to the aggregation of nodes, forming communities within the network. Community structure is one of the essential features of complex networks [15]. Each module or community is composed of closely connected individuals due to similar structural characteristics and positions. Community detection methods are specifically designed to partition the internal structure of complex networks with the goal of grouping network nodes into tightly connected communities. Compared to other traditional methods, community detection methods pay more attention to the patterns of connections between nodes, allowing for better capture of the local structural characteristics of the network. We utilize a typical community detection algorithm, the Louvain algorithm, to partition complex networks, which is described in detail below.

The Louvain algorithm [16] is grounded in multilevel optimization of modularity, offering the advantage of speed and accuracy in obtaining a hierarchical community structure with approximately linear time complexity. It takes a heuristic approach to maximize the local modularity of smaller communities, joining only if such aggregation leads to an increase in modularity. It is the preferred method for clustering (community detection) of complex networks [17] and was rated as one of the best community detection algorithms by [18]. Two key concepts are integral to the algorithm: modularity Q and modularity gain ΔQ [16], for which we provide the relevant formulas below.

- Modularity (Q)

$$Q = \sum_{C}[\frac{\sum_{in}}{2m} - (\frac{\sum_{tot}}{2m})^2], \tag{1}$$

where m denotes the total number of edges in the graph, $\sum_{in}$ denotes the sum of the weights of the edges interconnected within community C, and $\sum_{tot}$ denotes the sum of the weights of the edges connected to the nodes of community C, including the edges inside the community as well as the edges outside the community.

- Modularity gain (ΔQ)

$$\begin{aligned}\Delta Q(i \to C) &= [\frac{\sum_{in} + k_{i,in}}{2m} - (\frac{\sum_{tot} + k_i}{2m})^2] - \\ &[\frac{\sum_{in}}{2m} - (\frac{\sum_{tot}}{2m})^2 - (\frac{k_i}{2m})^2] \\ &= \frac{1}{2m}(k_{i,in} - \frac{\sum_{tot} k_i}{m}),\end{aligned} \tag{2}$$

where k_i denotes the sum of the weights of the edges connected to node i and $k_{i,in}$ denotes the sum of the weights of the edges of node i connected to the nodes in community C.

The main flow of the Louvain algorithm is as follows:

1. Initially, each node is regarded as a separate community;
2. For each node i, try to assign it to a neighbor community in turn and calculate the modularity gain ΔQ after assignment, find the assignment method with the maximum modularity gain and assign it if its $\Delta Q > 0$, otherwise leave it unchanged;
3. Repeat the steps in 2 until the communities in which all nodes are located no longer change;
4. Compress a community into a new node, convert the weights of edges interconnected by nodes within the community to the weights of the ring of the new node, and convert the weights of edges between communities to the weights of the edges between the new nodes;
5. Repeat the above steps until the results converge.

Through the Louvain algorithm, we distinguish the network structure hierarchically with a high degree of association between hosts within the community. Next, we use the delineated communities as the objective space to develop the definition of metrics for cyber mimic defense.

4. Metrics in the Objective Space

In this section, based on the objective space constructed by network partitioning, we develop multidimensional evaluation metrics to measure the effectiveness of cyber mimic defense technology.

4.1. Single Node

Definition 1 (Network topology). *We represent the network community as a binary group $NC_i = (N, E)$, where N is the set of all nodes (hosts) in the network, including switches, routers, firewalls, etc., and E is the set of connection relationships between these nodes.*

Definition 2 (Vulnerability set). *A collection of all possible vulnerabilities, especially zero-day vulnerabilities.*

$$VUL = VUL_1 \cup VUL_2 \cup \cdots \cup VUL_n,$$

where VUL_i represents the set of all vulnerabilities on node Ni (suppose n nodes in the community) and $VUL_i = \{vul_j | vul_j$ is a certain vulnerability on node $V_i\}$.

In CMD systems, the adjudication algorithm generally follows the "majority voting" principle. Consequently, when more than half of the executors possess the same symbiotic vulnerabilities and are successfully exploited by an attacker, they can potentially deceive the arbiter and allow the attack to evade detection, similar to an environment where CMD is not deployed. As a result, we propose the following hypothesis.

Assumption 1. *For a CMD system with (2l + 1) online executors, when there exists an (l + 1) order symbiotic vulnerability vul_t, it is considered that the node where the CMD system is deployed has vulnerability vul_t.*

Definition 3 (Vulnerability vector). *Construct vulnerability vector V_i for each node.*

$$\mathbf{V_i} = (v_1, v_2, \ldots, v_m)^T, m = |VUL|,$$

$$v_k = \begin{cases} 1 & vul_k \in VUL_i \\ 0 & vul_k \notin VUL_i \end{cases}, k = 1, 2, \cdots, m.$$

Definition 4 (Node–vulnerability matrix). *The indication of the corresponding relationship between nodes and vulnerabilities.*

$$\mathbf{NV_{n*m}} = (\mathbf{V_1}, \mathbf{V_2}, \ldots \mathbf{V_n})^T.$$

Definition 5 (CVSS vector). *Construct vulnerability score vector CVSS for each vulnerability.*

$$\mathbf{CVSS} = (cvss_1, cvss_2, \ldots, cvss_m)^T,$$

where $cvss_i(i = 1, 2, \ldots, m)$ represents one-tenth (For normalization) of the Common Vulnerability Scoring System (CVSS) score for the corresponding vulnerability.

Definition 6 (Importance vector). *Construct importance vector IM for each node considering the centrality (location of nodes in the community) and value (value of resources owned by nodes).*

$$\mathbf{IM} = (im_1, im_2, \ldots, im_n)^T,$$

$$im_k = w_1 \times centrality + w_2 \times value, \; k = 1, 2, \ldots, n.$$

where $w_i(i = 1, 2)$ represents the weight of the corresponding factors, $\sum w_i = 1$. Here, the weights and factors can be appropriately adjusted according to the actual situation.

In Definitions 2–5, the vulnerability set and CVSS score can be obtained from the open CVE vulnerability database. In Definition 6, the value depends on the property and resources owned by the nodes, and the centrality calculation method needs to be selected according to the actual network situation from degree centrality, betweenness centrality, closeness centrality, etc., and all of these concepts are defined.

Independent failure risk.

In CMD systems, the heterogeneity among different executors within the redundant structure is crucial and directly impacts the overall performance of the model. When multiple executors share the same vulnerability, it can lead to attacks escaping detection, rendering the cyber mimic defense strategy ineffective. From the perspective of individual nodes, we define the independent failure risk based on vulnerabilities, represented as an n-dimensional vector, where the i-th component represents the independent failure risk of node N_i. In the formula, the importance vector **IM** represents the likelihood of an attacker choosing to target a node, while **NV** × **CVSS** represents the likelihood of successfully compromising a node if targeted in an attack.

$$\mathbf{RI} = \mathbf{IM}^T \times \mathbf{NV} \times \mathbf{CVSS}. \quad (3)$$

4.2. Relationship between Nodes

Definition 7 (Executor set). *The set of executors in the CMD system, each capable of independently implementing service functions, is denoted as $A = \{A_1, A_2, \ldots\}$, where A_i represents a specific executor.*

Definition 8 (Higher-order symbiotic vulnerability). *Exploitable vulnerabilities that can achieve the same attack effect for m executors in executor set A ($m \geqslant 3$).*

Definition 9 (Adjacency matrix). *The adjacency matrix represents the adjacency between nodes (We assume undirected edges in the community, so it is a symmetric matrix).*

$$\mathbf{Adj} = (a_{ij})_{n\times n},$$

$$a_{ij} = \begin{cases} 1 & edge_{ij} \in E \\ 0 & edge_{ij} \notin E \end{cases}.$$

Common-mode index.

When multiple nodes share similar vulnerabilities, attackers can rapidly exploit these vulnerabilities on one node, potentially affecting multiple nodes in a similar or identical manner. This leads to the rapid horizontal spread of the attack's impact, resulting in irreversible consequences. Nodes within the same network community are closely interconnected and often exhibit similar or identical component structures in the real environment, such as accessory modules purchased from the same batch of manufacturers. Therefore, we introduce the concept of the common-mode index to measure the similarity within a community, thereby revealing potential security risks.

We define the common-mode index between node N_s and node N_t ($N_s, N_t \in N$) as follows.

$$I_{(N_s,N_t)} = \frac{\sum_{vul_k \in (VUL_s \cap VUL_t)} CVSS_{vul_k}}{\sum_{vul_k \in (VUL_s \cup VUL_t)} CVSS_{vul_k}}, \quad (4)$$

where $CVSS_{vul_k}$ indicates the CVSS score of the vulnerability vul_k to characterize the magnitude of the vulnerability's harm.

Considering the particularity of the CMD system, we need to make another consideration for the node where CMD is deployed. In CMD, multiple redundant executors independently run the output results and obtain the final results through adjudication, and the online executors are in a state of dynamic transformation. Due to these characteristics, the cognition of the vulnerability set of the node where CMD is deployed needs to be changed. After deploying CMD on the node, its vulnerability set is in a dynamic state. Accordingly, we give the definition of the common-mode index between the node N_{cmd} where CMD is deployed and the ordinary node N_x.

$$I_{(N_{CMD},N_x)} = \sum I_{(N_{CMD},N_x),t_i} \times \frac{t_i}{T}, \quad (5)$$

where $I_{(N_{cmd},N_x),t_i}$ denotes the common-mode index in a period t_i for a dynamically updated CMD vulnerability set and T denotes a period as long as possible to show the possible states of the executor set.

Based on the above definition, we integrate the common-mode index between nodes into matrix form, as shown below.

$$\mathbf{CM} = (c_{ij})_{n\times n},$$

$$c_{ij} = I_{(N_i,N_j)}. \quad (6)$$

Transfer probability.

When an adversary breaks through a host, they often use this host as a base and then launch attacks on other hosts to expand the control range and spread worms and viruses. Typically, the attack spreads from the compromised host to neighboring hosts, gradually infecting the entire network. Considering the attack transfer between neighboring nodes, we propose the concept of transfer probability.

We construct the transfer matrix to represent the single-step transfer probability.

$$\mathbf{P} = \begin{bmatrix} p_{11} & p_{12} & \cdots & p_{1n} \\ p_{21} & p_{22} & \cdots & p_{2n} \\ \vdots & \vdots & \ddots & \vdots \\ p_{n1} & p_{n2} & \cdots & p_{nn} \end{bmatrix}, \tag{7}$$

where n represents the number of nodes in the community and p_{ij} denotes the probability of an attacker moving from node i to node j in a single step, which is defined as follows.

$$p_{ij} = \begin{cases} 0 & a_{ij} = 0 \\ \frac{p_1}{d_i} & a_{ij} = 1 \; and \; t_{ij} = 0, \\ \frac{p_2}{d_i} & a_{ij} = 1 \; and \; t_{ij} = 1 \end{cases} \tag{8}$$

where p_1, and p_2, respectively, represent the success rate of transfer from node i to node j with or without common-mode vulnerability (prior knowledge is needed for machine learning in practical application), d_i denotes the degree of the i-th node, and a_{ij} is an element in the adjacency matrix **A**. If $a_{ij} = 1$ then it means there is an edge between node i and node j, $a_{ij} = 0$ means there is no edge between node i and node j, and t_{ij} is an element in $\mathbf{T_{n*n}}$, which is defined as follows.

$$\mathbf{T} = (t_{ij})_{n*n},$$

$$t_{ij} = \begin{cases} 1 & VUL_s \cap VUL_t \neq \varnothing \\ 0 & VUL_s \cap VUL_t = \varnothing \end{cases}.$$

From the above definition, we can know:

1. $0 \leqslant p_1 \leqslant p_2 \leqslant 1$
2. $\mathbf{P} = \mathbf{P^T}$

Before starting the definition of transfer probability, a related lemma and a theorem are given.

Lemma 1. *$\|P\|_1 < 1$, where $\|\cdot\|_1$ is the 1-norm of the matrix.*

We know that $\|P\|_1 = \max_{1 \leqslant i \leqslant n} \sum_{j=1}^{n} |p_{ij}|$, so to prove $\|P\|_1 < 1$, we have to prove $\max_{1 \leqslant i \leqslant n} \sum_{j=1}^{n} |p_{ij}| < 1$, that is, to prove $\forall i, \sum_{j=1}^{n} |p_{ij}| < 1$.

$\forall i$, we can prove that $\sum_{j=1}^{n} |p_{ij}| = \sum_{j=1}^{n} p_{ij} = \sum_{j=1}^{n} a_{ij} p_{ij} \leqslant \sum_{j=1}^{n} a_{ij} \frac{p_2}{d_i} = \frac{p_2}{d_i} \sum_{j=1}^{n} a_{ij} = \frac{p_2}{d_i} d_i = p_2$.

Therefore, $\|P\|_1 \leqslant p_2 < 1$. The lemma is proved.

Theorem 1. *If $\|P\| < 1$, then $I + P + P^2 + \cdots P^n + \cdots$ converges, and $I + P + P^2 + \cdots P^n + \cdots = (I - P)^{-1}$.*

Since $\|P\| < 1$, then $\|I\| + \|P\| + \|P\|^2 + \cdots + \|P\|^n + \cdots$ converges. And since the completeness of $(P_{n*n}, \|\cdot\|)$, then $I + P + P^2 + \cdots P^n + \cdots$ converges.

$$
\begin{aligned}
&(I-P)\left(I+P+P^2+\cdots P^n+\cdots\right)\\
=&\left(I+P+P^2+\cdots P^n+\cdots\right)\\
&-\left(P+P^2+\cdots P^n+\cdots\right)\\
=&I.
\end{aligned}
$$

Therefore, $I+P+P^2+\cdots P^n+\cdots=(I-P)^{-1}$. *The theorem is proved.*

Combining the basic transfer factor ε_0 and the multi-step transfer case, we define the transfer probability matrix as follows.

$$
\begin{aligned}
\mathbf{TP}&=\varepsilon_0\mathbf{Adj}+(\mathbf{P})+(\mathbf{P})^2+(\mathbf{P})^3+\cdots\\
&=\varepsilon_0\mathbf{Adj}+(\mathbf{I}-\mathbf{P})^{-1}-\mathbf{I}.
\end{aligned}
\tag{9}
$$

4.3. Entire Community

Comprehensive failure risk.

In the previous sections, we established quantitative evaluation metrics for individual nodes and between nodes. By combining these metrics, we defined the comprehensive community failure risk, which assesses the local effectiveness of deploying the cyber mimic defense strategy. We formulated it as a quadratic expression as shown below, where **RI** represents the individual node failure risk, and **CM**$\odot$**TP** represents the probability of attack spread.

$$RC=\mathbf{RI}^{\mathrm{T}}\times(\mathbf{CM}\odot\mathbf{TP})\times\mathbf{RI},\tag{10}$$

where $\odot$ denotes the Hadamard product, i.e., the multiplication of corresponding elements.

5. Simulation

In this section, we conduct simulation experiments and comparative analysis to validate the effectiveness and rationality of the above metrics.

Firstly, we use the NetworkX package in Python to generate a scale-free network structure with a large number of nodes and apply the Louvain algorithm to partition the network. After a limited number of iterations, the finite number of communities was successfully divided. We color different communities separately for visualization, and the example effect is shown in Figure 2.

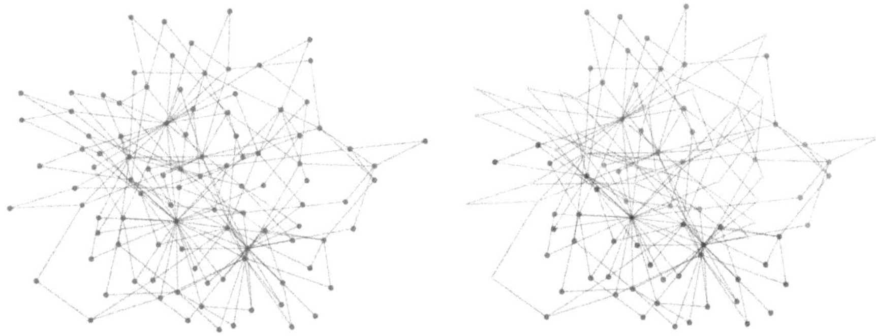

(**a**) Before division (**b**) After division

Figure 2. The example effect of community division.

Secondly, we construct the vulnerability set by selecting n vulnerabilities from the open CVE database and then generate each component of the vulnerability vector corresponding to each node with probability p. For nodes with CMD deployed, we add up the vulnerability vectors of each executor in the online executor set (assuming a total of $2l + 1$ online executors), and consider a component in the resulting sum vector as indicating the presence of a specific vulnerability if it is greater than l.

After completing the community partitioning and selecting the vulnerability set, we utilize the network topology structure, vulnerability information, and pre-defined asset values as inputs to calculate the metric values according to the formulas provided in Section 4. To validate the reasonableness of the proposed metrics, we conduct simulated attacks and compared the computed metric values with the results of the simulated attacks

Since the location of the attack initiation in the network is generally unknown, it can be regarded as a random event [19]. We simulate the attacks from a probabilistic perspective, where each simulation involves multiple attacks, and each attack is considered independent. A simulated attack can be divided into the following three phases: initial attack, horizontal spread, and clearance. (1) Initial attack: The attacker randomly selects a node and a vulnerability v from the vulnerability set to attack. If the chosen node possesses vulnerability v, the attack is considered successful; otherwise, it fails. (2) Horizontal spread: If the attacker successfully infiltrates the network in the initial attack, at each time step, it can attempt to spread to neighboring nodes. If a neighbor node lacks vulnerability v, the success rate of spreading to it is denoted by p_1; otherwise, it is set to p_2 (where $p_1 < p_2$). (3) Clearance: Considering that both regular nodes and CMD nodes have their own checking and clearing mechanisms, we assume that at each time step, there is a certain probability of the attacker being detected and cleared by the node's protection mechanism. For regular nodes, there is a probability of p_rec to find and clear exploitable vulnerabilities at each time step (p_rec is set based on the actual probability). For CMD nodes, if fewer than half of the executors have vulnerability v in the dynamic scheduling of each time step, the attacker will not achieve their goal under the CMD's ruling mechanism. This scenario is equivalent to the vulnerability being cleared. The simulated attack continues until it no longer spreads, and then this round of simulated attack is concluded.

In the simulated attacks mentioned above, the number of simulated attacks on nodes or communities can reflect their actual vulnerability to some extent. Combining the calculated failure risk of nodes and communities (i.e., independent and comprehensive failure risk) with the number of attacks, we draw a scatter plot on the two-dimensional coordinate system, as shown in Figure 3. After fitting the scatter points, it can be seen that the number of attacks is roughly proportional to the calculated failure risk. We also performed a comparative analysis of the transfer probabilities between nodes, part of which is visualized in Figure 4. The squares within the i-th row and j-th column of the figure represent the transfer probability from node i to node j, with color intensity denoting the probability magnitude Comparing the calculated and experimental results, we observe a close alignment. All of these indicate that our proposed metrics have excellent practical application value.

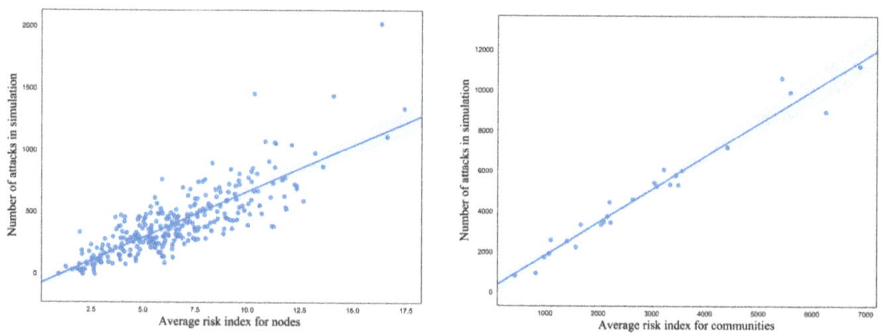

(a) Risk for nodes (b) Risk for communities

Figure 3. The relationship between failure risk and number of attacks.

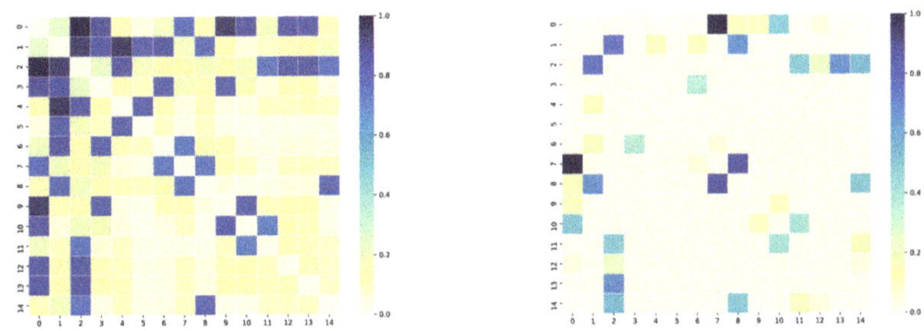

(**a**) Calculated values (**b**) Experimental values

Figure 4. Transfer probability between nodes.

In addition, considering the scale of the network and the estimated deviation of p_1 and p_2, we calculated the coefficients of correlation between the theoretical vulnerability and the actual vulnerability and summarized them in Tables 1–3. As shown in Table 1, we tested the effectiveness of evaluation metrics for nodes and communities under different network scales, and we can see that when the network scale gradually expands, our indicators fit well with the actual situation, especially for our community-based research ideas. In Tables 2 and 3, we consider the effect of the metrics when the estimated values p_1 and p_2 in the transition probability mentioned above exhibit some deviation. The majority of data in these tables exceed 0.7, signifying a robust correlation coefficient. This implies that despite potential estimation deviation, our metrics retain substantial error tolerance.

Table 1. Average coefficient of correlation for nodes and communities under different variables.

Number of Trials	Number of Nodes	Number of Simulated Attacks	Average Coefficient of Correlation for Nodes	Average Coefficient of Correlation for Communities
100	100	10,000	0.73	0.91
100	200	20,000	0.71	0.94
50	500	50,000	0.71	0.97
20	1000	100,000	0.70	0.97
20	2000	200,000	0.69	0.98

Finally, two comparative experiments are given, and the experimental data are shown in Table 4. We compared with two related models [13,20] to further substantiate our model's performance. The outcomes demonstrate that our method yields favorable results for calculating the correlation coefficients of nodes and communities within the intricate network environment. Specifically, our approach outperforms other methods in terms of nodes, and as the number of nodes progressively increases, the superiority of our model becomes particularly pronounced within the community context.

Table 2. Coefficient of correlation for nodes considering the estimated deviation of p_1 and p_2.

Coefficient of Correlation / Estimated Deviation of p_2 / Estimated Deviation of p_1	−5%	−4%	−3%	−2%	−1%	0%	1%	2%	3%	4%	5%
−5%	0.92	0.98	0.56	0.99	0.95	0.99	0.83	0.94	0.99	0.91	0.86
−4%	0.95	0.96	0.97	0.96	0.96	0.97	0.98	0.96	0.67	0.96	0.99
−3%	0.92	0.95	0.95	0.97	0.74	0.99	0.96	0.98	0.97	0.74	0.96
−2%	0.98	0.89	0.99	0.83	0.92	0.96	0.93	0.85	0.92	0.96	0.94
−1%	0.78	0.95	0.88	0.95	0.71	0.99	0.94	0.92	0.98	0.93	0.78
0%	0.96	0.95	0.87	0.98	0.88	0.99	0.84	0.90	0.97	0.91	0.97

Table 2. *Cont.*

Coefficient of Correlation / Estimated Deviation of p_1 \ Estimated Deviation of p_2	−5%	−4%	−3%	−2%	−1%	0%	1%	2%	3%	4%	5%
1%	0.95	0.91	0.93	0.93	0.93	0.95	0.99	0.92	0.91	0.96	0.91
2%	0.96	0.90	0.88	0.93	0.95	0.96	0.92	0.90	0.94	0.98	0.98
3%	0.92	0.98	0.97	0.83	0.97	0.96	0.89	0.96	0.87	0.95	0.98
4%	0.92	0.90	0.98	0.94	0.93	0.82	0.54	0.90	0.84	0.96	0.86
5%	0.94	0.93	0.94	0.92	0.82	0.95	0.97	0.96	0.97	0.95	0.96

Table 3. Coefficient of correlation for communities considering the estimated deviation of p_1 and p_2.

Coefficient of Correlation / Estimated Deviation of p_1 \ Estimated Deviation of p_2	−5%	−4%	−3%	−2%	−1%	0%	1%	2%	3%	4%	5%
−5%	0.66	0.78	0.57	0.77	0.74	0.82	0.72	0.75	0.77	0.73	0.82
−4%	0.72	0.72	0.80	0.77	0.76	0.78	0.81	0.72	0.73	0.80	0.77
−3%	0.79	0.84	0.76	0.76	0.80	0.79	0.69	0.65	0.74	0.74	0.75
−2%	0.77	0.61	0.79	0.78	0.70	0.73	0.68	0.71	0.69	0.70	0.75
−1%	0.71	0.74	0.62	0.73	0.64	0.80	0.72	0.78	0.72	0.74	0.66
0%	0.78	0.76	0.74	0.74	0.75	0.85	0.74	0.63	0.74	0.79	0.76
1%	0.74	0.73	0.77	0.74	0.71	0.70	0.78	0.74	0.78	0.76	0.76
2%	0.70	0.69	0.74	0.74	0.74	0.67	0.72	0.63	0.72	0.74	0.80
3%	0.83	0.67	0.77	0.73	0.79	0.70	0.77	0.76	0.76	0.76	0.84
4%	0.80	0.76	0.74	0.78	0.81	0.77	0.70	0.79	0.68	0.74	0.64
5%	0.78	0.77	0.72	0.49	0.65	0.81	0.79	0.80	0.63	0.72	0.78

Table 4. Average coefficient of correlation for nodes and communities for different models.

Number of Trials	Number of Nodes	Number of Simulated Attacks	Models	Average Coefficient of Correlation for Nodes	Average Coefficient of Correlation for Communities
			Our model	0.73	0.91
100	100	10,000	Model 1	0.35	0.66
			Model 2	0.65	0.98
			Our model	0.71	0.94
100	200	20,000	Model 1	0.39	0.72
			Model 2	0.60	0.98
			Our model	0.71	0.97
50	500	50,000	Model 1	0.48	0.82
			Model 2	0.52	0.97
			Our model	0.70	0.97
20	1000	100,000	Model 1	0.56	0.88
			Model 2	0.58	0.95
			Our model	0.69	0.98
20	2000	200,000	Model 1	0.55	0.92
			Model 2	0.52	0.96

6. Related Works

Security Strategy Measurement. The importance of evaluation metrics in evaluating the effectiveness of new security strategies that are constantly emerging cannot be overstated. Lingyu Wang et al. proposed a theoretical model based on zero-day security, combining the value of target assets and the shortest attack sequence to obtain k-zero-day security metrics [21]. Jin B. Hong et al. classified and proposed a series of performance metric definitions based on different characteristics of attack and defense behaviors, including attack cost, attack path exposure time, defense deployment cost, and downtime [22] Jin B. Hong et al. also used the hierarchical attack representation model and the importance measure to evaluate the effectiveness and scalability of MTD technology [23]. Hai Jin et al. proposed a security framework that automatically senses and updates in container-based cloud environments and builds a multidimensional attack graph model to analyze attack behavior [11]. Warren Connell et al. proposed a maximizing utility function approach to capture the trade off between security and performance [24]. Luis Muñoz-González et al. modeled attack graphs and used Bayesian inference to perform static and dynamic analysis [25]. Luis Muñoz-González et al. also proposed a Bayesian-based probabilistic graphical model to estimate the vulnerability and interconnection of system components and calculate the attack probability of target nodes to determine security [26]. Mengyuan Zhang et al. evaluated the network diversity based on the effective quantity of different resources, and the minimum and average attack effort, respectively [27]. These studies typically conduct evaluations from a global perspective. However, due to the significant asymmetry in network attacks and defenses, especially in complex networks, they are unable to identify specific regions where security benefits can be obtained.

Cyber Mimic Defense Measurement. As research on cyber mimic defense unfolds, how to evaluate the effectiveness of CMD deployment becomes a key issue. Fei Yu et al. conducted a series of experiments on basic, common-mode, and differential-mode attacks to obtain the defense success rate and analyzed the delay and throughput to reflect their performance loss [28]. Congqi Shen et al. proposed a decentralized multi-adjudicator arbiter approach to determine the defense effectiveness using the consistent convergence of subarbiters after data injection attacks [29]. Quan Ren et al. analyzed the applicability of cyber mimic defense in a software-defined network from the aspects of availability, response time, compromise tolerance, and performance [30]. Haiyang Yu et al. studied the effect of cyber mimic defense in a distributed system from the aspects of data reliability, fault repair, and security [31]. Chen Yu et al. analyzed the security and effectiveness of mimic DAA scheme [32]. Wei Liu et al. evaluated the mimic defense strategy in terms of storage limitation, throughput, and algorithm speed [33]. Yufeng Zhao et al. constructed a security quantification model from multiple angles, analyzed the different characteristics of cyber mimic defense architecture, and achieved a relatively complete security quantification method [34]. These studies primarily focus on measuring the security of cyber mimic defense system itself, and there is currently a lack of research on evaluating the effectiveness of deploying cyber mimic defense in large-scale networks.

7. Conclusions

In this paper, we propose a series of cyber mimic defense evaluation metrics by partitioning the complex network with the idea of the Louvain algorithm and mapping it to the objective space for finer-grained evaluation, incorporating common-mode index, transfer probability, and failure risk. Numerous simulation results demonstrate that our proposed metrics are highly reliable and can accurately reflect the effectiveness of cyber mimic defense technology deployed in complex networks. In future research, we will further refine the metrics for cyber mimic defense and integrate them with real-world scenarios. We believe that this work will inspire researchers in related fields and contribute to the improvement of the cyber mimic defense measurement.

Author Contributions: Conceptualization, Z.H. and Y.Y.; Methodology, Z.H., Y.Y., J.F. and H.Z.; Software, Z.H. and J.F.; Validation, Z.H., J.F. and H.Z.; Formal analysis, Z.H. and Y.Y.; Investigation, Z.H. and H.Z.; Resources, Z.H., Y.Y., J.F. and J.H.; Data curation, Z.H., Y.Y. and H.Z.; Writing—original draft, Z.H., Y.Y. and H.Z.; Writing—review & editing, Z.H., Y.Y. and J.H.; Visualization, Z.H.; Supervision, Y.Y.; Project administration, G.C.; Funding acquisition, G.C. All authors have read and agreed to the published version of the manuscript.

Funding: This research was funded by the National Key Research and Development Program of China (Grant No. 2020YFB1804604) and the Special Funds for Basic Scientific Research Operations of Central Universities (Grant No. 2242023k30053).

Institutional Review Board Statement: Not applicable.

Informed Consent Statement: Not applicable.

Data Availability Statement: Not applicable.

Conflicts of Interest: The authors declare no conflict of interest. The funders had no role in the design of the study; in the collection, analyses, or interpretation of data; in the writing of the manuscript; or in the decision to publish the results.

References

1. Zheng, Y.; Li, Z.; Xu, X.; Zhao, Q. Dynamic defenses in cyber security: Techniques, methods and challenges. *Digit. Commun. Netw.* **2022**, *8*, 422–435. [CrossRef]
2. Yang, J.; Chen, X.; Chen, S.; Jiang, X.; Tan, X. Conditional variational auto-encoder and extreme value theory aided two-stage learning approach for intelligent fine-grained known/unknown intrusion detection. *IEEE Trans. Inf. Forensics Secur.* **2021**, *16*, 3538–3553. [CrossRef]
3. Yousef, W.A.; Traoré, I.; Briguglio, W. UN-AVOIDS: Unsupervised and Nonparametric Approach for Visualizing Outliers and Invariant Detection Scoring. *IEEE Trans. Inf. Forensics Secur.* **2021**, *16*, 5195–5210. [CrossRef]
4. Tian, W.; Du, M.; Ji, X.; Liu, G.; Dai, Y.; Han, Z. Honeypot detection strategy against advanced persistent threats in industrial internet of things: A prospect theoretic game. *IEEE Internet Things J.* **2021**, *8*, 17372–17381. [CrossRef]
5. Giraldo, J.; El Hariri, M.; Parvania, M. Decentralized Moving Target Defense for Microgrid Protection against False-Data Injection Attacks. *IEEE Trans. Smart Grid* **2022**, *13*, 3700–3710. [CrossRef]
6. Hu, Y.; Xun, P.; Zhu, P.; Xiong, Y.; Zhu, Y.; Shi, W.; Hu, C. Network-based multidimensional moving target defense against false data injection attack in power system. *Comput. Secur.* **2021**, *107*, 102283. [CrossRef]
7. Sengupta, S.; Chowdhary, A.; Sabur, A.; Alshamrani, A.; Huang, D.; Kambhampati, S. A survey of moving target defenses for network security. *IEEE Commun. Surv. Tutor.* **2020**, *22*, 1909–1941. [CrossRef]
8. Negi, P.S.; Garg, A.; Lal, R. Intrusion detection and prevention using honeypot network for cloud security. In Proceedings of the 2020 10th International Conference on Cloud Computing, Data Science & Engineering (Confluence), Noida, India, 29–31 January 2020; IEEE: Piscataway, NJ, USA, 2020; pp. 129–132.
9. Wu, J. *Cyberspace Mimic Defense*; Springer: Berlin/Heidelberg, Germany, 2020.
10. Zheng, J.; Wu, G.; Wen, B.; Lu, Y.; Liang, R. Research on SDN-based mimic server defense technology. In Proceedings of the 2019 International Conference on Artificial Intelligence and Computer Science, Wuhan, China, 12–13 July 2019; pp. 163–169.
11. Jin, H.; Li, Z.; Zou, D.; Yuan, B. Dseom: A framework for dynamic security evaluation and optimization of mtd in container-based cloud. *IEEE Trans. Dependable Secur. Comput.* **2019**, *18*, 1125–1136. [CrossRef]
12. Li, H.; Hu, J.; Ma, H.; Huang, T. The architecture of distributed storage system under mimic defense theory. In Proceedings of the 2017 IEEE International Conference on Big Data (Big Data), Boston, MA, USA, 11–14 December 2017; IEEE: Piscataway, NJ, USA, 2017; pp. 2658–2663.
13. Picek, S.; Hemberg, E.; O'Reilly, U.M. If you can't measure it, you can't improve it: Moving target defense metrics. In Proceedings of the 2017 Workshop on Moving Target Defense, Dallas, TX, USA, 30 October 2017; pp. 115–118.
14. Barabási, A.L. Scale-free networks: A decade and beyond. *Science* **2009**, *325*, 412–413. [CrossRef] [PubMed]
15. Fortunato, S. Community detection in graphs. *Phys. Rep.* **2010**, *486*, 75–174.
16. Blondel, V.D.; Guillaume, J.L.; Lambiotte, R.; Lefebvre, E. Fast unfolding of communities in large networks. *J. Stat. Mech. Theory Exp.* **2008**, *2008*, P10008. [CrossRef]
17. Cohen-Addad, V.; Kosowski, A.; Mallmann-Trenn, F.; Saulpic, D. On the power of louvain in the stochastic block model. *Adv. Neural Inf. Process. Syst.* **2020**, *33*, 4055–4066.
18. Fortunato, S.; Lancichinetti, A. Community detection algorithms: A comparative analysis: Invited presentation, extended abstract. In Proceedings of the 4th International ICST Conference on Performance Evaluation Methodologies and Tools, Pisa, Italy, 20–22 October 2009.
19. Shameli-Sendi, A.; Louafi, H.; He, W.; Cheriet, M. Dynamic optimal countermeasure selection for intrusion response system. *IEEE Trans. Dependable Secur. Comput.* **2016**, *15*, 755–770. [CrossRef]
20. Yang, S.; Chen, W.; Zhang, X.; Liang, C.; Wang, H.; Cui, W. A graph-based model for transmission network vulnerability analysis. *IEEE Syst. J.* **2019**, *14*, 1447–1456. [CrossRef]
21. Wang, L.; Jajodia, S.; Singhal, A.; Cheng, P.; Noel, S. k-zero day safety: A network security metric for measuring the risk of unknown vulnerabilities. *IEEE Trans. Dependable Secur. Comput.* **2013**, *11*, 30–44. [CrossRef]
22. Hong, J.B.; Enoch, S.Y.; Kim, D.S.; Nhlabatsi, A.; Fetais, N.; Khan, K.M. Dynamic security metrics for measuring the effectiveness of moving target defense techniques. *Comput. Secur.* **2018**, *79*, 33–52. [CrossRef]
23. Hong, J.B.; Kim, D.S. Assessing the effectiveness of moving target defenses using security models. *IEEE Trans. Dependable Secur. Comput.* **2015**, *13*, 163–177. [CrossRef]
24. Connell, W.; Menasce, D.A.; Albanese, M. Performance modeling of moving target defenses with reconfiguration limits. *IEEE Trans. Dependable Secur. Comput.* **2018**, *18*, 205–219. [CrossRef]
25. Muñoz-González, L.; Sgandurra, D.; Barrère, M.; Lupu, E.C. Exact inference techniques for the analysis of Bayesian attack graphs. *IEEE Trans. Dependable Secur. Comput.* **2017**, *16*, 231–244. [CrossRef]

26. Muñoz-González, L.; Sgandurra, D.; Paudice, A.; Lupu, E.C. Efficient attack graph analysis through approximate inference. *arXiv* **2016**, arXiv:1606.07025.
27. Zhang, M.; Wang, L.; Jajodia, S.; Singhal, A.; Albanese, M. Network diversity: A security metric for evaluating the resilience of networks against zero-day attacks. *IEEE Trans. Inf. Forensics Secur.* **2016**, *11*, 1071–1086. [CrossRef]
28. Yu, F.; Wei, Q.; Geng, Y.; Wang, Y. Research on Key Technology of Industrial Network Boundary Protection based on Endogenous Security. In Proceedings of the 2021 IEEE 4th Advanced Information Management, Communicates, Electronic and Automation Control Conference (IMCEC), Chongqing, China, 18–20 June 2021; IEEE: Piscataway, NJ, USA, 2021; Volume 4, pp. 112–121.
29. Shen, C.; Chen, S.X.; Wu, C.M. A Decentralized Multi-ruling Arbiter for Cyberspace Mimicry Defense. In Proceedings of the 2019 International Symposium on Networks, Computers and Communications (ISNCC), Istanbul, Turkey, 18–20 June 2019; IEEE: Piscataway, NJ, USA, 2019; pp. 1–6.
30. Ren, Q.; Hu, T.; Wu, J.; Hu, Y.; He, L.; Lan, J. Multipath resilient routing for endogenous secure software defined networks. *Comput. Netw.* **2021**, *194*, 108134. [CrossRef]
31. Yu, H.; Li, H.; Yang, X.; Ma, H. On distributed object storage architecture based on mimic defense. *China Commun.* **2021**, *18*, 109–120. [CrossRef]
32. Yu, C.; Chen, L.; Lu, T. A Direct Anonymous Attestation Scheme Based on Mimic Defense Mechanism. In Proceedings of the 2020 International Conference on Internet of Things and Intelligent Applications (ITIA), Zhenjiang, China, 27–29 November 2020; IEEE: Piscataway, NJ, USA, 2020; pp. 1–5.
33. Liu, W.; Peng, Y.; Tian, Z.; Li, Y.; She, W. A Medical Blockchain Privacy Protection Model Based on Mimicry Defense. In Proceedings of the International Conference on Artificial Intelligence and Security, Hohhot, China, 17–20 July 2020; Springer: Berlin/Heidelberg, Germany, 2020; pp. 581–592.
34. Zhao, Y.; Zhang, Z.; Tang, Y.; Ji, X. A Security Quantification Method for Mimic Defense Architecture. In Proceedings of the 2021 IEEE 5th Advanced Information Technology, Electronic and Automation Control Conference (IAEAC), Chongqing, China, 12–14 March 2021; IEEE: Piscataway, NJ, USA, 2021; Volume 5, pp. 36–40.

applied sciences

MDPI

Article

Design and Implementation of Multi-Cyber Range for Cyber Training and Testing

Moosung Park [1,2], Hyunjin Lee [3], Yonghyun Kim [2], Kookjin Kim [1,4] and Dongkyoo Shin [1,4,*]

1 Department of Computer Engineering, Sejong University, Seoul 05006, Republic of Korea
2 R.O.K Agency for Defense Development, Seoul 05771, Republic of Korea
3 R.O.K Hanwha Systems, Seongnam 13524, Republic of Korea
4 Department of Convergence Engineering for Intelligent Drones, Sejong University, Seoul 05006, Republic of Korea
* Correspondence: shindk@sejong.ac.kr

Abstract: It is essential to build a practical environment of the training/test site for cyber training and weapon system test evaluation. In a military environment, cyber training sites should be continuously developed according to the characteristics of the military. Weapons with cyber security capabilities should be deployed through cyber security certification. Recently, each military has been building its own cyber range that simulates its battlefield environment. However, since the actual battlefield is an integrated operation environment, the cyber range built does not reflect the integrated battlefield environment that is interconnected. This paper proposes a configuration plan and operation function to construct a multi-cyber range reflecting the characteristics of each military to overcome this situation. In order to test the multi-cyber range, which has scenario authoring and operation functions, and can faithfully reflect reality, the impact of DDoS attacks is tested. It is a key to real-world mission-based test evaluation to ensure interoperability between military systems. As a result of the experiment, it was concluded that if a DDoS attack occurs due to the infiltration of malicious code into the military network, it may have a serious impact on securing message interoperability between systems in the military network. Cyber range construction technology is being developed not only in the military, but also in school education and businesses. The proposed technology can also be applied to the construction of cyber ranges in industries where cyber-physical systems are emphasized. In addition, it is a field that is continuously developing with the development of technology, such as being applied as an experimental site for learning machine learning systems.

Keywords: cyber training; cybersecurity test and evaluation; scenario authoring; cyber range

Citation: Park, M.; Lee, H.; Kim, Y.; Kim, K.; Shin, D. Design and Implementation of Multi-Cyber Range for Cyber Training and Testing. *Appl. Sci.* **2022**, *12*, 12546. https://doi.org/10.3390/app122412546

Academic Editors: Peter R.J. Trim and Yang-Im Lee

Received: 2 November 2022
Accepted: 6 December 2022
Published: 7 December 2022

1. Introduction

Cyberwar, even in the recent case of the Russia-Ukraine conflict, is represented in the form of a hybrid warfare scenario accompanied by regular warfare, and it is constantly carried out in peacetime. Accordingly, countries around the world are advancing cyber security technology to enhance their ability to carry out cyberwarfare and following development procedures that emphasize security in the development of weapons systems. The basis of cyber warfare performance will be the strengthening of the cyberspace. Therefore the training of personnel capable of carrying out defense and attacks will be the basis of cyber power. Realistic defense training is more efficient when conducted in a real-world environment. It is common to conduct such training in a virtual simulation environment because the nature of cyberwarfare can cause irreversible damage to the actual system in the training process. However, a training environment that lacks realism makes it difficult to expect practical results.

In order to attain a robust cyber defense capability, cybersecurity capabilities should be equipped from the time the weapon system is built, and while it is tested and evaluated

Jim Highsmith emphasized that technical debt incurred when things are not properly fixed exponentially increases over years [1].

A cyber range is a practice field that provides the ability to research, develop, test, or conduct cyber training on military capabilities in cyberspace [2]. In order to properly evaluate tests, it should be conducted in a realistic cyber range environment. In addition, cybersecurity capabilities and interoperability functions should be implemented prior to operational test and evaluation to ensure the success of system development [3]. In other words, in order to increase the effectiveness of cyber training, it is necessary to build a cyber range that resembles a realistic environment. Moreover, a cyber range that resembles a real system can also be used as a test evaluation site for weapon systems.

In a military system where accuracy and security are emphasized above all else, it is necessary to build a realistic cyber range and conduct a variety of training and test evaluations. Recently, each military has been building its own cyber range that simulates the battlefield environment of each military. However, since the actual battlefield is an integrated operation environment, the cyber range built does not reflect the integrated battlefield environment that is interconnected. In order to overcome this, the concept of a multi-cyber range proposed in this paper will be developed into a highly useful concept. This paper presents a tool to allow a single cyber range to realistically link multiple cyber training ranges and authoring scenarios that are connected between ranges based on integrated management measures between training ranges.

In this way, the construction of practical testbeds in various kinds of education and research is required, and it is a reality that is actively being researched and developed in schools and government institutions. Therefore, the main agenda of this paper, the design structure of connecting and expanding various ranges, will be indispensable for the development of cyber ranges.

The remainder of this paper is structured as follows. Section 2 describes related works. Section 3 proposes a practical cyber range structure that can also be used for cyber training and interoperability assessment, explains the scenario-building measures and related functional elements, and presents the results of the implementation. Section 4 conducts cyber warfare experiments in range-connected situations to verify that the built range can best represent a cyber threat/combat situation. Section 5 describes the contributions of this paper and the direction of future research.

2. Related Works

2.1. National Cyber Range (NCR)

DARPA in the U.S. has operated a cyber training range since 2009 and is moving it to the U.S. Test and Training Resource Management Center (TRMC) for further development to facilitate use in real-world training and test evaluations. For test spaces in the security area, L1 switches can be used to interface with the range to support training and test evaluations even in multi-level security environments [4]. In addition, during the weapon system acquisition lifecycle, all six stages of Cyber Test and Evaluation (T&E) were supported, and the range was developed to a level where test evaluation results were officially recognized.

The main capabilities of NCR are:

1. Multiple Independent Levels of Security (MILS) architecture enables simultaneous operation of multiple trials in different secret classes;
2. Quick emulation of complex operational environments;
3. Automation support for accurate repeated testing;
4. Support for different types and disciplines (test, training, research, etc.).

NCR serves as a cyber range that provides a mission-adaptive, hi-fidelity cyber environment for assessing independent and objective cyber testing and progressive cyberspace capabilities. It also integrates the test evaluation infrastructure of cyberspace through partnerships across the U.S. Department of Defense, the U.S. Department of Homeland Security, and industry and academia. The NCR facility is a special certified communication information facility that maintains a variety of hardware and software computing resources and provides a test environment that encompasses wired and wireless networks.

Vincent E. set out 11 limitations based on their experience using NCR and cited the need for further development [5]. Since NCR was developed, the use of case statistics in Table 1 shows that it has been applied to training and various tests. In other words, it shows that the role of the cyber range is not only for training and practice, but also for the development of weapon systems as a field for test evaluation.

Table 1. Number of NCR Uses by Sector.

	FY11	FY12	FY13	FY14	FY15	FY16
Cyberspace Capability DT&E	1	3	3	2	4	8
Cyberspace Capability OT&E		1	1		3	2
Cyberspace Capability M&S/R&D			5	7		
Training/Exercises		1	3	11	22	27
Mission Rehearsal		1	1		2	4
MDAP Cybersecurity DT&E				2	9	17

However, most of the limitations of NCR mentioned in [5] are management factors. It is an aspect caused by the need for many participants and it is difficult to systematically manage complex NCR resources. Problems that cannot be controlled when multiple ranges are connected and seem to require systematic automation of the management system.

2.2. Capability of Cyber Training System

A cyber training system is a system in which the training manager (White Team) prepares and controls the training environment, while the trainers, the cyber attacking group (Red Team) and the defenders (Blue Team), can train in the training environment. The cyber training system consists of a cyber battlefield environment construction function that simulates the actual battlefield environment as a cyber battlefield environment, a scenario authoring function that can produce various scenarios for training, and a training control function that can control, monitor, and evaluate training. The cyber training system is operated in the following order: training plan setting, training goal setting, writing of training scenario, training performance according to the scenario, training monitoring, evaluation and post-analysis of training results, and reporting of training results [6].

Usually, in the military, some units have established a cyber training range, which is used to train cyber warriors, and the ranges are mainly composed to mimic the Internet environment. As the aspects of cyber warfare become more complex, the level needed for training must also be advanced, especially in the area of defense, where tactical training of the concept of simulated combat needs to be carried out. The training scenario has the essential role of providing a user interface to design the training and mounting it into the training system. From this point of view, the training scenario should be able to include a number of factors that can increase the diversity and quality of the training. This is because the test evaluation should be carried out in the same environment as the environment in which the system will be operated, such that a complete mission-based test evaluation can be carried out. For example, a Distributed Denial of Service (DDoS) attack on an Army Corps server would cause problems with the transfer of data interlocked to the Joint Command, Control, Communication, Computer and Intelligence (C4I) system, which in turn would limit the Joint Chiefs of Staff's perception of the Army situation. Therefore, the mission of a Joint Operations War is bound to be affected.

Nikos Oikonomou proposed in [7] the need to connect and integrate services with the European cyber range due to the high cost of building and managing the cyber range, while Olivier Jacq proposed in [8] the need to build a Maritime cyber range through the Maritime's cyber risk assessment. In addition, Adamantini emphasized in [9] that it is difficult for a single organization to build and manage multi-domain ranges. Therefore, it is necessary to connect ranges from different organizations to achieve real-world fidelity, and, when connecting multiple ranges, use Virtual Private Network (VPN) technology to connect. In addition, outside of the military, ordinary schools and enterprises are also building cyber ranges, and are also evolving into services using cloud technology.

In addition, cyber ranges are also being used for security education, in various industries, and the construction of intelligent learning models. The cyber range was built to train procedures for analyzing/handling threats in real-world environments based on cyber threat scenarios in a more real-world cyber-physical environment rather than a theoretical approach to cybersecurity education [10]. The results of education at the university level proved how efficient it is to conduct it in a practical educational environment. The testbed was built for efficient learning of machine learning systems in the SCADA environment [11]. It is a well-known fact that the accuracy of a machine learning system is determined by the amount of training data that is required. To this end, cyber security researchers conducted an experiment to build a realistic cyber range to learn a machine learning system while carrying out a cyber-attack. By learning based on various cyber threat data that are difficult to obtain in real systems, the role of a cyber range in the development of intelligent models is being emphasized.

To develop a distributed intrusion detection system applied in an industrial control system environment, the test bed built a cyber range with a mix of physical equipment, simulation models, and emulated models [12]. This also drove the functional and performance accuracy of intrusion detection systems by building and testing in realistic environments.

SWaT is used to understand the impact of cyber and physical attacks on water treatment systems, to evaluate the effectiveness of attack detection algorithms, and to evaluate the effectiveness of defense mechanisms when a system is under attack [13]. Experience with testbeds has emphasized the importance of conducting research in an active and realistic environment.

Smyrlis, M. researched a model-based scenario authoring technology to improve user adaptability in cyber education. This study enabled the creation of customized training scenarios based on a comprehensive, model-based description of the organization and its security posture [14].

Ukwandu, E. examined and classified existing cyber ranges and testbeds. The latest trends detail the different dimensions of this classification and highlight the diminishing differentiation between application areas [15]. Chouliaras, N. et al. conducted a systematic survey of 10 cyber ranges developed over the past decade through structured interviews. The existing cyber range determined that there were many elements requiring improvement with new technological developments. They also mentioned that in the near future, digital twin technology will be applied to cyber range construction technology [16].

As such, it can be seen that cyber ranges are needed in many areas. In addition, cyber range technology incorporating artificial intelligence and IoT technology continues to be developed. The issue of emulating weapons systems in the military is also a very important issue and should be considered.

3. Multi-Cyber Range Structure for Training, Testing, and Evaluation

3.1. Structure of Range

It is necessary to establish an environment in which the battlefield management system environment is centered on supporting the Joint Chiefs of Staff and the tactical environment of each military branch can be comprehensively simulated to enable mission-based evaluation. As shown in Figure 1, each military branch shall establish a range of their own, and the Joint Chiefs of Staff shall design/build a light bulb range based on the joint command and control system to interconnect, train, and test functions.

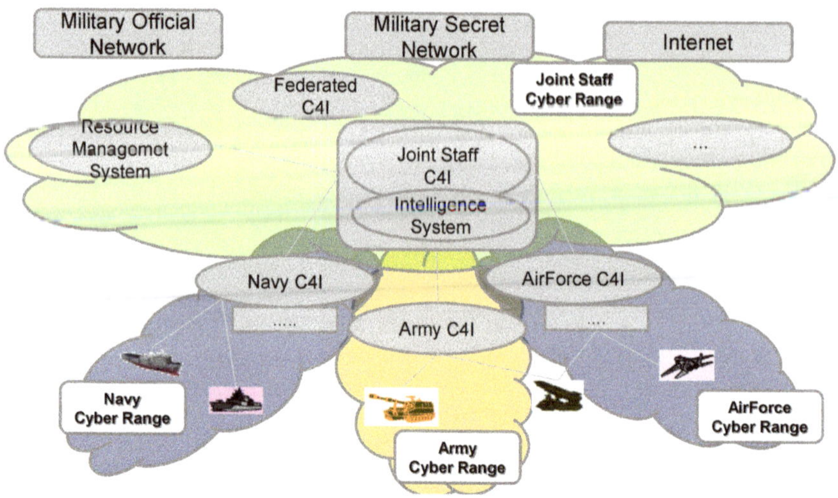

Figure 1. Operational architecture of theater level cyber range.

3.1.1. Networking for Federating Ranges

Centered on the Joint Chiefs of Staff Range, each army unit's cyber range must be configured around the battlefield management system where each subsystem is connected to the Live Virtual Constructive (LVC) concept in order to construct mission-based cyber training and a weapon system test evaluation environment. Therefore, each military cyber training range should develop a range environment around the battlefield management system, and it should be connected as in actual conditions.

Even in an actual environment, the C4I system of each group shares information situations based on Message Text Format (MTF) messages through an interoperating server, and the ability to interoperate information is an important element of the implementation of joint operation warfare. Therefore, in each cyber range, each battlefield management system should be simulated to support responses to interoperating messages. This can be simulated based on the Interface Control Document (ICD) between each system.

In addition, each battlefield management system synchronizes the battlefield situation through synchronization between centers with the concept of a distributed center, which is also an important function of the battlefield management system. Therefore, when simulating a battlefield management system, the synchronization between servers is also an important simulation object.

A Cross Domain Solution (CDS) is used to securely link areas with different secret ratings. The connection between ranges via CDS has the advantage of allowing trainers to train in a real-world environment by actually reflecting the operating environment of the battlefield management system, and to conduct interoperability assessments before an Operational Test (OT) that has not been carried out in the proposed range. However, a separate management channel is required for configuration management and scenario sharing between ranges, which can be used to establish a separate Range Management Channel, such as in Figure 2, using a VPN.

3.1.2. Architecture of Range Management Function

Focusing on the battlefield management system, the range configuration capability is similar to the actual operating system in sub-tactical systems, intranets, and the Internet and should be gradually expanded. Each cyber range has essential functions such as configuration management, scenario creation, and test data generation for independent operation.

However, for configuration and creation of scenarios over a range-to-range connection, a special channel is needed to control whether or not a separate range resource can be managed and supported. To this end, it is proposed to build a portal around the main range with the necessary functions to share and connect the status of resources to conduct training and test evaluation. The proposed management configuration is the same as in Figure 3.

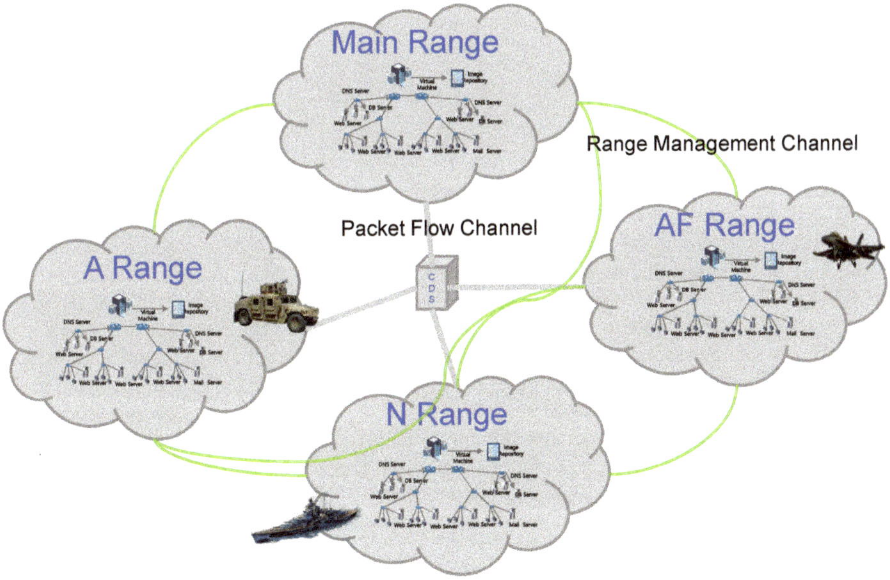

Figure 2. Networking architecture for connecting ranges.

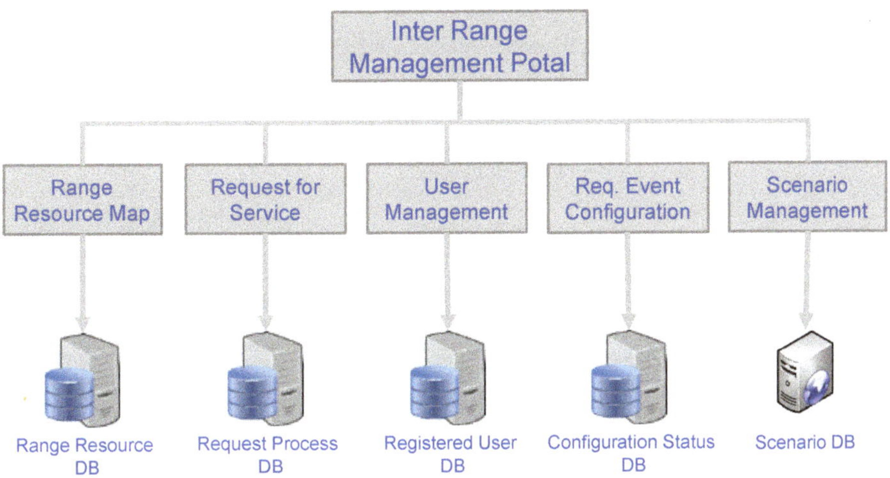

Figure 3. Multi-Cyber Range Management Function.

Each range shares the DB through a multi-cyber range management portal operated by the main range and cooperates with the range configuration. Each range lists the assets it has and presents the default configuration as a service template. Users who will utilize the assets of other ranges to conduct training and testing will apply for services using the service request management function, and refer to the default configuration templates provided by each range. In addition to the hardware and delivery functions that make up the range, a request form is written, including the traffic generation requirements and the coordination team (Red, Blue, White).

The Range Configuration Management module maps the available resources provided by each range based on the contents of the service request to configure the optimal range. The user management module is managed and executed by the user participating in the training and testing, and the requested support personnel at each event. Traffic Flow controls normal traffic between ranges by range, depending on the scenario in which it is written.

3.2. *Multiple Cyber Range Scenario Authoring and Traffic Flow Control*

3.2.1. Scenario Management

Scenario authoring at a single range is the basic procedure of starting with training/test information, constructing a configurable network topology at the range, and creating a normal/abnormal traffic distribution plan that meets the training/test intention. The training/test authoring tool facilitates recycling or extending existing utilization scenarios based on procedures performed at a single range.

In a multi-range environment, the network, traffic generation, and training/test participants and agents must be able to configure the training/test environment based on the resources allowed at each range. Hence, as shown in Figure 4, even in a single range scenario configuration procedure, it is necessary to have a multi-range configuration consultation procedure for each step. Figure 5 represents the scenario procedure in an existing single range. Figure 6 illustrates the procedure for configuring a scenario in a multi-range environment. For example, constructing a training/testing scenario in an environment where the Joint C4I System and the Army C4I System are interoperated, and an environment in which all ranges participate will be a theater-level test environment. Scenario authoring proceeds is done modularly in a single range without a negotiation process of the range. In the Figures 5 and 6, the purple line is linked to the detailed configuration function to help constructing the overall scenario by specifying the range to which the resource to be configured in the training or test belongs.

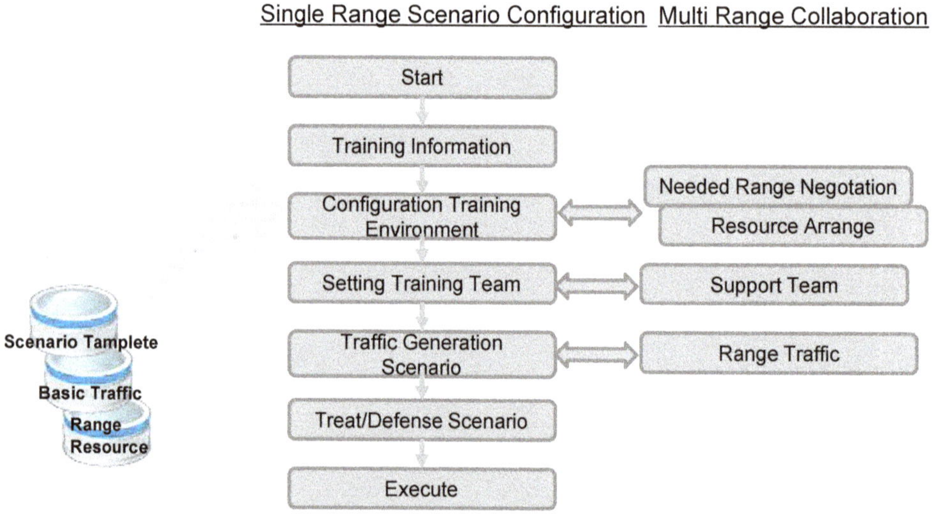

Figure 4. Procedure for Scenario Authoring.

Procedurally constructed scenarios are stored in the DB and recycled in the future, or selected as the default template so that the scenario can evolve. Based on the basic template scenario and the historical scenario, additional resource configuration should be able to configure the licensed resources in a drag and drop manner, as shown in Figure 7, and all resources are managed by tagging their range affiliations.

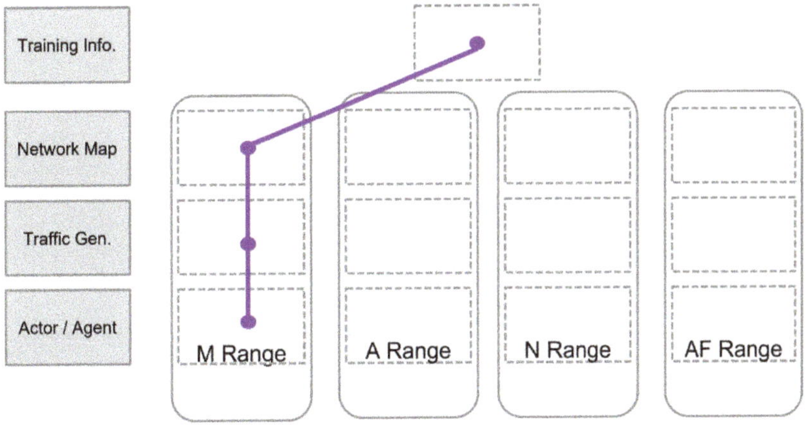

Figure 5. Scenario Authoring at Single Range.

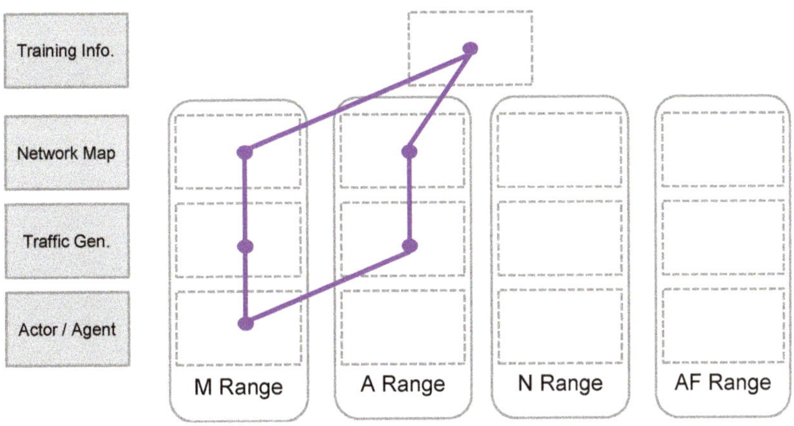

Figure 6. Scenario Authoring through Multi-Range.

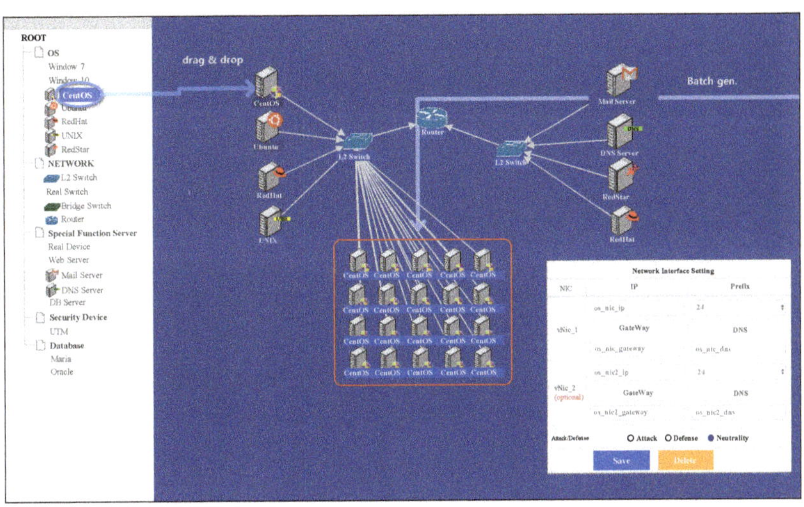

Figure 7. Network Map Configuration Function.

3.2.2. Normal Traffic Generation

Normal traffic generation and reproduction is a very important factor in the reproduction of the actual environment, and at the same time as the establishment of a practical training and test evaluation environment. In a related study [17], network traffic was generated in three ways: probabilistic generation, replication of actual network traffic, and the use of instruction lists for applications in the test network. Bieniasz, J. et al. proposed a new approach to generating datasets for cyber threat research on multi-node systems [18]. This has been made useful in fields where information concealment technology is applied. Traffic in military systems is likely to generate traffic with regular statistics depending on wartime/peacetime situations. Therefore, the method of replicating and generating actual network traffic according to the situation is considered the most efficient. The traffic used by a cyber range requires the process of building a dataset, as shown in Figure 8 [19].

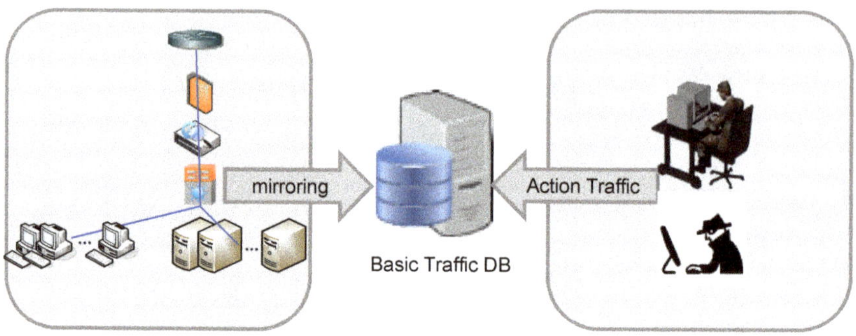

Figure 8. Collect real operation traffic.

In order to develop a plan for traffic generation, a traffic generating node must first be set up in the network topology. The traffic generating node should be specified on the basis of what actually occurs at the server node, but, according to the scenario, a special node can be set up to generate the collected traffic, and the traffic generated by the terminal node may be generated by mixing the use of the terminal traffic template by designation. The terminal traffic template selects the terminal traffic template to be used in the basic dataset DB, and grasps the traffic distribution terminal through the traffic distribution terminal information in the header. In addition, the traffic distribution process can be reproduced through the number of messages and traffic type information.

In the battlefield management system, direct traffic between the terminals is limited, so assuming the traffic between the terminal and the server, it is possible to reproduce simple traffic. Normal traffic is managed in a Packet Capture (PCAP) file and used as basic data by managing the data set collected during peacetime/training, and the server included in the configuration of the network map is essentially designated as traffic generation equipment and operated. Traffic generated at the other terminal should be separately specified to generate traffic. In addition, since the characteristics of training and test evaluation are set at the time to be reproduced, and not the current time, traffic generation should also be designed so that a multiplier generation or hold function can be given a timer function.

In order to establish a practical environment, traffic flow reflecting the characteristics of the military battlefield management system needs to be efficient to perform with the concept of replay based on the information collected. Attack traffic generation is often replayed by building existing case data into a basic data set, but it is difficult to judge it as actual traffic due to differences in training and testing environments. Therefore, if possible, traffic is naturally generated by attack agents or Red Team actions that are applied to training and testing, so a separate attack traffic generation is not necessary for real-world environment configuration.

Traffic generating nodes according to the scenario are specified as shown in Figure 9 to configure a normal training environment. TA1 and TA2 nodes can be considered as acting as interoperating servers that make up each system, and the TS1 node can be assumed to be a node that generates information as they move. However, when configured in a scenario that requires a separate small amount of traffic generation for training purposes, traffic generation agents such as mail behavior and Internet usage behavior are utilized without using large basic traffic.

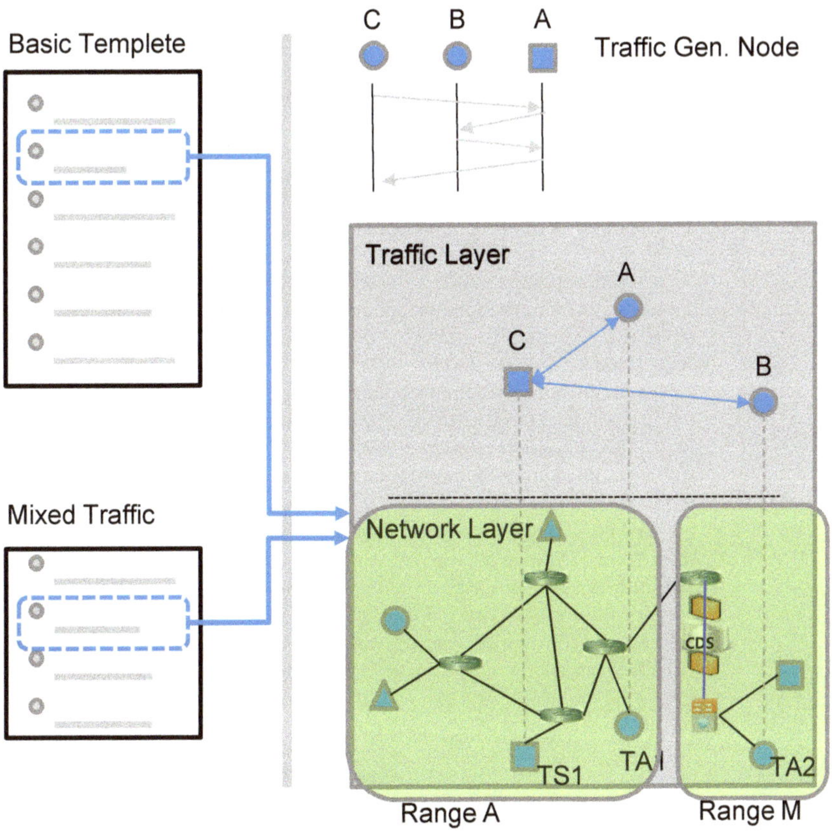

Figure 9. Mapping traffic generation node.

3.2.3. Automatic Scoring

The training is basically based on MITRE's Adversarial Tactics, Techniques, and Common Knowledge (ATT&CK) and consists of the Red Team's combat action in the seventh phase of the Cyber Kill Chain. We proposed a way to set the desired time by phase/combat action and set the desired time, as it is continuously supplemented during training. When integrating the desired time for each phase/combat action into the score for each phase designated by the manager into the score for each phase of the battle, it shall be possible to evaluate whether the task given at the time of the desired time is actually achieved by reflecting compliance with the desired time.

The phase generation function should record relevant information around the phase name and time, such as in Figure 10, and if the desired time is specified for each combat action, the training score can be automatically calculated according to Equation (1). However, the combat performance score reflects the manual score by the training instructor.

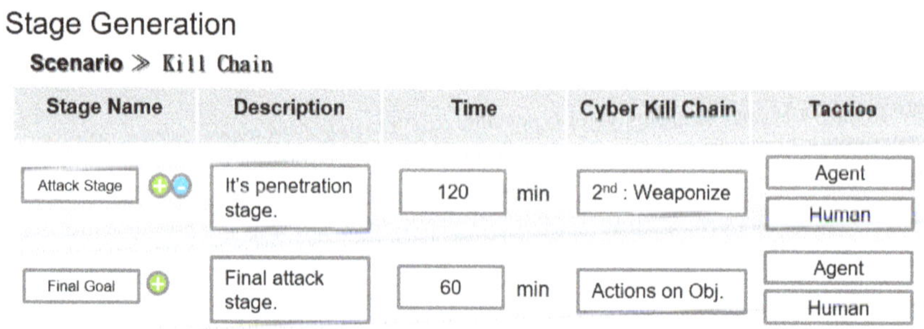

Figure 10. Phase generation function.

$$\text{Phase Score} = \sum(\text{Combat Action Score } \times (\text{Desired Time/ExecuteTime})) \quad (1)$$

As a result of the training, the scores for each trainer/team are automatically calculated based on whether the combat action performed was achieved, and the timeliness of the mission is evaluated to reflect the desired time. The Combat Action Score is calculated by monitoring CPU usage and file system/process/network changes. However, it is necessary to control the training time by stipulating that the performance time for each combat action shall not exceed 1.5 times the desired time. Monitoring user behavior for automatic evaluation is limited because it is calculated based on some system change information. Therefore, this automatic calculation feature is suitable for defensive training against known attacks and is not suitable for free attack/defense training of Red and Blue teams.

As a result of the training, the scores for each trainer/team are automatically calculated based on whether the combat action performed was achieved, and the timeliness of the mission is evaluated to reflect the desired time. However, it is necessary to control the training time by stipulating that the performance time for each combat action shall not exceed 1.5 times the desired time. The situation that occurs between trainings is controlled/analyzed through a visualization tool that shows the range configuration topology, a detailed event list, and the results of the attack/defense behavior analysis, such as in Figure 11.

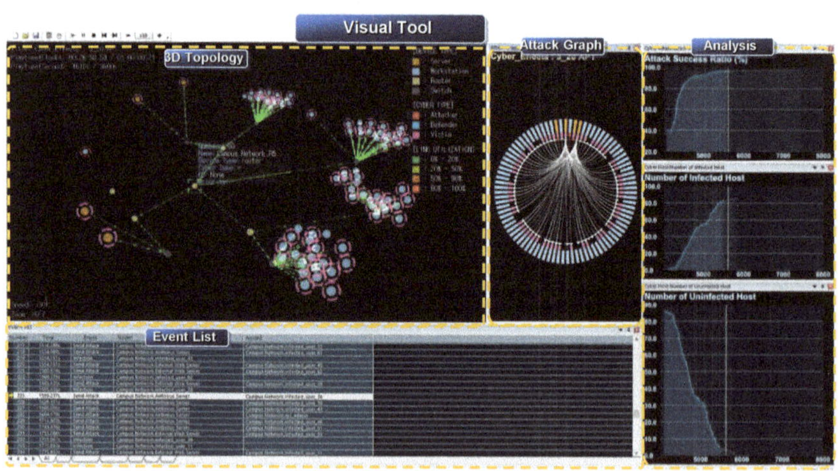

Figure 11. Training situation visualization.

4. Cyber Warfare Experiment with Range Connection

The importance of information sharing in modern warfare is, needless to say, a key element of battlefield operations. Training and evaluation of cybersecurity is emphasized as being mission-oriented [20], and mission-oriented assessments require that all environments in which the system operates are reproduced in order for a proper mission-oriented assessment to be achieved. Therefore, since the battlefield management system of each army is operated in conjunction with the tactical system operated by each military, and at the level of the Joint Chiefs of Staff, it has a hierarchical structure in which the battlefield situation is synthesized in conjunction with the command and control system of each army. Thus, the joint chiefs of staff and the cyber range of each army must be linked to the training and test evaluation to achieve practical training and test evaluation.

The characteristics of the battlefield management system are built and operated in a distributed environment, and the guarantee of traffic for synchronization between distributed servers is an important factor in matching the battlefield situation. In cyber defense, training, security testing, and evaluation between battlefield management systems, the match of the battlefield situation is achieved through the exchange of messages between each system. Other major generated traffic is situational information input and inquiry by the user. Therefore, ensuring the flow of Enterprise Application Integration (EAI) traffic between server sites and message traffic between interoperating servers is an important evaluation indicator for accomplishing the task. Therefore, based on the Information Exchange Requirements (IER) between the battlefield management systems, the success of training and test and evaluation can be analyzed around the flow of interoperating data.

For the experiment, when two ranges were configured, as shown in Figure 12, and a DDoS attack in the form of User Datagram Protocol (UDP) flooding occurred on the Corps server according to the malicious behavior of an insider in the A Range network, as shown in Figure 13. The effect of the IER between the joint C4I system interoperating servers in conjunction with the Corps server was experimented with and the limitation of sharing the battlefield situation by cyberattack was investigated.

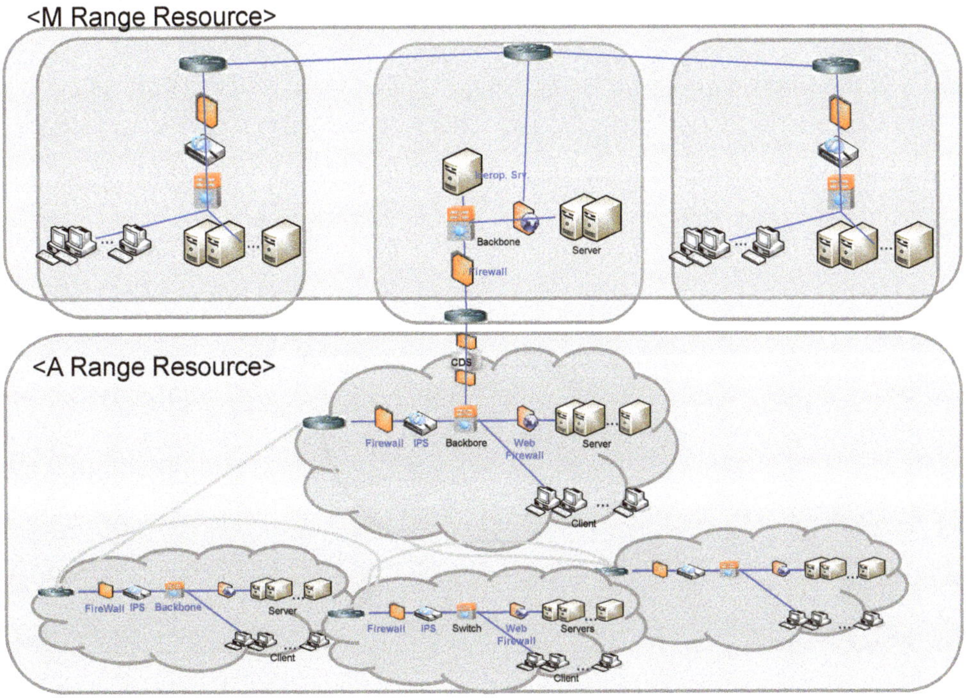

Figure 12. Multi-range configuration example (Joint Staff C4I 3, Army C4I 4 SITE).

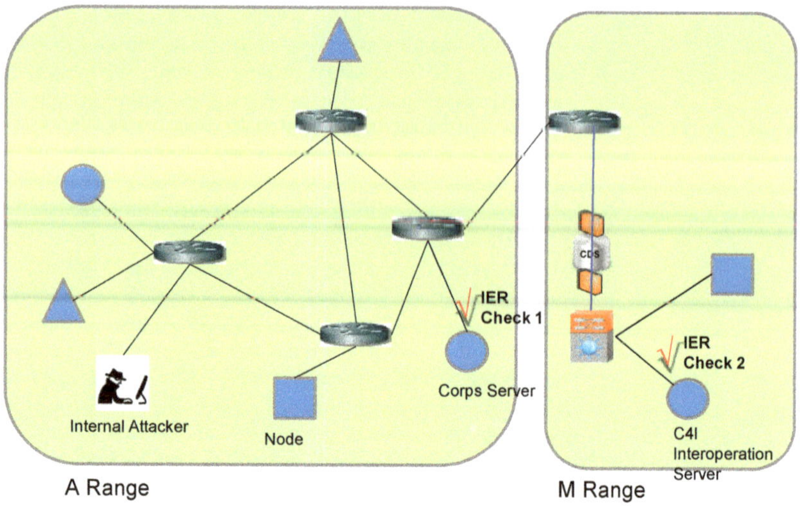

Figure 13. Test environment for IER process.

There are attacker terminals and a number of C4I terminals inside the Army network, and an IER occurs between the Corps server and the Joint Chiefs of Staff C4I interoperating server. The attacker terminal hijacks the antivirus server inside the Army network to configure the Command and Control (C&C) server, and the terminal is infected through the antivirus patch. The infected terminal generates a DDoS attack on the legion server, affecting the transmission quality of the IER.

The experimental environment is shown in Table 2. The experiment was performed in a total of eight scenarios, and the DDoS attack traffic characteristics by scenario are shown in Table 3. Specific items measured by the scenario are shown in the table.

Table 2. Experiment Characteristics.

Item	Value	Information
	IER Information	
IER delay limit	3 s	
IER traffic size	1500 bytes	Exponential distribution
IER interval	0.1 s	Exponential distribution
	DDoS Attack Information	
# of DDoS participating terminals	Max. 82	Increased cyberattack progress
Attack start time	300 s	
Attack duration	1000 s	
Attack interval	0.1 s	
Attack traffic size	1~15 Kbits	Various per scenarios

Table 3. DDoS Attack Load for Each Scenario.

Scenario #	1	2	3	4	5	6	7	8
DDoS traffic size (kbits)	15	14	13	12	9	6	3	1
Attack interval (s)	0.1	0.1	0.1	0.1	0.1	0.1	0.1	0.1
Average DDoS Attack Traffic (Mbps)	11.7	10.9	10.1	9.4	7.0	4.7	2.4	0.8

1. IER delay limit: The delay limit that IER should be received by in order to not to affect a military operation. It is a concept similar to service level agreement (SLA).
2. IER size and inter-arrival time: These factors are used to define traffic characteristics of IER. Average traffic volume of IER can be calculated by IER size/IER interval.
3. DDoS Attack Information: These parameters describe the characteristics of a DDoS attack by an attacker; # of DDoS participating terminals means the number of terminals that generate the DDoS attack traffic. DDoS attack maintains during the attack duration after the attack start time. Attack interval and DDoS traffic size mean the attack traffic generation interval and the traffic size per a DDoS attack, respectively. Thus, we can calculate the traffic volume of the DDoS attack by DDoS traffic size/attack inter-arrival time.
4. End-to-end IER transmission delay: Automation support for accurate repetitive testing of the time it takes from the generation of an IER on the Corps C4I server until the Joint Chiefs of Staff interlocking server receives the IER.
5. IER received ratio: The percentage of IERs sent that are successfully received.
6. IER success ratio: The percentage of IERs received that arrive within the IER delay limit (the percentage of IERs that satisfy timeliness).
7. IER failure ratio: The ratio of received IERs to those who arrive after three times the IER latency limit time.
8. IER perished ratio: The percentage of IERs received that exceed the IER delay limit time but arrive within three times the IER delay limit time.

Figure 14 shows the number of cyberattack infected terminals over time. Terminals participating in DDoS attacks are variable depending on the time of the vaccine patch. The patch time is variable depending on the scenario, but the attack start time dramatically increases, and the infection occurs even during the course of a DDoS attack.

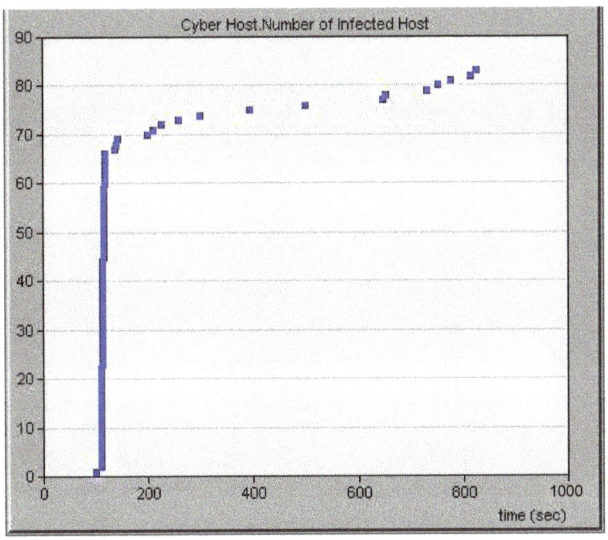

Figure 14. Number of devices participating in a cyberattack according to simulation time.

Figure 15 shows the end-to-end IER transmission delay according to the simulation time by scenario. The figure shows that the delay in end-to-end IER transmission increases during the time of the DDoS attack. In particular, scenarios 1 and 2 show that DDoS traffic exceeds the link load (the link on the legion server is 10 Mbps, creating a bottleneck for experimentation), resulting in a dramatic increase in transmission delays. Not only can DDoS attack traffic be affected by end-to-end transmission delays even when the traffic is less than the link load, but it takes a certain amount of time to process the IERs that have been queued up even after the end of the attack.

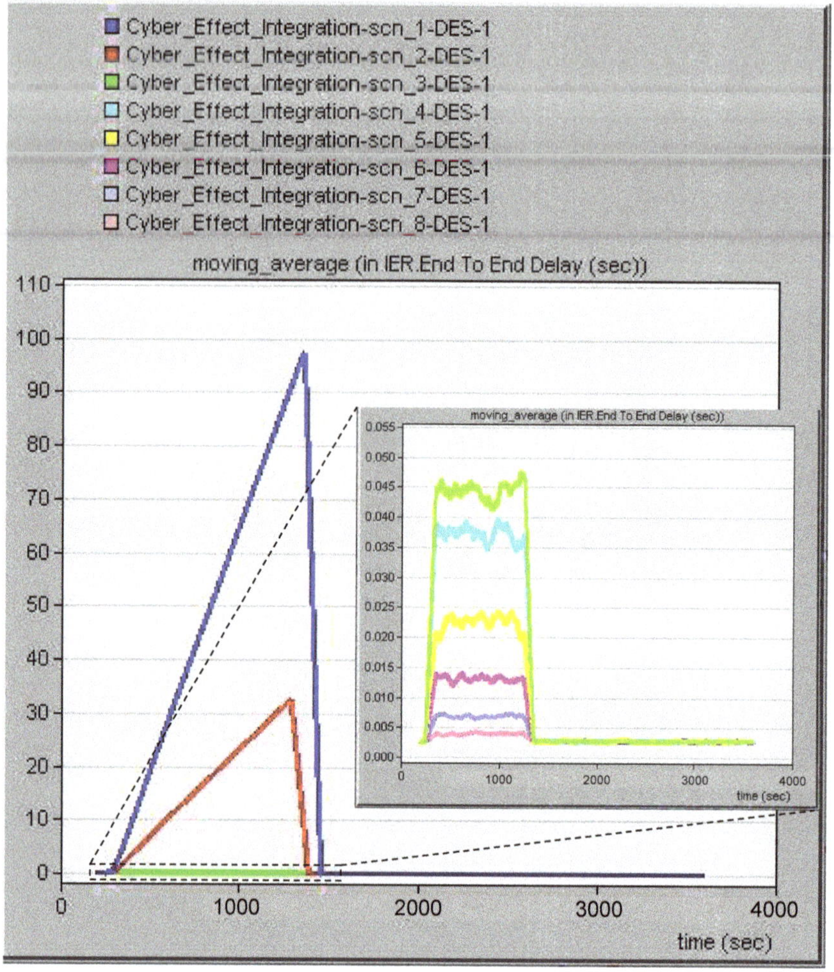

Figure 15. End-to-end IER transmission delay according to the DDoS attack volume.

Figure 16 is an illustration of the transmission characteristics of the IER according to the scenario: (a) is the reception rate of the IER, (b) is the success rate of the IER transmission, (c) is the IER transmission failure rate, and (d) is the IER delay reception rate. (a) shows that in all scenarios, except scenario 8, the IER reception rate decreases during the DDoS attack and then increases again when the DDoS attack ends. However, (b), (c), and (d) show that the IER did not satisfy the timeliness required and exceeded it by 20%, even in situations where the DDoS attack was low relative to the link load (scenarios 7 and 8). Therefore, a cyberattack by an internal attacker can have a serious impact on the battlefield management system.

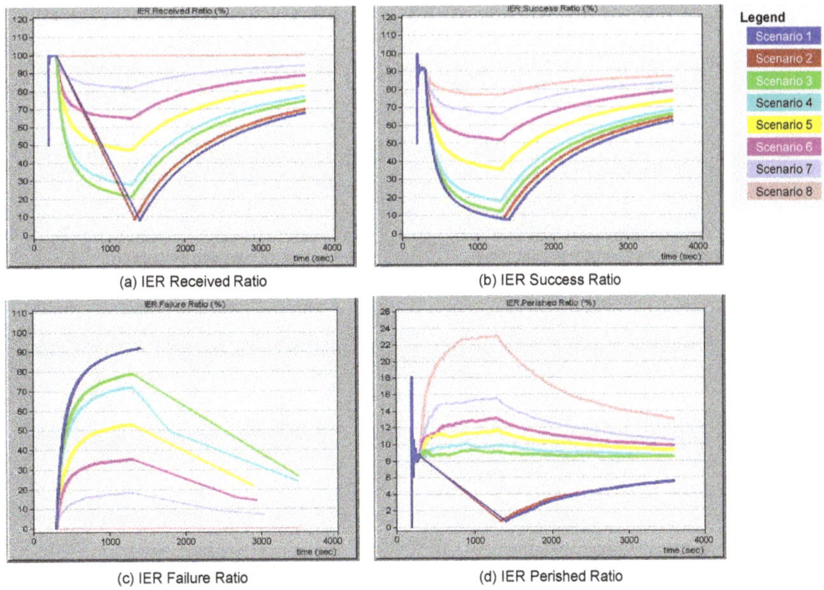

Figure 16. IER transmission characteristics under DDoS attack load.

The scenario applied in this paper is presented as a way to simulate or analyze the impact of a cyberattack on military operations in the event of a cyberattack on a military network. To respond to such an attack, the response team can mitigate a DDoS attack by restricting the total traffic or the amount of traffic circulating on individual nodes by checking IPS's anomaly detection policy in the path of the attacked node in the short term, and adjust the ACL of the real firewall or constructive model environment firewall to block access to the nodes participating in the DDoS attack. It can also analyze the command delivery information of the nodes involved in a DDoS attack to perform a response, such as blocking the connection of the C&C server that delivered the attack command, and can master these tips of action through the cyber range.

5. Conclusions

A military cyber range configuration should not only be configured as a training ground for cybersecurity education, but also as a training ground where the cyberspace guarded by the military can be realistically configured to carry out effective defensive operations. It should also serve as a testing ground for cyber ranges to conduct inorganic system development net weather security tests and interoperability tests.

To this end, the range of each Army, which was previously established in the form of a cyber defense training field, was developed around the battlefield management system of each Army, and the function of forming a scenario was developed by proposing a method of forming a range jointly by connecting each Army range together with the Joint Chiefs of Staff. Through this, we made it possible to conduct practical cyber defense training and proposed a test site for interoperability test evaluation during Development Tests (DT) in the development of weapons systems. As in the case of NCR in the U.S., we also need to make continuous progress through in-depth simulation of the actual battlefield management system based on the multi-range configuration presented in this paper. This research will ensure that, in the future, various tactical weapon systems can be combined with the battlefield management system, and the interoperability evaluation and security test of the new weapon system can be carried out.

We designed and built a method of combining the basic configuration results of the cyber range for training with several ranges. For actual interoperability test evaluation, it is necessary to develop a method for virtualizing and operating each application system. Additionally, the method of operation in conjunction with the tactical system over a wireless link needs to be addressed.

The advantages of the multi-cyber range proposed in this paper are as follows. First, a single range can be extended to form a training environment. Second, by creating an integrated test environment that looks like reality, mission-based impact assessment is possible. However, the disadvantage of this proposed technology is that it can further complicate the management element. As we have seen in the case of NCR, the operation of a cyber range has issues that require a great deal of management and engagement in terms of personnel. To overcome this, agent technology with various AI technologies is needed. In the future, automatic preferences, automatic traffic generation, and automatic attack agents should be developed to meet user needs.

Technically, we believe that this operating concept can be developed into a training and T&E system with greater realism and visibility by combining digital twin and metaverse technologies.

Author Contributions: Conceptualization, M.P.; Funding acquisition, D.S.; Methodology, K.K. and H.L.; Design of Scenario, Y.K.; Supervision, D.S.; Validation, H.L.; Writing-original draft, M.P.; Writing-review and editing, D.S. All authors have read and agreed to the published version of the manuscript.

Funding: This work was supported by a National Research Foundation of Korea (NRF) grant funded by the Korea government (MSIT) (No. 2022R1F1A1074773).

Institutional Review Board Statement: Not applicable.

Informed Consent Statement: Not applicable.

Data Availability Statement: Not applicable.

Conflicts of Interest: The authors declare no conflict of interest.

Abbreviations

The following abbreviations are used in this manuscript:

NCR	National Cyber Range
TRMC	Test and Training Resource Management Center
T&E	Test and Evaluation
MILS	Multiple Independent Levels of Security
DDoS	Distributed Denial of Service
C4I	Command, Control, Communication, Computer and Intelligence
VPN	Virtual Private Network
LVC	Live Virtual Constructive
MTF ICD	Message Text Format Interface Control Document
CDS	Cross Domain Solution
OT	Operational Test
OS	Operating System
DNS	Domain Name System
GUI	Graphic User Interface
PMS	Patch Management System
IER	Information Exchange Requirements
CPE	Common Platform Emulation
CVE	Common Vulnerability Enumeration
ACL	Access Control List
PCAP ATT&CK	Packet Capture Adversarial Tactics, Technique & Common Knowledge
EAI UDP	Enterprise Application Integration User Datagram Protocol
C&C DT	Command and Control Development Test

References

1. Bloom, J. *The Financial Implication of Technical Debt*; CAST Software Ltd.: New York, NY, USA, 22 February 2011.
2. Damodaran, S.K.; Smith, K. *CRIS Cyber Range Lexicon*; Version 1.0 (Report 59-0001); MIT Lincoln Laboratory: Lexington, KY, USA, 2015.
3. Hutchison, S.J. *Shift Left! Test Earlier in the Life Cycle*; Defense Acquisition University: Fort Belvoir, VA, USA, 2013.
4. Oikonomou, N.; Mengidis, N.; Spanopoulos-Karalexidis, M.; Voulgaridis, A.; Merialdo, M.; Raisr, L.; Hanson, K.; Vallee, P.L.; Tsikrika, T.; Vrochidis, S.; et al. ECHO Federated Cyber Range: Towards Next-Generation Scalable Cyber Ranges. In Proceedings of the 2021 IEEE International Conference on Cyber Security and Resilience (CSR), Rhodes, Greece, 26–28 July 2021.
5. Urias, V.E.; Stout, W.M.S.; Van Leeuwen, B.; Lin, H. Cyber Range Infrastructure Limitations and Needs of Tomorrow: A Position Paper. In Proceedings of the 2018 International Carnahan Conference on Security Technology (ICCST), Montreal, QC, Canada, 22–25 December 2018.
6. Pridmore, L.; Lardieri, P.; Hollister, R. National Cyber Range (NCR) Automated Test Tools: Implications and Application to Network-Centric Support Tools. In Proceedings of the 2010 IEEE AUTOTESTCON, IEEE, Orlando, FL, USA, 13–16 September 2010.
7. Ferguson, B.; Tall, A.; Olsen, D. National Cyber Range Overview. In Proceedings of the 2014 IEEE Military Communications Conference, Baltimore, MD, USA, 6–8 October 2014.
8. Jacq, O.; Salazar, G.P.; Parasuraman, K.; Kuusijarvi, J.; Gkaniatsou, A.; Latsak, E.; Amditis, A. The Cyber-MAR Project: First Results and Perspectives on the Use of Hybrid Cyber Ranges for Port Cyber Risk Assessment. In Proceedings of the 2021 IEEE International Conference on Cyber Security and Resilience (CSR), Rhodes, Greece, 26–28 July 2021.
9. Peratikou, A.; Louca, C.; Shiaeles, S.; Stavrou, S. On Federated Cyber Range Network Interconnection. In *Lecture Notes in Networks and Systems, Proceedings of the 12th International Networking Conference. INC 2020. Plymouth, UK, 5 January, 2020*; Springer: Berlin/Heidelberg, Germany, 2021.
10. Cruz, T.; Simões, P. Down the Rabbit Hole: Fostering Active Learning through Guided Exploration of a SCADA Cyber Range. *Appl. Sci.* **2021**, *11*, 9509. [CrossRef]
11. Teixeira, M.A.; Salman, T.; Zolanvari, M.; Jain, R.; Meskin, N.; Samaka, M. SCADA System Testbed for Cybersecurity Research Using Machine Learning Approach. *Future Internet* **2018**, *10*, 76. [CrossRef]
12. Cruz, T.; Rosa, L.; Proença, J.; Maglaras, L.; Aubigny, M.; Lev, L.; Jiang, J.; Simoes, P. A cybersecurity detection framework for supervisory control and data acquisition systems. *IEEE Trans. Ind. Inform.* **2016**, *12*, 2236–2246. [CrossRef]
13. Mathur, A.; Tippenhauer, N. SWaT: Secure Water Treatment Testbed for Research and Training in the Design of Industrial Control Systems. In Proceedings of the IEEE Computer Society International Conference on Computers, Software & Applications, Vienna, Austria, 11 April 2016.
14. Smyrlis, M.; Somarakis, I.; Spanoudakis, G.; Hatzivasilis, G.; Ioannidis, S. CYRA: A Model-Driven CYber Range Assurance Platform. *Appl. Sci.* **2021**, *11*, 5165. [CrossRef]
15. Ukwandu, E.; Farah, M.A.B.; Hindy, H.; Brosset, D.; Kavallieros, D.; Atkinson, R.; Tachtatzis, C.; Bures, M.; Andonovic, I.; Bellekens, X. A Review of Cyber-Ranges and Test-Beds: Current and Future Trends. *Sensors* **2020**, *20*, 7148. [CrossRef] [PubMed]
16. Chouliaras, N.; Kittes, G.; Kantzavelou, I.; Maglaras, L.; Pantziou, G.; Ferrag, M.A. Cyber Ranges and TestBeds for Education, Training, and Research. *Appl. Sci.* **2021**, *11*, 1809. [CrossRef]
17. Vishwanath, K.V.; Vahdat, A. Swing: Realistic and Responsive Network Traffic Generation, *IEEE/ACM. Trans. Netw.* **2009**, *17*, 712–725.
18. Bieniasz, J.; Szczypiorski, K. Dataset Generation for Development of Multi-Node Cyber Threat Detection Systems. *Electronics* **2021**, *10*, 2711. [CrossRef]
19. Botta, A.; Dainotti, A.; Pescapé, A. A tool for the generation of realistic network workload for emerging networking scenarios. *Comput. Netw.* **2012**, *56*, 3531–3547. [CrossRef]
20. Heinbockel, W.; Noel, S.; Curbo, J. Mission Dependency Modeling for Cyber Situational Awareness. In *NATO IST-148 Symposium on Cyber Defence Situation Awareness*; NATO: Sofia, Bulgaria, 3 October 2016.

applied sciences

MDPI

Article

Squill: Testing DBMS with Correctness Feedback and Accurate Instantiation

Shihao Wen [1], Peng Jia [1,*], Pin Yang [1] and Chi Hu [2]

1 School of Cyber Science and Engineering, Sichuan University, Chengdu 610207, China; wenshihao@stu.scu.edu.cn (S.W.); yangpin@scu.edu.cn (P.Y.)
2 China Academy of Engineering Physics, Mianyang 621900, China; huchi16@nudt.edu.cn
* Correspondence: pengjia@scu.edu.cn

Abstract: Database Management Systems (DBMSs) are the core of management information systems. Thus, detecting security bugs or vulnerabilities of DBMSs is an essential task. In recent years, grey-box fuzzing has been adopted to detect DBMS bugs for its high effectiveness. However, the seed scheduling strategy of existing fuzzing techniques does not consider the seeds' correctness, which is inefficient in finding vulnerabilities in DBMSs. Moreover, current tools cannot correctly generate SQL statements with nested structures, which limits their effectiveness. This paper proposes a fuzzing solution named Squill to address these challenges. First, we propose correctness-guided mutation to utilize the correctness of seeds as feedback to guide fuzzing. Second, Squill embeds semantics-aware instantiation to correctly fill semantics to SQL statements with nested structures by collecting the context information of AST nodes. We implemented Squill based on Squirrel and evaluated it on three popular DBMSs: MySQL, MariaDB, and OceanBase. In our experiment, Squill explored 29% more paths and found 3.4× more bugs than the existing tool. In total, Squill detected 30 bugs in MySQL, 27 in MariaDB, and 6 in OceanBase. Overall, 19 of the bugs are fixed with 9 CVEs assigned. The results show that Squill outperforms the previous fuzzer in terms of both code coverage and bug discovery.

Keywords: coverage-based grey-box fuzzing; database testing; vulnerability

Citation: Wen, S.; Jia, P.; Yang, P.; Hu, C. Squill: Testing DBMS with Correctness Feedback and Accurate Instantiation. *Appl. Sci.* **2023**, *13*, 2519. https://doi.org/10.3390/app13042519

Academic Editor: Luis Javier García Villalba

Received: 22 November 2022
Revised: 10 February 2023
Accepted: 13 February 2023
Published: 15 February 2023

1. Introduction

Database management systems (DBMSs) are widely used worldwide as the core of modern information systems. Like other complicated computer applications, the security and reliability of DBMSs face severe challenges. Malicious attacks on DBMSs, such as remote code execution or denial of service, will seriously harm the information system. Therefore, it is of great significance to efficiently detect DBMS vulnerabilities to improve their robustness and the security of the information system built on them.

Black-box fuzzing, or generation-based fuzzing, has been extensively used in finding DBMS bugs, such as SQLsmith [1] and SQLancer [2–4]. Security researchers have found a considerable number of bugs using this technique. A black-box fuzzer treats the program as a black box and is unaware of internal program structure [5]. It randomly generates a large number of SQL statements and executes them in the DBMS. The current input is saved for subsequent analysis when unexpected behavior occurs, such as a crash. The disadvantage of black-box fuzzing has been thoroughly discussed by the academic circle, which is inefficiency. Since the generation of SQL statements is entirely random, considering the complexity of the DBMS, most of the inputs generated by the black-box fuzzer will be difficult to trigger the deep program logic, in which bugs often hide. Despite inefficiency, this technique still has a wide range of uses. Since black-box fuzzing does not require the source code of the DBMS, it can test some commercial DBMSs that are not open source.

Researchers have studied grey-box fuzzing actively in recent years. The main difference between grey-box fuzzing and black-box fuzzing is that the former leverages instrumentation to glean information about the program [5], such as code coverage. With an initial seed queue, the grey-box fuzzer performs a series of mutations on seeds to generate new inputs and saves the inputs that trigger a new state (or crash) of the program for future mutation. Therefore, compared with black-box fuzzer, grey-box fuzzer can explore the deep states of the program gradually. The well-known AFL [6] collects the code coverage of the program during fuzzing by instrumentation, and DBMS vendors have applied it to DBMS testing. For example, SQLite used AFL as a standard part of the testing strategy until it was superseded by better fuzzers [7]. However, since fuzzer [8–10], like AFL, was not initially designed for DBMS fuzzing, the SQL statements generated by AFL often have syntactic or semantic errors, making it hard to trigger the deep logic of DBMSs (such as the optimizer). Squirrel [11], a recent work focusing on DBMS fuzzing, has solved this problem to some extent, making it the state-of-art grey-box DBMS fuzzer. It introduces the structure-aware mutator for SQL statements into AFL. After mutation, it fills inputs with new semantics to improve the syntactic and semantic correctness.

In recent years, many new solutions have been proposed for grey-box fuzzer to improve fuzzing efficiency. An important one is improving the seed scheduling strategy. However, less attention has been paid to the seed scheduling strategy in the DBMS fuzzing area. In DBMS fuzzing, different seeds have different correctness, and seeds with different correctness contribute differently to fuzzing. Hence, scheduling seeds by speed and size, the seed scheduling strategy in the existing grey-box DBMS fuzzer, is inefficient. Another challenge in grey-box DBMS fuzzing is the semantics filling of SQL statements. In order to make the SQL statement generated by mutation pass the semantic check of DBMSs, Squirrel proposes a method called Semantics-Guided Instantiation to fill the SQL skeleton with concrete semantics. However, the instantiation method of Squirrel does not perform well on SQL statements with nested structures due to design issues. A significant reason is that Squirrel cannot distinguish between nodes with the same type but at different levels. The problem of instantiation makes Squirrel hard to generate complex SQL statements, limiting its effectiveness in finding DBMS bugs.

In this paper, we implement a grey-box fuzzer, Squill, to address the challenges faced in current DBMS fuzzing. As the particularity of DBMS fuzzing scenarios, we propose correctness-guided mutation, which utilizes the correctness of SQL statements as feedback to guide fuzzing. We design two heuristic methods to improve the fuzzing efficiency by collecting the correctness (valid, syntax-error, semantics-error) of each seed. First, we prioritize mutating valid seeds because of their effectiveness in generating new paths and crashes. Second, we give some seeds with syntactic or semantic errors more opportunities to participate in mutation as material to activate interesting SQL structures in them more rapidly. In addition, we propose semantics-aware instantiation, which has the ability to guarantee the semantic correctness of the inputs with nested structures. We design a new instantiation stage in which we fill the nodes with semantics according to the predetermined constraints. During instantiation, we traverse each node of the AST in turn and parse according to the node type. While traversing, we collect the context information of each node so that we can distinguish nodes of the same type but at different levels and assign different dependencies to them. For example, with the context information of a node, we can distinguish whether it is at the beginning of a SELECT statement or a subquery in FROM clause and treat it differently.

We implemented Squill based on Squirrel. To understand the effectiveness of Squill, we evaluated it on three popular databases: MySQL [12], MariaDB [13], and OceanBase [14]. Squill successfully found 63 memory error issues, including 30 bugs in MySQL, 27 bugs in MariaDB, and 6 bugs in OceanBase. We have reported all of our findings to the developers of the appropriate DBMS. At the time of paper writing, 19 bugs have been fixed, and 9 CVE numbers have been assigned due to the danger of these vulnerabilities. Our evaluation shows that correctness-guided mutation helps to improve the efficiency of fuzzers in path exploration and bug finding. We also compare our work with the current state-of-the-art tool, Squirrel. After 24 h of testing, Squill found 15, 17, and 2 bugs in each of the three DBMSs, while Squirrel found only 3, 7, and 0 bugs. Furthermore, results show that semantics-aware instantiation outperforms the instantiation of Squirrel in the correct semantic filling of complex SQL statements.

In this paper, we first introduce Squirrel's mutation and instantiation method. Then we illustrate the necessity of scheduling seeds according to correctness through experiments and illustrate the drawbacks of Squirrel's instantiation method with examples. In addition, we introduce our solutions Squill to these two problems, including correctness-guided mutation and semantics-aware instantiation. Eventually, we prove the effectiveness of Squill through experiments.

In conclusion, this paper makes the following contributions:

- We investigated the drawbacks of the current seed scheduling strategy and the problem of Squirrel's instantiation method. We conclude that seeds should be scheduled based on correctness, and a new instantiation method that can correctly generate semantics for SQL statements with nested structures is demanded.
- We propose correctness-guided mutation, which utilizes the correctness of seed execution as feedback to guide fuzzing and improve efficiency. Moreover, we propose semantics-aware instantiation to address the challenge of correct semantics generation for SQL statements with nested structures. We implement Squill, a coverage-guided DBMS fuzzer that applies the two solutions above.
- We evaluated Squill on several real-world DBMSs and found 63 bugs.The results show that Squill outperforms the previous fuzzer in terms of both code coverage and bug discovery. We have released the source code of Squill at https://github.com/imbawenzi/Squill (accessed on 22 November 2022).

2. Background

Our proposed solution, Squill, is built on the state-of-the-art DBMS fuzzer, Squirrel. In this section, we first present an overview of Squirrel. We also introduce the challenges that current grey-box DBMS fuzzing faces and illustrate the motivation of Squill.

2.1. Overview of Squirrel

Squirrel is a recent work that aims to detect memory errors in DBMSs. Based on AFL, Squirrel modifies the mutation component so that the fuzzer can guarantee the syntactic correctness of SQL statements when mutating. As the input may be a combination of multiple parts from different SQL statements, there is a considerable probability for its semantics to be wrong. After mutation, Squirrel fills the skeleton of the SQL query with concrete operands (such as table name) through query instantiation to improve the semantic correctness.

A fuzzing loop of Squirrel starts with an empty database and inputs a set of SQL statements into DBMS, which generally include CREATE, INSERT, UPDATE, and SELECT statements. After Squirrel completes one execution, it will empty the database. Squirrel will add the input to the seed queue when it triggers new code coverage. So that Squirrel can mutate based on previous seeds, triggering the deep logic of DBMSs, compared with black-box DMBS fuzzer.

2.1.1. Mutation of Squirrel

Squirrel implements a SQL parser that converts SQL statements into AST. The mutation of the seeds (SQL statements) is based on the AST. Each node has an associated type (or grammar type), such as `SelectStmt` for the root node of a SELECT statement. Squirrel proposes three new mutation operators, including insertion, deletion, and replacement of an AST node. There is an AST subtree library in Squirrel, which we call the mutation material library. Squirrel will convert the original input and new seeds into AST and add all subtrees of these AST to the mutated material library. When performing a replace or insert mutation, Squirrel randomly selects a subtree whose root node has the same type as the target node from the mutation material library to mutate. In this way, Squirrel can maintain SQL statements in a structural manner and guarantee syntactic correctness during mutation. In the AST parser, Squirrel additionally assigns a refined data type, used in the instantiation, to nodes with semantics, such as table name.

2.1.2. Instantiation of Squirrel

The new SQL statement generated by mutation is a syntax-correct skeleton with semantics stripped. Squirrel fills it with concrete values in the process called instantiation. For data definition nodes, such as table name and column name in the CREATE statement, Squirrel directly generates concrete data to fill the node and record it. For other nodes, Squirrel will first construct the dependency graph of nodes according to the preset dependency rules of different refined data types. For the node in the graph with more than one parent, Squirrel randomly picks one to establish the edge. After that, the dependency graph is filled from top to bottom to complete the semantics filling of each node.

Figure 1 is an instantiation example of a SELECT statement in Squirrel, where `x*` is the placeholder for the semantics to be filled, and `v*` represents the semantics after filling. For the CREATE statement, we assume that the semantics has been assigned. The SELECT statement has two types of nodes that need to be instantiated, the node whose refined data type is `kDataColumnName` and the node whose refined data type is `KDataTableName`. Squirrel specifies the dependencies between these two refined data types. That is, the column name depends on the table name. Since the column name `x1` can come from both table `x3` and table `x4`, and `x2` is the same, the dependency graph in the figure can be constructed. Then Squirrel randomly selects a parent for each node, assuming that `x1`, `x2` depend on `x3`. Finally, the dependency graph is filled from top to bottom. For the table name, Squirrel randomly selects one from the existing tables(`v1` and `v5`). For the column name, Squirrel randomly selects a column name from the table it depends on. At this point, the SELECT statement is filled with semantics.

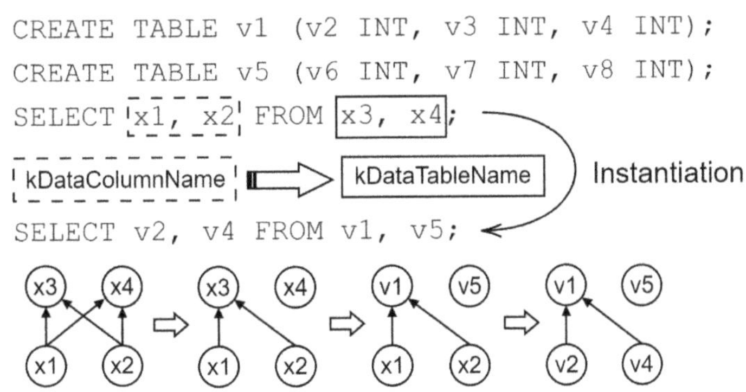

Figure 1. An instantiation example of Squirrel.

2.2. *Motivation*

2.2.1. Correctness Feedback

In grey-box fuzzing, fuzzers usually collect some information to guide fuzzing. For example, AFL collects seeds' size and execution speed and prioritizes mutating the smaller and faster seeds. Some studies [15–17] have shown that information, such as the rareness of branches, the number of memory reads or writes, and the number of branches that seed changed can guide fuzzer to perform better. In DBMS fuzzing, there is a noticeable difference in the correctness of the seeds. For example, Listing 1 shows some SQL statements with different correctness. An intuitive assumption is that seeds with different correctness contribute differently to fuzzing.

Listing 1. SQL statements with different correctness.

```
--- Valid
SELECT row_number() OVER w, v1 FROM v2 WINDOW w AS (PARTITION BY
v3 ORDER BY v4);

--- Semantics-error
SELECT row_number() OVER w, v1 FROM v2;
--- ERROR: Window name 'w' is not defined.

--- Syntax-error
SALECT row_number() OVER w, v1 FROM v2 WINDOW w AS (PARTITION BY
v3 ORDER BY v4);
--- ERROR: MySQL server version for the right syntax to use
near 'SALECT\ldots'
```

To verify our hypothesis, we conducted experiments on Squirrel to evaluate the contribution of seeds with different correctness. The result is demonstrated in Figure 2. According to the correctness of seeds, we divided the seeds into three types: valid (or semantics-correct), syntax-error, and semantics-error. We counted the seeds number of each type in a DBMS fuzzing process, as shown in Figure 2a. The abscissa indicates the total number of seeds in the process of fuzzing, and the ordinate indicates the number of different correctness seeds during the period. We found that most of the seed increments come from valid seeds. In other words, most of the paths explored by fuzzing were the program logic of DBMS after the syntactic and semantic check. It is because only inputs that are syntactic and semantic correct can proceed to the following phases, such as optimization and execution, triggering new code coverage.

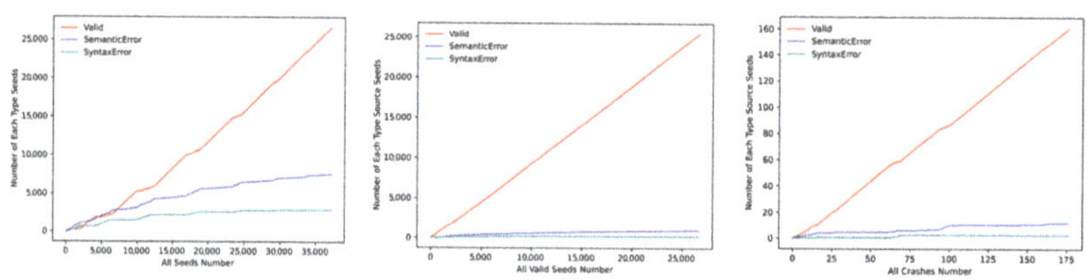

(**a**) Correctness of seeds (**b**) Source seed type of valid seeds (**c**) Source seed type of crashes

Figure 2. Contributions of seeds with different correctness in a DBMS fuzzing process.

We also counted the correctness of the valid seed's source seed in this fuzzing, as shown in Figure 2b. The abscissa indicates the total number of valid seeds in the fuzzing process, and the ordinate indicates the number of different correctness valid seed's source seeds during the period. A seed's source seed means that the seed was generated by the mutation based on its source seed. It can be seen that the majority of valid seeds are mutated from valid seeds. Considering the proportion of valid seeds in all seeds, it shows that valid seeds have a greater probability of generating valid seeds than seeds with syntactic and semantic errors. It is because if seeds with syntactic and semantic errors want to generate valid seeds, they need to mutate the wrong structures into correct ones, which is more difficult.

Moreover, we counted the correctness of the crash source seed, as shown in Figure 2c. The abscissa indicates the total number of crashes in the fuzzing process, and the ordinate indicates the number of different correctness crash source seeds during the period. The result shows that valid seeds are more likely to generate crash inputs than seeds with syntactic and semantic errors, as crashes often hide in the deep logic of the DBMS. To cause a crash, the input needs to pass the DBMS's syntactic and semantic check so that the DBMS can execute it. Therefore, the input that causes a crash is often valid.

Motivation. According to the analysis above, we can conclude that seeds with different correctness have different contributions to fuzzing. Hence, the seed schedulers in existing fuzzers, which schedules seeds by speed and size, are not efficient. Ideally, valid seeds should be mutated prior to invalid seeds because of their effectiveness in generating new paths and crashes. Therefore, a better seed scheduling strategy is demanded.

2.2.2. Limitation of Squirrel's Instantiation

The instantiation method of Squirrel works well on simple SQL statements. However, when faced with complex SQL statements, this method shows its limitation. In this paper, we define complex SQL statements as long SQL statements with nested structures, such as subqueries. When Squirrel translates SQL statements into AST, it will initialize the node containing semantics with a corresponding refined data type. It means that when recursive parsing, such as a subquery, nodes at different levels will be assigned with the same refined data type, for Squirrel parses them with the same grammar. However, there may be dependencies between nodes at different levels. Therefore, an error occurs when using the refined data type to determine the dependencies between nodes in nested structures. From another point of view, this problem is caused by Squirrel defining the dependency between nodes in the syntax analysis stage, in which the information about SQL statements is not enough to construct a complicated dependency.

Suppose there are SQL statements shown in Figure 3, which are similar to that in Figure 1, except the SELECT statement has a subquery. For descriptive convenience, the subquery does not have an alias here. Repeating the instantiation described in Section 2.1.2, the refined data type of x1, x2, x3, and x4 is kDataColumnName. Hence, they all depend on the table name nodes x5 or x6 in the same statement. Assuming that `x1` depends on `x5`, and `x2`, `x3`, `x4` depend on `x6`, the dependency graph in the figure can be constructed and filled. We can see that `x1` is filled with an invalid column name v2 that does not exist in the subquery result because `x1` comes from table v1 while x3, x4 come from table v5. Even if there is only one subquery, Squirrel still has a high probability of filling in the wrong semantics, let alone in the case of multiple subqueries.

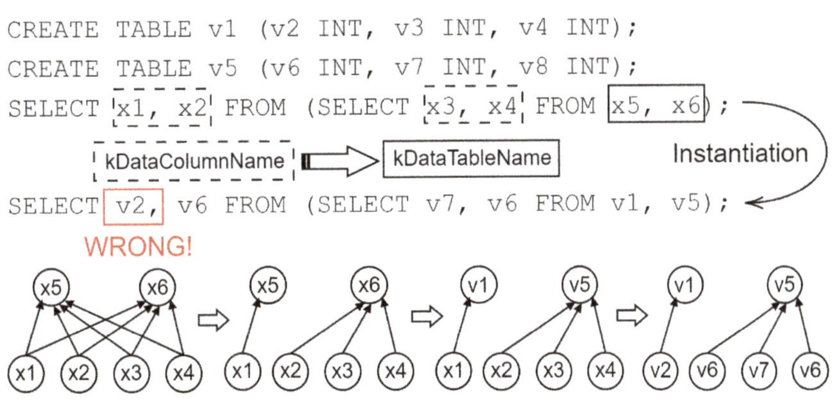

Figure 3. Squirrel's instantiation of SQL statements with a subquery.

In fact, `x1` and `x2` should depend on `x3` and `x4`, as `x1` and `x2` should come from the result of subquery in the FROM clause. Squirrel cannot do that by defining more data relation rules because it initializes both column name nodes in the subquery and the main SELECT statement with the same refined data type. During instantiation, it appears to Squirrel that these nodes are all the same. In this example, Squirrel has no way of distinguishing between `x1` and `x3` and has difficulty establishing a dependency that makes x1 depend on x3 and x4. Because of the above problem, in practice, Squirrel will discard the input with multiple subqueries for their low semantics-correct rate after instantiation.

Motivation. A new instantiation method that can correctly generate semantics for SQL statements with nested structures is demanded. In order to achieve this goal, the new instantiation method should not rely on the refined data type defined in the AST translator to construct the dependency graph.

3. Design of Squill

We propose two practical solutions to address the above challenges. First, we provide correctness-guided mutation, which contains two heuristic methods, utilizing the correctness of seeds as feedback to improve the efficiency of fuzzing (Section 3.1). Second, we introduce semantics-aware instantiation (Section 3.2). During instantiation, we collect the context information of nodes. So we can know the level of the node according to the context information and build dependencies across levels when traversing to a nested structure.

Figure 4 shows an overview of Squill, where the white components are the original Squirrel, and our design is marked in grey. Squill follows the general flow of grey-box DBMS fuzzing, which mainly includes mutation, instantiation, and fuzzing. First, Squill selects the next seed to mutate from the seed queue. Squill will preferentially select the seeds with syntactic and semantic correctness. Then, the seed is translated into AST. The mutator randomly performs replacement, insertion, and deletion mutations on the AST. Squill adds an interesting material library to participate in the mutation. During the instantiation phase, new inputs generated by mutation will be filled with semantics to maintain semantic correctness. We design a new instantiator to address the challenge of correct semantics generation for SQL statements with nested structures. In the end, Squill will take these test cases as input to the DBMS, detect whether the DBMS has crashed, and add the input that triggers new code coverage to the seed queue.

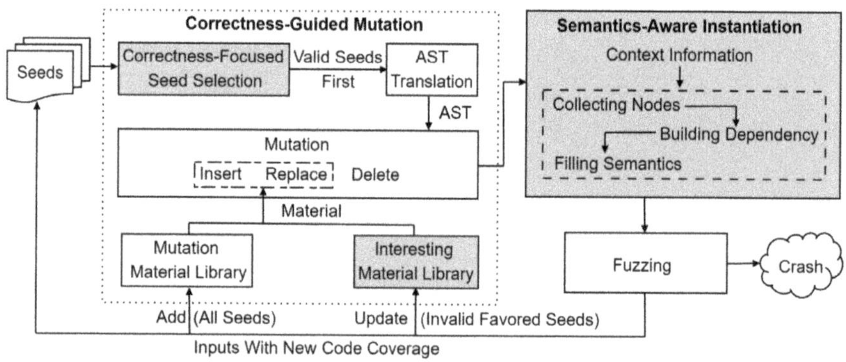

Figure 4. Overview of Squill.

3.1. Correctness-Guided Mutation

Since seeds with different correctness contribute differently to fuzzing, the fuzzer should not treat them equally. In this section, we propose two correctness-guided heuristic methods to improve the efficiency of fuzzing in path exploration and bug finding.

3.1.1. Correctness-Focused Seed Selection

In DBMS fuzzing, most of the seed increments come from valid seeds. Valid seeds can trigger deeper logic of DBMS than those with syntactic and semantic errors, exploring more paths. In addition, valid seeds have a higher probability of generating valid seeds, producing more crashes. Based on the conclusion above, we propose a correctness-focused seed selection strategy. We mutate valid seeds first, then seeds with semantics-error, and finally seeds with syntax-error. Because mutating valid seeds is more likely to generate valid seeds, leading to more path exploration and bug finding. The process of seed selection is shown in Algorithm 1.

Algorithm 1 Correctness-focused seed selection.

```
1: // Run when fuzzer generates new seed
2: function UPDATE_BITMAP_SCORE(new_seed)
3:     for i from 0 to MAP_SIZE do
4:         // trace_bits is the bitmap of current seed
5:         if trace_bits[i] is not 0 then
6:             if top_rated[i] is not 0 then
7:                 if new_seed has better correctness than top_rated[i] then
8:                     top_rated[i] = new_seed;
9:                 end if
10:                // Compare speed and size with top_rated[i]
11:                // when they have same correctness
12:                if new_seed has the same correctness as top_rated[i] then
13:                    if new_seed is faster and smaller than top_rated[i] then
14:                        top_rated[i] = new_seed;
15:                    end if
16:                end if
17:            else
18:                top_rated[i] = new_seed;
19:            end if
20:        end if
21:    end for
22: end function
```

We implement correctness-focused seed selection based on AFL's original seed selection mechanism. AFL updates `top_rated` whenever it finds a new seed `top_rated` is an array that has the same size as the bitmap, where each value records the seed with the highest score on the corresponding edge in the bitmap. The faster, smaller seed will have a higher score. Then, AFL uses a greedy algorithm to select a minimum subset of seeds that contain all edges in the bitmap from seeds recorded in `top_rated`. Seeds in this subset are marked as `favored`. The `favored` seeds have a higher probability of mutating. AFL uses this mechanism to reduce the seed queue and improve the efficiency of fuzzing.

In the original seed selection mechanism, AFL prioritizes fuzzing faster and smaller seeds. This mechanism tends to select more syntax-error or semantics-error seeds to participate in fuzzing. The seeds with syntactic or semantic errors usually have a faster execution speed, as they terminate at the syntactic and semantic check phase of DBMS. The subsequent phases of DBMS, such as the optimization and storage phase, are often time-consuming. In our method, we preferentially update seeds with better correctness into `top_rated`, as shown in Algorithm 1 Line 7–9. We define that valid seeds are better than semantics-error seeds, which are better than syntax-error seeds. We first compare the correctness of seeds and then consider their execution speed and size only if they have the same correctness (Line 12–16). It ensures that valid seeds are mutated preferentially.

3.1.2. Mutation with Interesting Material Library

Valid seeds usually trigger deep program logic. In contrast, seeds with syntactic or semantic errors (in other words, invalid) often terminate at the early phase of DBMS, such as the syntactic and semantic check. It means that the optimizer and executor of DBMS do not actually process these invalid SQL statements. So some SQL structures in the syntax-error or semantics-error seed are unactivated, as subsequent phases of DBMS do not actually process them. That is, although some SQL structures can trigger new DBMS logic, they are not actually executed because they are in an invalid seed. We call these SQL structures interesting structures here. For example, the query in Line 4 of Listing 1 is an invalid input, which uses a not existing window `w`. The valid one is shown in Line 2. The function `row_number()` in Line 4 may be an interesting structure. It is not actually executed since it is in a semantics-error SQL statement that does not pass the semantic check of DBMS.

Therefore, we designed a method to filter out these interesting structures and activate them, as shown in Algorithm 2. We maintain an interesting material library, which contains subtrees of all current `favored` and invalid seeds (Line 15–17). When the fuzzer needs a material (subtree) from the mutation material library to participate in mutation, it has a certain probability of obtaining the material from the interesting material library (Line 21–30). The variable `probability` in Line 25 is an input parameter, which is set to 5 by default.

Algorithm 2 Mutation with interesting material library.

```
1:  // Run when top_rated changed
2:  function CULL_QUEUE(void)
3:      temp_bitmap[MAP_SIZE] = 0;
4:      set_empty(interesting_library);
5:      for i from 0 to MAP_SIZE do
6:          if top_rated[i] is not 0 and temp_bitmap[i] is 0 then
7:              // Favored means mutating first
8:              top_rated[i] is favored;
9:              // Record the bitmap of top_rated[i] to temp_bitmap
10:             for j from 0 to MAP_SIZE do
11:                 if top_rated[i].bitmap[j] is not 0 then
12:                     temp_bitmap[j] = 1;
13:                 end if
14:             end for
15:             if top_rated[i] is invalid then
16:                 add_into_interesting_library(top_rated[i]);
17:             end if
18:         end if
19:     end for
20: end function
21: // Run when insertion or replacement
22: function GET_IR_FROM_LIBRARY(type)
23:     // Get a random number from 1 to 100
24:     rand_int = get_rand_int(100);
25:     if rand_int<probability then
26:         get_from_interestring_library(type);
27:     else
28:         get_from_all_library(type);
29:     end if
30: end function
```

We utilize the `favored` mechanism in AFL to select seeds that may contain interesting structures. After correctness-focused seed selection, the `favored` and invalid seed must trigger the program state (edge) that valid seeds have not triggered. For example, some SQL structures in these seeds might trigger a unique logic of the DBMS parser. When these SQL structures are actually executed, it is likely to bring path exploration or bug finding in the optimizer or executor of DBMS. Since it is difficult to generate valid seeds from the mutation of seeds with syntactic and semantic errors, we give the mutation material (subtrees) of these seeds more opportunities to participate in mutation, making interesting structures executed in valid seeds after insertion or replacement.

3.2. Semantics-Aware Instantiation

We design an instantiation algorithm to address the challenge of correct semantics generation for SQL statements with nested structures. In instantiation, while traversing AST nodes in the order of SQL statements, we parse nodes according to the node's type and context information (such as the type of parent and adjacent nodes). In this way, we can distinguish nodes of the same type but at different levels, as their context information is different. For example, the type of the parent node of a main SELECT statement and a subquery is distinct. With this information, we can construct a series of detailed constraints on nodes based on prior knowledge (the relationship between semantics in SQL statements) and then fill them with semantics correctly according to these constraints. For example, for a table name node in a CREATE statement, we can know whether it comes from a CREATE TABLE statement or a CREATE TRIGGER statement according to the context information when parsing it. For the former, we will fill it with a newly generated unique table name. For the latter, we will randomly assign a table name to it from the currently existing table name (created in the previous SQL statement).

We divide semantics into simple and complicated semantics, depending on the complexity of constraints. When traversing to a node, if we can instantly assign semantics to it without error, we call the semantics that the node has as simple semantics. The dependency constraints of nodes with such semantics are relatively simple, usually across statements. For example, the table name dropped in the DROP statement is from tables created in the previous CREATE statements. When filling a DROP statement with semantics, the previous CREATE statement has been traversed and instantiated. At this point, the existing table names are determined, which can be instantly assigned to the table name node in the DROP statement. When traversing to a node, if we cannot instantly assign semantics to it but need to wait until the entire SQL statement is parsed and fill it with consideration of the semantics of other nodes, we call the semantics that the node has as complicated semantics. For example, the column name in a SELECT clause depends on one of the tables in the FROM clause, which means that the former should be a column of the latter, and we need to instantiate the latter before the former.

3.2.1. Instantiation of Simple Semantics

Simple semantics mainly exist in CREATE, DROP, and ALTER statements, as well as nodes that do not have dependencies, such as function names. In instantiation, Squill maintains a data structure called the information table that stores the current database information, which mainly contains the table name and column name of the created tables. This information table also stores information of indexes, views, and triggers. When instantiating CREATE, DROP, and ALTER statements, we perform creating, deleting, and modifying operations in the information table correspondingly, such as in real DBMS. For the CREATE statement, we generate and assign a unique table name and column name (or index name) to the corresponding node. We record this information in the information table described above. For the ALTER statement, we will randomly choose a table name from the currently existing table name. Whether it is to modify, delete, or add a column name, Squill randomly assigns a column name from the table chosen above and modifies the corresponding information in the information table. Similarly to the DROP statement, we randomly assign a table name and delete it in the information table. For nodes without dependency, including function, integer, and floating point number, Squill will randomly assign a predefined value to them. In addition, the alias node will be assigned a unique name when traversed.

3.2.2. Instantiation of Complicated Semantics

Instantiation of complicated semantics is performed in SQL statements with column-table dependency, including SELECT, INSERT, and UPDATE statements. It is performed within one SQL statement, as the dependency between column and table is not across statements. For example, there are two independent SELECT statements. The column name in the former and the table name in the latter are irrelevant. The instantiation of complicated semantics includes three stages, collecting nodes, building dependency, and filling semantics, as shown in Algorithm 3. Note that since an input of Squill is composed of multiple SQL statements, the instantiation of complicated semantics is often performed multiple times for an input.

Algorithm 3 Instantiation of complicated semantics.

```
1: ColumnList ← Column[];
2: VirtualTableList ← VirtualTable[];
3: DependencyList ← map(Column, VirtualTable);
4: function INSTANTIATION(root)
5:     // A. Collecting Nodes
6:     for each ColumnNode in SelectTarget, Where... do
7:         // Convert node to Column
8:         ColumnList.add(ColumnParser(ColumnNode));
9:     end for
10:    for each TableNode in From, InsertTable, UpdateTable do
11:        // Convert node to VirtualTable
12:        VirtualTableList.add(TableParser(TableNode));
13:    end for
14:    for each SubQueryNode do
15:        // Process subquery recursively
16:        Instantiation(SubQueryNode);
17:        if SubQueryNode is in From then
18:            VirtualTableList.add(SubQueryParser(SubQueryNode));
19:        end if
20:    end for
21:
22:    // B. Building Dependency
23:    for each Column in ColumnList do
24:        DependencyList[Column]=RandChoose(VirtualTableList);
25:    end for
26:
27:    // C. Filling Semantics
28:    for each (Column, VirtualTable) in DependencyList do
29:        if VirtualTable is not filled then
30:            FillVirtualTable(VirtualTable);
31:        end if
32:        FillColumn(Column);
33:    end for
34:    FillVirtualTableNotInDependencyList();
35: end function
```

A. Collecting Nodes. While traversing AST, we collect the node with complicated semantics based on the type and context information of the current node and store it in the corresponding data structure (Line 5–20). For the SELECT statement, we collect column name nodes in select target, function parameter, WHERE, GROUP BY, ORDER BY, and WINDOW clauses, storing them in the data structure called `Column` (Line 6–9). `Column` stores not only the column name node but also the alias node and the table name node corresponding to the column name node, if they exist. With the help of context information, we can describe a column abstractly. Similarly, we collect the column name node in the insert and update the target clause of the INSERT and UPDATE statement.

For the table name, since we want to treat the result of the subquery as a table, we define a data structure called `VirtualTable` to represent a table. `VirtualTable` includes a table name node, an array of `Column`, and the alias node of the table, which can describe a table or the result of a subquery. For the SELECT statement, we collect the table name node in the FROM clause. For INSERT and UPDATE statements, we collect their target table name nodes (Line 10–13). For subqueries, we process them recursively, instantiating them from inner to outer (Line 14–20). For the subquery in FROM clause of the SELECT statement, we treat the result of it as a table in the subsequent dependency construction. For the subquery in other clauses, such as in the WHERE clause, we instantiate it like a SELECT statement since there is no external dependency within it.

Figure 5 shows the data structure that contains nodes in the main `SELECT` statement, where `Column_x4` and `Column_x5` are the data structure `Column` which contains nodes `x4` and `x5`. `VirtualTable` describes a table (`x10`) or the result of a subquery (`s1`) by filling different fields (`TableNameNode` or `ColumnList`). When parsing, we recursively processed subqueries, which means that the subquery and the main `SELECT` statement will be parsed with the same function, and the subquery will be instantiated before the outer query. Therefore, the corresponding data structure (such as `Column_x4` and `Column_x5`) is created while parsing the subquery.

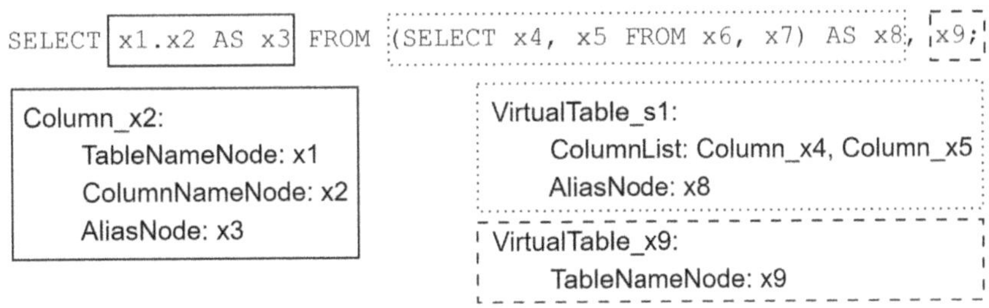

Figure 5. An example of collecting nodes.

B. Building Dependency. After creating the corresponding data structure, we construct the dependency between `Column` and `VirtualTable` (Line 22–25). Obviously, after processing, the dependency is very clear, which is that all `Column` depends on the `VirtualTable` in `FROM` clause. We randomly select a `VirtualTable` for each `Column` to depend on and record the dependency.

C. Filling Semantics. For each `Column-VirtualTable` dependency recorded, we fill nodes in it with semantics (Line 27–34). We fill `VirtualTable` first, and then the `Column` which depends on it. If the `VirtualTable` describes a table, we randomly assign the table name node in it with a table name from currently existing table names. The column name node in the `Column` which depends on the `VirtualTable` will be assigned a random column name from the table selected. The table name node in the `Column` will be filled with the same table name in `VirtualTable`, if it exists. If the `VirtualTable` describes the result of a subquery, we do not need to instantiate nodes of the `Column` in it, as they have been filled with semantics in the instantiation of the subquery (Line 16). The column name node in the `Column` will be filled with a column name in a random one of the `Column` in the `VirtualTable`. The table name node in the `Column` will be filled with the alias of the `VirtualTable`. At last, we fill semantics of the `VirtualTable` that is not depended on by any `Column` (Line 34).

Example. Figure 6 is an example of the instantiation of complicated semantics in which the SQL statement is the same as that in Section 2.2.2. Suppose that the first two CREATE statements have been instantiated, where the table names and column names have been generated, filled in nodes, and recorded in the information table. For the SELECT statement, there are two instantiation processes, one is the instantiation of the subquery, and the other is the instantiation of the main SELECT statement. Since the process is similar, here we focus on the instantiation of the main SELECT statement. Assume that the instantiation result of the subquery is as in `step1`. This SELECT statement contains two `Column` and a `VirtualTable`, where the `VirtualTable` describes the result of a subquery. Obviously, both `Column_x1` and `Column_x2` depend on `VirtualTable_s1`. We can construct a dependency graph as shown in the figure. Compared with Squirrel, the dependency graph here is more abstract. The dependency graph in Squirrel is constructed with AST nodes, while the dependency graph in Squill consists of abstract data structures, such as Column and VirtualTable. As `VirtualTable_s1` contains the result of a subquery, assuming that we randomly choose `Column_v7` for `Column_x1`, and `Column_v6` for `Column_x2`, we can fill nodes in them with semantics based on the dependencies. The result after filling in semantics is shown in `step2`. Compared with Squirrel's method, semantics-aware instantiation can effectively handle the SQL statement with nested structures like subquery.

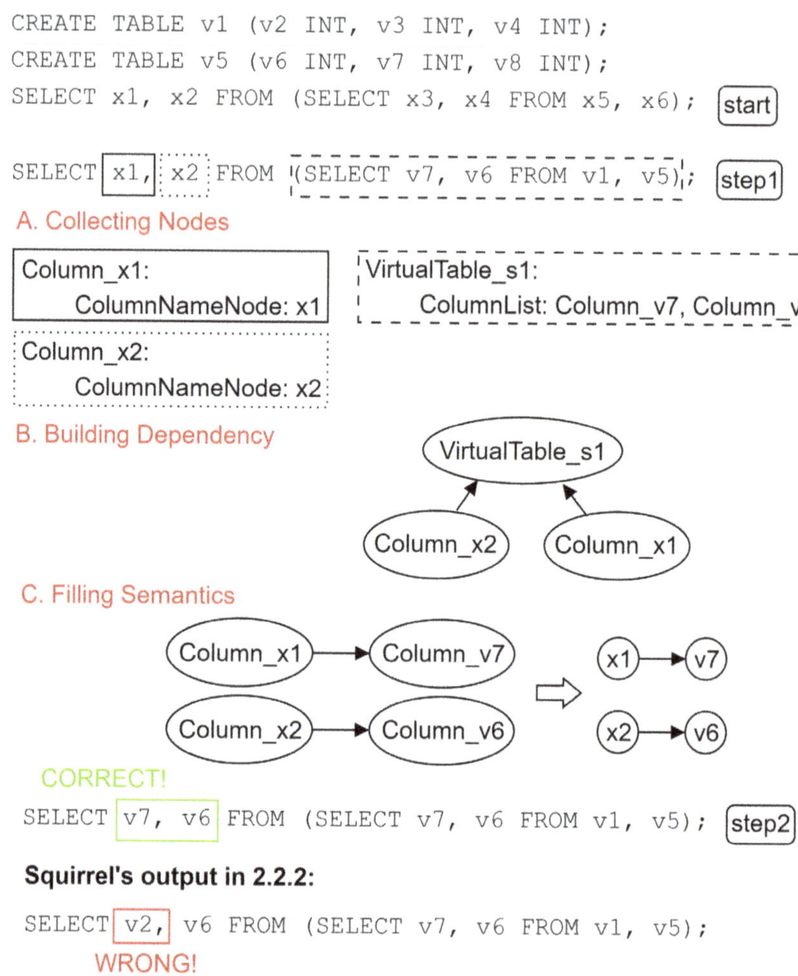

Figure 6. An example of instantiation of complicated semantics.

4. Implementation

Squill is implemented based on Squirrel. Since Squirrel is at the top of AFL, we implement the correctness-guided mutation based on the seed selection mechanism of AFL. In the implementation, we judge the correctness of the input according to the error code returned by the DBMS after executing. Additionally, the interesting material library has the same structure as the mutation material library in Squirrel, where the main difference between them is that the former stores subtrees of seeds that are invalid and `favored` while the latter stores subtrees of all seeds. We implement a new instantiation stage after mutation to replace the instantiator of Squirrel for its fundamental limitation in design. We improve the grammar of the AST parser since there are omissions and errors in Squirrel's grammar, and we remove the code that defines the refined data type.

5. Evaluation

We applied our tool Squill on real-world DBMSs to verify its effectiveness. The evaluation was designed to answer the following questions:

Q1. Can Squill detect bugs from well-tested DBMSs? (Section 5.1).
Q2. Can Squill perform better than existing tools? (Section 5.2).
Q3. How does correctness-guided mutation help fuzzing? (Section 5.3).
Q4. What is the contribution of semantics-aware instantiation? (Section 5.4).

We selected three popular real-world DBMSs for evaluation, including MySQL, MariaDB, and OceanBase. We mainly compared Squill with Squirrel, as Squirrel had been shown to outperform other mutation-based fuzzers, such as AFL, and generation-based fuzzers, such as SQLsmith. We did not compare Squill with SQLRight [18] and SQLancer, because their target is the logic bug of DBMSs, while Squill, like Squirrel, focuses on the memory error of DBMSs. We perform the experiments on three computers with Ubuntu 18.04 system, Intel(R) Core(TM) i7-10700 (2.90 GHz) CPU, and 32 GB memory. We used the llvm mode of AFL to instrument the DBMS. Because of the large codebase of DBMSs, we set the bitmap size to 256 K and used a 20% ratio instrumentation. The DBMS versions in the experiment are all the latest, including MySQL 8.0.29, MariaDB 10.10.0, and OceanBase 3.1.4. In the experiments, due to resource bottleneck, we ran one DBMS and a fuzzer on each machine for 24 h at a time and repeated three times. Squill and Squirrel used the same seed and initial library in experiments.

5.1. DBMS Bugs

In total, Squill found 63 bugs, including 30 bugs from MySQL, 27 from MariaDB, and 6 from OceanBase. The details of these bugs are shown in Table 1. We have reported all bugs to the developers of the appropriate DBMS. At the time of paper writing, 19 of all bugs have been fixed, with 9 CVEs assigned. The type of bugs found by Squill are listed in the second column of Table 1. Specifically, Squill found 10 bugs related to buffer overflows and use-after-free.

Table 1. Bugs detected by Squill.

ID	Type	Description	Status	Reference
MySQL 8.0.27				
1	SEGV	Common_table_expr::clone_tmp_table()	Fixed	CVE-2022-21509
2	HOF	make_join_readinfo()	Fixed	CVE-2022-21526
3	SEGV	Item_field::fix_outer_field()	Fixed	CVE-2022-21527
4	SEGV	push_new_name_resolution_context()	Fixed	CVE-2022-21528
5	SEGV	QEP_shared_owner::table()	Fixed	CVE-2022-21529
6	SEGV	Item_field::used_tables()	Fixed	CVE-2022-21530
7	SEGV	QEP_shared_owner::idx()	Fixed	CVE-2022-21531
8	HOF	compare_fields_by_table_order()	Fixed	CVE-2022-21438
9	AF	Query_expression::accumulate_used_tables()	Fixed	CVE-2022-21459
10	AF	MoveCompositeIteratorsFromTablePath()	Fixed	BUG106045
11	SEGV	Query_block::next_query_block()	Fixed	BUG106047
12	AF	temptable::Handler::position()	Verified	BUG106048
13	SEGV	Item_subselect::exec()	Verified	BUG106050
14	SEGV	Bitmap::merge()	Verified	BUG106051
15	SEGV	TABLE::empty_result_table()	Verified	BUG106058
16	AF	SubqueryWithResult::single_query_block()	Verified	BUG106061
17	AF	TABLE_LIST::create_materialized_table()	Verified	BUG106055

Table 1. *Cont.*

ID	Type	Description	Status	Reference
MySQL 8.0.29				
18	HUAF	Item_field::used_tables_for_level()	Verified	BUG108241
19	AF	handler::ha_index_next_same()	Verified	BUG108242
20	AF	Bounds_checked_array::operator[]()	Verified	BUG108243
21	SEGV	KEY::records_per_key()	Verified	BUG108244
22	AF	add_key_fields()	Verified	BUG108246
23	AF	add_key_field()	Verified	BUG108247
24	AF	Query_block::get_derived_expr()	Verified	BUG108248
25	AF	Item_func_case::find_item()	Verified	BUG108249
26	SBOF	Query_expression::prepare()	Verified	BUG108251
27	AF	Item_func_in::val_int()	Verified	BUG108252
28	SEGV	Item_ref::walk()	Verified	BUG108253
29	AF	copy_contexts()	Verified	BUG108254
30	SBOF	Item_func::fix_fields()	Verified	BUG108255
MariaDB 10.3.35				
31	SEGV	update_depend_map_for_order()	Verified	MDEV-28501
32	SEGV	st_select_lex::next_select()	Verified	MDEV-28502
33	SEGV	get_addon_fields()	Verified	MDEV-28503
34	SEGV	With_element::get_name()	Verified	MDEV-28504
35	SEGV	sub_select()	Verified	MDEV-28505
36	AF	find_field_in_table_ref()	Verified	MDEV-28506
37	SEGV	Item_field::fix_outer_field()	Verified	MDEV-28507
38	AF	create_tmp_table()	Fixed	MDEV-28508
39	SEGV	Bitmap::merge()	Verified	MDEV-28509
40	SEGV	get_sort_by_table()	Verified	MDEV-28510
41	SEGV	Item_subselect::init_expr_cache_tracker)	Verified	MDEV-28614
42	AF	handler::ha_rnd_next()	Verified	MDEV-28615
43	SEGV	Item_ref::fix_fields()	Verified	MDEV-28616
44	SEGV	TABLE_LIST::set_check_materialized()	Fixed	MDEV-28617
45	SEGV	Item_equal::val_int()	Verified	MDEV-28618
46	SEGV	Window_funcs_sort::setup()	Verified	MDEV-28619
47	SEGV	Item_subselect::get_cache_parameters()	Verified	MDEV-28620
48	AF	Item_subselect::exec()	Verified	MDEV-28621
49	SEGV	Item_exists_subselect::exists2in_processor()	Verified	MDEV-28622
50	AF	resolve_ref_in_select_and_group()	Verified	MDEV-28623
51	AF	Item_field::fix_fields()	Verified	MDEV-28624
MariaDB 10.10.0				
52	SBOF	st_select_lex_unit::set_unique_exclude()	Verified	MDEV-29358
53	HUAF	Field::is_null()	Verified	MDEV-29359
54	SEGV	grouping_field_transformer_for_where()	Verified	MDEV-29360
55	SBOF	resolve_references_to_cte()	Verified	MDEV-29361
56	AF	Item_singlerow_subselect::val_int()	Verified	MDEV-29362
57	HUAF	calc_group_buffer()	Verified	MDEV-29363
OceanBase 3.1.4				
58	AF	ObInsertResolver::resolve_insert_values()	Fixed	issues 986
59	HOF	ABitSet::myffsl()	Fixed	issues 987
60	SEGV	ObLatchMutex::try_lock()	Fixed	issues 988
61	AF	ObSelectStmtPrinter::print_with()	Fixed	issues 989
62	SEGV	sql::ObExpr::count()	Fixed	issues 995
63	SEGV	ObMergeJoinOp::ChildRowFetcher::next()	Fixed	issues 1000

HUAF: heap-use-after-free. SBOF: stack-buffer-overflow. HOF: heap-buffer-overflow. SEGV: segmentation violation. AF: assertion failure

Case Study. Squill detected a bug in MariaDB (ID 44 in Table 1, PoC in Listing 2), which can cause a DBMS crash by a null pointer accessing. This bug happened in INSERT...SELECT statements whose WHERE condition contains an IN/ANY/ALL predicand with a special GROUP clause, which can be eliminated and contains a subquery over a mergeable derived table referencing the updated table. It is caused by the incorrect access to the derived table which has been eliminated. The bug can cause a similar crash when executing a single-table DELETE statement with EXISTS subquery whose WHERE condition is like this. Executing this kind of query will cause a crash of DBMS in the preparation phase. The stability of the DBMS is critical, as it is usually the infrastructure for some information systems which require high availability, such as business systems in banks. Denial-of-service attacks based on such vulnerabilities can make the DBMS crash, resulting in serious consequences.

Listing 2. A PoC of ID 44 in Table 1.

```
CREATE TABLE v0 ( v1 BOOLEAN, v2 INT, v3 INT );
CREATE TABLE v4 ( v5 INT NOT NULL, v6 INT, v7 INT );
INSERT INTO v4 ( v7 ) VALUES ( ( ( TRUE ,v5 ) NOT IN
( SELECT ( - 49 ) AS v8, -128 FROM v0 GROUP BY ( TRUE, v3 )
NOT IN ( SELECT v5, ( SELECT v2 FROM ( WITH v9 AS ( SELECT v7
FROM ( SELECT NOT v5 <= 'x' , FROM v4 GROUP BY v7 ) AS v10 )
SELECT v7, ( v7 = 67 OR v7 > 'x' ) FROM v4 ) AS v11 NATURAL JOIN
v0 WHERE v7 = v3 ) AS v12 FROM v4 ), v2 ) OR v5 > 'x' ) );
```

5.2. Comparison with Existing Tools

We evaluate Squill and Squirrel on three real-world DBMSs, MySQL, MariaDB, and OceanBase, to help us better understand the performance of Squill. As shown in Figure 7, we compare the capability of bug finding and path exploration between the two tools. The number details are listed in Table 2. More program paths explored and more bugs found per unit of time means better fuzzer performance. We also compared the type of bugs they found, which represents how harmful the bug is. Since Squirrel will drop long inputs with multiple subqueries, we disable the length and the subquery check of Squirrel, denoted as $\text{Squirrel}_{!check}$.

Table 2. The number of paths and bugs explored by each fuzzer in 24 h.

DBMS	Squill		Squirrel		$\text{Squirrel}_{!check}$	
	Paths	Bugs	Paths	Bugs	Paths	Bugs
MySQL	32,827	15	46,232	3	26,387	6
MariaDB	33,904	17	62,465	7	23,835	10
OceanBase	13,081	2	16,684	0	10,630	0

In statistics, we deduplicate crashes to the corresponding bug since a bug often causes hundreds of crashes and summarize the number of bugs by the hour. Due to the multithreading feature of DBMSs, the unique crash mechanism of AFL is hard to deduplicate DBMS crashes accurately. For MySQL and MariaDB, we deduplicate crashes according to the report output by ASan [19]. For OceanBase, we use GDB [20] to debug each crash after fuzzing and deduplicate according to the information, such as the call stack of functions, at the time of the DBMS crash.

Path Exploration. Figure 7a–c show the number of paths explored by Squill, Squirrel, and $\text{Squirrel}_{!check}$ over time in MySQL, MariaDB, and OceanBase. As we can see, Squirrel explored more paths than Squill and $\text{Squirrel}_{!check}$. It is because Squirrel drops long inputs with multiple subqueries for their low semantics-correct rate after instantiation. The input generated by Squirrel is very short and simple and with a fast execution speed. However, with semantics-aware instantiation, we do not need to limit the number of subqueries in inputs generated by Squill. Thus Squill can generate long and complex inputs, which means a slow execution speed. Faster execution usually means more paths. So we tested Squirrel$_{!check}$, which can also generate long and complex inputs, to evaluate Squill more comprehensively. Compared with Squirrel$_{!check}$, Squill explored 24% more paths in MySQL, 40% in MariaDB, and 23% in OceanBase. Moreover, Squill and Squirrel found many more paths on MySQL and MariaDB than OceanBase. We think this may be caused by the feature of OceanBase as a distributed database and the bad grammar compatibility of the fuzzer with OceanBase.

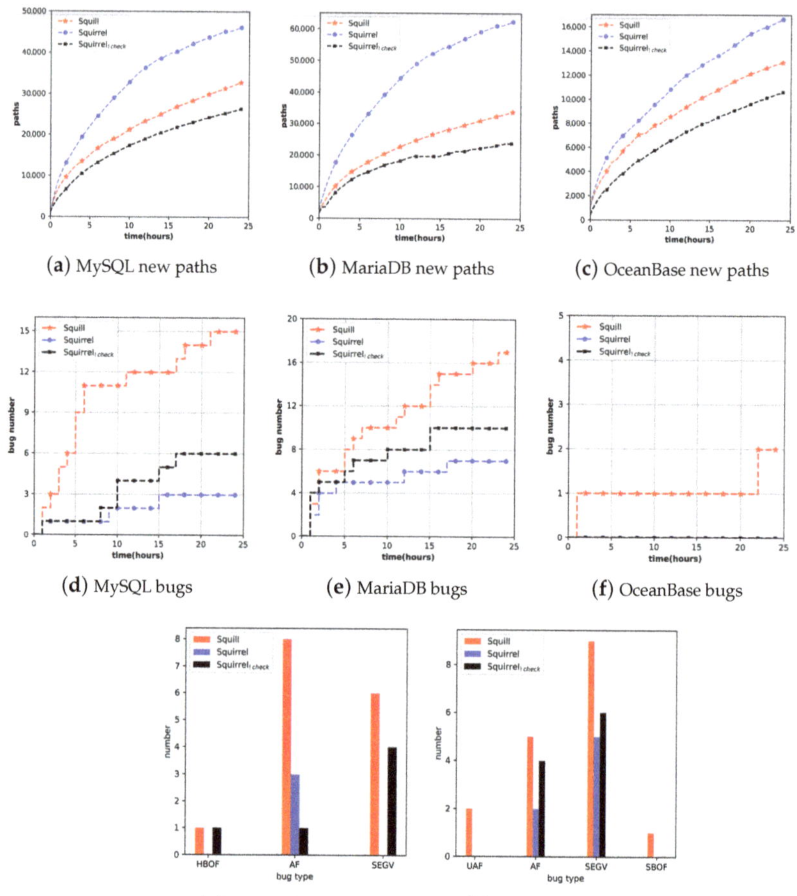

(**a**) MySQL new paths (**b**) MariaDB new paths (**c**) OceanBase new paths

(**d**) MySQL bugs (**e**) MariaDB bugs (**f**) OceanBase bugs

(**g**) MySQL bugs type (**h**) MariaDB bugs type

Figure 7. Comparison with existing tools.

Bug Finding. Figure 7d–f show the number of bugs found by Squill, Squirrel, and Squirrel$_{!check}$ over time in MySQL, MariaDB, and OceanBase. In total, Squill found 3.4x and 2x more bugs than Squirrel and Squirrel$_{!check}$, which shows the effectiveness of Squill in bug finding. Note that Squill and Squirrel$_{!check}$ found more bugs than Squirrel in MySQL and MariaDB. The result proves that there is no fundamental reason that maximizing the number of paths (or seeds) is directly connected to finding bugs [21]. Figure 7g,h show the type of bugs found by Squill, Squirrel, and Squirrel$_{!check}$. The main types of bugs are assertion fails and SEGV. It shows that Squill found a total of four buffer-related errors, while Squirrel and Squirrel$_{!check}$ only found one.

Overall, Squill outperforms Squirrel in finding memory error bugs of real-world DBMSs. Because Squill has the ability to generate valid complex SQL while Squirrel cannot. Moreover, Squill embeds correctness-guided mutation, which can improve the efficiency of fuzzing. Squill can also explore more paths than $\text{Squirrel}_{!check}$, which shows the effectiveness of Squill.

5.3. Contribution of Correctness-Guided Mutation

To understand the contribution of different factors in correctness-guided mutation, we disable each factor to perform unit tests in MySQL and measure various aspects of the fuzzing process. In addition to the capabilities of bug finding and path exploration, we also compare the correctness of the input. Figure 8 shows the result, where $\text{Squill}^{!seed}_{!lib}$ means we disable both correctness-focused seed selection and mutation with interesting material library, and $\text{Squill}_{!lib}$ means we only disabled interesting material library. Since the implementation of the mutation with interesting material library relies on correctness-focused seed selection, we do not disable the latter and keep the former.

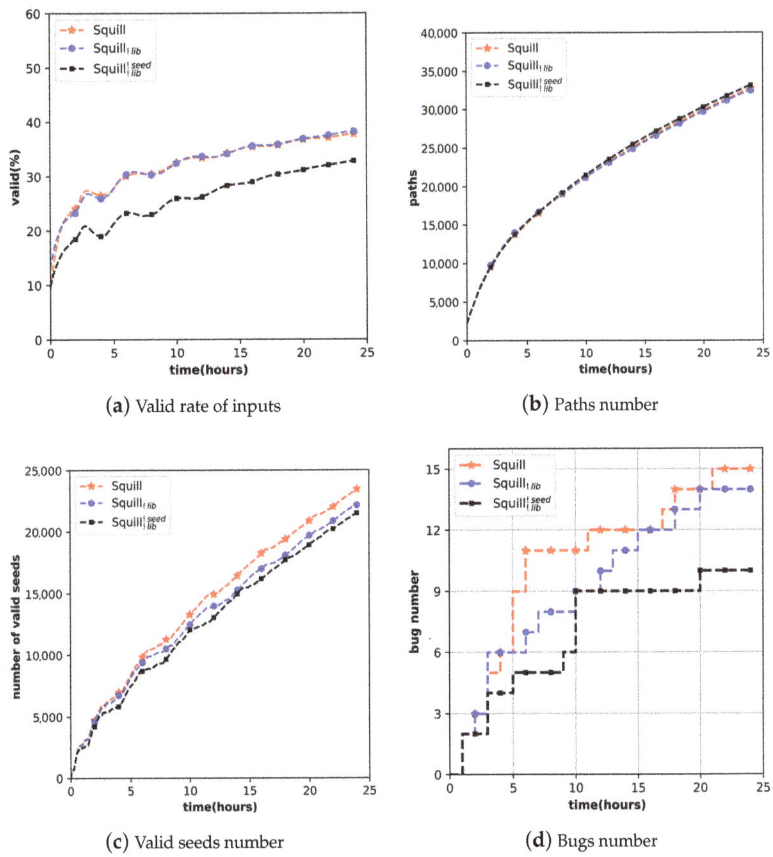

(**a**) Valid rate of inputs (**b**) Paths number

(**c**) Valid seeds number (**d**) Bugs number

Figure 8. Contributions of correctness-guided mutation. The experiment is performed on MySQL.

Correctness of Inputs. Figure 8a shows the valid rate of inputs when fuzzing, which means the proportion of valid inputs in all inputs. Higher valid rate of inputs when fuzzing is better, because we want the input to pass the validity check of the DBMS. The result is $\text{Squill} \approx \text{Squill}_{!lib} > \text{Squill}^{!seed}_{!lib}$, where $\text{Squill}_{!lib}$ is 6% higher than $\text{Squill}^{!seed}_{!lib}$. The result shows that the correctness-focused seed selection improves the ability of the fuzzer to generate more valid inputs because of its strategy to prioritize mutating seeds which is valid.

Path Exploration. Figure 8b shows the number of paths explored by each fuzzer. Squill, Squill$_{!lib}$, and Squill$^{!seed}_{!lib}$ are almost equal in the number of paths, and Squill$^{!seed}_{!lib}$ is slightly higher than the other two. Due to that Squirrel$^{!seed}_{!lib}$ generates more syntactically and semantically incorrect inputs, as shown in Figure 8a, its inputs are executed faster, leading to more paths. In addition, we count the number of valid seeds generated during fuzzing, as shown in Figure 8c. A seed represents a path since only if an input triggers a new path will it be saved into the seed queue as a seed. Therefore, the number of valid seeds reflects the capability of exploring the path which passes the syntactic and semantic check of DBMS. The result is Squill > Squill$_{!lib}$ > Squill$^{!seed}_{!lib}$, where Squill is approximately 9% higher than Squill$^{!seed}_{!lib}$, and Squill$_{!lib}$ is about 3% higher than Squill$^{!seed}_{!lib}$. The result shows that both two mechanisms can help fuzzing in path exploration.

Bug Finding. Figure 8d shows the number of bugs found by Squill with each setting, where the original Squill achieves the best results. Squill and Squill$_{!lib}$ found 15 and 14 bugs in MySQL, while Squill$^{!seed}_{!lib}$ only found 10. The results show that the correctness-guided mutation plays an important role in bug finding.

Overall, both mechanisms of correctness-guided mutation improve the effectiveness of Squill in path exploration and bug finding, where correctness-focused seed selection improves the ability of Squill to generate more valid inputs and mutation with interesting material library helps Squill explore more DBMS states after the syntactic and semantic check.

5.4. Contribution of Semantics-Aware Instantiation

In this section, we evaluate semantics-aware instantiation introduced in Section 3.2. We perform instantiation method of Squill and Squirrel to instantiate SQL statements of the same dataset and input the SQL statements with semantics to DBMS. We evaluate the instantiation of Squill by comparing the correctness of these inputs. The higher valid rate of input after instantiation is, the better instantiation method is. In the end, we illustrate the advantages of Squill instantiation through a practical example.

The dataset contains all valid seeds in one MySQL fuzzing of Squill because we want to compare the two methods' capability to instantiate some critical inputs and ensure that these inputs can be correctly instantiated. We normalize the seeds before adding them to the dataset, that is, removing the semantics in them. Due to the design of translating AST to string, the input generated by Squill has a very tiny probability that it cannot be parsed by itself (same with Squirrel). Moreover, there are differences between the grammar of Squill and Squirrel, and Squirrel cannot parse some inputs of Squill. So we remove the seeds that both Squill and Squirrel cannot parse. The evaluation results are shown in Table 3.

Table 3. The comparison between instantiation of Squill and Squirrel.

Fuzzer	Seed Size	Valid	Invalid	Total	Valid Rate
Squill	**<1 kb**	5246	373	5619	93.36%
	1∼1.5 kb	15,810	1328	17,138	92.25%
	>1.5 kb	8280	676	8956	92.45%
	total	29,336	2377	31,713	92.5%
Squirrel	**<1 kb**	4304	1315	5619	76.6%
	1∼1.5 kb	9747	7391	17,138	56.87%
	>1.5 kb	4419	4537	8956	49.34%
	total	18,470	13,243	31,713	58.24%

The results show that the instantiation of Squill (92.5% valid rate) outperforms Squirrel's (58.24% valid rate). In addition, we make separate statistics according to the file size of the seeds. The file size of seeds corresponds to the length of the SQL statement, which we think is positively related to the complexity of the SQL statement. Long SQL statement usually means more complicated dependencies between nodes and more nested structures, such as subquery. With the increased complexity of SQL statements (file size), the valid rate of Squirrel's instantiation is significantly reduced, while the correct rate of Squill's instantiation changes less. This shows the advantage of Squill's instantiation in processing complex SQL statements. Because of the randomness in semantics filling, the valid rate of Squill's instantiation is not 100%, though the input was instantiated correctly before.

Listing 3 is a PoC of ID 23 in Table 1. It can be seen that there is a nested structure containing subqueries in the SELECT statement in Line 3. This kind of nested structure is pervasive in SQL statements generated by mutation, which may be closely related to overflow vulnerabilities. It is difficult for Squirrel to instantiate such type of structure since Squirrel is hard to build correct dependencies between subqueries, such as the semantics of the first two v4 positions in Line 3.

Listing 3. A PoC of ID 23 in Table 1.

```
CREATE TABLE v0 ( v1 NUMERIC UNIQUE, v2 BIGINT ) ;
CREATE TABLE v3 ( v4 INT, v5 INT ) ;
SELECT 1 FROM v3 GROUP BY ( SELECT v4 FROM
( SELECT v4 FROM ( SELECT v4 FROM v3 UNION SELECT v1 FROM v0 )
AS v7 WHERE (v4 = 0 AND v4 = -1 AND v4 = 67 ) ) AS v9);
```

6. Discussion

In this section, we discuss several limitations of our current implementation and possible future directions.

Universality of Fuzzer. In this paper, the instantiation of Squill is based on the grammar of MySQL, which has low universality. So we chose MariaDB and OceanBase, which are compatible with MySQL grammar, for evaluation. The cost of migrating this approach to other DBMSs is slightly higher than Squirrel. Moreover, the universality of the method is also very important [22,23]. In the future, we plan to achieve the universality of the fuzzer by implementing an instantiation method that satisfies the intersection of most SQL grammars and then writing extensions for each DBMS based on this universal method.

Mode of Input. Both Squill and Squirrel start with an empty database, and the input is a combination of CREATE, INSERT, and SELECT statements. We observed that most of the seeds that triggered new code coverage were mutated in SELECT statements. Changing the data inserted and the table structure created usually does not bring new paths. We think there is room for optimization. For example, we can construct a series of tables with complex structure and data as the initial database and only input SELECT statements in fuzzing. This can save the overhead of table creation and data insertion of each input.

Mutation Operator. Squill and Squirrel use the same mutation operators, including insertion, deletion, and replacement of AST nodes. We think there are other mutation operators suitable for DBMS fuzzing scenarios. For example, the random recursive mutation operator mentioned in Nautilus [24] randomly selects a recursive tree and repeats the recursion 2^n times. Such mutation operators may help trigger buffer overflow vulnerabilities of DBMSs.

Fuzzing Partial. Most of the vulnerabilities detected by Squill and Squirrel are located in the parser and optimizer components of the DBMS. It means the main target of the current DBMS fuzzer is the parser and optimizer rather than the executor of the DBMS. However, the storage process in the executor is time-consuming, as it involves the disk IO. So one optimization idea is to separate the parser and optimizer by analyzing the source code of the DBMS. Fuzzing these separated-out functions can significantly reduce the overhead during the execution phase of the DBMS, improving the efficiency of fuzzing.

7. Related Work

In this section, we discuss the recent DBMS testing technologies related to Squill.

Black-box DBMS Fuzzing. Black-box fuzzing, or generation-based fuzzing, has been widely used to detect DBMS bugs. With a specific predefined schema, continuously generating a large number of SQL statements into the DBMS to trigger abnormal behaviors (usually crashes) of the DBMS is one method of black-box DBMS fuzzing. Sqlsmith [1] is a representative of this kind of black-box DBMS fuzzer. Based on AST, it randomly generates SQL query statements for the initial database through a series of highly customized rules. In addition, differential testing is another standard method used to detect DBMS vulnerabilities in black-box DBMS fuzzing. Rags [25] and Sparkfuzz [26] send the same SQL query to different DBMSs and detect correctness bugs by comparing the differences in the results. Sqlancer [2–4] constructs different SQL statements of functionally equivalent through several different patterns and inputs them into the same DBMS. If the results are different, the DBMS might have a logical bug. Similarly, AMOEBA [27] constructs query pairs that are semantically equivalent to each other and then compares their response time on the same database system to detect performance bugs. The main difference between Squill and the works above is that Squill is a grey-box fuzzer with feedback like code coverage. Compared with blind fuzzing, fuzzing with feedback can comprehensively explore program states and trigger the deep logic of DBMSs.

Grey-box DBMS Fuzzing. In recent years, grey-box or mutation-based fuzzing has shown its effectiveness in memory error bug detection [28–37]. AFL [6], which is an important milestone in the area of software security testing [38], has been applied to DBMS fuzzing. However, the fuzzer, like AFL, performs poorly in generating structural inputs, such as SQL statements. Though there are many works trying to address this challenge, such as Zest [39], GRIMOIRE [40], and Nautilus [24]. Their ability to generate syntactically and semantically correct SQL queries is still not good enough due to the strict syntactic and semantic requirements of the DBMS. The recent work Squirrel [11] focuses on the DBMS fuzzing scenarios. Through a customized parser based on Bison [41] and Flex [42], Squirrel translates SQL statements into AST and mutates based on the AST to guarantee the syntax correctness of the inputs. After mutation, Squirrel fills the newly generated inputs with semantics to increase their semantic correctness. There are many works based on Squirrel. With its industry-oriented design, Ratel [43] improves the feedback precision in DBMS fuzzing and enhances the robustness of input generation. SQLRight [18] combines differential testing and mutation-based fuzzing to detect logic bugs of the DBMS. Squill is also based on Squirrel, using the correctness of seeds as feedback to guide fuzzing. Moreover, Squill introduces an instantiation method that can generate correct semantics for SQL statements with nested structures.

8. Conclusions

In this paper, we design and implement Squill to find memory errors in DBMSs. We introduce the correctness of seeds into DBMS fuzzing as feedback and propose two methods: correctness-focused seed selection and mutation with interesting material library. Additionally, we investigate the challenge of semantics filling in DBMS fuzzing and design a new instantiation method to address this challenge. We evaluated Squill on popular real-world DBMSs and found 30 bugs in MySQL, 27 in MariaDB, and 6 in OceanBase, with 9 CVEs assigned. The evaluation showed that Squill could find more bugs in DBMSs than existing tools.

Author Contributions: Conceptualization, P.J.; Data curation, P.Y.; Formal analysis, S.W.; Funding acquisition, P.J.; Methodology, S.W.; Software, S.W.; Supervision, C.H.; Validation, P.Y.; Visualization, S.W.; Writing—original draft, S.W.; Writing—review and editing, P.J. All authors have read and agreed to the published version of the manuscript.

Funding: This research was funded by National Key R&D projects of China OF FUNDER grant number 2021YFB3101803.

Institutional Review Board Statement: Not appliable.

Informed Consent Statement: Not appliable.

Data Availability Statement: Not applicable.

Conflicts of Interest: The authors declare no conflicts of interest.

References

1. Seltenreich, A. SQLSmith. Available online: https://github.com/ansel/sqlsmith (accessed on 22 November 2022).
2. Rigger, M.; Su, Z. Detecting Optimization Bugs in Database Engines via Non-Optimizing Reference Engine Construction. In Proceedings of the 28th ACM Joint Meeting on European Software Engineering Conference and Symposium on the Foundations of Software Engineering, ESEC/FSE 2020, Virtual Event, 8–13 November 2020; Association for Computing Machinery: New York, NY, USA, 2020; pp. 1140–1152. [CrossRef]
3. Rigger, M.; Su, Z. Testing Database Engines via Pivoted Query Synthesis. In Proceedings of the 14th USENIX Conference on Operating Systems Design and Implementation, OSDI'20, Virtual, 4–6 November 2020; USENIX Association: Berkeley, CA, USA, 2020.
4. Rigger, M.; Su, Z. Finding Bugs in Database Systems via Query Partitioning. *Proc. ACM Program. Lang.* **2020**, 4, 1–30. [CrossRef]
5. Wikipedia. Fuzzing. Available online: https://en.wikipedia.org/wiki/Fuzzing (accessed on 22 November 2022).
6. Zalewski, M. American Fuzzy Lop. Available online: https://github.com/google/AFL (accessed on 22 November 2022).
7. Consortium, S. How SQLite Is Tested. Available online: https://www.sqlite.org/testing.html (accessed on 22 November 2022).
8. LLVM. LibFuzzer. Available online: https://www.llvm.org/docs/LibFuzzer.html (accessed on 22 November 2022).
9. Google. Honggfuzz. Available online: https://github.com/google/honggfuzz (accessed on 22 November 2022).
10. Fioraldi, A.; Maier, D.; Eißfeldt, H.; Heuse, M., AFL++: Combining Incremental Steps of Fuzzing Research. In Proceedings of the 14th USENIX Conference on Offensive Technologies, Berkeley, CA, USA, 11 August 2020; USENIX Association: Berkeley, CA, USA, 2020.
11. Zhong, R.; Chen, Y.; Hu, H.; Zhang, H.; Lee, W.; Wu, D. SQUIRREL: Testing Database Management Systems with Language Validity and Coverage Feedback. In Proceedings of the CCS '20: 2020 ACM SIGSAC Conference on Computer and Communications Security, Virtual Event, 9–13 November 2020; Ligatti, J., Ou, X., Katz, J., Vigna, G., Eds.; ACM: New York, NY, USA, 2020; pp. 955–970. . [CrossRef]
12. Oracle. MySQL Server. Available online: https://github.com/mysql/mysql-server (accessed on 22 November 2022).
13. Foundation, M. MariaDB. Available online: https://github.com/MariaDB/server (accessed on 22 November 2022).
14. Group, A. OceanBase. Available online: https://github.com/oceanbase/oceanbase (accessed on 22 November 2022).
15. Lemieux, C.; Sen, K. FairFuzz: A Targeted Mutation Strategy for Increasing Greybox Fuzz Testing Coverage. In Proceedings of the 33rd ACM/IEEE International Conference on Automated Software Engineering, ASE 2018, Montpellier, France, 3–7 September 2018; Association for Computing Machinery: New York, NY, USA, 2018; pp. 475–485. [CrossRef]
16. Wang, Y.; Jia, X.; Liu, Y.; Zeng, K.; Bao, T.; Wu, D.; Su, P. Not All Coverage Measurements Are Equal: Fuzzing by Coverage Accounting for Input Prioritization. In Proceedings of the 27th Annual Network and Distributed System Security Symposium, NDSS 2020, San Diego, CA, USA, 23–26 February 2020; The Internet Society: Washington, DC, USA, 2020.
17. Yue, T.; Wang, P.; Tang, Y.; Wang, E.; Yu, B.; Lu, K.; Zhou, X. EcoFuzz: Adaptive Energy-Saving Greybox Fuzzing as a Variant of the Adversarial Multi-Armed Bandit. In Proceedings of the 29th USENIX Conference on Security Symposium, SEC'20, San Diego, CA, USA, 12–14 August 2020; USENIX Association: Berkeley, CA, USA, 2020.
18. Liang, Y.; Liu, S.; Hu, H. Detecting Logical Bugs of DBMS with Coverage-based Guidance. In Proceedings of the 31st USENIX Security Symposium (USENIX Security 22), Boston, MA, USA, 10–12 August 2022; USENIX Association: Boston, MA, USA, 2022; pp. 4309–4326.
19. Google. AddressSanitizer. Available online: https://github.com/google/sanitizers/wiki/AddressSanitizer (accessed on 22 November 2022).
20. Foundation, F.S. GDB: The GNU Project Debugger. Available online: http://www.sourceware.org/gdb/ (accessed on 22 November 2022).
21. Klees, G.; Ruef, A.; Cooper, B.; Wei, S.; Hicks, M. Evaluating Fuzz Testing. In Proceedings of the 2018 ACM SIGSAC Conference on Computer and Communications Security, CCS '18, Toronto, ON, Canada, 15–19 October 2018; Association for Computing Machinery: New York, NY, USA, 2018; pp. 2123–2138. [CrossRef]
22. Yan, L.; Ahmad, M.W.; Jawarneh, M.; Shabaz, M.; Raffik, R.; Kishore, K.H.; Azeem, I. Single-Input Single-Output System with Multiple Time Delay PID Control Methods for UAV Cluster Multiagent Systems. *Secur. Commun. Netw.* **2022**, 2022, 3935143. [CrossRef]
23. Gao, H.; Kareem, A.; Jawarneh, M.; Ofori, I.; Raffik, R.; Kishore, K.H. Metaheuristics Based Modeling and Simulation Analysis of New Integrated Mechanized Operation Solution and Position Servo System. *Math. Probl. Eng.* **2022**, 2022, 1466775. [CrossRef]
24. Aschermann, C.; Frassetto, T.; Holz, T.; Jauernig, P.; Sadeghi, A.; Teuchert, D. NAUTILUS: Fishing for Deep Bugs with Grammars. In Proceedings of the 26th Annual Network and Distributed System Security Symposium, NDSS 2019, San Diego, CA, USA, 24–27 February 2019; The Internet Society: Washington, DC, USA, 2019.
25. Slutz, D. *Massive Stochastic Testing of SQL*; Technical Report MSR-TR-98-21; Publisher: Burlington, MA, USA, 1998.

26. Ghit, B.; Poggi, N.; Rosen, J.; Xin, R.; Boncz, P. SparkFuzz: Searching Correctness Regressions in Modern Query Engines. In Proceedings of the Workshop on Testing Database Systems, DBTest '20, Portland, OR, USA, 19 June 2020; Association for Computing Machinery: New York, NY, USA, 2020. [CrossRef]
27. Liu, X.; Zhou, Q.; Arulraj, J.; Orso, A. Automatic Detection of Performance Bugs in Database Systems Using Equivalent Queries. In Proceedings of the 44th International Conference on Software Engineering, ICSE '22, Pittsburgh, PA, USA, 25–27 May 2022; Association for Computing Machinery: New York, NY, USA, 2022; pp. 225–236. [CrossRef]
28. Gan, S.; Zhang, C.; Qin, X.; Tu, X.; Li, K.; Pei, Z.; Chen, Z. CollAFL: Path Sensitive Fuzzing. In Proceedings of the 2018 IEEE Symposium on Security and Privacy (SP), San Francisco, CA, USA, 20–24 May 2018; pp. 679–696. [CrossRef]
29. Manès, V.J.M.; Kim, S.; Cha, S.K. Ankou: Guiding Grey-Box Fuzzing towards Combinatorial Difference. In Proceedings of the ACM/IEEE 42nd International Conference on Software Engineering, ICSE '20, Seoul, Republic of Korea, 27 June–19 July 2020; Association for Computing Machinery: New York, NY, USA, 2020; pp. 1024–1036. [CrossRef]
30. Lyu, C.; Ji, S.; Zhang, C.; Li, Y.; Lee, W.H.; Song, Y.; Beyah, R. MOPT: Optimized Mutation Scheduling for Fuzzers. In Proceedings of the 28th USENIX Conference on Security Symposium, SEC'19, Berkeley, CA, USA, 14–16 August 2019; USENIX Association: Berkeley, CA, USA, 2019; pp. 1949–1966.
31. Zhou, C.; Wang, M.; Liang, J.; Liu, Z.; Jiang, Y. Zeror: Speed up Fuzzing with Coverage-Sensitive Tracing and Scheduling. In Proceedings of the 35th IEEE/ACM International Conference on Automated Software Engineering, ASE '20, Melbourne, Australia, 21–25 September 2020; Association for Computing Machinery: New York, NY, USA, 2020; pp. 858–870. [CrossRef]
32. Aschermann, C.; Schumilo, S.; Blazytko, T.; Gawlik, R.; Holz, T. REDQUEEN: Fuzzing with Input-to-State Correspondence. In Proceedings of the 26th Annual Network and Distributed System Security Symposium, NDSS 2019, San Diego, CA, USA, 24–27 February 2019; The Internet Society: Washington, DC, USA, 2019.
33. Chen, P.; Chen, H. Angora: Efficient Fuzzing by Principled Search. In Proceedings of the 2018 IEEE Symposium on Security and Privacy (SP), San Francisco, CA, USA, 20–24 May 2018; pp. 711–725. [CrossRef]
34. Park, S.; Xu, W.; Yun, I.; Jang, D.; Kim, T. Fuzzing JavaScript Engines with Aspect-preserving Mutation. In Proceedings of the 2020 IEEE Symposium on Security and Privacy (SP), San Francisco, CA, USA, 18–21 May 2020; pp. 1629–1642. [CrossRef]
35. Yun, I.; Lee, S.; Xu, M.; Jang, Y.; Kim, T. QSYM: A Practical Concolic Execution Engine Tailored for Hybrid Fuzzing. In Proceedings of the 27th USENIX Security Symposium (USENIX Security 18), Vancouver, BC, Canada, 16–18 August 2017; USENIX Association: Baltimore, MD, USA, 2018; pp. 745–761.
36. Pham, V.T.; Böhme, M.; Santosa, A.E.; Căciulescu, A.R.; Roychoudhury, A. Smart Greybox Fuzzing. *IEEE Trans. Softw. Eng.* **2021**, *47*, 1980–1997. [CrossRef]
37. Wüstholz, V.; Christakis, M. Harvey: A Greybox Fuzzer for Smart Contracts. In Proceedings of the 28th ACM Joint Meeting on European Software Engineering Conference and Symposium on the Foundations of Software Engineering, ESEC/FSE 2020, Virtual Event, 8–13 November 2020; Association for Computing Machinery: New York, NY, USA, 2020; pp. 1398–1409. [CrossRef]
38. Fioraldi, A.; Maier, D.; Zhang, D.; Balzarotti, D. LibAFL: A framework to build modular and reusable fuzzers. In Proceedings of the CCS 2022, 29th ACM Conference on Computer and Communications Security, Los Angeles, CA, USA, 7–11 November 2022; ACM: New York, NY, USA, 2022.
39. Padhye, R.; Lemieux, C.; Sen, K.; Papadakis, M.; Le Traon, Y. Semantic Fuzzing with Zest. In Proceedings of the 28th ACM SIGSOFT International Symposium on Software Testing and Analysis, ISSTA 2019, Beijing, China, 15–19 July 2019; Association for Computing Machinery: New York, NY, USA, 2019; pp. 329–340. [CrossRef]
40. Blazytko, T.; Aschermann, C.; Schlögel, M.; Abbasi, A.; Schumilo, S.; Wörner, S.; Holz, T. GRIMOIRE: Synthesizing Structure while Fuzzing. In Proceedings of the 28th USENIX Security Symposium (USENIX Security 19), Santa Clara, CA, USA, 14–16 August 2019; USENIX Association: Santa Clara, CA, USA, 2019; pp. 1985–2002.
41. Foundation, F.S. Gnu Bison. Available online: https://www.gnu.org/software/bison (accessed on 22 November 2022).
42. Paxson, V. Flex. Available online: https://github.com/westes/flex (accessed on 22 November 2022).
43. Wang, M.; Wu, Z.; Xu, X.; Liang, J.; Zhou, C.; Zhang, H.; Jiang, Y. Industry Practice of Coverage-Guided Enterprise-Level DBMS Fuzzing. In Proceedings of the 43rd International Conference on Software Engineering: Software Engineering in Practice, ICSE-SEIP '21, Madrid, Spain, 22–30 May 2021; IEEE Press: Hoboken, NJ, USA, 2021; pp. 328–337. [CrossRef]

applied sciences

Article

Not All Seeds Are Important: Fuzzing Guided by Untouched Edges

Chen Xie [1], Peng Jia [1,*], Pin Yang [1], Chi Hu [2], Hongbo Kuang [1], Genzuo Ye [1] and Xuanquan Hong [1]

[1] School of Cyber Science and Engineering, Sichuan University, Chengdu 610207, China; xiechen1@stu.scu.edu.cn (C.X.); yangpin@scu.edu.cn (P.Y.); hongbokuang@stu.scu.edu.cn (H.K.); yegenzuo@stu.scu.edu.cn (G.Y.); leehung@stu.scu.edu.cn (X.H.)
[2] China Academy of Engineering Physics (CAEP), Mianyang 621900, China; huchi16@nudt.edu.cn
* Correspondence: pengjia@scu.edu.cn

Abstract: Coverage-guided greybox fuzzing (CGF) has become the mainstream technology used in the field of vulnerability mining, which has been proven to be effective. Seed scheduling, the process of selecting seeds from the seeds pool for subsequent fuzzing iterations, is a critical component of CGF. While many seed scheduling strategies have been proposed in academia, they all focus on the explored regions within programs. In response to the inefficiencies of traditional seed scheduling strategies, which often allocate resources to ineffective seeds, we introduce a novel seed scheduling strategy guided by untouched edges. The strategy generates the optional seed set according to the information on the untouched edges. We also present a new instrumentation method to capture unexplored areas and guide the fuzzing process toward them. We implemented the prototype UntouchFuzz on top of American Fuzzy Lop (AFL) and conducted evaluation experiments against the most advanced seed scheduling strategies. Our results demonstrate that UntouchFuzz has improved in code coverage and unique vulnerabilities. Furthermore, the method proposed is transplanted into the fuzzer MOpt, which further proves the scalability of the method. In particular, 13 vulnerabilities were found in the open-source projects, with 7 of them having assigned CVEs.

Keywords: vulnerability mining; greybox fuzzing; seed scheduling

Citation: Xie, C.; Jia, P.; Yang, P.; Hu, C.; Kuang, H.; Ye, G.; Hong, X. Not All Seeds Are Important: Fuzzing Guided by Untouched Edges. *Appl. Sci.* **2023**, *13*, 13172. https://doi.org/10.3390/app132413172

Academic Editors: Peter R.J. Trim and Yang-Im Lee

Received: 24 October 2023
Revised: 5 December 2023
Accepted: 5 December 2023
Published: 12 December 2023

1. Introduction

Fuzzing is a prevalent and effective automated software testing method that has already found numerous vulnerabilities in real-world applications [1–7]. Fuzzers efficiently explore the input space of the program under test, operating at nearly raw execution speeds, with the aim of identifying specific inputs that can provoke program crashes or anomalous behaviors. However, the input space of most real-world programs is so large that it is difficult to fully explore. Moreover, vulnerabilities are sparse in an application, with only certain specific inputs capable of triggering vulnerabilities [8].

American Fuzzy Lop (AFL) [9], one of the most popular and widely used coverage-guided greybox fuzzers in both academia and industry, is an efficient fuzzing tool for file applications and has already discovered many high-risk vulnerabilities across various projects. AFL employs a mutation-based fuzzing approach by mutating the binary data of the seed file to find test cases that improve coverage or trigger crashes. Research [10] indicates that the performance of mutation-based fuzzers depends on seed scheduling, essentially determining the prioritization of which seed to mutate.

The main challenge in seed scheduling is to determine which seeds in the corpus are more likely to explore new code space in the program when mutated. From the perspective of code coverage, the main role of seed scheduling is to prioritize those seeds that are more promising to trigger new code coverage after being mutated. AFL, for example, utilizes a greedy algorithm to maintain an optimal seed set that covers all explored edges. However, seeds related to validation check edges are also added to the seed set by mistake, and fuzzing these seeds will waste a lot of computational overhead [11].

In the field of seed scheduling, many scholars have conducted in-depth research. Classic seed scheduling strategies mainly guide seed scheduling and prioritization through the distribution of edge coverage or path coverage throughout the fuzzing process. For example, AFLFast [12] gives priority to those seeds likely to trigger low-frequency paths. FairFuzz [13] prioritizes those triggering rare edges, while EcoFuzz [14] prioritizes those based on the seed's self-transition probability. However, the distribution of coverage information mentioned above is related to the control flow graph (CFG) of the program. Consequently, the performance of such fuzzers can vary across programs with different CFGs.

To decouple the strong correlation between seed scheduling strategies and target programs, SLIME [15] classifies seeds by constructing multiple property queues and employs the upper confidence bound variance (UCB-V) algorithm to select the optimal seed queue and then fuzzing the seeds in that queue. Furthermore, Alphuzz [16] considers seed interdependencies and uses a Monte Carlo search tree approach for seed scheduling.

However, the existing methods mentioned above neglect to focus on the unexplored regions within the control flow graph (CFG) of the program. For instance, consider a seed s whose execution path does not contain any untouched edges, but a coverage-guided fuzzer still marks s as a favored seed. On one hand, this seed is likely to cover branches related to validation checks, while those checks are often hard to solve. On the other hand, when this seed initially joins the seed queue, it contains untouched edges. However, as fuzzing proceeds, other seeds may explore these untouched edges, resulting in a seed execution path without untouched edges. Our insight is that if a seed does not contain any untouched edges, there is little sense in prioritizing mutations on it.

In response to the aforementioned challenges, we propose a seed scheduling strategy based on untouched edges extracted from underlying CFG. Unlike traditional seed scheduling strategies that pay more attention to explored regions, our approach gives precedence to seeds that incorporate untouched edges, which are then subjected to fuzzing as a priority. We instrument the target program to collect data on edge coverage and untouched edge information. Subsequently, we revise the untouched edge coverage information for all seeds within the queue. Ultimately, we select an optimal minimal subset of seeds that covers all untouched edges from the seed pool. Furthermore, our scheduling strategy allocates more energy to seeds with more low-frequency untouched edges in the queue. Our insight is that low-frequency untouched edges imply a high probability of being explored, while high-frequency untouched edges are likely to be hard-to-solve edges. Therefore, we enhance the effectiveness of seed scheduling by allocating extra energy to seeds associated with more low-frequency untouched edges, further encouraging in-depth exploration of these areas.

The main contributions of this paper are summarized as follows:

(1) We propose a new seed scheduling strategy that efficiently selects seeds based on unexplored regions within program execution paths. This approach prevents wasting resources on ineffective seeds, a common issue in traditional scheduling methods.
(2) We applied the new seed scheduling strategy to a new greybox fuzzing tool named UntouchFuzz. To our knowledge, UntouchFuzz is the first fuzzer to utilize unexplored area information as the basis for seed scheduling. The source code is available at https://github.com/bladchan/untouchFuzz.git (accessed on 22 October 2023).
(3) We evaluated UntouchFuzz on 12 programs, demonstrating its effectiveness when compared to four AFL-based seed schedulers.

The rest of the paper is organized as follows. Section 2 discusses the background. Section 3 illustrates our motivation with an example. Section 4 shows our design of UntouchFuzz and its technical details. Implementation details are listed in Section 5. Section 6 shows the evaluation results. Section 7 discusses several limitations of our implementation. Section 8 concludes this paper.

2. Background

2.1. Techniques

In this section, we provide the background on coverage-guided greybox fuzzing and focus on the coverage acquisition method and seed scheduling in the classical fuzzer AFL.

2.1.1. Coverage-Guided Greybox Fuzzing

Coverage-guided greybox fuzzing continuously generates test cases by employing coverage feedback loops and preserves seeds that yield new coverage. Specifically, coverage-guided greybox fuzzing includes four main stages [17]:

(1) Seed Scheduling: Effective seeds are chosen from a pool of seeds based on a scheduling strategy, where effectiveness refers to the fact that new code can be explored more easily by mutating that seed.
(2) Energy Scheduling: Appropriate mutation counts (energy) are assigned based on the attributes of the chosen effective seeds.
(3) Seed Mutation: Within the allocated energy, various mutation operations are performed on the selected seeds to generate new test cases.
(4) Seed Selection/saving: Each generated test case is executed on the target program, and seeds are evaluated based on corresponding coverage information. If a test case enhances the coverage of the target program, it is chosen as a new seed. Through this feedback loop, the coverage of the target program under test continues to increase and it is more likely to generate test cases that trigger new bugs.

2.1.2. Lightweight Instrumentation

The key idea of the coverage-guided greybox fuzzing method lies in coverage acquisition data, as the coverage feedback mechanism propels the entire process of coverage-guided greybox fuzzing forward. AFL, as one of the state-of-the-art coverage-guided fuzzing tools, employs a lightweight instrumentation technique to capture transitions between program basic blocks [18]. AFL assigns a unique random ID to each basic block in CFG. The transition between two basic blocks, i.e., the edge, is defined as in Equation (1). In particular, $edge_i$ represents the ID of the edge, $prev_{bb}$ refers to the ID of the previous basic block, and cur_{bb} refers to the ID of the basic block where the current transition occurs. Note that $prev_{bb}$ is shifted one bit to the right in order to distinguish different transition orders between two basic blocks.

$$edge_i = (prev_{bb} \gg 1) \oplus cur_{bb} \quad (1)$$

AFL maintains a default 64 KB bitmap. Each byte in the bitmap is utilized to log the transition count associated with the byte's index in the bitmap, as illustrated in Figure 1. In this way, AFL can effectively track and count the edges covered by different inputs in the program.

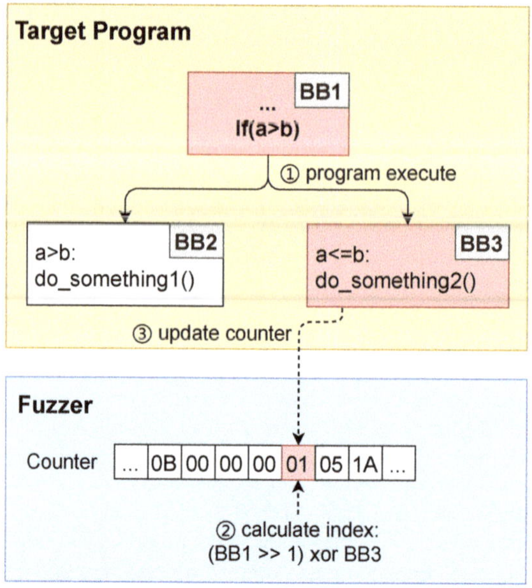

Figure 1. Details of AFL edge coverage collection.

2.1.3. Seed Scheduling

AFL employs a genetic algorithm to preserve test cases that cover new edges or hit new edge counts and utilizes a greedy algorithm to generate a favored seed subset from the seed queue, as described in Algorithm 1.

Algorithm 1 Generating the favored seed subset

Input: top_{rated} maintained by AFL
Output: A favored seed set S_{fav}
1 $memset(temp_v, 255, MAP\_SIZE)$
2 $S_{fav} \leftarrow \emptyset$
3 **for** $i = 0$ *to* MAP_SIZE **do**
4   **if** $top_{rated}[i]$ *and* $(temp_v[i \gg 3]\ \&\ (1 \ll (i\ \&\ 7)))$ **then**
5   **for** $j = (MAP\_SIZE/8 - 1) \rightarrow 0$ **do**
6             $temp_v \leftarrow temp_v \& \sim top_{rated}[i].trace_{mini}[j]$
7   **end**
8         $S_{fav} \leftarrow S_{fav} \cup top_{rated}[i]$
9   **end**
10 **end**
11 **return** S_{fav}

In the initial step, the algorithm maintains an array, $temp_v$, to keep a record of edges currently covered by favored seeds. Then, in line 3, the algorithm iterates through the highest-scoring seed of each edge. Notably, the highest-scoring seeds are dynamically maintained by AFL during the fuzzing process based on criteria such as seed size and execution speed.

Subsequently, the algorithm determines whether the seed has already been covered by other seeds within the favored seed set in lines 4–7. If the edge remains uncovered, we update edge coverage information in $temp_v$. Finally, in line 8, the seed is added to the favored seed set.

Once this favored seed set is obtained, AFL prioritizes these seeds and allocates more energy for mutation. It is important to acknowledge that AFL assumes that if a seed triggers new edges, fuzzing that seed will likely trigger more edges. However, this assumption has certain limitations when dealing with unexplored regions, as we will discuss in detail in Section 3.

2.2. Related Work

In this section, we discuss the closely related works.

Coverage-guided greybox fuzzing. Coverage-guided greybox fuzzing is one of the most effective techniques for finding vulnerabilities and bugs, garnering significant attention from both academia and industry. Coverage-based greybox fuzzers typically adopt the coverage information to guide different program path explorations.

Since a coverage guidance engine is a key component for the greybox fuzzers, much effort has been devoted to improving their coverage. For example, REDQUEEN [19], GREYONE [20] and PATA [21] employ lightweight taint analysis to penetrate some paths protected by magic bytes comparisons. Driller [22], T-Fuzz [23] and QSYM [3] incorporate symbolic execution engines to delve into deeper program codes. Angora [24] adopts a gradient descent technique to resolve path constraints to break some hard comparisons. MemFuzz [25], MemLock [26], and ovAFLow [27] augment evolutionary fuzzing by additionally leveraging information about memory accesses or memory consumption performed by the target program. CollAFL [28] proposes a coverage-sensitive fuzzing approach to mitigate path collisions. Furthermore, AFLGo [29], Hawkeye [30], Beacon [31], and SelectFuzz [32] utilize alternative metrics for directing fuzzing toward user-specified target sites in the program.

Seed scheduling. In this paper, we focus on improving the seed scheduling component in a fuzzer. With the seed set, seed scheduling is essential for addressing two key issues: (1) which seed to select for the next round and (2) the time budget for the selected seed. In practice, instead of time budget, most fuzzers optimize the number of mutations performed on the selected seeds, i.e., energy scheduling.

AFLFast [12] models path transitions as Markov chain [33], efficiently guiding fuzzing to explore undiscovered path transitions. FairFuzz [13] leverages a targeted mutation strategy to prioritize the exploration of rare branches. EcoFuzz [14] adopts the Variant of the Adversarial Multi-armed Bandit Model (VAMAB) model to prioritize and allocate more energy to the seeds with lower self-transition probabilities. The insight behind it is that the low self-transition probability indicates the high probability of discovering new paths after mutating. Alphuzz [16] models the seed scheduling problem as a Monte Carlo tree search (MCTS) problem. Its key observation is that the relationships among seeds are valuable for seed scheduling. SLIME [15] prioritizes seeds based on reward estimated by a customized upper confidence bound variance-aware (UCB-V) algorithm on different property seed queues, adaptively allocating energy to the seed with different properties.

3. Motivating Example

We use a program's control flow graph in Figure 2 to illustrate our motivation. The initial seed A is considered to be a quality seed, i.e., the seed is capable of executing the main logic codes of the program. Figure 2a shows the execution path of seed A, with red circles denoting the basic blocks covered by the seed. When the coverage-guided greybox fuzzer mutates seed A, it can easily produce seed B and seed C. Both of these seeds cover edges associated with validation checks at the control flow graph level, i.e., BB1→BB10 and BB2→BB10.

The appearance of these two new edges is interpreted by the fuzzer as an increase in edge coverage, leading to their inclusion in the seed queue. Furthermore, according to the previous description in Section 2.1.3, AFL marks both seed B and seed C as favored seeds during seed scheduling since they cover new edges. However, when attempting to mutate these seeds, they encounter challenges in producing descendant seeds that explore deeper code areas.

We further assume that after mutating seed A, the fuzzer generates not only seed B and seed C but also seed D. Seed D's execution path is shown in Figure 2d. The difference between seed A and seed D lies in the fact that seed D explores basic block BB6, an unexplored area of seed A's execution path. In this case, both two seeds are effective for the exploration of basic block BB5. However, if seed E, which covers basic block BB5 is produced later, seed A and seed D become ineffective as they make limited contributions to exploring basic blocks BB8 and BB9. Nevertheless, AFL still regards seed A and seed D as favored because they both cover new edges.

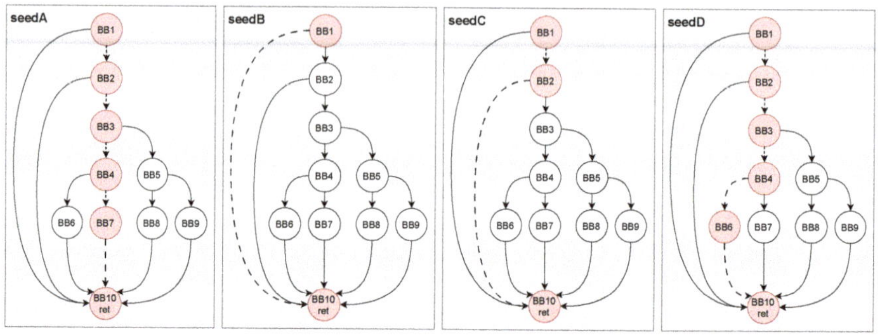

(**a**) seed A (**b**) seed B (**c**) seed C (**d**) seed D

Figure 2. The CFGs of the motivating example.

As discussed in Section 2.1.3, AFL's seed scheduling algorithm selects favored seeds based on covered edges. However, it lacks tracking information for unexplored areas at the CFG level. AFL's perspective hinges on the assumption that if a seed covers new edges, it is more likely to explore unexplored regions. This perspective relies on a crucial precondition: the new edges must be adjacent to unexplored areas; otherwise, these edges are likely related to validation checks or have already been explored by other seeds. Particularly in the context of expansive and complex programs, AFL expends substantial fuzzing resources on ineffective seeds, impeding the prioritization of genuinely effective seeds and slowing down the convergence of the fuzzing process. Therefore, this paper addresses the issue in AFL's seed scheduling algorithm by introducing a new mechanism to track unexplored regions.

4. Design of UntouchFuzz

4.1. Overview

Figure 3 shows the overview of UntouchFuzz. In comparison to traditional coverage-guided greybox fuzzers, UntouchFuzz introduces an additional bitmap for tracking untouched edges. The untouched edge instrumentation mechanism updates this bitmap during program execution. Seed scheduling is then performed based on the information from this bitmap, prioritizing the mutations of favored seeds generated by scheduler.

To further explain, UntouchFuzz starts with an initial corpus as a seed set. By using the seed scheduling mechanism, the fuzzer selects a seed from the seed set for mutations. The number of mutations is determined by the energy scheduling mechanism, based on seed attributes. The fuzzer then executes the AFL-instrumented program with mutated test cases. If the program causes a crash, the test case is preserved on the local disk. If it covers new edges, the test case is preserved as a seed and added to the seed queue. Meanwhile, the coverage-increasing seed is provided to the program instrumented for untouched edge tracking, collecting information on untouched edges to guide the next seed scheduling process.

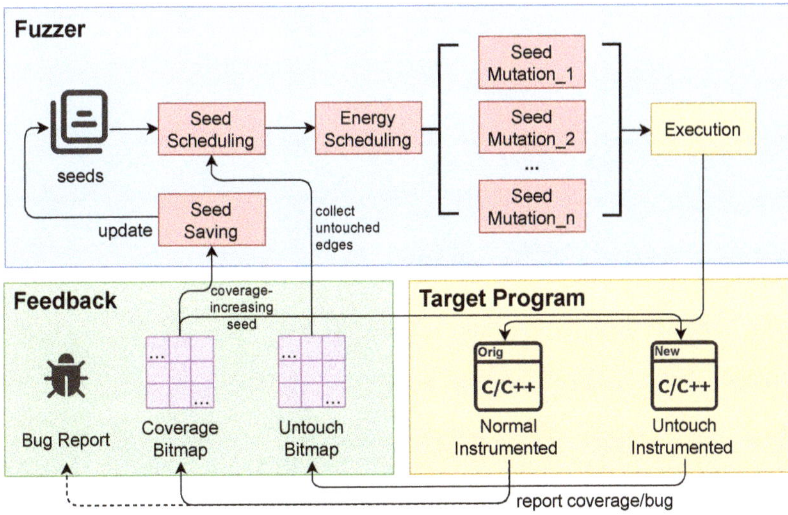

Figure 3. Overview of UntouchFuzz.

Furthermore, UntouchFuzz allocates more energy to seeds with more low-frequency untouched edges in the seed queue. This allocation aims to encourage these seeds to make more attempts at breaking through these low-frequency untouched edges. In the following sections, we will discuss the methods for collecting information on untouched edges in Section 4.2, introduce the seed scheduling algorithm based on untouched edges in Section 4.3, and outline slight improvements to energy scheduling in Section 4.4.

4.2. Untouched Edges Tracking

As mentioned in Section 2.1.2, AFL employs a lightweight instrumentation technique to track program edge coverage. In essence, AFL's instrumentation assigns a random ID to each basic block within the target program's CFG. During program execution, the instrumentation codes calculate the corresponding edge index based on Equation (1) and use this index to update the coverage bitmap.

To ensure minimal impact on program execution speed, AFL utilizes a lightweight XOR operation for computing coverage indices. Similarly, the instrumentation codes responsible for gathering untouched edge information should also be lightweight to minimize disruption to program execution.

We introduce our instrumentation approach with a practical example. Figure 4 provides a snippet of branches within the program's CFG. The hexadecimal values in green boxes represent random IDs allocated by the AFL instrumentation for each basic block. According to Equation (1), the edge index for the transition from basic block BB1 to BB2 is calculated as $(0xabcd \gg 1) \oplus 0x1234 = 0x47d2$, while the edge index for the transition from basic block BB1 to BB3 is calculated as $(0xabcd \gg 1) \oplus 0x5678 = 0x039e$.

Suppose a particular seed triggers a transition from basic block BB1 to BB2, leaving the transition from BB1 to BB3 unexplored. In this scenario, we label edge $0x039e$ as an untouched edge within the execution path of the seed. A straightforward method for capturing untouched edges is to insert instrumentation codes within basic block BB2. Such codes update a byte of untouched edge bitmap by using $0x039e$ as a static index pre-allocated during compilation. This approach works efficiently when a basic block has only one predecessor, but confusion arises when a basic block has multiple incoming edges.

Consider the seed's execution path: BB0→BB1→BB2→BB1. Basic block BB1 has two incoming edges: $0xe693$ from BB0 and $0xa2d7$ from BB2. The corresponding unexplored edges are $0xb6a5$ and $0xb2a1$. Employing the aforementioned instrumentation codes, distinguishing between these two untouched edges becomes a challenging endeavor.

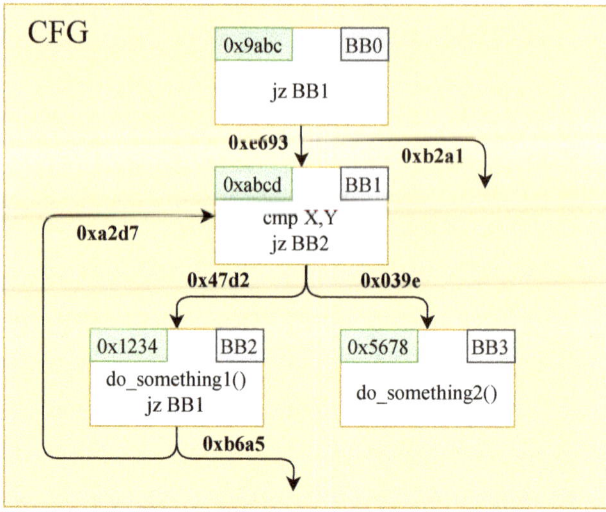

Figure 4. Example of an AFL-instrumented program's CFG.

Upon analyzing the above example, it becomes evident that static pre-allocation of untouched edge IDs is impractical when a basic block has multiple predecessors. To address this challenge, we propose a dynamic method for obtaining untouched edge IDs based on the properties of XOR operations. Specifically, we employ a global variable named "$\_\_afl\_bb\_ids$" to maintain the XOR value of the IDs of two basic blocks at branch transition point. When a transition between basic blocks occurs, we perform an XOR operation on the ID of the transitioned-to basic block with "$\_\_afl\_bb\_ids$" to retrieve the ID of the other unexplored basic block. Furthermore, we define the equation for calculating untouched edge IDs as follows:

$$edge_{untouch} = (prev_{bb} \gg 1) \oplus (\_\_afl\_bb\_ids \oplus cur_{bb}) \tag{2}$$

$$where,\ \_\_afl\_bb\_ids = cur_{bb} \oplus untouch_{bb} \tag{3}$$

Illustrated with the control flow graph in Figure 4, consider the branch transition point within basic block BB1, which leads to two transitions to basic blocks BB2 and BB3. At BB1, we update the value of "$\_\_afl\_bb\_ids$" by performing an XOR operation on the IDs of BB2 and BB3, resulting in $0x1234 \oplus 0x5678 = 0x444c$. When a transition occurs from basic block BB1 to BB2, we can recover the ID of the unexplored basic block BB3 by using "$\_\_afl\_bb\_ids$": "$\_\_afl\_bb\_ids \oplus ID_{BB2} = 0x444c \oplus 0x1234 = 0x5678$". Subsequently, we calculate the untouched edge ID between BB1 and BB3 using Equation (1): "$(0xabcd \gg 1) \oplus 0x5678 = 0x039e$". The calculated value is then utilized to update the bitmap information for the untouched edges.

Algorithm 2 delineates the process of instrumenting untouched edges. Initially, in lines 2–6, the algorithm employs AFL's native edge coverage-based instrumentation, concurrently capturing the random IDs allocated to each basic block. Subsequently, from lines 8 to 27, the algorithm performs to instrument for untouched edges.

To elucidate further, lines 10–12 of the algorithm determine whether the current "untouch_inst1" is invoked. "untouch_inst1" appends instrumentation codes to the beginning of each basic block. The primary function of these codes lies in fetching the value of the global variable "$\_\_afl\_bb\_ids$" within the program. It subsequently calculates the ID for the untouched edge per Equation (2) and leverages this ID to update the untouched edge bitmap "$\_\_afl\_untouch\_ptr$".

Algorithm 2 Instrumentation for untouched edges

```
   Input: A control flow G = (BB, E)
   Output: A new control flow G' = (BB', E')
1  BB_IDs ← ∅, G' ← ∅
2  for (BB, _) ∈ G do
3      // perform AFL's raw instrumentation
4      BB, ID ← afl_inst(BB) // ID is assigned by AFL for each basic block
5      BB_IDs ← BB_IDs ∪ (BB, ID)
6  end
7  // traverse all basic blocks (our modification)
8  for (BB, E) ∈ G do
9      // if the current basic block is not the start basic block of G
10     if BB ≠ G.firstBB() then
11         BB ← untouch_inst1(BB)
12     end
13     __afl_bb_ids ← 0
14     // traverse all successor basic blocks of the current basic block
15     for BB_Succ ∈ Successors(BB) do
16         for (BB_t, BB_ID) ∈ BB_IDs do
17             // find the ID of the successor basic block
18             if BB_Succ = BB_t then
19                 ID ← BB_ID
20                 break
21             end
22         end
23         __afl_bb_ids ← __afl_bb_ids ⊕ ID
24     end
25     BB ← untouch_inst2(BB, __afl_bb_ids)
26     G' ← G' ∪ (BB, E)
27 end
28 return G'
```

Moving to lines 15–24, the algorithm traverses the two successor basic blocks of the current basic block, performing XOR operation on the IDs allocated to these two basic blocks. Finally, in line 25, the function "untouch_inst2" is invoked. The primary aim of the code is to update the value of the global variable "*__afl_bb_ids*" with the control flow graph, setting it to the outcome of the XOR operation mentioned earlier. At this point, all steps of the untouched edge instrumentation algorithm have been completed.

At the low assembly level, a basic block always has exactly two successor basic blocks. However, when we move up to higher-level compiler intermediate languages, it is not guaranteed that a basic block has always exactly two successor basic blocks, especially in programs that contain Switch statements. Specific solutions to this issue will be provided in Section 5. Additionally, it is worth noting that the instrumentation approach proposed still leads to the problem of edge index collisions, where different edges in the CFG are assigned with the same index value [28].

4.3. Seed Scheduling Based on Untouched Edges

In the preceding Section 4.2, we introduced the method for obtaining untouched edge information. In this section, we delve into seed scheduling based on the collected untouched edges. Similar to AFL, we maintain an array, '*untouch_top_rated*', with a size of MAP_SIZE to store the most favored seed for each untouched edge. We also employ a greedy algorithm, specifically the minimum covering set algorithm [34], to generate an optimal seed set that contains all currently untouched edges.

Algorithm 3 describes the seed scheduling algorithm based on untouched edges. Initially, in line 1, the algorithm feeds the newly added seed, denoted as 's', to the instrumented program 'P'' with untouched edges. Subsequently, it acquires the edge coverage bitmap and the untouched edge bitmap of that seed. Following this, in lines 2–27, the algorithm iterates through each index value in the bitmap.

Algorithm 3 Seed scheduling based on untouched edges

Input: New seed s, program P' instrumented by Algorithm 1, untouch_top$_{rated}$
Output: A favored seed set

```
1  trace_bits, untouch_bits ← execute(P', s)
2  for i = 0 → MAP_SIZE do
3      if trace_bits[i] then
4          if untouch_bits[i] then
5              untouch_bits[i] = 0 // handle cases when executing in loops
6          end
7          if virgin_untouch[i] then
8              virgin_untouch[i] = 255 // the edge is explored
9          end
10     end
11     if untouch_bits[i]&&virgin_untouch[i] ≠ 255 then
12         virgin_untouch[i] = 1 // The edge is not explored by other seeds
13     end
14     if virgin_untouch[i] = 255 then
15         untouch_top_rated[i] ← NULL
16         continue
17     end
18     if untouch_bits[i] ≠ NULL then
19         if untouch_top_rated[i] then
20             fav_factor ← s' execution time × s' file size
21             if fav_factor > untouch_top_rated[i]' execution time × its file size then
22                 continue
23             end
24         end
25         untouch_top_rated[i] = s
26     end
27 end
28 return coverMinSet(untouch_top_rated)
```

Specifically, in lines 3–10, the algorithm first checks whether the edge has been both marked as a touched edge and an untouched edge in the two bitmaps. If this condition is true, it suggests that the edge is likely in a loop structure. Due to the repeated edges covered in loops, it is possible that edges untouched in a previous iteration of the loop are now covered in the current iteration. To mitigate this effect, the algorithm sets the value of the untouched edge bitmap corresponding to this edge's index to 0.

The algorithm also maintains a global array called '*virgin_untouch*'. The indices of this array correspond to edge IDs, and the array values indicate the status of the respective untouched edges: 0 denotes an untouched edge that has not been covered by the execution path of any seed, 1 denotes that a seed's execution path includes this untouched edge, and 255 denotes an untouched edge that has been covered by the execution path of another seed, i.e., it has been "explored". If the current edge is covered by seed 's', and the corresponding value in the '*virgin_untouch*' array is not 0, the algorithm updates the value to 255.

Moving on to lines 11–13, if the edge is an untouched edge for the current seed 's' and has not been "explored" by other seeds in history, the algorithm updates the value in the '*virgin_untouch*' array to 1. In lines 14–17, if the corresponding '*virgin_untouch*' value for this edge is 255, indicating that the edge is no longer untouched, the algorithm sets the '*untouch_top_rated*' value for this edge to NULL. The '*untouch_top_rated*' maintains the best seed for each untouched edge.

In lines 18–26 of the algorithm, the current seed s is compared with the metrics of the best seed for this untouched edge. We continue to use AFL's default metrics, which is the seed's execution speed multiplied by its file size. Then, at line 28, the algorithm invokes the $coverMinSet()$ function, which generates a minimal seed set containing all untouched edges found, following the logic described in Algorithm 1.

In the end, the algorithm outputs the optimal seed set based on untouched edges. UntouchFuzz prioritizes testing seeds from this selected seed set.

4.4. Energy Scheduling Optimization

The goal of energy scheduling is to allocate energy efficiently to the chosen seeds for optimal mutations. It is essential to strike the right balance in energy allocation. Allocating excessive energy can lead to a significant waste of fuzzing resources on a single seed. Conversely, insufficient energy allocation may underutilize a seed's potential to explore new paths, as discussed in reference [14].

In this paper, we obtain the number of seeds in the seed set for each untouched edge included. Following this, we sort the number of seeds for these untouched edges in ascending order and select the top $\beta\%$ (40% is the default in this paper) as rare untouched edges. Then, we calculate the difference between the number of current seed's rare untouched edge '$rare_s[i]$' and the maximum number of seed 'max_s' among all rare untouched edges then calculate the average distance value $dist_s$ based on Equation (4).

$$dist_s = \frac{\sum_i^n (max_s - rare_s[i])}{n} \tag{4}$$

Assuming the original energy allocated to the seed was p, it is now assigned a new energy of $(1 + \alpha \times dist_s) \times p$, where α is the default value of 0.3. In our insight, if a certain untouched edge appears frequently in the seed set, it implies a high probability that the selected seed contains this untouched edge and suggests that this particular untouched edge is likely to be difficult to explore. Therefore, if a seed includes many high-frequency untouched edges, it should be allocated less energy. Conversely, if the seed contains numerous low-frequency untouched edges, it should be allocated more energy.

To elaborate on our insight, Figure 5 illustrates it through an example. Assuming that the current phase has fuzzed for a while and the seed set contains four seeds, with two global untouched edges: BB1→BB10 and BB5→BB8. According to the seed scheduling algorithm outlined in Section 4.3, seed B is identified as a favored seed due to the presence of untouched edge BB1→BB10 in its execution path and superior seed attribute. Similarly, seed D is designated as a favored seed because of the inclusion of the new untouched edge BB5→BB8 in its execution path.

Subsequently, the fuzzing process prioritizes the mutation of seed B and seed D. As previously mentioned, the untouched edge BB1→BB10 appears in the execution paths of all four seeds, indicating it is not a rare untouched edge. In contrast, BB5→BB8 is considered as a rare untouched edge since it appears only in the execution path of seed D. Consequently, we can make a reasonable conjecture that the untouched edge BB1→BB10 is likely associated with a hard-to-solve constraint, whereas BB5→BB8 is likely to represent a more manageable constraint. To increase the likelihood of covering edge BB5→BB8, we allocate more energy to the mutation of seed D based on the previously outlined energy allocation mechanism. It is essential to note that the provided example is simplified for explanatory purposes, while the real-world program will be complex.

However, in AFL, when new seeds are discovered, the fuzzer doubles the energy allocated to the seeds. Hence, we do not intentionally reduce the energy but instead provide seeds with a higher initial energy value to the seeds, aiming to fully unleash the potential of seeds to discover new paths.

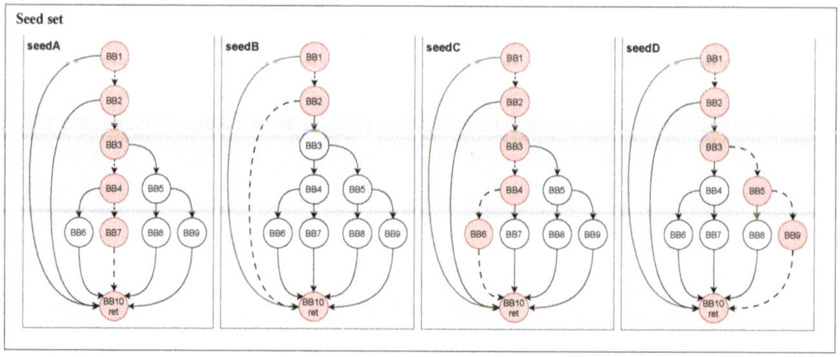

Figure 5. Example of our insight.

5. Implementation

We implemented a prototype of UntouchFuzz on the top of AFL 2.57b, comprising two key components: instrumentation based on untouched edges and the main fuzzing loop. Next, we discuss a few important implementation details.

For the instrumentation, we employed the LLVM framework [35] to instrument the target program's codes, collecting information about untouched edges. However, LLVM IR contains SwitchInst instructions, which can lead to situations where basic blocks with SwitchInst instructions have multiple successor basic blocks. To address this, we utilized a pass available in the AFL++ [6] instrumentation tools that splits switch statements. By applying this pass, all SwitchInst instructions in the target program's LLVM IR are transformed into if. . . else. . . structures, converting basic blocks that originally had multiple successor basic blocks into those with only two successor basic blocks. This enables us to effectively implement the instrumentation method described in Section 4.2.

Before entering the main fuzzing loop, we launched an instrumented program using another fork server for untouched edge instrumentation in UntouchFuzz. Here, we considered that instrumentation due to untouched edges might impact the program's execution speed. Therefore, we only run the untouched edge-instrumented version of the program when seeds are added to the queue. In other scenarios, we run the AFL's native instrumented version of the program. However, it is important to note that to maintain instrumentation consistency, we applied the aforementioned switch statement, splitting pass to the AFL's native instrumentation.

Regarding the fuzzing main loop, we introduced an $update\_untouch\_score()$ function to maintain the best seed for each untouched edge. Additionally, we made modifications to the $cull\_queue()$ and $fuzz\_one()$ functions in AFL to implement seed scheduling and energy allocation. Other logic in the fuzzing main loop remained unchanged.

6. Evaluation

In this section, we evaluated the effectiveness of UntouchFuzz and answer the following questions:

- RQ1: How effective is UntouchFuzz at improving coverage faster when compared with other seed scheduling strategies?
- RQ2: Can UntouchFuzz discover more unique crashes with respect to other seed scheduling strategies?
- RQ3: How does the seed scheduling based on untouched edges perform in other fuzzers?
- RQ4: Can UntouchFuzz detect new vulnerabilities in real-world programs?

6.1. Experiment Settings

(1) Baseline Seed Scheduling Strategies: Starting from the minimum coverage set, rare paths, new paths, and same-prefix coverage, the seed scheduling method proposed in this paper was compared with native AFL [9], AFLFast [12], EcoFuzz [14], and Alphuzz [16] seed scheduling strategies. It is important to note that these five tools differ only in their seed scheduling mechanisms while all other components remain the same. FairFuzz [13] was not compared due to its custom mutator, which aims to obtain mutation byte masks. When fuzzing large-scale seeds, the custom mutator might lead to starvation in subsequent seeds. Conversely, if seeds are small, the additional coverage gained from deterministic mutation might be unfair compared to fuzzers without deterministic mutation. Previous work [36] removed this custom mutation phase in experiments, but this deviated from FairFuzz's original intent. Therefore, we do not select FairFuzz for comparison.

Additionally, the multi-property queue seed scheduling method SLIME [15] is not compared due to the additional instrumentation for the target program. This instrumentation affects program execution speed and differs significantly from the instrumentation used by the aforementioned tools. The tool we implemented can share the same instrumentation program with these tools, ensuring no experimental differences in comparison.

(2) Benchmark Programs: We selected 12 real-world binary programs for testing based on their popularity, testing frequency, and diversity of categories. As shown in Table 1, these 12 binary programs include popular binary utilities (such as readelf), image parsing and processing libraries (such as libjpeg-turbo, exiv2), audio parsing tools (such as mp3gain), document processing libraries (such as xpdf, libxml2), and network packet parsing tools (such as tcpdump). Since certain vulnerabilities (i.e., buffer overflows) do not affect the program execution, we apply Address Sanitizer (ASAN) [37] to capture memory errors.

Table 1. Twelve real-world programs for evaluation.

Program	Library and Version	Input Type	Commands
djpeg	libjpeg-turbo-2.1.91	jpg	@@
exiv2	exiv2-0.26	jpg	@@/dev/null
pdftotext	xpdf-4.0.0	pdf	@@
tcmdump	tcmdump-4.8.1	bin	-e -vv -nr @@
mp3gain	mp3gain-1.5.2	mp3	@@
mp42aac	Bento4-1.5.1-628	mp4	@@/dev/null
tiffcp	libtiff-3.9.7	tiff	-i -E 1 -H 10 -V 10 -S 8:4 -R 270 @@ ./output.tif
readelf	binutils-2.28	elf	-a @@
nm-new	binutils-2.28	elf	-A -a -l -S -s –special-syms –synthetic –with-symbol-versions -D @@
xmllint	libxml-2.98	xml	@@
bsdtar	libarchive-3.2.0	tar	-xf @@/dev/null
mujs	SQLite-3.8.9	text(js)	@@

(3) Initial Corpus: The initial seed corpus used comes from datasets provided by Mopt [38] and uniFuzz [39], with some contributions from the open-source community.

(4) Experimental Environment: The experiments were conducted on a server with a 56-core Intel Xeon CPU, 128 GB of memory, and running Ubuntu 22.04. Deterministic mutations were disabled as it is less effective compared to AFL's havoc mutation strategy [40]. To reduce the impact of randomness, we ran each benchmark program for 24 h, repeating the process 10 times and taking the arithmetic mean as the final result [41].

(5) Experimental Metrics: We evaluated the proposed method against four fuzzers with different seed scheduling strategies, considering edge coverage, edge coverage over time, and the number of unique crashes. Additionally, the proposed guided mechanism for untouched edges was transplanted into MOpt to assess its performance across different

fuzzers. MOpt was chosen due to its focus on optimizing mutation operators while keeping the seed scheduling mechanism unaltered.

6.2. RQ1: Code Coverage Improving

Code coverage is a crucial metric for evaluating the performance of fuzzing techniques [17]. In general, the more code a fuzzer can cover in the target program, the higher the probability of discovering hidden vulnerabilities. As explained in Section 2.1.2, AFL [9] employs a 64KB bitmap to collect coverage information, with each byte in the bitmap representing the number of hits for a particular edge ID. AFL maps program branches to the bitmap by using a hash function. If a branch is explored, the byte at the index corresponding to its edge ID in the bitmap is updated. AFL maintains a simplified bitmap in real time and stores it on local disk, which allows us to assess code coverage based on this simplified bitmap.

Table 2 presents a comparison of UntouchFuzz with four other state-of-the-art fuzzers in terms of edge coverage. The data in the table represent the arithmetic average edge coverage across ten fuzzing tests. The results in Table 2 demonstrate that UntouchFuzz outperforms EcoFuzz and Alphuzz in edge coverage and slightly surpasses AFL and AFLFast. UntouchFuzz achieves better coverage than the other four baseline seed scheduling strategies in 11 out of the 12 programs tested (all except for mujs). Overall, the proposed method is effective in improving coverage. While the improvement percentages on AFL and AFLFast (2.16% and 3.31%) are relatively modest, these results are consistent with findings from previous research [40], which suggests that differences among fuzzers are minimized when using the single havoc strategy. Nonetheless, our seed scheduling strategy still has an impact on the direction of fuzzing evolution by concentrating mutation energy on more effective seeds, thus enhancing overall program coverage.

Table 2. Arithmetic mean edge coverage comparison.

	Default	**RarePath**	**NewPath**	**SamePrefix**	**UntouchedEdge**
Fuzzer	**AFL**	**AFLFast**	**EcoFuzz**	**Alphuzz**	**UntouchFuzz**
djpeg	3998	3850	3646	3613	**4101**
exiv2	12,152	12,104	11,310	11,958	**12,304**
pdftotext	15,160	15,120	11,408	14,510	**15,280**
tcpdump	18,208	17,252	16,134	17,640	**18,556**
mp3gain	1375	1376	1364	1370	**1382**
mp42aac	**3263**	3230	3178	3197	**3263**
tiffcp	5841	5801	5405	5692	**6659**
readelf	10,336	10,503	9100	10,029	**10,647**
nm-new	5455	5463	5530	5185	**5601**
xmllint	10,406	10,332	10,121	10,172	**10,470**
bsdtar	5250	5227	5109	5169	**5318**
mujs	9170	**9243**	8892	9084	9210
Total	100,614 (−2.16%)	99,501 (−3.31%)	91,197 (−12.71%)	97,619 (−5.30%)	**102,791**

Figure 6 shows the evolution of edge coverage over time, with samples taken every hour. Evidently, on the seven target programs, exiv2, pdftotext, tcpdump, tiffcp, readelf, nm-new, and bsdtar, UntouchFuzz has a significantly higher edge coverage growth rate than the other four baseline seed scheduling strategies. However, for the remaining five target programs, the fuzzers converge quickly due to their smaller scale and lower complexity, negating the advantage demonstrated by UntouchFuzz.

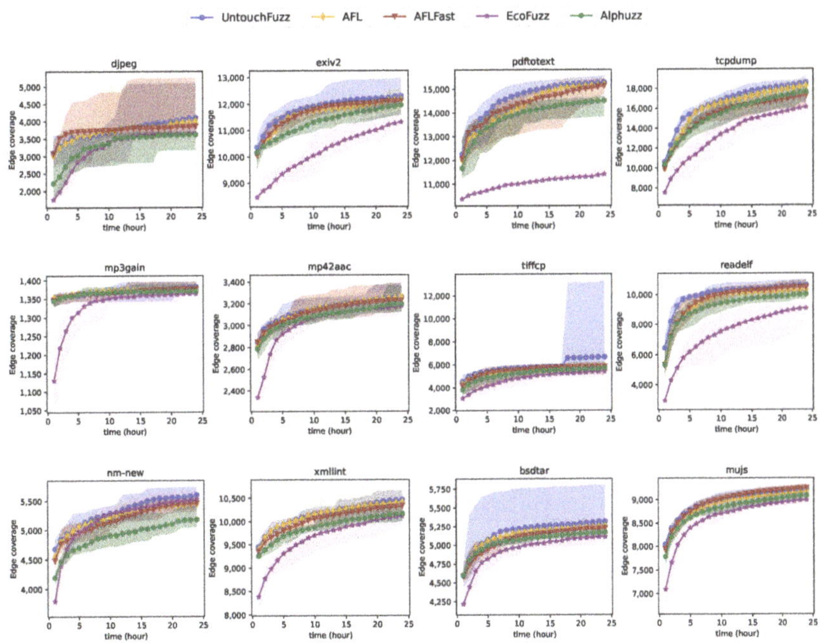

Figure 6. Edge coverage over time in five fuzzers.

6.3. RQ2: Unique Crashes

To further validate the effectiveness of the proposed method in vulnerability discovery, we conducted a statistical analysis of unique crashes. Table 3 presents a comparison of UntouchFuzz with four baseline fuzzers on the number of unique crashes, where the data in the table represent the total number of unique crashes discovered over ten rounds of testing. It is worth noting that, as in our experiments, we used the latest version of the djpeg program and did not find any valid crashes, so its results are not listed in Table 3.

As shown in the experimental results in Table 3, UntouchFuzz outperforms the other four baseline seed scheduling strategies in total unique crash discoveries, discovering the highest number of unique crashes on six of the programs under test. However, on the remaining five target programs, UntouchFuzz does not achieve the best results, but the difference between it and the best-performing approach is not significant. We attribute this to the randomness of mutation and differences in fuzzing evolution.

It is essential to note that the experimental results in Table 3 are directly taken from the unique crash metric of AFL-based fuzzers. However, the number of unique crashes may not accurately reflect the actual number of unique vulnerabilities because there is often a many-to-one relationship between them, which means that multiple crashes may correspond to a single vulnerability. For example, suppose crash one's triggering path is A →B→D, and crash two's triggering path is A→C→D, and both crashes occur at the same location in basic block D. From AFL's perspective, since the triggering paths of these two crashes are different, they are both considered unique crashes. However, from a root cause analysis perspective, both crashes are due to codes in basic block D, and thus, these two crashes should be categorized as the same vulnerability. Moreover, the differences in these two crash paths are likely related to changes in the input, where bytes in the input can affect the execution of subsequent basic blocks following basic block A. To obtain a more accurate count of unique vulnerabilities, we conducted deduplication of unique crashes based on the function call stack information provided by ASAN, selecting the top three functions and removing duplicate crashes. Table 4 presents the results after crash deduplication.

Table 3. Comparison on unique crashes.

	Default	RarePath	NewPath	SamePrefix	UntouchedEdge
Fuzzer	**AFL**	**AFLFast**	**EcoFuzz**	**Alphuzz**	**UntouchFuzz**
exiv2	503	476	301	529	**546**
pdftotext	3725	3424	122	2310	**3937**
tcpdump	1917	2008	1738	**2090**	2088
mp3gain	**1230**	1219	986	1128	1226
mp42aac	**502**	480	262	243	467
tiffcp	3495	3553	2317	3235	**3642**
readelf	19	2	1	5	**83**
nm-new	1550	1450	1435	641	**1662**
xmllint	4331	**4377**	3313	4063	4251
bsdtar	341	559	978	76	**603**
mujs	471	**524**	71	423	469
Total	18,084 (−4.92%)	18,072 (−4.99%)	11,524 (−64.65%)	14,743 (−28.70%)	**18,974**

Table 4. Results after crash deduplication.

	Default	RarePath	NewPath	SamePrefix	UntouchedEdge
Fuzzer	**AFL**	**AFLFast**	**EcoFuzz**	**Alphuzz**	**UntouchFuzz**
exiv2	102	96	59	99	**105**
pdftotext	159	166	25	100	**176**
tcpdump	515	488	435	**571**	519
mp3gain	65	66	**68**	61	67
mp42aac	22	26	**28**	24	24
tiffcp	381	406	318	336	**418**
readelf	3	2	1	2	**6**
nm-new	**165**	156	160	92	158
xmllint	17	**24**	15	19	19
bsdtar	37	37	31	12	**39**
mujs	20	19	19	10	**25**
Total	1486 (−4.17%)	1486 (−4.17%)	1159 (−34.25%)	1326 (−17.35%)	**1556**

The data in Table 4 reveal that various seed scheduling strategies exhibit distinct performances across different programs, with UntouchFuzz achieving the highest number of unique vulnerabilities in six programs. Overall, UntouchFuzz outperforms the other four baseline seed scheduling strategies based on the total number of discovered vulnerabilities. Furthermore, compared to EcoFuzz and Alphuzz, UntouchFuzz demonstrates a more stable performance across the programs.

6.4. RQ3: Scalability

To assess the scalability of the proposed approach, we integrated our method into the MOpt fuzzer, naming the modified tool "MOpt-u." Table 5 provides a comparison of edge coverage between UntouchFuzz, AFL, MOpt-u, and the original MOpt fuzzer. The results in Table 5 demonstrate that UntouchFuzz and MOpt-u outperform the original, unmodified fuzzers in terms of edge coverage across all 12 benchmark programs. This further substantiates the capability of the untouched edge-guided mechanism to enhance the performance of the original fuzzers.

Table 5. Comparison between AFL, MOpt and UntouchFuzz, MOpt-u.

Fuzzer	AFL	UntouchFuzz	Mopt	Mopt-u
djpeg	3998	**4101**	4073	**4397**
exiv2	12,152	**12,304**	12,404	**12,480**
pdftotext	15,160	**15,280**	15,322	**15,341**
tcpdump	18,208	**18,556**	18,467	**18,734**
mp3gain	1375	**1382**	1378	**1380**
mp42aac	**3263**	**3263**	3352	**3409**
tiffcp	5841	**6659**	5949	**6132**
readelf	10,336	**10,647**	11,128	**11,252**
nm-new	5455	**5601**	5573	**5674**
xmllint	10,406	**10,470**	10245	**10,408**
bsdtar	5250	**5318**	5270	**5273**
mujs	9170	**9210**	9121	**9157**
Total	100,614 (−2.16%)	102,791	102,282 (−1.31%)	**103,637**

6.5. RQ4: New Vulnerabilities

We used UntouchFuzz to find new vulnerabilities in open-source projects on GitHub. We reported these vulnerabilities to the respective projects. The details are in Table 6, confirming the effectiveness of UntouchFuzz in real-world scenarios.

Table 6. New vulnerabilities found by UntouchFuzz.

Project	Version	CVE/Issue	Type	Fixed Status
LIEF	v0.12.1	CVE-2022-40922	segmentation fault	✓
LIEF	v0.12.1	CVE-2022-40923	segmentation fault	✓
LIEF	v0.12.1	CVE-2022-43171	heap buffer overflow	✓
LIEF	v0.12.1	CVE-2022-43172	segmentation fault	✓
LIEF	v0.12.1	github-issue-785	allocator oom	✓
PcapPlusPlus	v22.11	CVE-2023-31991	heap buffer overflow	✓
libfyaml	v0.7.12	CVE-2023-31992	use after free	✓
libfyaml	v0.7.12	CVE-2023-31993	stack buffer overflow	✓
libfyaml	v0.7.12	github-issue-56	stack-buffer-overflow	✓
sxmlc	v4.5.2	github-issue-24	segmentation fault	✓
sxmlc	v4.5.2	github-issue-25	segmentation fault	✓
configor	v0.9.18	github-issue-97	infinite loop	✓
tom11	v3.7.1	github-issue-199	heap buffer overflow	✓

7. Discussion

In this section, we discuss several limitations of our current implementation:

(1) Our method is implemented on the top of AFL and not on AFL++. The choice to not implement it on AFL++ was due to the fact that AFL++ already integrates the seed scheduling mechanism from AFLFast and other advanced technologies. Porting the untouched edge guidance mechanism into AFL++ would have been complex. We plan to implement our approach to AFL++ in future work. Additionally, it is important to note that our method may not apply to base fuzzers outside of AFL, such as libFuzzer [42] or honggfuzz [1].

(2) Coverage bitmap collisions are a common issue in AFL-based fuzzers. AFL-based fuzzers typically use a fixed-size 64KB bitmap to collect coverage information, which can be adjusted via configuration. The fixed bitmap size might result in different edges being assigned the same edge ID, causing coverage bitmap collisions. Prior research attempts have aimed to optimize this issue by modifying instrumentation [28,43,44]. However, with the increase in the size of the target programs, expanding the bitmap to solve collision issues might not lead to significant performance improvements.

(3) We conducted fuzzing tests on 12 mainstream benchmark programs, running each program 10 times for 24 h per test. The experimental results demonstrate that, under the given initial corpus conditions, our tool effectively selects better seeds for fuzzing and guides the fuzzer toward maximizing program coverage. However, differences between the initial corpus and the target program can impact the performance of coverage-guided greybox fuzzers [10,15,45], causing variations in the evolutionary process and resulting in different outcomes.

8. Conclusions

In this paper, we concluded that the existing seed scheduling methods neglect to focus on the unexplored regions within the program's control flow graph. In response to this issue, we presented a greybox fuzzer guided by untouched edges, UntouchFuzz. We developed a lightweight instrumentation technique to track untouched edges. Furthermore, we designed a seed scheduling strategy based on untouched edges inspired by a minimal coverage sets algorithm. The strategy prioritizes seeds that include all untouched edges. Additionally, we made minor adjustments to the energy scheduler to align with the new seed scheduling method. In evaluation, UntouchFuzz outperformed the other fuzzers on code coverage and the number of vulnerabilities, further proving the untouched guidance mechanism proposed in this paper. To foster future research in this area, we have made our fuzzer open source. Further, future research could combine symbolic execution techniques with the untouched guidance mechanism for better fuzzing results. We plan to implement our mechanism into AFL++ and investigate the influence of bitmap collisions in our mechanism.

Author Contributions: Conceptualization, P.J.; data curation, C.X.; formal analysis, G.Y.; funding acquisition, P.J.; investigation, H.K.; methodology, C.X.; software , C.X.; supervision, P.Y. and C.H.; validation, X.H.; visualization, C.X.; writing—original draft, C.X.; writing—review and editing, C.X. and P.J. All authors have read and agreed to the published version of the manuscript.

Funding: This research was funded by National Key R&D projects of China OF FUNDER grant number 2021YFB3101803. This research was supported in part by the National Natural Science Foundation of China under Grant U2133208 and the Sichuan Youth Science and Technology Innovation Team under Grant 2022JDTD0014.

Institutional Review Board Statement: Not applicable.

Informed Consent Statement: Not applicable.

Data Availability Statement: The source code of fuzzer proposed in this research is available at https://github.com/bladchan/untouchFuzz.git (accessed on 22 October 2023).

Conflicts of Interest: The authors declare no conflict of interest.

References

1. Swiecki, R.; Gröbert, F. Honggfuzz. 2016. Available online: http://code.google.com/p/honggfuzz (accessed on 2 October 2023).
2. Schumilo, S.; Aschermann, C.; Gawlik, R.; Schinzel, S.; Holz, T. kAFL: Hardware-Assisted feedback fuzzing for OS kernels. In Proceedings of the 26th USENIX Security Symposium (USENIX Security 17), Vancouver, BC, Canada, 16–18 August 2017; pp. 167–182.
3. Yun, I.; Lee, S.; Xu, M.; Jang, Y.; Kim, T. QSYM: A practical concolic execution engine tailored for hybrid fuzzing. In Proceedings of the 27th USENIX Security Symposium (USENIX Security 18), Baltimore, MD, USA, 15–17 August 2018; pp. 745–761.
4. Pham, V.T.; Böhme, M.; Santosa, A.E.; Căciulescu, A.R.; Roychoudhury, A. Smart greybox fuzzing. *IEEE Trans. Softw. Eng.* **2019**, *47*, 1980–1997. [CrossRef]
5. Zheng, Y.; Davanian, A.; Yin, H.; Song, C.; Zhu, H.; Sun, L. FIRM-AFL:High-Throughput greybox fuzzing of IoT firmware via augmented process emulation. In Proceedings of the 28th USENIX Security Symposium (USENIX Security 19), Santa Clara, CA, USA, 14 August 2019; pp. 1099–1114.
6. Fioraldi, A.; Maier, D.; Eißfeldt, H.; Heuse, M. AFL++: Combining incremental steps of fuzzing research. In Proceedings of the 14th USENIX Workshop on Offensive Technologies (WOOT 20), Boston, MA, USA, 11 August 2020.
7. Pham, V.T.; Böhme, M.; Roychoudhury, A. AFLNet: A greybox fuzzer for network protocols. In Proceedings of the 2020 IEEE 13th International Conference on Software Testing, Validation and Verification (ICST), Porto, Portugal, 24–28 October 2020; pp. 460–465.

8. Zhu, X.; Wen, S.; Camtepe, S.; Xiang, Y. Fuzzing: A survey for roadmap. *ACM Comput. Surv. (CSUR)* **2022**, *54*, 1–36. [CrossRef]
9. Zalewski, M. American Fuzzy Lop (AFL) Fuzzer. 2017. Available online: http://lcamtuf.coredump.cx/afl/technical_details.txt (accessed on 2 October 2023).
10. Herrera, A.; Gunadi, H.; Magrath, S.; Norrish, M.; Payer, M.; Hosking, A.L. Seed selection for successful fuzzing. In Proceedings of the 30th ACM SIGSOFT International Symposium on Software Testing and Analysis, 2021, Virtual, 11–17 July 2021; pp. 230–243.
11. Zhang, K.; Xiao, X.; Zhu, X.; Sun, R.; Xue, M.; Wen, S. Path transitions tell more: Optimizing fuzzing schedules via runtime program states. In Proceedings of the 44th International Conference on Software Engineering, Pittsburgh, PA, USA, 21–29 May 2022; pp. 1658–1668.
12. Böhme, M.; Pham, V.T.; Roychoudhury, A. Coverage-based greybox fuzzing as markov chain. In Proceedings of the 2016 ACM SIGSAC Conference on Computer and Communications Security, Vienna, Austria, 24–28 October 2016; pp. 1032–1043.
13. Lemieux, C.; Sen, K. Fairfuzz: A targeted mutation strategy for increasing greybox fuzz testing coverage. In Proceedings of the 33rd ACM/IEEE International Conference on Automated Software Engineering, Montpellier, France, 3–7 September 2018; pp. 475–485.
14. Yue, T.; Wang, P.; Tang, Y.; Wang, E.; Yu, B.; Lu, K.; Zhou, X. EcoFuzz: Adaptive Energy-Saving greybox fuzzing as a variant of the adversarial Multi-Armed bandit. In Proceedings of the 29th USENIX Security Symposium (USENIX Security 20), Virtual, 12–14 August 2020; pp. 2307–2324.
15. Lyu, C.; Liang, H.; Ji, S.; Zhang, X.; Zhao, B.; Han, M.; Li, Y.; Wang, Z.; Wang, W.; Beyah, R. SLIME: Program-sensitive energy allocation for fuzzing. In Proceedings of the 31st ACM SIGSOFT International Symposium on Software Testing and Analysis, Virtual, 18–22 July 2022; pp. 365–377.
16. Zhao, Y.; Wang, X.; Zhao, L.; Cheng, Y.; Yin, H. Alphuzz: Monte carlo search on seed-mutation tree for coverage-guided fuzzing. In Proceedings of the 38th Annual Computer Security Applications Conference, Austin, TX, USA, 5–9 December 2022; pp. 534–547.
17. Wang, J.; Duan, Y.; Song, W.; Yin, H.; Song, C. Be sensitive and collaborative: Analyzing impact of coverage metrics in greybox fuzzing. In Proceedings of the 22nd International Symposium on Research in Attacks, Intrusions and Defenses (RAID 2019), Beijing, China, 23–25 September 2019; pp. 1–15.
18. Wang, M.; Liang, J.; Zhou, C.; Jiang, Y.; Wang, R.; Sun, C.; Sun, J. RIFF: Reduced Instruction Footprint for Coverage-Guided Fuzzing. In Proceedings of the 2021 USENIX Annual Technical Conference (USENIX ATC 21), Virtual, 14–16 July 2021; pp. 147–159.
19. Aschermann, C.; Schumilo, S.; Blazytko, T.; Gawlik, R.; Holz, T. REDQUEEN: Fuzzing with Input-to-State Correspondence. In Proceedings of the NDSS, San Diego, CA, USA, 24–27 February 2019; Volume 19, pp. 1–15.
20. Gan, S.; Zhang, C.; Chen, P.; Zhao, B.; Qin, X.; Wu, D.; Chen, Z. GREYONE: Data flow sensitive fuzzing. In Proceedings of the 29th USENIX Security Symposium (USENIX Security 20), Virtual, 12–14 August 2020; pp. 2577–2594.
21. Liang, J.; Wang, M.; Zhou, C.; Wu, Z.; Jiang, Y.; Liu, J.; Liu, Z.; Sun, J. Pata: Fuzzing with path aware taint analysis. In Proceedings of the 2022 IEEE Symposium on Security and Privacy (SP), San Francisco, CA, USA, 23–26 May 2022; pp. 1–17.
22. Stephens, N.; Grosen, J.; Salls, C.; Dutcher, A.; Wang, R.; Corbetta, J.; Shoshitaishvili, Y.; Kruegel, C.; Vigna, G. Driller: Augmenting fuzzing through selective symbolic execution. In Proceedings of the NDSS, San Diego, CA, USA, 21–24 February 2016; Volume 16, pp. 1–16.
23. Peng, H.; Shoshitaishvili, Y.; Payer, M. T-Fuzz: Fuzzing by program transformation. In Proceedings of the 2018 IEEE Symposium on Security and Privacy (SP), San Francisco, CA, USA, 21–23 May 2018; pp. 697–710.
24. Chen, P.; Chen, H. Angora: Efficient fuzzing by principled search. In Proceedings of the 2018 IEEE Symposium on Security and Privacy (SP), San Francisco, CA, USA, 21–23 May 2018; pp. 711–725.
25. Coppik, N.; Schwahn, O.; Suri, N. Memfuzz: Using memory accesses to guide fuzzing. In Proceedings of the 2019 12th IEEE Conference on Software Testing, Validation and Verification (ICST), Xi'an, China, 22–27 April 2019; pp. 48–58.
26. Wen, C.; Wang, H.; Li, Y.; Qin, S.; Liu, Y.; Xu, Z.; Chen, H.; Xie, X.; Pu, G.; Liu, T. Memlock: Memory usage guided fuzzing. In Proceedings of the ACM/IEEE 42nd International Conference on Software Engineering, Seoul, Republic of Korea, 27 June–19 July 2020; pp. 765–777.
27. Zhang, G.; Wang, P.F.; Yue, T.; Kong, X.D.; Zhou, X.; Lu, K. ovAFLow: Detecting Memory Corruption Bugs with Fuzzing-Based Taint Inference. *J. Comput. Sci. Technol.* **2022**, *37*, 405–422. [CrossRef]
28. Gan, S.; Zhang, C.; Qin, X.; Tu, X.; Li, K.; Pei, Z.; Chen, Z. Collafl: Path sensitive fuzzing. In Proceedings of the 2018 IEEE Symposium on Security and Privacy (SP), San Francisco, CA, USA, 21–23 May 2018; pp. 679–696.
29. Böhme, M.; Pham, V.T.; Nguyen, M.D.; Roychoudhury, A. Directed greybox fuzzing. In Proceedings of the 2017 ACM SIGSAC Conference on Computer and Communications Security, Dallas, TX, USA, 30 October– 3 November 2017; pp. 2329–2344.
30. Chen, H.; Xue, Y.; Li, Y.; Chen, B.; Xie, X.; Wu, X.; Liu, Y. Hawkeye: Towards a desired directed grey-box fuzzer. In Proceedings of the 2018 ACM SIGSAC Conference on Computer and Communications Security, Toronto, ON, Canada, 15–19 October 2018; pp. 2095–2108.
31. Huang, H.; Guo, Y.; Shi, Q.; Yao, P.; Wu, R.; Zhang, C. Beacon: Directed grey-box fuzzing with provable path pruning. In Proceedings of the 2022 IEEE Symposium on Security and Privacy (SP), San Francisco, CA, USA, 23–26 May 2022; pp. 36–50.
32. Luo, C.; Meng, W.; Li, P. Selectfuzz: Efficient directed fuzzing with selective path exploration. In Proceedings of the 2023 IEEE Symposium on Security and Privacy (SP), San Francisco, CA, USA, 22–25 May 2023; pp. 2693–2707.

33. Norris, J.R. *Markov Chains*; Cambridge University Press: Cambridge, UK, 1998.
34. Rebert, A.; Cha, S.K.; Avgerinos, T.; Foote, J.; Warren, D.; Grieco, G.; Brumley, D. Optimizing seed selection for fuzzing. In Proceedings of the 23rd USENIX Security Symposium (USENIX Security 14), Anaheim, CA, USA, 9–11 August 2023; pp. 861–875.
35. Lattner, C.; Adve, V. LLVM: A compilation framework for lifelong program analysis & transformation. In Proceedings of the International Symposium on Code Generation and Optimization, CGO 2004, Palo Alto, CA, USA, 20–24 March 2004; pp. 75–86.
36. She, D.; Shah, A.; Jana, S. Effective seed scheduling for fuzzing with graph centrality analysis. In Proceedings of the 2022 IEEE Symposium on Security and Privacy (SP), San Francisco, CA, USA, 23–26 May 2022; pp. 2194–2211.
37. Serebryany, K.; Bruening, D.; Potapenko, A.; Vyukov, D. AddressSanitizer: A fast address sanity checker. In Proceedings of the 2012 USENIX Annual Technical conference (USENIX ATC 12), Boston, MA, USA, 13–15 June 2012; pp. 309–318.
38. Lyu, C.; Ji, S.; Zhang, C.; Li, Y.; Lee, W.H.; Song, Y.; Beyah, R. MOPT: Optimized mutation scheduling for fuzzers. In Proceedings of the 28th USENIX Security Symposium (USENIX Security 19), Santa Clara, CA, USA, 14–16 August 2019; pp. 1949–1966.
39. Li, Y.; Ji, S.; Chen, Y.; Liang, S.; Lee, W.H.; Chen, Y.; Lyu, C.; Wu, C.; Beyah, R.; Cheng, P.; et al. UNIFUZZ: A Holistic and Pragmatic Metrics-Driven Platform for Evaluating Fuzzers. In Proceedings of the 30th USENIX Security Symposium (USENIX Security 21), Vancouver, BC, Canada, 11–13 August 2021; pp. 2777–2794.
40. Wu, M.; Jiang, L.; Xiang, J.; Huang, Y.; Cui, H.; Zhang, L.; Zhang, Y. One fuzzing strategy to rule them all. In Proceedings of the 44th International Conference on Software Engineering, Pittsburgh, PA, USA, 21–29 May 2022; pp. 1634–1645.
41. Klees, G.; Ruef, A.; Cooper, B.; Wei, S.; Hicks, M. Evaluating fuzz testing. In Proceedings of the 2018 ACM SIGSAC Conference on Computer and Communications Security, Toronto, ON, Canada, 15–19 October 2018; pp. 2123–2138.
42. Serebryany, K. Continuous fuzzing with libfuzzer and addresssanitizer. In Proceedings of the 2016 IEEE Cybersecurity Development (SecDev), Boston, MA, USA, 3-4 November 2016; p. 157.
43. Ahmed, A.; Hiser, J.D.; Nguyen-Tuong, A.; Davidson, J.W.; Skadron, K. BigMap: Future-proofing Fuzzers with Efficient Large Maps. In Proceedings of the 2021 51st Annual IEEE/IFIP International Conference on Dependable Systems and Networks (DSN), Taipei, Taiwan, 21–24 June 2021; pp. 531–542.
44. Hsu, C.C.; Wu, C.Y.; Hsiao, H.C.; Huang, S.K. Instrim: Lightweight instrumentation for coverage-guided fuzzing. In Proceedings of the Symposium on Network and Distributed System Security (NDSS), Workshop on Binary Analysis Research, San Diego, CA, USA, 18–21 February 2018; p. 40.
45. Lee, M.; Cha, S.; Oh, H. Learning Seed-Adaptive Mutation Strategies for Greybox Fuzzing. In Proceedings of the 2023 IEEE/ACM 45th International Conference on Software Engineering (ICSE), Melbourne, Australia, 15–16 May 2023; pp. 384–396.

applied sciences

MDPI

Article

Can Windows 11 Stop Well-Known Ransomware Variants? An Examination of Its Built-in Security Features

Yousef Mahmoud Al-Awadi [1], Ali Baydoun [1,*] and Hafeez Ur Rehman [1,2]

1 School of Computing and Data Sciences, Oryx Universal College with Liverpool John Moores University, Doha 34110, Qatar; 101769@oryx.edu.qa (Y.M.A.-A.); hafeez.r@oryx.edu.qa (H.U.R.)
2 Department of Computer Science, National University of Computer and Emerging Sciences, Islamabad 44000, Pakistan
* Correspondence: ali.b@oryx.edu.qa

Abstract: The ever-evolving landscape of cyber threats, with ransomware at its forefront, poses significant challenges to the digital world. Windows 11 Pro, Microsoft's latest operating system, claims to offer enhanced security features designed to tackle such threats. This paper aims to comprehensively evaluate the effectiveness of these Windows 11 Pro, built-in security measures against prevalent ransomware strains, with a particular emphasis on crypto-ransomware. Utilizing a meticulously crafted experimental environment, the research adopted a two-phased testing approach, examining both the default and a hardened configuration of Windows 11 Pro. This dual examination offered insights into the system's inherent and potential defenses against ransomware threats. The study's findings revealed that Windows 11 Pro does present formidable defenses. This paper not only contributes valuable insights into cybersecurity, but also furnishes practical recommendations for both technology developers and end-users in the ongoing battle against ransomware. The significance of these findings extends beyond the immediate evaluation of Windows 11 Pro, serving as a reference point for the broader discourse on enhancing digital security measures.

Keywords: cybersecurity; malware; ransomware; Windows 11; cyberattack

Citation: Al-Awadi, Y.M.; Baydoun, A.; Ur Rehman, H. Can Windows 11 Stop Well-Known Ransomware Variants? An Examination of Its Built-in Security Features. *Appl. Sci.* **2024**, *14*, 3520. https://doi.org/10.3390/app14083520

Academic Editors: Peter R. J. Trim and Yang-Im Lee

Received: 2 January 2024
Revised: 17 February 2024
Accepted: 20 February 2024
Published: 22 April 2024

1. Introduction

Our increasing reliance on technology has led to growth not only in its benefits but also in the threats it poses, as cyber-attacks occur at an alarming rate of once every 39 s [1]. Among these threats, ransomware particularly has proven to be one of the most damaging [2,3]. Ransomware stands out as an especially menacing adversary. Originating as simple scareware that tricked users into paying, ransomware has evolved into a sophisticated tool of cyber extortion, encrypting victims' data and demanding a ransom for its release [4]. This evolution underscores the increasing complexity and sophistication of cyber threats as most economic, commercial, cultural, social, and governmental activities are now being carried out in cyberspace, making them potential targets [5]. These escalating trends underline the need for robust cybersecurity measures embedded within modern operating systems like Microsoft Windows 11. In the face of this escalating threat, tech giants like Microsoft have risen to the challenge. With the introduction of Windows 11, Microsoft has ushered in advanced security features to mitigate such cyber onslaughts, particularly those posed by ransomware [6]. However, as these features are novel, their real-world efficacy against a spectrum of ransomware threats remains to be thoroughly assessed. This study is set to make a valuable contribution to the field of cybersecurity by assessing the effectiveness of Windows 11's built-in security features against a selection of the most widespread ransomware variants, focusing specifically on the MortalKombat ransomware variant. The results of this research could have important implications for a wide range of stakeholders, including individual users and organizations, developers, policymakers, academics, and professionals.

2. Literature Review

This review delves into the intricate details of ransomware and the robust security features incorporated in Windows 11 Pro. In doing so, it aims to provide a comprehensive context for the subsequent examination of the effectiveness of Windows 11 Pro's built-in security features against ransomware. This review comprehensively covers the important aspects of the ransomware variants we selected in the research and the robust security features incorporated in Windows 11 Pro.

2.1. Ransomware Overview

Ransomware is malware utilized to lock users out of their systems or encrypt their data, making it inaccessible [7]. Victims are then presented with ransom demands, primarily in the form of messages to make them aware of the encryption and instruct how the ransom should be paid for them to receive the decryption key [8]. Ransomware may impact different data or files on victims' devices [9]. The ransomware attack process can generally be broken down into five phases:

1. Delivery: Involves the delivery of the malicious payload to the targeted network by using spam emails, social engineering, etc. [10].
2. Deployment: Involves the extraction of a second payload from the initial malware, which can bypass detection by disguising itself as benign or routine network traffic [11].
3. Destruction: Begins searching for specific file types (pdf, docx, jpeg, etc.) across all accessible volumes in the system to encrypt them. The ransomware then communicates with the attacker's command and control server (C&C) to retrieve the encryption keys [12].
4. Dealing: Generates a ransom demand on the victim's screen. It includes instructions on how to pay the ransom and how the encrypted files can be retrieved after payment. However, it is worth noting that paying the ransom does not guarantee that the decryption keys will be provided or that the files can be recovered successfully [12].
5. Exfiltration and Persistence: Siphons off sensitive data, which can be used as additional leverage against victims or sold on the dark web, further monetizing the attack. Additionally, some ransomware variants aim to maintain persistence in the infected systems, often using methods such as creating or modifying registry keys, scheduled tasks, or injecting code into other processes [13].

2.2. Ransomware Variants

There are many notable ransomware variants in the extended literature, including WannaCry, CryptoLocker, Locky, Hive, Maze, Conti, MortalKombat, Cerber, etc. [14–16]. For the research's credibility and relevance, a representative set of well-known ransomware variants was selected for inspection. Table 1 illustrates the three main ransomware variants presented in the literature.

Table 1. Well-Known Ransomware Variants.

Ransomware Variant	Year of Discovery	Origin	Key Features	Encryption Method	Decryptor Available
Hive [14,17]	2021	Unclear	RaaS, Wiper capabilities, Double extortion methods	Proprietary	Version 5 variant only
MortalKombat [16,18]	2023	Unknown	Xorist ransomware family offshoot	TEA (Tiny Encryption Algorithm)	Yes
Cerber [19,20]	2016	Unknown	RaaS, Geographic targeting mechanism	AES and RSA-2048	Yes

These variants were chosen based on multiple criteria, including their prevalence in recent cyberattacks, their notoriety within the cybersecurity community, and their varied modes of operation, ensuring a comprehensive testing spectrum.

2.3. Windows 11 Security Characteristics

When Microsoft Windows 11 Pro was released in January 2022, Microsoft claimed that Windows 11 Pro brought several security enhancements compared to its predecessor, Windows 10. It introduces robust security features designed to guard against advanced, persistent threats, including ransomware, effectively integrating with a broader cybersecurity approach. Table 2 depicts the main security features of Microsoft Windows 11 Pro.

Table 2. Windows 11 Pro Security Characteristics.

Security Characteristic	Description
Virtualization-Based Security (VBS)	Uses hardware virtualization features to isolate key security processes, enhancing protection against attacks. VBS protects critical system components and sensitive data from ransomware [21].
Memory Integrity	Protects the memory from attacks, specifically from kernel-mode code injection techniques. It can significantly impede ransomware's ability to infiltrate and control system processes [22].
Mandatory Trusted Platform Module (TPM) 2.0	Provides hardware-based security by storing cryptographic keys and other sensitive data in an isolated and secure environment within the device, thereby preventing ransomware from accessing these crucial elements [23].
UEFI Secure Boot	Fortifies the boot process against unauthorized changes, ensuring that only trusted software is executed during the booting process and thereby preventing the loading of malicious bootloaders that could compromise the system [24].
Microsoft Pluton	Uses chip-to-cloud security technology designed to provide a new level of hardware security. It enhances protection against physical attacks and prevents the theft of credential and encryption keys [25].
Hardware-enforced Stack Protection	Adding another layer of defense against ransomware, it enhances system protection against Return Oriented Programming (ROP) attacks, a common exploitation method used by ransomware to hijack a program's control flow [26].
Regular Updates	Facilitates faster, less disruptive updates by minimizing the size of updates by up to 40% [27]. This encourages users to keep their systems updated with the latest security patches, which are vital in the fight against ransomware [27].

2.4. Related Studies

In the extensive realm of ransomware literature, there is a dire need for specific studies that delve deeper into the intricacies of modern defense mechanisms. While there is a plethora of information on general anti-ransomware strategies, a clear gap emerges when seeking an in-depth analysis of Windows 11's unique security features [28]. This study enthusiastically seeks to navigate this space, offering a keen focus on these features in the specific context of defending against crypto-ransomware. Notably, while contributions like those from [29] provide a broad-brush picture of the ransomware landscape and Windows 11's defenses, they stop short of presenting tangible tests or simulations. Recognizing this shortfall, our research endeavors to bridge this gap by creating a robust, controlled testing environment on a Windows 11 Pro virtual machine. The journey does not stop there. The literature space seems to skim over the detailed assessment of Windows 11's proactive security features, especially in the throes of real-world ransomware scenarios. Our research seeks to offer this granularity, presenting a meticulous review of the effectiveness of these defenses. In [30], the authors ventured into a comparative realm, equating paid antivirus solutions with Windows Defender in the broader malware detection arena. However, a noticeable gap exists in the depth of exploration regarding how the nuanced features of Windows Defender measure up against the formidable challenge posed by ransomware. While the study aptly contrasts paid antivirus solutions with Windows Defender in the broader context of malware detection, it falls short in providing a comprehensive examination of the specific capabilities and limitations of Windows Defender in addressing the ransomware menace.

3. Methodology

This section presents the broad strategy of the research, including the experimental design, the specifics of the experimental environment, and the details of the system and network configurations. The research design for this study is rooted in a practical and experimental approach, which is deemed most appropriate for a comprehensive evaluation of Windows 11 Pro's security features against ransomware threats.

3.1. Experimental Environment Setup

3.1.1. Experimental Environment

The experimental environment is established on a dedicated physical machine running Windows 11 Pro, reflecting a typical user configuration. The machine is equipped with 6 Cores, 16 GB RAM, and a 512 GB hard disk, carefully configured to mirror real-world, up-to-date systems as closely as possible.

3.1.2. Network Isolation

This is facilitated through a router that also behaves as a switch, creating a specialized environment exclusively for testing. Central to this setup is the gateway firewall housed within the router. It is armed with a specific rule that denies any connection attempts between the test environment and other devices or networks on the local grid.

3.1.3. User Data Simulation

To improve the realism of the test environment and align it more closely with potential real-world targets, a diverse array of file types is included on the physical machine running Windows 11 Pro. These file types cover various scripts, text documents, Word and Excel files, and PNG and JPEG images, to name a few, of various sizes. Additionally, some well-known software that can be found on a typical machine will be installed, such as Google Chrome, Google Drive, and Zoom, among others. By including a variety of file types in our test environment, the study caters to the potential behaviors of diverse ransomware strains.

3.2. Windows 11 Pro Configuration

These configurations, both default and hardened, aim to provide a comprehensive understanding of Windows 11 Pro's capabilities against ransomware attacks under different security postures.

3.2.1. Default Configuration

The first phase of testing maintains the default security configurations within Windows 11 Pro. This environment represents a baseline, simulating a common user system that relies on the standard settings provided by the operating system. In the default environment, Windows 11 Pro's Windows Defender is configured with the following features illustrated in Table 3.

Table 3. Windows 11 Pro Default Security Configuration.

Security Feature	Description
Real-time protection [31]	- Uses heuristics and behavior analysis to detect unknown threats. - Monitors file and program activity, alerting users to suspicious behaviors and blocking potentially harmful actions.
Cloud-delivered protection [32]	- Leverages Microsoft's vast cloud infrastructure to quickly identify and respond to emerging threats. - Uses machine learning models and big data analysis to predict and counteract new malware strains.
Automatic sample submission [32]	- Uses a secure channel to send suspicious files to Microsoft's labs. - The analysis helps in refining heuristics and improving detection algorithms.
Tamper protection [33]	- Secures against root-level attacks that attempt to disable security features. - Integrates with system-level permissions, ensuring only authorized users can modify security settings.
Firewall [34]	- Uses stateful inspection to monitor active connections. - Implements packet filtering to analyze network data and block malicious traffic.
Smart App Control [35]	- Uses a combination of local heuristics and cloud-based analysis to evaluate app behaviors. - Integrates with Microsoft's app reputation database to determine the trustworthiness of applications.
Reputation-based protection [36]	- Employs URL filtering and reputation checks to block access to malicious sites. - Uses a continuously updated database of known phishing sites, malware domains, and other online threats.
Memory access protection [37]	- Implements hardware-based virtualization to isolate key processes. - Uses Virtualization-Based Security (VBS) to protect critical parts of the OS from tampering.
Device encryption [38]	- Uses BitLocker technology to encrypt the entire hard drive. - Employs the Trusted Platform Module (TPM) to store encryption keys securely.
Exploit protection [39]	- Control Flow Guard (CFG): Protects against memory corruption vulnerabilities by checking the legitimacy of target addresses during indirect calls. - Data Execution Prevention (DEP): Blocks execution of code from data pages, preventing buffer overflow attacks.

3.2.2. Hardened Configuration

Following the initial testing, the Windows 11 Pro testing machine will be wiped clean, and specific enhancements and configurations will be applied to fortify the system against ransomware attacks. This "hardened" environment showcases how a more security-conscious user might configure their system, emphasizing particular features and settings to enhance resilience against malware and other threats. For the hardened environment, Windows 11 Pro's Windows Defender is configured with enhanced features. Table 4 shows the hardened configuration of the Windows 11 Pro testing machine.

Table 4. Windows 11 Pro Hardened Configuration.

Security Feature	Description
Controlled folder access	- Windows employs a feature called Controlled Folder Access in Windows 11 to implement this protective measure. It protects specific folders from unauthorized changes, acting as a defense layer against threats like ransomware. By default, it ensures the security of user directories, including but not limited to Documents, Pictures, Videos, and others.
Account protection	- Enhances the security of sign-in options.
Potentially unwanted app blocking	- Actively prevents potentially unwanted applications (PUAs) from being installed.
Memory integrity	- A part of Core isolation in device security, it is turned on to enhance memory protection.
Exploit protection adjustments	- Settings like forcing randomization for images (Mandatory ASLR) are activated.
Smart App Control	- Adjusted from 'Evaluation' mode to 'On' mode. In this mode, it actively assesses any app initiated on Windows, blocking it if deemed harmful or unwanted.
Isolated browsing	- While not installed by default, for the purpose of this study, it will be configured to provide an additional layer of protection during web browsing.

3.3. Rationale for Selecting the Ransomware

A representative set of known ransomware variants were chosen for testing. These samples should reflect the diversity of ransomware threats that real-world users might encounter. the scope was narrowed down to focus on three specific ransomware variants. This shift in the number of tested ransomware variants is aimed at delivering a more detailed and focused evaluation of Windows 11's resilience and detection capabilities.

These samples were chosen based on multiple criteria, including their prevalence in recent cyberattacks, their notoriety within the cybersecurity community, and their varied modes of operation, ensuring a comprehensive testing spectrum. This selection aims to provide a realistic representation of the threats that everyday users and businesses might face in real-world scenarios.

In the context of this research, MortalKombat, Hive, and Cerber ransomware variants were chosen for experimentation. This choice stems from their classification as recent ransomware variants and their deliberate focus on users of Windows operating systems. Furthermore, it is important to highlight the scarcity of literature addressing the definitions and operations of these ransomware variants, adding to the significance of this research endeavor.

3.4. Testing Procedures

The tests are conducted on a designated physical machine running Windows 11 Pro, specifically set up to evaluate Windows Defender's effectiveness against ransomware variants. This physical environment replicates a real-world scenario, allowing for a more authentic analysis. Figure 1 outlines the penetration testing procedures for assessing Windows Defender against ransomware variants. After setting up the testing environments, we introduce selected ransomware variants and meticulously observe the system's reactions. The figure highlights a structured approach that is designed to generate experimental data, offering insights into the effectiveness of Windows 11 Pro's security.

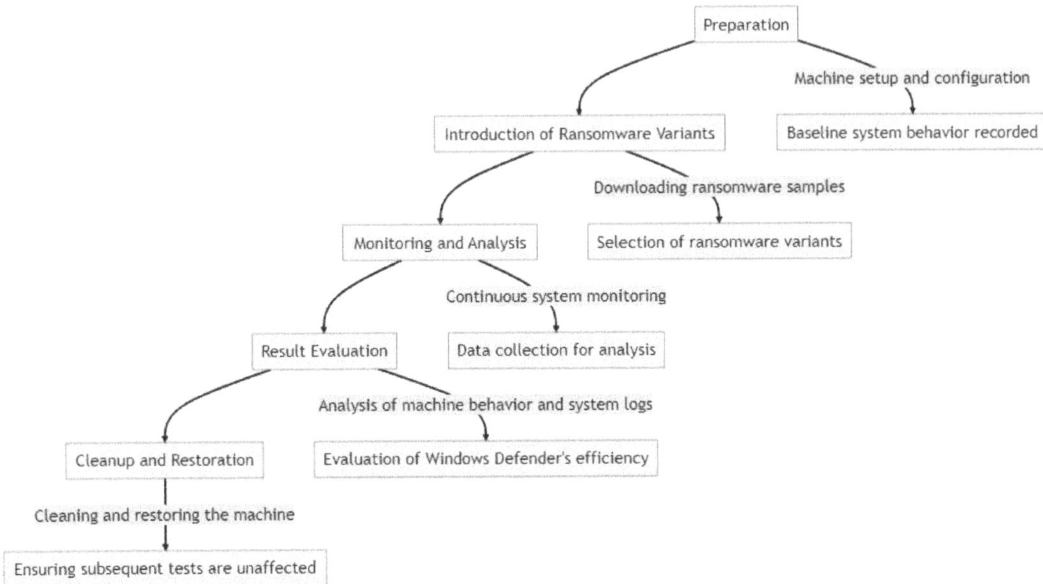

Figure 1. Flow Chart of Penetration Testing Procedures for Evaluating Windows Defender against Ransomware Variant.

3.4.1. Testing Procedures in the Default Environment

The default environment represents the baseline configuration, where the systems and networks operate without any specialized hardening or advanced security measures. The testing procedures conducted within this environment aim to evaluate the vulnerability of typical systems against ransomware attacks.

- Experiment Setup

In the default environment, all operating systems, applications, and network devices were configured to their out-of-the-box settings. Standard user privileges were granted, and no additional firewalls, intrusion detection systems, or endpoint protection mechanisms were deployed.

- Attack Simulation

Utilizing a controlled isolated network, a ransomware attack was simulated using a known ransomware strain. The efficiency and effectiveness of the attacks were measured based on the time to infiltration, the extent of encryption, and the ability to detect the attack.

- Data Collection and Analysis

Logs, alerts, and forensic data were collected during the attack simulation. The default environment's response was analyzed to identify potential weaknesses and entry points exploited by the ransomware, providing insights into commonly targeted vulnerabilities.

3.4.2. Testing Procedures in the Hardened Environment

The hardened environment refers to the configuration where the systems and networks are specifically fortified against potential ransomware attacks through the implementation of enhanced security measures.

- Experiment Setup

The hardened environment employed a combination of advanced firewalls, intrusion detection and prevention systems (IDPS), endpoint protection platforms (EPP), and restricted user privileges. All configurations were meticulously crafted to align with industry best practices for ransomware mitigation.

- Attack Simulation

Ransomware attack simulations were conducted under the same isolated conditions, utilizing the same strains as in the default environment. The focus here was to evaluate the resilience of the hardened environment by assessing the ability of ransomware to penetrate the enhanced security layers and by examining any thwarted attempts to communicate or propagate.

- Data Collection and Analysis

Comprehensive logs, alerts, and forensic data were collected to analyze the efficiency and effectiveness of the hardening measures. The analysis emphasized the points of resistance, identifying the specific mechanisms that successfully thwarted or mitigated the ransomware attacks. Comparative analysis with the default environment provided actionable insights into the value and impact of the applied hardening techniques.

3.5. Detection Procedure

Understanding how effectively the inherent security mechanisms in Windows 11, primarily Windows Defender, can detect the chosen ransomware variants is a central part of this study. Unlike penetration testing, this phase is tailored to evaluate the system's ability to recognize a malicious intrusion before serious harm occurs. Figure 2 depicts the detection testing process stages.

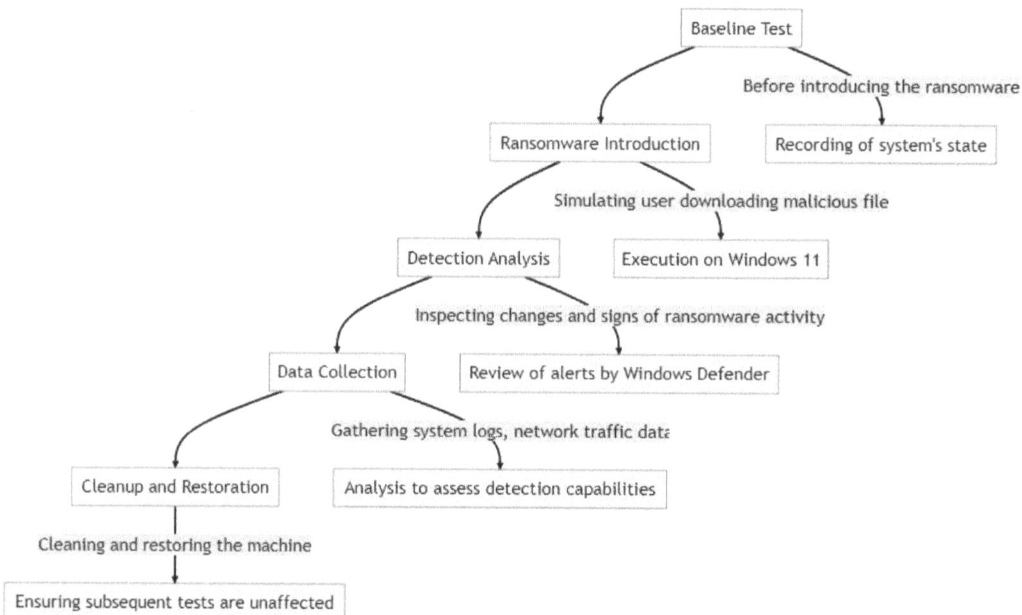

Figure 2. Flow Chart of Ransomware Detection Testing Process.

3.6. Testing Tools

For an accurate assessment of the malware's behavior and Windows Defender's response, various monitoring and filtering techniques were implemented. These strategies included:

- Sysinternals Suite: A collection of system utilities to monitor system behavior.
- Process Monitor: Observes real-time file system, registry, and thread activity.

- Windows Event Logs: Analysis of system logs to detect suspicious activity.
- Windows Defender Logs: Review of the logs specific to Windows Defender to evaluate its response to the malware.

4. Experimental Results

In this section, we present the outcomes of our assessment of Windows 11 Pro's defenses against three prominent ransomware variants: Cerber, Hive, and MortalKombat. The test was executed in two distinct environments: a Windows 11 Pro default configuration and a hardened configuration with enhanced security settings of Windows 11 Pro (As specified in the Section 3). The following section will delve into the system's responses, supplemented by detailed log insights, to provide a comprehensive understanding of Windows 11 Pro's capabilities and defense mechanisms.

4.1. MortalKombat

Default Environment Outcome

Upon unzipping the MortalKombat ransomware in the default environment, Microsoft Defender instantly detected the threat and took immediate action. The malware was detected as *Trojan:Win32/Vindor!pz*. Figure S1 illustrates the system behaviors as soon as the MortalKombat ransomware was unpacked.

Key Observations:

Subsequent to its detection, the ransomware was immediately quarantined, ensuring no further malicious actions could take place. The threat was isolated, and the system reported, "No additional actions required", signifying a successful containment of the threat.

Despite the user potentially attempting to execute the ransomware post-extraction, Windows Defender ensured that the ransomware remained non-functional. It is notable that there was no activity observed, underscoring the immediate action taken by Windows Defender, which prevented any further malicious operations from taking place. Figure S2 represents the Windows Defender logs, highlighting the severity of the ransomware and the status and the action taken.

4.2. Hardened Environment Testing Outcome

In the hardened environment, the download process for the MortalKombat ransomware was interrupted even before completion. This immediate halt can be seen in the log entries, as shown in Figure S3.

Such an immediate response can be attributed to the enhanced security layers in this configuration. The layered defenses, specifically features like Smart App Control, actively assess every file or application being introduced into the system. Given the intelligence of Smart App Control, it likely recognized the download as a potential threat even before the download was completed, thereby halting the process. Figure S4 illustrates the detection process of Smart App Control. The incident also highlights the efficacy of the integrated Threat Intelligence feeds that continuously provide the security infrastructure with real-time updates on emerging threats and attack vectors. This up-to-date knowledge enables the security components to proactively anticipate potential threats and respond swiftly.

Additionally, the utilization of Behavioral Analysis within the security framework contributed significantly to thwarting the attack. By establishing a baseline of normal system behavior, any deviations from this pattern trigger alerts. This technology effectively identified the ransomware's attempts to encrypt files and initiate unauthorized connections, even though the specific signature of the MortalKombat ransomware might not have been previously known. Therefore, in summary, the collaborative nature of these security layers, each complementing the other's strengths, collectively created a multi-faceted defense mechanism that not only intercepted the MortalKombat ransomware but also actively deterred its progress. This scenario exemplifies the crucial importance of a proactive and comprehensive security strategy, especially in the face of ever-evolving cyber threats.

The incident serves as a testament to the dedication of the cybersecurity team and the effectiveness of their strategic planning and technological implementation. A process creation log for sdbinst.exe (a Windows file used to address compatibility issues [40]) indicated that the Application Compatibility Database Installer was invoked. This is a typical behavior to ensure compatibility when new software or applications are introduced. The initiation of SecurityHealthHost.exe suggests the activation of the Windows Security Health Host. This service assesses the overall security and health of the system. Its activation here might be in response to the potential threat detected during the download. Another notable process creation was smartscreen.exe, which is the Windows Defender SmartScreen. This is an integral part of Windows' security infrastructure, designed to warn users about potentially malicious websites and downloads.

A registry value was set by SecurityHealthHost.exe. This could be a notification or alert pertaining to the detected threat, showcasing how the system responds internally by setting flags or notifications. The log entry showing a change in the file creation time for the ransomware by MsMpEng.exe (Windows Defender's core process) suggests that Defender interacted with the file, possibly during its scanning or quarantining process.

4.3. Cerber

4.3.1. Default Environment Outcome

Upon extracting the Cerber ransomware in the default environment, Windows Defender instantly sprang into action even if the user tried to execute Cerber before Windows Defender took action. The attached appendices (Figures S5 and S6) provide all the screenshots associated with these actions. The Windows log also indicates the Windows Defender behaviors in detecting the Cerber ransomware. Figure S7 shows the logs of Windows Defender when it detected the Cerber ransomware. The subsequent section outlines the main observations from testing and executing the Cerber ransomware variant on the Windows 11 Machine.

Key Observations:

Origin: The threat was detected on the test machine, meaning the antivirus did not rely on cloud-based checks for this detection.

Type: The detection type was "Concrete", which suggests that the antivirus software was certain about the malicious nature of the file.

Source: The detection source being "System" means the system processes or the operating system itself flagged it.

User: The threat was executed or encountered under the system authority, which means it might have had elevated permissions.

Process Name: The process that initiated or was infected by the ransomware is not known, as per the log.

Action Taken: The Windows Defender antivirus took the action to "Quarantine" the file, effectively isolating it to prevent any harm. No further actions were required from the user's end as the threat was neutralized. The operation was successful, as indicated by the error description.

While the threat was automatically quarantined, a potential improvement could be allowing the notification to await user acknowledgment, ensuring they are informed about the detected threat. Currently, the notification is temporary and might be missed by the user.

The Windows Defender protection history offers insights, but we must ensure users can easily access this information. It is essential to prioritize user experience while maintaining stringent security.

4.3.2. Hardened Environment Outcome

Upon attempting to download the Cerber ransomware in the hardened environment, the process was interrupted even before the completion of the download. This immediate halt can be attributed to the log entries, shown in Figure S8. The instant disruption is indicative of the comprehensive defense mechanisms that the hardened configuration offers, with multiple layers working in tandem to ensure the highest level of security.

The in-transit disruption happened only if the user tried to download the malicious file using Microsoft Edge, which reveals that the security feature that came into play is part of Microsoft Edge. If the user tried to download a malicious file using the most popular browser, Google Chrome, he/she would be given the option to keep the file, like any other executable (exe) file (Figure S9).

In the above case, it was observed that Real-time protection, which actively detected the ransomware in the default environment, took a backseat in the hardened mode. Instead, the responsibility for ransomware detection transitioned to Smart App Control as soon as the user attempted to execute the application (Figure S10). This shift in roles between the two features is not just an incidental behavior but reflects a deeper strategy in Windows 11 Pro's defense mechanisms.

The move from the default to the hardened setting showcased a marked shift in the dynamics of Windows Defender's behavior. While Real-time protection served as the initial line of defense in the default setting, actively identifying and handling threats, the hardened environment saw Smart App Control stepping up. The logic behind this shift can be attributed to the enhanced security layers in the hardened setting. Even as Real-time protection continues its vigil, scanning files in real-time, Smart App Control operates more at the application execution juncture.

In sum, the collaboration and prioritization of Real-time protection and Smart App Control in the hardened setting underline Windows 11 Pro's adaptability and strategic defense. It is a clear demonstration of the OS's dedication to equipping users with flexible and potent protection against modern threats.

4.4. *Hive*

4.4.1. Default Environment Outcome

Upon attempting to download the Hive ransomware in the default environment, Windows Defender instantly detected the malicious content and did not even allow the completion of the download. This immediate detection signifies that Windows Defender identified the ransomware pattern in real-time as the file was being downloaded. Figures S11 and S12 show the Windows Defender logs in the default testing environment when attempting to download the Hive ransomware. The key testing observations from executing the Hive ransomware variant on Windows 11 can be summarized as follows:

Key Observations:

The notification indicating the threat is a system-generated alert and does not necessitate any user acknowledgment. It serves to inform the user about the blocked content but does not require any further action on their part. The identified threat is termed as "Trojan:Script/Sabsik.TE.A!ml". This classification indicates that Hive contains malicious scripts commonly associated with Trojan activities. The detection source, "Downloads and attachments", reinforces the idea that Defender halted the ransomware during the download process itself.

Name of the Threat: Trojan:Script/Sabsik.TE.A!ml

This specifies the type and variant of malware detected. In this case, it is a Trojan, which is malicious software that masquerades as legitimate software. This is a unique identifier assigned to this specific threat by Microsoft Defender.

Severity: Severe.

This indicates the potential harm the detected threat can cause to the system. "Severe" indicates it can inflict significant damage or unauthorized access.

Category: Trojan.

This confirms the type of malware detected.

Detection Origin: Internet.

This means the malware was detected during a download or while interacting with online content.

Detection Source: Downloads and attachments.

This indicates that the malware was detected as part of a downloaded file or an email attachment.

Action: Quarantine.

This is the action Defender took upon detecting the threat. "Quarantine" means the suspicious file was isolated to prevent it from causing harm.

Action Status: No additional actions required.

After quarantining the threat, Windows Defender determined no further actions were needed. The threat has been contained.

4.4.2. Hardened Environment Outcome

In the hardened configuration, the behavior observed was consistent with that of the default environment. Once again, Hive could not complete its download process due to Windows Defender's intervention. This outcome reinforces the robustness of the hardened configuration and its ability to block known threats even before they land on the system.

Key Observations:

Hive seems to be distinct in its immediate detection during download, unlike some other ransomware. This could be attributed to its signature or behavioral patterns being prominently recognized as malicious.

The utilization of a different browser, Chrome, resulted in a different user experience. Chrome presented a warning but still gave an option to proceed, suggesting that the blocking mechanism in Edge (Figure S13) is more rigid, possibly due to tighter integration with Windows Defender or the OS's inherent security features. This variation highlights the importance of browser-level security and its coordination with the OS.

The detection source being "Internet" in both logs implies that the ransomware was detected during its transit from the web server to the local machine, halting its download.

Hive's detection by Windows Defender during the download phase, irrespective of the configuration, is a testament to the potency of modern anti-malware solutions. Several factors could be contributing to this agile response:

Prominence: Hive's prominence in the cyber threat landscape might have led to its prioritization. Being a known ransomware variant, it is plausible that security solutions have been fine-tuned to detect it with heightened sensitivity.

Signature Recognition: Every piece of malware has a signature—a unique set of characteristics that define it. Hive's signature might be particularly distinct or overtly malicious, making it an easy target for real-time scanning engines like Windows Defender.

Behavioral Patterns: Modern cybersecurity solutions have evolved beyond mere signature-based detections. Behavioral analysis plays a significant role in identifying threats. There is a possibility that even during its initial download stages, Hive exhibits certain behaviors or patterns that act as red flags for detection systems.

Cloud-backed Analysis: The integration of cloud-backed analysis in security solutions like Windows Defender provides an additional layer of rapid threat detection. If Hive has been flagged, analyzed, and documented in a cloud repository recently, its details would be fresh in the database. This can lead to an almost instantaneous detection as soon as it begins downloading, given the real-time synchronization between the local machine and the cloud database. The prompt detection of Hive underscores the importance of keeping security solutions updated. As ransomware variants evolve, so do detection mechanisms, and the tussle continues. The immediate recognition of Hive serves as a reassuring indicator of Windows Defender's capabilities, especially when configured for maximum security. Table 5 below shows the testing results for the three said ransomware variants.

Table 5. Testing comparison between different ransomware variants.

MortalKombat Ransomware		Cerber Ransomware		Hive Ransomware	
Default	**Hardened**	**Default**	**Hardened**	**Default**	**Hardened**
After unzipping the MortalKombat ransomware, Windows 11 takes immediate quarantine action, and the logs indicate the following actions: - Name of the Threat: Trojan:Win32/Vindor!pz - Severity: Severe - Action: Quarantine - Action Status: No additional action required Despite the user potentially attempting to execute the ransomware post-extraction, Windows Defender ensured that the ransomware remained non-functional.	- The download process was interrupted - Proactively anticipated a potential ransomware threat - Behavioral Analysis effectively identified the ransomware's attempts to encrypt files and initiate unauthorized connections, even though the specific signature of the MortalKombat ransomware might not have been previously known.	Upon extracting the Cerber ransomware in the default environment, Windows Defender instantly took the following actions: - Name of the Threat: Ransom:Win32/Avaddon.P!MSR - Severity: Concrete - Action: Quarantine - Action Status: No further action is required Even if the user tries to execute Cerber, Windows Defender takes prompt blocking action.	- Upon attempting to download the Cerber ransomware in the hardened environment, the downloading process was interrupted even before the completion of the download. - Smart App Control took immediate action as soon as the user attempted to execute the Cerber ransomware.	Upon attempting to download the Hive ransomware in the default environment, Windows Defender instantly detected the malicious content and did not even allow the completion of the download. - Name of the Threat: - Trojan:Script/Sabsik.TE.A!ml - Severity: Severe - Category: Trojan - Action: Quarantine - Action Status: No further action is required Defender halted the ransomware during the download process itself.	- Hive could not complete its download process due to Windows Defender's intervention. This outcome reinforces the robustness of the hardened configuration and its ability to block known threats even before they land on the system. - Hive seems to be distinct in its immediate detection during download, unlike some other ransomware. This could be attributed to its signature or behavioral patterns being prominently recognized as malicious.

5. Discussion

The testing of the MortalKombat, Hive, and Cerber ransomware variants on Windows 11 Pro offers a comprehensive insight into the operating system's defense mechanisms, both in default and hardened configurations. Here, we provide a discussion of the responses to these ransomware variants to understand the overall robustness and strategy of Windows 11 Pro's security features.

The hardened environment demonstrated Smart App Control's dominance in ransomware detection, even overshadowing Real-time protection. When the ransomware variants were introduced, Smart App Control, backed by its cloud intelligence, probably identified the file's hash or metadata matching with known threats, resulting in an immediate halt of the download.

While Real-time protection is designed to scan and detect malicious files actively, the Smart App Control feature is more reactive, awaiting an execution attempt. The advantage is that even if a user momentarily disables Real-time protection, Smart App Control remains vigilant.

In essence, the behavior exhibited against the ransomware variants reinforces the strategic interplay between Real-time protection and Smart App Control, offering users a robust and dynamic defense mechanism.

- Detection Timing:

Detection happened after unzipping in the default and in pre-download completion in the hardened environment.

- Security Feature Engagement:

In the default setting, Real-time protection was the primary defense mechanism. In the hardened mode, Smart App Control took the reins, reflecting the strategic shift in defense based on the configuration.

- User Interaction and Notifications:

Both variants triggered notifications which, though temporary, were effective in informing the user about the detected threat.

- Sysmon Observations:

Sysmon logs revealed that various system processes were invoked, reflecting the operating system's response to a potential threat.

- Adaptability of Defense Mechanisms:

The testing of these ransomware variants showcased Windows 11 Pro's dynamic response based on the configuration. While the default setting relies heavily on Real-time protection, the hardened mode introduces a layered defense strategy, emphasizing the role of Smart App Control.

While the threat was automatically quarantined, a potential improvement could be allowing the notification to await user acknowledgment, ensuring they're informed about the detected threat. Currently, the notification is temporary, which might be missed by the user.

Relevance to End-Users: The Windows Defender protection history offers insights, but we must ensure users can easily access this information. It is essential to prioritize user experience while maintaining stringent security.

The hardened configuration in Windows 11 Pro acknowledges that many modern threats are not solely dependent on the presence of malicious files, but rather on the behavior of applications and processes. Attackers often exploit legitimate applications to execute malicious actions, making it challenging for traditional antivirus methods to detect such activities solely based on file signatures or patterns.

Interpreting Windows Defender's Defensive Dynamics

Several key observations and patterns were discerned. Windows Defender showcased its efficacy in both default and hardened configurations, effectively neutralizing the ransomware attempts and underscoring its robust capability to identify and combat a range of threats. The juxtaposition between the default and hardened configurations was enlightening. While the default configuration leaned heavily on Real-time protection, detecting and quarantining ransomware either post-download or during execution attempts, the hardened environment exhibited more proactive defense mechanisms. The download processes were often interrupted, a testament to the layered security approach in this advanced setting. An integral part of the analysis was recognizing the role of browser-based protections. The granular insights from the Sysmon logs offered an in-depth view into the processes and system alterations instigated by ransomware and the corresponding responses triggered by Windows Defender. In many scenarios, these logs displayed an absence of malicious activity, a clear indication of Windows Defender's swift and effective interventions. A fascinating dynamic was evident in the interplay of Windows Defender features. Particularly in the hardened setting, while Real-time protection remained the frontline defense in the default mode, Smart App Control assumed a more dominant role during the application execution phase. Conclusively, this comprehensive exploration underscores that Windows 11 Pro's security features, irrespective of default or hardened configurations, present a formidable defense against ransomware threats.

6. Conclusions

The digital age has brought with it a plethora of opportunities, but it has also introduced numerous challenges, not least of which is the threat of ransomware. This paper set out to explore the capabilities of Windows 11 Pro in mitigating well known ransomware variants threats and to evaluate its embedded security features thoroughly. Through systematic testing in both default and hardened configurations, this research has shown that while Windows 11 Pro offers significant defenses against ransomware, there is always room for improvement. As ransomware evolves, so too must the defenses against it. The security features in Windows 11 Pro, such as Controlled Folder Access and Smart App Control, provide robust protection, but their efficacy is closely tied to user understanding and interaction. Thus, striking a balance between user-friendly features and robust security measures remains a challenge. An important takeaway from this study is the realization that security is not just about having advanced features; it is about the constant adaptation and updating of these features in response to emerging threats. Moreover, user awareness and behavior play a pivotal role in system security. Even the most sophisticated features can be compromised if not used correctly or if users are not vigilant.

Supplementary Materials: The following supporting information can be downloaded at: https://www.mdpi.com/article/10.3390/app14083520/s1, Figure S1. Evident of detecting the MortalKombat ransomware by Windows Defender; Figure S2. Windows Defender logs after detection the MortalKombat variant; Figure S3. Windows Defender logs showing the detection of the MortalKombat ransomware download; Figure S4. Smart App Control popup when attempting to run the MortalKombat executable; Figure S5. Windows Defender detecting Cerber upon extraction; Figure S6. Windows Defender prevented the user from manually executing the Cerber exe; Figure S7. Windows Defender logs showing the detection of the Cerber ransomware; Figure S8. The Windows Defender logs entries in the hardened environment indicating the detection of Cerber ransomware; Figure S9. Chrome giving the user the option to allow executables; Figure S10. Smart App Control popup when attempting to run the Cerber executable; Figure S11. Windows Defender Logs showing the detection of Hive Ransomware in the Default testing environment; Figure S12. Windows Defender Logs showing action taken to protect local machine from Hive Ransomware; Figure S13. Hive Ransomware being detected in transit in Microsoft Edge browser.

Author Contributions: Conceptualization, A.B. and H.U.R.; methodology, A.B. and Y.M.A.-A.; software, Y.M.A.-A.; validation, Y.M.A.-A., A.B. and H.U.R.; formal analysis, A.B. and H.U.R.; investigation, Y.M.A.-A.; resources, A.B. and H.U.R.; data curation, Y.M.A.-A.; writing—original draft preparation, Y.M.A.-A., A.B. and H.U.R.; writing—review and editing, A.B. and H.U.R.; visualization, Y.M.A.-A., A.B. and H.U.R.; supervision, A.B.; project administration, A.B.; funding acquisition, Y.M.A.-A., A.B. and H.U.R. All authors have read and agreed to the published version of the manuscript.

Funding: We are extremely thankful to the Qatar National Library for supporting the Open Access publication charges of this publication.

Institutional Review Board Statement: Not applicable.

Informed Consent Statement: Not applicable.

Data Availability Statement: The original contributions presented in the study are included in the article/supplementary material, further inquiries can be directed to the corresponding author.

Conflicts of Interest: The authors declare no conflict of interest.

References

1. Lavieille, P.; Atlas, I.A.H. IsoEx: An explainable unsupervised approach to process event logs cyber investigation. *arXiv* **2023**, arXiv:2306.09260.
2. Eren, M.E.; Bhattarai, M.; Rasmussen, K.; Alexandrov, B.S.; Nicholas, C. MalwareDNA: Simultaneous Classification of Malware, Malware Families, and Novel Malware. In Proceedings of the 2023 IEEE International Conference on Intelligence and Security Informatics (ISI), Charlotte, NC, USA, 2–3 October 2023; pp. 1–3.
3. Aldauiji, F.; Batarfi, O.; Bayousef, M. Utilizing cyber threat hunting techniques to find ransomware attacks: A survey of the state of the art. *IEEE Access* **2022**, *10*, 61695–61706. [CrossRef]
4. Razaulla, S.; Fachkha, C.; Markarian, C.; Gawanmeh, A.; Mansoor, W.; Fung, B.C.; Assi, C. The Age of Ransomware: A Survey on the Evolution, Taxonomy, and Research Directions. *IEEE Access* **2023**, *11*, 40698–40723. [CrossRef]
5. Li, Y.; Liu, Q. A comprehensive review study of cyber-attacks and cyber security; Emerging trends and recent developments. *Energy Rep.* **2021**, *7*, 8176–8186. [CrossRef]
6. Numminen, A. Windows Technical Hardening against the Most Prevalent Threats. 2023. Available online: http://urn.fi/URN:NBN:fi:jyu-202305293330 (accessed on 19 December 2023).
7. Humayun, M.; Jhanjhi, N.; Alsayat, A.; Ponnusamy, V. Internet of things and ransomware: Evolution, mitigation and prevention. *Egypt. Inform. J.* **2021**, *22*, 105–117. [CrossRef]
8. Ryan, M. *Ransomware Revolution: The Rise of a Prodigious Cyber Threat*; Springer: Berlin/Heidelberg, Germany, 2021.
9. McIntosh, T.; Kayes, A.; Chen, Y.-P.P.; Ng, A.; Watters, P. Ransomware mitigation in the modern era: A comprehensive review, research challenges, and future directions. *ACM Comput. Surv. CSUR* **2021**, *54*, 1–36. [CrossRef]
10. O'Kane, P.; Sezer, S.; Carlin, D. Evolution of ransomware. *IET Netw.* **2018**, *7*, 321–327. [CrossRef]
11. Lee, W. *Malware and Attack Technologies Knowledge Area Issue*; CyBOK: Bristol, UK, 2021.
12. Moussaileb, R.; Cuppens, N.; Lanet, J.-L.; Bouder, H.L. A survey on windows-based ransomware taxonomy and detection mechanisms. *ACM Comput. Surv. (CSUR)* **2021**, *54*, 1–36. [CrossRef]
13. Gittins, Z.; Soltys, M. Malware persistence mechanisms. *Procedia Comput. Sci.* **2020**, *176*, 88–97. [CrossRef]
14. Kim, G.; Kim, S.; Kang, S.; Kim, J. A method for decrypting data infected with hive ransomware. *J. Inf. Secur. Appl.* **2022**, *71*, 103387. [CrossRef]
15. Kurniawan, A.; Riadi, I. Detection and analysis cerber ransomware based on network forensics behavior. *Int. J. Netw. Secur.* **2018**, *20*, 836–843.
16. Jeffrey, C. Hackers Hit US Windows Systems with "Mortal Kombat" Ransomware. Available online: https://www.techspot.com/news/97608-hackers-hit-us-windows-systems-mortal-kombat-ransomware.html (accessed on 4 December 2023).
17. Jethva, B.; Traoré, I.; Ghaleb, A.; Ganame, K.; Ahmed, S. Multilayer ransomware detection using grouped registry key operations, file entropy and file signature monitoring. *J. Comput. Secur.* **2020**, *28*, 337–373. [CrossRef]
18. Wright, R. Bitdefender Releases Decryptor for MortalKombat Ransomware. Available online: https://www.techtarget.com/searchsecurity/news/365531919/Bitdefender-releases-decryptor-for-MortalKombat-ransomware (accessed on 4 December 2023).
19. Kara, I.; Aydos, M. Static and dynamic analysis of third generation cerber ransomware. In Proceedings of the 2018 International Congress on Big Data, Deep Learning and Fighting Cyber Terrorism (IBIGDELFT), Ankara, Turkey, 3–4 December 2018; pp. 12–17.
20. Adamov, A.; Carlsson, A. The state of ransomware. Trends and mitigation techniques. In Proceedings of the 2017 IEEE East-West Design & Test Symposium (EWDTS), Novi Sad, Serbia, 29 September–2 October 2017; pp. 1–8.
21. Microsoft. Virtualization-Based Security (VBS). Available online: https://learn.microsoft.com/en-us/windows-hardware/design/device-experiences/oem-vbs (accessed on 4 December 2023).

22. Microsoft. Enable Virtualization-Based Protection of Code Integrity. Available online: https://learn.microsoft.com/en-us/windows/security/hardware-security/enable-virtualization-based-protection-of-code-integrity (accessed on 1 December 2023).
23. Microsoft. Windows Hello Biometrics in the Enterprise. Available online: https://learn.microsoft.com/en-us/windows/security/identity-protection/hello-for-business/hello-biometrics-in-enterprise (accessed on 15 November 2023).
24. Microsoft. Secure Boot. Available online: https://learn.microsoft.com/en-us/windows-hardware/design/device-experiences/oem-secure-boot (accessed on 21 November 2023).
25. Microsoft. New Security Features for Windows 11 Will Help Protect Hybrid Work. Available online: https://www.microsoft.com/en-us/security/blog/2022/04/05/new-security-features-for-windows-11-will-help-protect-hybrid-work/ (accessed on 1 December 2023).
26. Microsoft. Understanding Hardware-Enforced Stack Protection. Available online: https://techcommunity.microsoft.com/t5/windows-os-platform-blog/understanding-hardware-enforced-stack-protection/ba-p/1247815 (accessed on 12 December 2023).
27. Microsoft. Windows 11: The Optimization and Performance Improvements. Available online: https://techcommunity.microsoft.com/t5/microsoft-mechanics-blog/windows-11-the-optimization-and-performance-improvements/ba-p/2733299 (accessed on 30 November 2023).
28. Khan, I.U.K.U.; Ouaissa, M.; Ouaissa, M.; Abou El Houda, Z.; Ijaz, M.F. *Cyber Security for Next-Generation Computing Technologies*; CRC Press: Boca Raton, FL, USA, 2024.
29. Pogonin, D.; Korkin, I. Microsoft Defender Will Be Defended: MemoryRanger Prevents Blinding Windows AV. *arXiv* **2022**, arXiv:2210.02821.
30. Santos, D. Comparison of Paid Subscription vs. Freeware Software on Antivirus Program. 2021. Available online: http://hdl.handle.net/10790/6828 (accessed on 18 December 2023).
31. Microsoft. Stay Protected with Windows Security. Available online: https://support.microsoft.com/en-us/windows/stay-protected-with-windows-security-2ae0363d-0ada-c064-8b56-6a39afb6a963 (accessed on 8 November 2023).
32. Microsoft. Cloud Protection and Sample Submission at Microsoft Defender Antivirus. Available online: https://learn.microsoft.com/en-us/microsoft-365/security/defender-endpoint/cloud-protection-microsoft-antivirus-sample-submission?view=o365-worldwide (accessed on 2 December 2023).
33. Microsoft. Exploit Protection Reference. Available online: https://learn.microsoft.com/en-us/microsoft-365/security/defender-endpoint/exploit-protection-reference?view=o365-worldwide (accessed on 29 November 2023).
34. Microsoft. Windows Firewall: New and Upcoming Features for 2023. Available online: https://techcommunity.microsoft.com/t5/windows-events/windows-firewall-new-and-upcoming-features-for-2023/ev-p/3971637 (accessed on 12 November 2023).
35. Microsoft. What Is Smart App Control? Available online: https://support.microsoft.com/en-au/topic/what-is-smart-app-control-285ea03d-fa88-4d56-882e-6698afdb7003 (accessed on 5 November 2023).
36. Microsoft. Protect Your PC from Potentially Unwanted Applications. Available online: https://support.microsoft.com/en-us/windows/protect-your-pc-from-potentially-unwanted-applications-c7668a25-174e-3b78-0191-faf0607f7a6e (accessed on 16 November 2023).
37. Microsoft. Kernel DMA Protection (Memory Access Protection) for OEMs. Available online: https://learn.microsoft.com/en-us/windows-hardware/design/device-experiences/oem-kernel-dma-protection (accessed on 23 November 2023).
38. Microsoft. Device Encryption in Windows. Available online: https://support.microsoft.com/en-us/windows/device-encryption-in-windows-ad5dcf4b-dbe0-2331-228f-7925c2a3012d (accessed on 16 November 2023).
39. Microsoft. Enable Exploit Protection. Available online: https://learn.microsoft.com/en-us/microsoft-365/security/defender-endpoint/enable-exploit-protection?view=o365-worldwide (accessed on 1 December 2023).
40. Microsoft. Using the Sdbinst.exe Command-Line Tool. 2023. Available online: https://learn.microsoft.com/en-us/windows/deployment/planning/using-the-sdbinstexe-command-line-tool (accessed on 2 January 2024).

applied sciences

MDPI

Article

Secure Control of Linear Controllers Using Fully Homomorphic Encryption

Jingshan Pan [1,2,3,4,5], Tongtong Sui [4], Wen Liu [3,5], Jizhi Wang [2,4,5,*], Lingrui Kong [4], Yue Zhao [4] and Zhiqiang Wei [1]

1 College of Information Science and Engineering, Ocean University of China, Qingdao 266100, China; panjsh@sdas.org (J.P.); weizhq@sdas.org (Z.W.)
2 Shandong Provincial Key Laboratory of Computer Networks, Shandong Computer Science Center (National Supercomputing Center in Jinan), Jinan 250014, China
3 Jinan Institute of Supercomputing Technology, Jinan 250301, China; liuwen@jnist.cn
4 Faculty of Computer Science and Technology, Qilu University of Technology (Shandong Academy of Science), Jinan 250102, China; sui_tongtong@163.com (T.S.); konglingrui125@163.com (L.K.); zy13296440360@163.com (Y.Z.)
5 Quancheng Laboratory, Jinan 250100, China
* Correspondence: wangjzh@sdas.org

Abstract: In actual operation, there are security risks to the data of the network control system, mainly in the form of possible eavesdropping of signals in the transmission channel and parameters in the controller leading to data leakage. In this paper, we propose a scheme for encrypting linear controllers using fully homomorphic encryption, which effectively removes these security risks and substantially improves the security of networked control systems. Meanwhile, this paper uses precomputation to handle data encryption, which eliminates the encryption time and solves the drawback of fully homomorphic encryption that it is difficult to apply due to the efficiency problem. Compared to previous schemes with precomputation, for the first time, we propose two methods to mitigate the problem of the slight security degradation caused by precomputation, which makes our scheme more secure. Finally, we provide numerical simulation results to support our scheme, and the data show that the encrypted controller achieves normal control and improves safety and efficiency.

Keywords: encrypted controller; networked control system; fully homomorphic; BFV encryption

Citation: Pan, J.; Sui, T.; Liu, W.; Wang, J.; Kong, L.; Zhao, Y.; Wei, Z. Secure Control of Linear Controllers Using Fully Homomorphic Encryption. *Appl. Sci.* **2023**, *13*, 13071. https://doi.org/10.3390/app132413071

Academic Editors: Andrea Prati, Peter R.J. Trim and Yang-Im Lee

Received: 18 October 2023
Revised: 27 November 2023
Accepted: 4 December 2023
Published: 7 December 2023

1. Introduction

As control technology, network communication technology, and computer technology advance, the network control system (NCS) also advances steadily. It is a feedback control system which realizes a closed-loop control through the control network. The key benefits of a network control system include less system connection, a high dependability, a flexible structure, a simple system extension, and the possibility to implement resource sharing for information. This has led to its widespread application in key infrastructure such as water, transportation, and power. However, NCSs are not completely secure [1], if a malicious user has invaded the controller without authorization, it can lead to the leakage of important information of the control system, which can make infrastructure failures such as power plants sustain huge failures and losses [2–4]. Therefore, the security of network control systems is becoming increasingly important and has attracted the attention of researchers.

The traditional antieavesdropping method is communication encryption, as shown in Figure 1a, which is to encrypt the data sampled by the sensor to hide the data. This is equivalent to putting a lock on the data, and it is difficult for a malicious attacker without a corresponding key to open the lock and eavesdrop on the data. However, this also prevents the controller from operating on the locked data. It is necessary to decrypt the data into plaintext after transmission to the controller, and then the computation of the plaintext data is completed in the controller. Then, the computed signal is encrypted by the controller and transmitted to the actuator to perform decryption. However, in this process, this conventional communication encryption not only requires two encryptions and decryptions, but the data in the controller are in plaintext as well, and thus does not protect this part of the data from eavesdropping. Kogiso et al. [5] proposed the concept of encrypted controller in 2015 to make up for the deficiency of communication encryption. The ideal encryption controller can directly calculate the encrypted data. As shown in Figure 1b, the data exist in ciphertext throughout the network control loop, and a malicious attacker would have no way to get at it. In this way, the encryption controller greatly improves the security of the data in the NCS compared to the NCS without the encrypted controller.

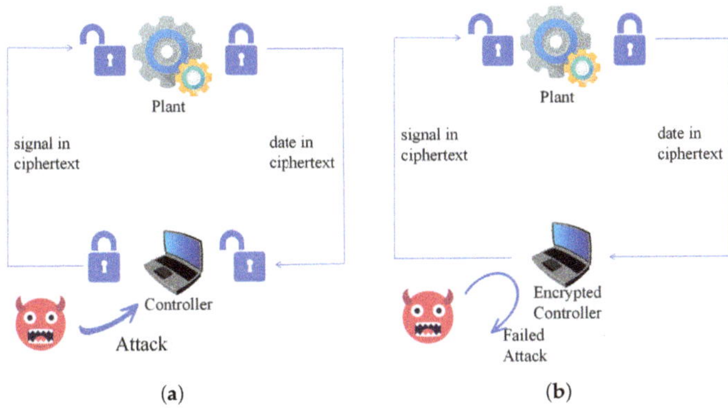

Figure 1. The NCS (**a**) without encrypted controller; (**b**) with encrypted controller.

However, the concrete encryption controller scheme proposed by Kogiso et al. [5] was implemented by RSA [6] and ElGamal [7] partially homomorphic encryption. Subsequent research studies on encrypted controllers have also mostly been conducted on partially homomorphic encryption controllers, including a series of research studies based on encrypted controllers [8–11] with Paillier [12] encryption and ElGamal encryption. Homomorphic encryption allows direct operations on ciphertext. Partially homomorphic encryption is homomorphic encryption that supports only one operation in addition or multiplication. Therefore, it may not be possible to complete the operation in the encrypted controller using partially homomorphic encryption. This leads to the fact that operations inside the encrypted controller are often guaranteed at the expense of data security. Therefore, can we implement the encrypted controller using a fully homomorphic encryption scheme? For the first time, a fully homomorphic encrypted controller was shown by Kim et al. [13]. But the scheme faced two problems. On the one hand, they suggested running multiple controllers and a catch-up mechanism to solve the problem because the controller could not function while the bootstrapping of encrypted variables was performed. On the other hand, the finite life of the encrypted variable was reduced with the operation. To solve this problem, a tree-based algorithm was introduced. But this also led to an increased complexity of the control systems.

In this paper, we propose a encrypted controller scheme that is more secure and efficient while less complex. The highlights of this paper are as follows. Firstly, we adopt a fully homomorphic encryption scheme [14,15] proposed by Brakerski, Fan, and Vercauteren to encrypt the controller in this paper, which we usually refer to as BFV encryption. Secondly, we use the method of generating tables by precomputation [16] to improve encryption efficiency. Thirdly, we propose to continuously update the table to improve security. Finally, we describe the attack scenarios [17–19] in this study and discuss the security of the scheme.

Here is the structure for the rest of the article. In Section 2, we present the related work. In Section 3, we introduce the mathematical symbols and some basic knowledge. In

Section 4, we introduce our proposed encrypted controller using BFV encryption and give two methods to improve security and analyze the security of the scheme. In Section 5, a numerical example is given to demonstrate that the encrypted controller can implement regular control and to verify that the precomputation saves time. We discuss the findings, implications, and some limitations of this paper in Section 6, and in Section 7, we summarize the paper, presenting the advantages of the scheme and outlining the current shortcomings.

2. Related Work

Homomorphic encryption is a form of encryption that is capable of performing computations on encrypted data, and its research can be traced back to the idea of homomorphic encryption proposed by Rivest in [20]. The idea can be described as the ability to directly perform functions on the ciphertext without knowing the private key under an encryption scheme with homomorphic properties. For a long time thereafter, partially homomorphic encryption, where only one of the operations of addition and multiplication can be performed on the ciphertext, developed by leaps and bounds, e.g., RSA and ElGamal are multiplicative homomorphic encryption schemes, and Paillier is an additive homomorphic encryption scheme. But there has been no breakthrough in homomorphic encryption schemes that can support both additive and multiplicative operations. It was not until 2009 that Gentry [21] first proposed a fully homomorphic encryption scheme based on an ideal lattice, which allowed anyone without a private key to perform any valid computable function on the encrypted data. According to different construction ideas, fully homomorphic encryption can be roughly classified into three categories: the first category is the fully homomorphic schemes constructed based on the hard problem on an ideal lattice, which is represented by the scheme proposed by Gentry in [21] and its improvement [22]; the second category is the fully homomorphic scheme constructed based on the (R)LWE problem, which is represented by the scheme proposed by Brakershi et al. [23,24], which has improved efficiency compared to the first category, and the fully homomorphic encryption proposed in [14,15] used in this paper belongs to this category; the third category is the fully homomorphic encryption scheme that does not require any key exchange, and this category of schemes is represented by the scheme proposed by Gentry et al. in [25]. As we all know, fully homomorphic encryption is more secure but inefficient compared to partially homomorphic encryption. The efficiency of fully homomorphic encryption has been greatly improved in recent years, such as GSW encryption [25] and BGV encryption [24] in the second and third categories; both of them are more efficient fully homomorphic encryption schemes, with an encryption time reaching the *ms* level. However, this time-consuming aspect of the control system cannot be ignored. The two methods we propose improve efficiency and ensure that the security of the program is not compromised. In terms of efficiency, the time spent on encryption is completely eliminated compared to existing encryption schemes. This results in a significant increase in efficiency compared to existing fully homomorphic encryption schemes. In terms of security, the table used for encryption is constantly updated, so the scheme still maintains the high security of homomorphic encryption.

The above is the research work on homomorphic encryption, and with the rise of NCSs, there has been a focus on using homomorphic encryption as a tool to improve the security of networked control systems. Homomorphic encryption was first used for NCSs in [5], where two partially homomorphic encryption schemes, RSA and ElGamal, were used in the method. Paillier was subsequently proposed to be used for NCSs. Recent studies [26–29] have proposed many ways to further improve and optimize these schemes, such as maintaining stability and performance. Among them, ref. [26] proposed to update the key pair and ciphertext by simple update rules and modulo operations at each sampling cycle, which brought some inspiration and reference to this paper. In this paper, we apply in-cycle updating of plaintext–ciphertext pairs in a fully homomorphic encrypted controller scheme to improve the security of the scheme.

The application of fully homomorphic encryption to NCSs has been late and rare because of efficiency issues in real-world applications. Ref. [13] considers the application of fully homomorphic encryption to NCSs to alleviate the extra overhead and quantization errors caused by quantization recovery. Subsequently, ref. [16] proposed to use a non-strictly fully homomorphic encryption scheme for encryption and performed optimization. We refer to the method of [16] and propose a new fully homomorphic encryption scheme. The scheme performs well in terms of security and efficiency compared to existing schemes using fully homomorphic encryption. Specifically, the security is comparable to [13] and much higher than [16], and the efficiency is much higher than [13,16]. In addition, compared with [13], our scheme does not require multiple controllers, so the control system is simpler, which is more favorable for applications in practice. In addition to this, another popular control scheme involving optimization is model predictive control, and [30–32] consider a model predictive control scheme for related linear systems. For some of the current challenges, ref. [33] outlines them accordingly. However, since our scheme already involves a large amount of computation, we do not use this technique in this paper, but in the future, we will consider using model predictive control for our scheme.

3. Preliminaries

The paper makes use of the following notions.

In this paper, we use bold uppercase letters (e.g., $\mathbf{A}, \mathbf{B}$) to denote matrices and, similarly, lowercase letters (e.g., $\mathbf{a}, \mathbf{b}$) to denote column vectors. We use $\mathbb{R}$ to denote the set of real numbers; thus, $z \in \mathbb{R}$ is a real number. If z_1 is the closest integer to z, then we denote z_1 by $\lfloor z \rceil$, which means it is the only integer in the half open interval $(z - 1/2, z + 1/2]$.

We identify $\mathbb{Z} \bigcap (-q/2, q/2]$ as a representation of $\mathbb{Z}_q$ for an integer q and use $[z]_q$ or $r_p(z)$ to indicate the interval into which the integer z modulo q is reduced. To represent the sampling of x based on a distribution D, we use the notation $x \leftarrow D$. When D is a finite set, it means sampling from the uniform distribution over D.

Along with the explanations of the aforementioned symbols, we also provide some definitions of the fundamental terms that will be used throughout the remainder of this paper.

3.1. RLWE Problem

The RLWE problem is the underlying mathematically difficult problem of securing cryptographic methods. Before introducing the RLWE problem, it is necessary to familiarize oneself with some of the notations in the definition that follows.

Let $\Phi_M(X)$ be the Mth cyclotomic polynomial of degree $N = \phi(M)$ for a positive integer M. Let $R = \mathbb{Z}[X]/(\Phi_M(X))$ be the ring of integers in the $Q[X]/(\Phi_M(X))$ number field. For the residue ring of R modulo an integer q, we write $R_q = R/qR$. We write $R_q^\vee = R^\vee/qR^\vee$, where $R^\vee$ is the dual fractional ideal of R. For a positive integer modulus of $q \geq 2$, $s \in R_q^\vee$, $r \in (\mathbb{R}^+)^N$, and an error distribution of $\chi := \lfloor \Psi_r \rceil_{R^\vee}$.

Definition 1 ([34]). *(Ring learning with errors (RLWE) distribution) We define $A_{N,q,\chi}(s)$ as the* **RLWE** *distribution that is formed by uniformly sampling $a \leftarrow R_q$ at random, $e \leftarrow \chi$ and returning $(a, a \cdots + e) \in R_q \times R_q^\vee$.*

Definition 2 ([35]). *((Decision) RLWE) The (decision) RLWE, denoted by* $\mathbf{RLWE}_{N,q,\chi}(\mathcal{D})$, *is the problem of distinguishing arbitrarily many independent samples chosen according to $A_{N,q,\chi}(s)$ for a random choice of s sampled from the distribution $\mathcal{D}$ over $R^\vee$ from the same number of uniformly random and independent samples from $R_q \times R_q^\vee$.*

3.2. Fully Homomorphic Encryption

Fully homomorphic encryption plays an important role in this paper, and it is defined as follows.

Definition 3. *A fully homomorphic encryption scheme* **FHE** = (**Gem**, **Enc**, **Dec**, **Eval**) *is described as follows:*

- $\mathbf{Gen}(1^\lambda) \to (pk, sk, evk)$: *input security parameter* λ, *output* (pk, sk, evk) *where pk is a public key, sk is a secret key, and evk is an evaluation key.*
- $\mathbf{Enc}(m, pk) \to c$: *input message m and public key pk, output ciphertext c.*
- $\mathbf{Dec}(c, sk) \to m'$: *input ciphertext c and secret key sk, compute and output the plaintext* m'.
- $\mathbf{Evaluation}(f, c_1, c_2, \cdots, c_N, evk) \to c^*$: *input a set of ciphertexts* $(c_1, c_2, \cdots, c_N)$, *function f, and evaluation key evk. Compute and output the evaluation ciphertext* c^*.

Fully homomorphic encryption can be performed on data in the form of ciphertexts of arbitrary complexity, which we describe more intuitively in Figure 2 below. A set of data $(m_1, m_2, \cdots, m_N)$ is encrypted with the algorithm **Enc** to obtain a set of ciphertexts $(c_1, c_2, \cdots, c_N)$, and an arbitrary computation f is performed on the ciphertext set to obtain c^*. The value of c^* after decryption should be the same as the value of the calculation done directly on the plaintext.

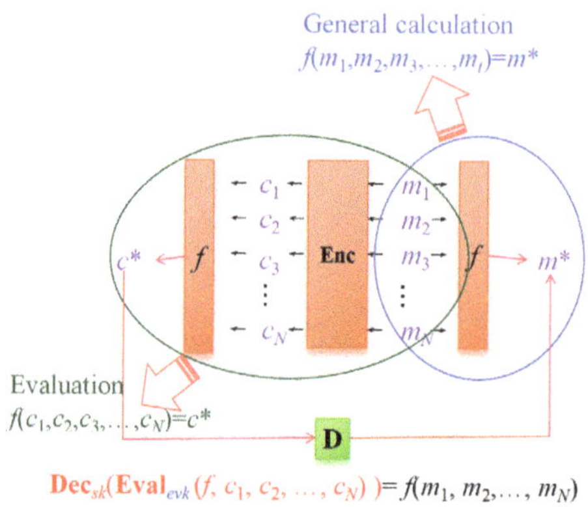

Figure 2. Graphical representation of fully homomorphic encryption

3.3. BFV Encryption

It is assumed that the security-parameter-related noise distribution χ is a discrete Gaussian distribution on the ring R, and that the uniform random noise distribution χ' is also on the ring R. The seven probabilistic polynomial time (PPT) algorithms (**SecretKeyGen**, **PublicKeyGen**, **EvaluateKeyGen**, **Enc**, **Dec**, **Add**, **Mult**) used in BFV encryption are as follows:

- $\mathbf{SecretKeyGen}(1^\lambda)$: input security papameter λ, sample $\mathbf{s} \leftarrow R_2$, and output secret key noted as $sk = \mathbf{s}$.
- $\mathbf{PublicKeyGen}(sk)$: input secret key, sample $\mathbf{a} \leftarrow R_q$, $\mathbf{e} \leftarrow \chi$, and output public key

$$pk = ([-(\mathbf{a} \cdot \mathbf{s} + \mathbf{e})]_q, \mathbf{a}).$$

- **EvluateKeyGen**:
  - Version 1: parameters (sk, T): for $i = 0, \cdots, l = \lfloor log_T(q) \rfloor$, sample $\mathbf{a_i}, R_q, \mathbf{e_i} \leftarrow \chi$, perform the following operation, and return

$$rlk = [([-(\mathbf{a_i} \cdot \mathbf{s} + \mathbf{e}_i) + T^i \cdot \mathbf{s}^2]_q, \mathbf{a}_i) : i \in [0 \cdots l]].$$

  - Version 2: parameters (sk, p): sample vectors $\mathbf{a} \leftarrow R_{p \cdot q}$, $\mathbf{e} \leftarrow \chi'$, and then return

$$rlk = ([-(\mathbf{a} \cdot \mathbf{s} + \mathbf{e} + p \cdot \mathbf{s}^2]_{p \cdot q}, \mathbf{a}).$$

- $\mathbf{Enc}(pk, \mathbf{m})$: to encrypt a message $m \in R_t$. We set $\mathbf{p}_0 = pk[0], \mathbf{p}_1 = pk[1]$ and sample $\mathbf{u} \in R_2, \mathbf{e}_1, \mathbf{e}_2 \in \chi$, then return the final ciphertext

$$ct = ([\mathbf{p}_0 \cdot \mathbf{u} + \mathbf{e}_1 + \Delta \cdot \mathbf{m}]_q, [\mathbf{p}_1 \cdot \mathbf{u} + \mathbf{e}_2]_q).$$

- **Dec**(sk, ct): set $\mathbf{c}_0 = ct[0], \mathbf{c}_1 = ct[1]$ and compute and output the result of the decryption

$$m' = [\lfloor \frac{t \cdot [\mathbf{c}_0 + \mathbf{c}_1 \cdot \mathbf{s}]_q}{q} \rceil]_t.$$

- **Add**(ct_1, ct_2): return

$$([ct_1[0] + ct_2[0]]_q, [ct_1[1] + ct_2[0]]_q),$$

  where $ct_1 = \mathbf{Enc}(pk, \mathbf{m_1}), ct_2 = \mathbf{Enc}(pk, \mathbf{m_2})$.
- **Mult**(ct_1, ct_2): compute

$$\mathbf{c}_0 = [\lfloor \frac{t \cdot (ct_1[0] \cdot ct_2[0])}{q} \rceil]_q$$

$$\mathbf{c}_1 = [\lfloor \frac{t \cdot (ct_1[0] \cdot ct_2[1] + ct_1[1] \cdot ct_2[0]}{q} \rceil]_q$$

$$\mathbf{c}_2 = [\lfloor \frac{t \cdot (ct_1[1] \cdot ct_2[1])}{q} \rceil]_q$$

  - Relinearization version 1: rewrite $\mathbf{c}_2$ equivalently to be based on T, i.e., write $\mathbf{c}_2 = \sum_{i=0}^{l} \mathbf{c}_2^{(i)} T^i$ with $\mathbf{c}_2^{(i)} \in R_T$ and set

$$\mathbf{c}_0' = [\mathbf{c}_0 + \sum_{i=0}^{l} rlk[i][0] \cdot \mathbf{c}_2^{(i)}]_q \text{ and } \mathbf{c}_1' = [\mathbf{c}_1 + \sum_{i=0}^{l} rlk[i][1] \cdot \mathbf{c}_2^{(i)}]_q.$$

    Return $(\mathbf{c}_0', \mathbf{c}_1')$.
  - Relinearization version 2: compute

$$(\mathbf{c}_{2,0}, \mathbf{c}_{2,1}) = ([\lfloor \frac{\mathbf{c}_2 \cdot rlk[0]}{p} \rceil]_q, [\lfloor \frac{\mathbf{c}_2 \cdot rlk[1]}{p} \rceil]_q),$$

    and return $([\mathbf{c}_0 + \mathbf{c}_{2,0}]_q, [\mathbf{c}_1 + \mathbf{c}_{2,1}]_q)$.

For a better understanding, the following Figure 3 represents the whole process of fully homomorphic encryption.

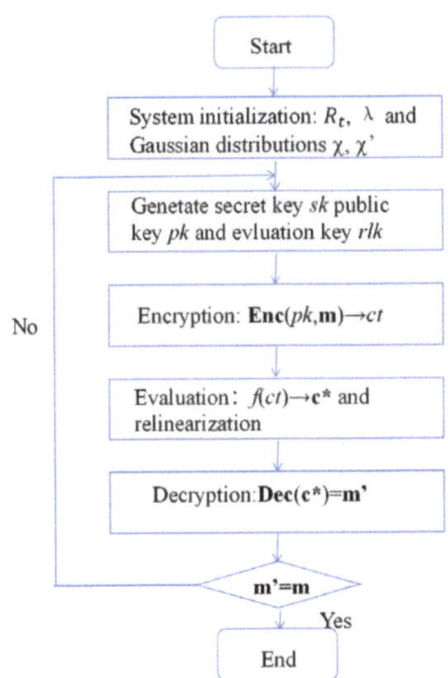

Figure 3. The process of fully homomorphic encryption.

3.4. Encrypted Controller

The discrete-time linear controller case that is under consideration in this work is summarized in the following form:

$$f : \begin{cases} \mathbf{x}(t+1) = \mathbf{A}\mathbf{x}(t) + \mathbf{B}\mathbf{y}(t) \\ \mathbf{u}(t) = \mathbf{C}\mathbf{x}(t) + \mathbf{D}\mathbf{y}(t), \end{cases} \tag{1}$$

where $\mathbf{y}(t) \in \mathbb{R}^m$ is a controller input (or a plant output), $\mathbf{u}(t) \in \mathbb{R}^l$ is a controller output (or a plant input), and t is a step. $\mathbf{A}, \mathbf{B}, \mathbf{C}$, and $\mathbf{D}$ are controller parameter values. The following is an equivalent rewriting of Equation (1):

$$\begin{bmatrix} \mathbf{x}(t+1) \\ \mathbf{u}(t) \end{bmatrix} = f(\mathbf{\Phi}, \xi(t)) = \mathbf{\Phi}\xi(t), \tag{2}$$

where the parameter $\mathbf{\Phi}$ and the input ξ are represented in the following form:

$$\mathbf{\Phi} := \begin{bmatrix} A & B \\ C & D \end{bmatrix} \in \mathbb{R}^{\alpha \times \beta}, \qquad \xi := \begin{bmatrix} x \\ y \end{bmatrix} \in \mathbb{R}^{\beta}.$$

with $\alpha := n + l$ and $\beta := n + m$.

Definition 4 ([27]). *For an NCS, we assume that given a linear controller f in (1) for an NCS, the controller's input y and output u are encrypted using the encryption algorithm $E = (\mathbf{Gen}, \mathbf{Enc}, \mathbf{Dec})$. If a map f_E exists such that the equation*

$$f_E(\mathbf{Enc}(k_p, \overline{\mathbf{\Phi}}), \mathbf{Enc}(k_p, \overline{\xi})) = \mathbf{Enc}(k_p, \overline{f}(\mathbf{\Phi}, \xi)) \tag{3}$$

holds, then we call f_E the encrypted controller of f. Here, $\overline{\mathbf{\Phi}} \in \mathcal{M}^{\alpha \times \beta}$, $\overline{\xi} \in \mathcal{M}^{\beta}$, and $\overline{f}(\cdot) \in \mathcal{M}^{\alpha}$ are the plaintexts, rounded to ensure that each component can be represented as an element of the information space.

4. Control System with Encrypted Controller

In this section, we encrypt the controller using the BFV encryption scheme to obtain Scheme 1, and we precompute to save time. Precomputation speeds up the encrypted control system's operation and reduces the amount of time required for encryption. However, the appearance of precomputation changes the underlying encryption algorithm from random encryption to deterministic encryption, which reduces the security of the BFV scheme. As a result, we suggest two schemes to strengthen scheme 1's security from two different angles.

4.1. Encrypted Controller Using BFV Encryption Scheme

First, we encrypt the controller using the BFV encryption method and follow the method in the literature [16] to adopt precomputation to save the time of encryption. We describe the process for this scheme based on Figure 4 in the following. We have drawn the flowchart as Figure 5 to help understand.

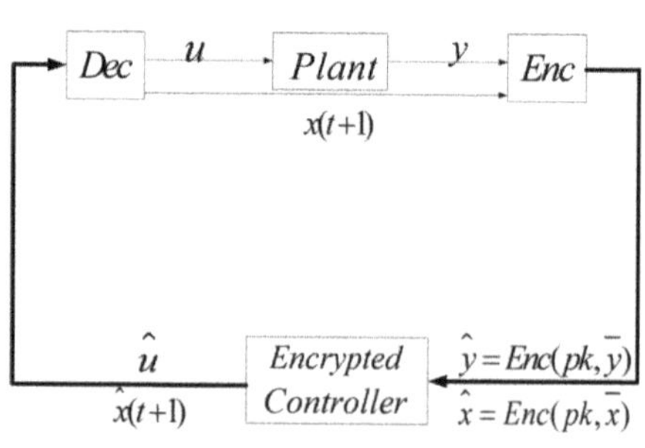

Figure 4. The schematic diagram of a networked control system with BFV encryption.

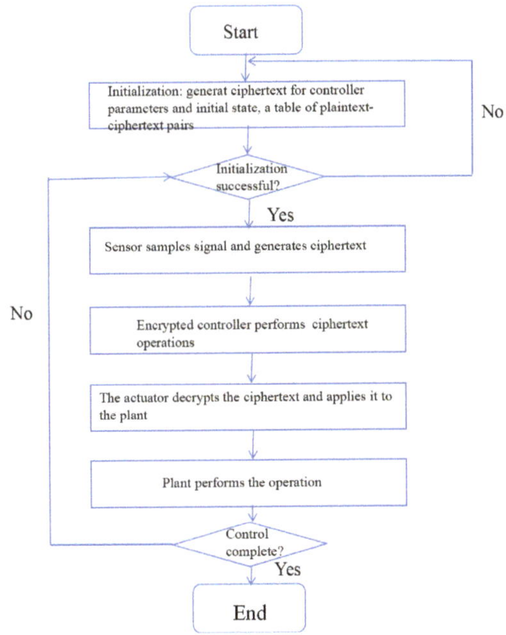

Figure 5. The flowchart of a networked control system with BFV encryption.

Scheme 1:

We describe each step in detail based on the above flowchart:

- The parameters $\boldsymbol{A}, \boldsymbol{B}, \boldsymbol{C}$, and $\boldsymbol{D}$ as well as the controller's initial state $\boldsymbol{x}(0)$, are encrypted to produce its ciphertext $\hat{\boldsymbol{A}}, \hat{\boldsymbol{B}}, \hat{\boldsymbol{C}}, \hat{\boldsymbol{D}}, \hat{\boldsymbol{x}}(0)$ in the controller's design, and this ciphertext is transmitted to the controller.
- The sensor collects the signal $\boldsymbol{y}$ and then uses the plaintext index to look up the table generated by precomputation to obtain the corresponding ciphertext $\hat{\boldsymbol{y}} = \mathbf{Enc}(\boldsymbol{y})$, thus realizing the encryption process, which is then passed to the encrypted controller.
- The encryption controller performs homomorphic operations after obtaining the ciphertext signal $\hat{\boldsymbol{y}}$. The BFV homomorphism operation states that (1) really operates in the cipher space after being encrypted as:

$$\begin{aligned} \hat{\boldsymbol{x}}(t+1) &= \hat{\boldsymbol{A}} \times \hat{\boldsymbol{x}}(t) + \hat{\boldsymbol{B}} \times \hat{\boldsymbol{y}}(t) \\ \hat{\boldsymbol{u}}(t) &= \hat{\boldsymbol{C}} \times \hat{\boldsymbol{x}}(t) + \hat{\boldsymbol{D}} \times \hat{\boldsymbol{x}}(t) \end{aligned} \tag{4}$$

  The ciphertext $\hat{\boldsymbol{x}}(t+1)$ of the state and the ciphertext $\hat{\boldsymbol{u}}(t)$ of the output are passed to the actuator after the homomorphic operation is completed.
- The actuator block decrypts the controller's output cipher to obtain $\boldsymbol{u}(t)$ and applies it to the plant, decrypts the $\hat{\boldsymbol{x}}(t+1)$ to obtain the state $\boldsymbol{x}(t+1)$ and passes it to the sensor.
- The state $\boldsymbol{x}(t+1)$ is encrypted by the sensor and sent to the encrypted controller.

Remark 1. *(About the table generated by the precomputation)*

1. *Based on the plant, the control function, and the initial condition, we can identify the range of* $\boldsymbol{x}$ *and* $\boldsymbol{y}$.
2. *We run the control system, create plaintext–ciphertext pairs* $(m, \hat{m})$ *by encryption, and place them in a table. The final generated table contains all pairs between* $\boldsymbol{y}$ *and* $\boldsymbol{x}$.
3. *The table can determine the corresponding ciphertext when the sensor gathers the value* $\boldsymbol{y}$ *of the plant and encrypts it.*

Remark 2. *(About state $\mathbf{x}(t)$) To avoid the problem that the noise of the ciphertext $\mathbf{x}(t)$ increases significantly after homomorphic addition and homomorphic multiplication in the encrypted controller, the following measures are taken. The ciphertext of state $\mathbf{x}(t+1)$ after homomorphic encryption is sent to the actuator for decryption and then returned to the cryptographic controller for a new round of homomorphic computation after sensor encryption.*

With regard to Figure 5, we explain the illustration in detail by means of the following example. We want to control a conveyor belt to move a table through the control system. Then, at the beginning, we need to initialize the settings. The transmission speed has an upper limit, so we generate a plaintext–ciphertext table for the range of the speed. After successful initialization, we can start the subsequent operation. Suppose we input speed v; the axis begins to rotate and moves the table. At this point, the following workflow begins. The sensor obtains the axis rotation speed v_1, consults the table to generate the ciphertext value c_{v_1}, and sends it to the encrypted controller. The encrypted controller carries out operations in the form of ciphertext to generate new states and signals. Then, the ciphertext of signals is output to the actuator to decrypt and control the axis to accelerate or decelerate its rotation. The cycle repeats itself until the completion of the transmission of the table.

In scheme 1, the controller input and parameters are ciphertext since the encrypted controller allows homomorphic additions and homomorphic multiplications. As shown in the Figure 4, in the whole control process, from the sensor sampling data and encrypting until the actuator decrypts the ciphertext, the data in the transmission channel and encrypted controller are all ciphertext, which greatly improves the security compared with the previous partially homomorphic encrypted controller. The table is generated in advance, so that the ciphertext can be obtained only by looking up the table according to the plaintext index, saving the time of encryption. However, we know that BFV encryption is a random encryption scheme, and the precomputation causes the search table to obtain the same ciphertext from a plaintext m, that is to say, from random encryption to deterministic encryption. This process reduces the security of the scheme. Therefore, we propose two approaches below to remedy this deficiency.

4.2. Security Enhancement

In the previous section, we saw that although the precomputation apparently improves the efficiency, it also reduces the security of the scheme to a certain extent. Therefore, we propose two approaches below to solve this problem from two aspects. In the following scheme, we use the method of periodically updating the table to enhance security.

We just present a general idea here; the control system's average computation time for each iteration is provided in Section 5 below. Additionally, specific values may be substituted.

4.2.1. Periodic Update Table

The two methods we propose to improve security have no difference with scheme 1 in the general process framework, except for the specific operations in the second step related to obtaining the ciphertext in the second step. Since each ciphertext has only one fixed corresponding ciphertext in a table, we consider updating the table generated by the precalculation regularly. The most intuitive way to solve this problem is to ensure that each cycle of the control system has a new table, but in general, the control system takes much longer to compute each iteration than it takes to generate an estimated table. In this way, to complete the table update, the computing power of the precalculation process must be greatly improved, and the precalculation time must be guaranteed to be less than the control cycle. In this paper, we do not consider excessive requirements on the hardware of the control system; we hope to complete the security improvement through a "natural" method. Therefore, we do the next best thing and consider updating the table regularly.

First, we assume that the control system in scheme 1 takes approximately a ms per iteration to calculate, while the time spent on generating a precomputed encryption table is about b ms. Therefore, a table can be updated after about b/a iterations of the control system. In order to ensure that the table can be updated, we set an update every $([b/a]+1)$ cycle.

4.2.2. Provide Multiple Tables

From another perspective, if we randomly select one of the encrypted tables to search for plaintext–ciphertext pairs each time, the problem that there is only one ciphertext value corresponding to the plaintext m can be avoided.

We assume that α tables are generated when the plaintext–ciphertext pairs are generated in the initial precomputation, and a table is randomly selected from α tables and then retrieved according to the plaintext index during each encrypted table lookup. To prevent excessive storage burden, the value of α cannot be too large. But if α is too small, randomness is not enough. At the same time, we consider incorporating the idea of Section 4.2.1 into it and completing the update of a table after b/a iterations of the control system. It may take many cycles to complete the update of α tables, but this is not important, because each encrypted table lookup is a random table selected from α. This practice only further enhances the security of the scheme on this basis.

4.3. Attack Scenario and Security

NCSs are at risk of eavesdropping attacks because plants and controllers communicate with each other over network links. Our proposed network control system with encrypted controllers is well protected against eavesdropping attacks, and we briefly describe it here. We consider the following attack scenarios.

In our model, the attacker $\mathcal{A}$ has mainly the following described capabilities.

1. Adversary $\mathcal{A}$ can collect data within the communication channel through an eavesdropping attack.
2. Adversary $\mathcal{A}$ can collect data within the controller through an eavesdropping attack.

Note: In addition to the capabilities listed above, the decryptor, encryptor, and actuator of the control system cannot be compromised by an attacker $\mathcal{A}$.

We say that the scheme is not resistant to eavesdropping attacks if the attacker $\mathcal{A}$ can obtain the controller parameters A, B, C, and D or signals y in polynomial time; otherwise, we say that it is eavesdropping-resistant.

The security of the control system in this attack scenario is analyzed. Attacker $\mathcal{A}$ collects data in the controller and communication channel by eavesdropping. In this scenario, the data in the controller and the data in the communication channel are in the form of ciphertext, which is encrypted using the BFV encryption scheme. In order to obtain useful data, the attacker $\mathcal{A}$ needs to reduce the ciphertext to plaintext. The BFV encryption scheme is based on the difficult problem of RLWE and hence cannot obtain useful plaintext data in polynomial time. Therefore, our scheme is resistant to eavesdropping attacks.

4.4. Comparison of Four Schemes

In terms of safety and efficiency, we contrast the scheme proposed in this study with a number of traditional schemes that have been previously offered. For their specific processing time, we refer to the literature [16], and the average processing time in this paper will be given in the next section.

It can be seen from Table 1 that in the previous partially homomorphic encryption schemes, the security of data transmitted in the channel during the process from sensor to actuator and data in the controller cannot be ensured at the same time. This suggests that security is not ideal. Subsequent BGN encryption schemes can well avoid this problem. The BFV encrypted controller scheme proposed in this paper and BGN encrypted controller can ensure that the data inside the controller and in the transmission channel are ciphertext. In terms of efficiency, the two partially homomorphic encryption schemes are efficient, and BGN is relatively inefficient. But this time can also be suitable for the control system's sampling cycle. However, by accelerating precomputation, the scheme using the BFV technique suggested in this study achieves an efficiency that is almost identical to partially homomorphic encryption. On the other hand, considering the homomorphic operation inside the controller, the homomorphic multiplication of BGN can only be performed once, but partially homomorphic encryption and the BFV scheme proposed in this paper do not have this defect.

Table 1. Analysis of three control systems.

	Data Security	Homomorphic Operation	Average Processing Time [1]
With RSA encryption	Only the data inside the controller	Many times	$(10+a)$ [2] ms
With Paillier encryption	Only the data in the channel	Many times	16 ms
With BGN encryption	All data are secure	Add multiple times and multiply once	96 ms
With BFV encryption	All data are secure	Many times	24 ms

[1] The data are obtained with a key length of 512 bits; [2] "a" represents the additional communication time required.

Through the above analysis, the BFV scheme is excellent in both security and efficiency.

5. Numerical Example

In this section, we first give a concrete numerical example of a control system, then simulate with our scheme to obtain a series of results, and subsequently analyze the obtained graphs to support our scheme.

The control system is made up of the following discrete-time linear plant and the following kind of linear controller, according to the numerical example in [5]. $p_1(t)$ and $p_2(t)$ are the internal states of the plant, and they satisfy:

$$\begin{aligned} \begin{bmatrix} p_1(t+1) \\ p_2(t+1) \end{bmatrix} &= \begin{bmatrix} 0.99998 & 0.0197 \\ -0.0197 & 0.97025 \end{bmatrix} \begin{bmatrix} p_1(t) \\ p_2(t) \end{bmatrix} + \begin{bmatrix} 0.0000999 \\ 0.0098508 \end{bmatrix} u(t), \\ y(t) &= \begin{bmatrix} 1 & 0 \end{bmatrix} \begin{bmatrix} p_1(t) \\ p_2(t) \end{bmatrix}, \end{aligned} \tag{5}$$

where the initial states are $p_1(0) = 1$ and $p_2(0) = 0$, and the linear controller's internal states are $x_1(t)$ and $x_2(t)$, satisfying:

$$\begin{aligned} \begin{bmatrix} x_1(t+1) \\ x_2(t+1) \end{bmatrix} &= \begin{bmatrix} 1 & 0.0063 \\ 0 & 0.3678 \end{bmatrix} \begin{bmatrix} x_1(t) \\ x_2(t) \end{bmatrix} + \begin{bmatrix} 0 \\ 0.0063 \end{bmatrix} y(t), \\ u(t) &= \begin{bmatrix} 10 & -99.9 \end{bmatrix} \begin{bmatrix} x_1(t) \\ x_2(t) \end{bmatrix} - 3y(t), \end{aligned} \tag{6}$$

where the initial states are $x_1(0) = 0$ and $x_2(0) = 0$.

Numerical Results

The BFV encryption was implemented through Microsoft's Simple Encrypted Arithmetic Library (SEAL). The following diagram was obtained by calling SEAL to simulate the encryption of a specific number of cases.

Figure 6 shows the simulation results corresponding to the time of input y and output u. The control input response shows some minor quantization errors, but such quantization errors are so small that they can be ignored. As can be seen from Figure 6, the closed-loop system's control performance and stability can be realized with the help of the BFV encryption controller.

Figure 7 depicts the time change for computing the iterations of the controlled system following the BFV encryption of the controller. Figure 7a represents the calculation time of each iteration of the control system without BFV encryption using precomputation, with an average time of 32.40 ms; Figure 7b represents the time after precomputation, with an average time of 23.99 ms. As can be seen from the comparison of the two pictures, it is estimated that about 35% of the time will be saved, which is still considerable.

The suggested encryption control system's histogram of ciphertext is displayed in Figure 8. It can be assumed that the ciphertext in the suggested cryptosystem follows a discrete uniform distribution because the histogram distribution is nearly flat.

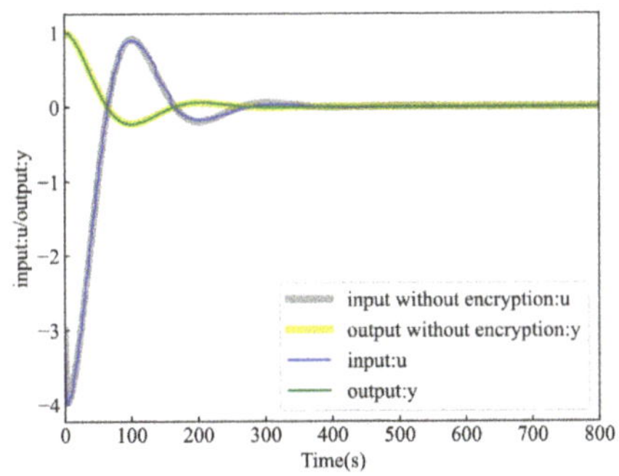

Figure 6. Comparison of output/input with and without the proposed cybersecurity enhancement: BFV encryption.

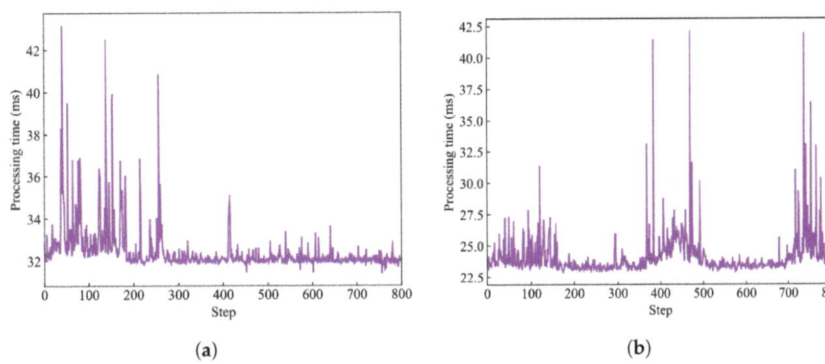

Figure 7. Time variation at each iteration calculation by the encrypted controller, (**a**)without precomputation; (**b**)with precomputation.

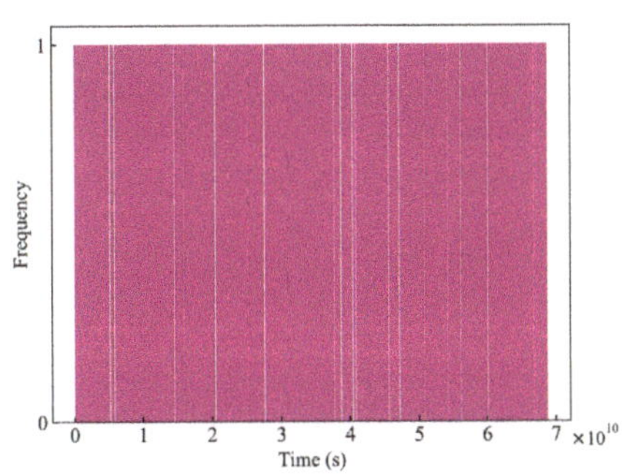

Figure 8. Histogram of the first element of the controller gain in ciphertext.

6. Discussion

In this paper, we proposed a scheme for encrypting the controller using fully homomorphic encryption. We verified that this encrypted controller could achieve a normal control and that the efficiency and security of the encrypted controller were improved using numerical examples in Section 5.

The study in this paper has the following implications. First, in terms of security, our scheme ensures that both the data in the controller and in the channel are not eavesdropped on, so the data security in the whole network control system is improved. This compensates for the lack of security in previous homomorphic encrypted controller schemes [5,13,16]. Second, in terms of efficiency, we use precomputation to alleviate the latency problem caused by fully homomorphic encryption, which reduces the iteration time of the control system and also improves the response time of the networked control system with encrypted controllers. Finally, our scheme has a simple control system and does not require more encrypted controllers compared to the scheme proposed by Kim et al. [13].

This paper also suffers from the following limitations. On the one hand, our scheme places some demands on the computational power of the device, as it requires constant computation to generate tables for encrypted access. On the other hand, homomorphic encryption, especially fully homomorphic encryption, is still difficult to apply in realistic scenarios. Only simulation results are considered in this paper to validate the scheme, which may be problematic in further practical applications specifically. Related issues will be further investigated in future work.

7. Conclusions and Future Work

7.1. Conclusions

In this paper, we proposed a scheme which effectively improved the security of an NCS by encrypting the controller using fully homomorphic encryption. Specifically, all data in the system could be well secured from eavesdropping and recording. We further reduced the time spent on encryption in the scheme by precomputation and improved the efficiency of the encryption controller. In addition, for the security of the network control system, we further proposed two methods to improve the security. An efficient and secure NCS is of great practical significance.

7.2. Future Work

The scheme proposed in this paper can provide some technical guarantee for the data security of NCSs, but this scheme needs to be improved continuously, and in the following aspects, further research needs to be conducted. Firstly, the scheme proposed in this paper is still only in the simulation stage, and further research is needed for future consideration of applications in a real environment. Secondly, in practice, the shorter the iteration time of the control system, the better; therefore, further improvement in efficiency or the design of more efficient schemes should be considered in the future.

Author Contributions: Conceptualization, J.P.; methodology, J.P.; software, T.S.; validation, L.K. and Y.Z.; formal analysis, W.L.; investigation, T.S. and J.W.; resources, J.P.; data curation, T.S.; writing—original draft preparation, J.P.; writing—review and editing, W.L. and Z.W.; supervision, J.W.; funding acquisition, Z.W. All authors have read and agreed to the published version of the manuscript.

Funding: This work was supported in part by the National Key Research and Development Program of China (No. 2023YFF0905300), the Key R&D Plan of Shandong Province (No. 2022CXGC020106), Major Innovation Project of Science, Education and Industry of Shandong Academy of Sciences (No. 2022JBZ01-01), Innovation Capacity Improvement Project for Small and Medium-sized Technology-based Enterprises of Shandong Province (No. 2023TSGC0641).

Data Availability Statement: Data is contained within the article.

Conflicts of Interest: The authors declare no conflict of interest.

References

1. Yohanandhan, R.V.; Elavarasan, R.M.; Manoharan, P.; Mihet-Popa, L. Cyber-Physical Power System (CPPS): A Review on Modeling, Simulation, and Analysis with Cyber Security Applications. *IEEE Access* **2020**, *8*, 151019–151064. [CrossRef]
2. Yampolskiy, M.; Andel, T.R.; Mcdonald, J.T.; Glisson, W.B.; Yasinsac, A. Intellecutal protection in additive layer manufacturing: Requirements for secure outsouring. In Proceedings of the 4th Program Protection and Reverse Engineering Workshop, New Orleans, LA, USA, 9 December 2014.
3. Wall, D.S.; Yar, M. Intellecutal property crime and the Internet: Cyber-piracy and 'stealing' information intangibles. In *Handbook of Internet Crime*; Wall, D.S., Yar, M., Eds.; Willan: London, UK, 2010; pp. 255–272.
4. Mclaughlin, S. On dynamic malware payloads aimed at programmable logic controllers. In Proceedings of the 6th USENIX Conference on Hot Topics in Security, San Francisco, CA, USA, 9 August 2011; p. 10.
5. Kogiso, K.; Fujita, T. Cyber-Security Enhancement of Networked Control Systems Using Homomorphic Encryption. In Proceedings of the IEEE Conference on Decision and Control, Osaka, Japan, 15–18 December 2015.
6. Rivest, R.L.; Shamir, A.; Adleman, L. A method for obtaining digital signatures and public-key cryptosystem. *Commun. ACM* **1978**, *21*, 120–126. [CrossRef]
7. ElGamal, T. A public key cryptosystem and a signature scheme based on discrete logarithms. *IEEE Trans. Inf. Theory* **1985**, *31*, 469–472. [CrossRef]
8. Farokhi, F.; Shames, I.; Batterham, N. Secure and private control using semi-homomorphic encryption. *Control Eng. Pract.* **2017**, *67*, 13–20. [CrossRef]
9. Lin, Y.; Farokhi, F.; Shames, I.; Nesic, D. Secure control of nonlinear systems using semi-homomorphic encryption. In Proceedings of the IEEE Conference on Decision and Control, Miami, FL, USA, 17–19 December, 2018.
10. Murguia, C.; Farokhi, F.; Shames, I. Secure and private implementation of dynamic controllers using semi-homomorphic encryption. *IEEE Trans. Autom. Control* **2020**, *65*, 3950–3957. [CrossRef]
11. Kosieradzki, S.; Zhao, X.; Kawase, H.; Qiu, Y.; Kogiso, K.; Ueda, J. Secure teleoperation control using somewhat homomorphic encryption. *IFAC-PapersOnLine* **2020**, *55*, 593–600. [CrossRef]
12. Paillier, P. Public-key cryptosystems based on composite degree residuosity classes. In *Advances in Cryptology—Eurocrypt'99, Proceedings of the International Conference on the Theory and Application of Cryptographic Techniques, Prague, Czech Republic, 2–6 May 1999*; Stern, J., Ed.; Springer: Berlin/Heidelberg, Germany, 1999; pp. 223–238.
13. Kim, J.; Lee, C.; Shim, H.; Cheon, J.H.; Kim, A.; Kim, M.; Song, Y. Encrypting controller using fully homomorphic encryption for security of cyber-physical systems. *IFAC-PapersOnLine* **2020**, *49*, 175–180. [CrossRef]
14. Brakerski, Z. Fully Homomorphic Encryption without Modulus Switching from Classical GapSVP. 2012. Available online: https://eprint.iacr.org/2012/078 (accessed on 13 June 2023).
15. Fan, J.; Vercauteren, F. Somewhat Practical Fully Homomorphic Encryption. Cryptology Eprint Archive. 2012. Available online: https://eprint.iacr.org/2012/144 (accessed on 9 June 2023).
16. Pan, J.; Sui, T.; Liu, W.; Wang, J.; Kong, L.; Zhao, Y. Secure Control Using Homomorphic Encryption and Efficiency Analysis. *Secur. Commun. Netw.* **2023**, *2023*, 6473497 . [CrossRef]
17. Wang, D.; Li, W.; Wang, P. Measuring Two-Factor Authentication Schemes for Real-Time Data Access in Industrial Wireless Sensor Networks. *IEEE Trans. Ind. Inform.* **2018**, *14*, 4081–4092. [CrossRef]
18. Wang, Q.; Wang, D. Understanding Failures in Security Proofs of Multi-Factor Authentication for Mobile Devices. *IEEE Trans. Inf. Forensics Secur.* **2022**, *18*, 597–612. [CrossRef]
19. Yu, Y.; Xu, G.; Wang, X. Provably Secure NTRU Instances over Prime Cyclotomic Rings. In *Public-Key Cryptography—PKC 2017, Proceedings of the 20th IACR International Conference on Practice and Theory in Public-Key Cryptography, Amsterdam, The Netherlands, 28–31 March 2017*; Lecture Notes in Computer Science; Fehr, S., Ed.; Springer: Berlin/Heidelberg, Germany, 2017; pp. 409–434.
20. Rivest, R.; Adleman, L.; Deryouzos, M. On data banks and privacy homomorphisms. In *Foundations of Secure Computation*; Academic Press, Inc.: Orlando, FL, USA, 1978; pp. 169–180.
21. Gentry, C. Fully homomorphic encryption using ideal lattices. In Proceedings of the Forty-First Annual ACM Symposium on Theory of Computing, Bethesda, MD, USA, 31 May 2009; pp. 169–178.
22. Garg, S.; Gentry, C.; Halevi, S.; Raykova, M.; Sahai, A.; Waters, B. Candidate indistinguishability obfuscation and functional encryption for all circuits. In Proceedings of the 2013 IEEE 54th Annual Symposium on Foundations of Computer Science, Berkeley, CA, USA, 26–29 October 2013; pp. 40–19.
23. Brakerski, Z.; Vaikuntanathan, V. Efficient fully homomorphic encryption from (standard) LWE. In Proceedings of the 2011 IEEE 52nd Annual Symposium on Foundations of Computer Science, Palm Springs, CA, USA, 22–25 October 2011.
24. Brakerski, Z.; Gentry, C.; Vaikuntanathan, V. (Leveled) fully homomorphic encryption without bootstrapping. In Proceedings of the 3rd Innovations in Theoretical Computer Science Conference, Cambridge, MA, USA, 8–10 January 2012; pp. 309–325.
25. Gentry, C.; Sahai, A.; Waters, B. Homomorphic Encryption from Learning with Errors: Conceptually-Simpler, Asymptotically-Faster, Attribute-Based. Available online: https://eprint.iacr.org/2013/340 (accessed on 9 June 2013).
26. Teranishi, K.; Kogiso, K.; Shimada, N. Stability-guaranteed dynamic ElGamal cryptosystem for encrypted control systems. *IET Control Theory Appl.* **2020**, *14*, 2242–2252. [CrossRef]
27. Kogiso, K. Upper-Bound Analysis of Performance Degradation in Encrypted Control System. In Proceedings of the 2018 Annual American Control Conference, Milwaukee, WI, USA, 27–29 June 2018.

28. Tran, J.; Farokhi, F.; Cantoni, M.; Shames, I. Implementing homomorphic encryption based secure feedback control. *Control Eng. Pract.* **2020**, *97*, 104350.1–104350.12. [CrossRef]
29. Shoukry, Y.; Gatsis, K.; Alanwar, A.; Pappas, G.J.; Seshia, S.A.; Srivastava, M.; Tabuada, P. Privacy-aware quadratic optimization using partially homomorphic encryption. In Proceedings of the 2016 IEEE 55th Conference on Decision and Control (CDC), Las Vegas, NV, USA, 12–14 December 2016.
30. Darup, M.S.; Redder, A.; Quevedo, D.E. Encrypted cloud based MPC for linear systems with input constraints. *IFAC-PapersOnLine* **2018**, *51*, 535–542. [CrossRef]
31. Alexandru, A.B.; Morari, M.; Pappas, G.J. Cloud-Based MPC with Encrypted Data. In Proceedings of the 2018 IEEE Conference on Decision and Control (CDC), Miami, FL, USA, 17–19 December 2018.
32. Darup, M.S. Encrypted MPC based on ADMM real-time iterations. *IFAC-PapersOnLine* **2020**, *53*, 3508–3514. [CrossRef]
33. Darup, M.S.; Alexandru, A.B.; Quevedo D.E.; Pappas, G.J. Encrypted Control for Networked Systems: An Illustrative Introduction and Current Challenges. *IEEE Control Syst. Mag.* **2021**, *41*, 58–78. [CrossRef]
34. Song, C.; Huang, R. Secure Convolution Neural Network Inference Based on Homomorphic Encryption. *Appl. Sci.* **2023**, *13*, 6117. [CrossRef]
35. Lyubashevsky, V.; Peikert, C.; Regev, O. On Ideal Lattices and Learning with Errors over Rings. *Commun. ACM* **2013**, *60*, 43.1–43.35. [CrossRef]

applied sciences

MDPI

Article

Risk-Based Cybersecurity Compliance Assessment System (RC2AS)

Afnan Alfaadhel [1], Iman Almomani [1,2,*] and Mohanned Ahmed [1]

1 Security Engineering Lab, Computer Science Department, Prince Sultan University, Riyadh 11586, Saudi Arabia; 221421249@psu.edu.sa (A.A.); mqasem@psu.edu.sa (M.A.)
2 Computer Science Department, King Abdullah II School of Information Technology, The University of Jordan, Amman 11942, Jordan
* Correspondence: i.momani@ju.edu.jo or imomani@psu.edu.sa

Abstract: Cybersecurity attacks are still causing significant threats to individuals and organizations, affecting almost all aspects of life. Therefore, many countries worldwide try to overcome this by introducing and applying cybersecurity regularity frameworks to maintain organizations' information and digital resources. Saudi Arabia has taken practical steps in this direction by developing the essential cybersecurity control (ECC) as a national cybersecurity regulation reference. Generally, the compliance assessment processes of different international cybersecurity standards and controls (ISO2700x, PCI, and NIST) are generic for all organizations with different scopes, business functionality, and criticality level, where the overall compliance score is absent with no consideration of the security control risk. Therefore, to address all of these shortcomings, this research takes the ECC as a baseline to build a comprehensive and customized risk-based cybersecurity compliance assessment system (RC2AS). ECC has been chosen because it is well-defined and inspired by many international standards. Another motive for this choice is the limited related works that have deeply studied ECC. RC2AS is developed to be compatible with the current ECC tool. It offers an offline self-assessment tool that helps the organization expedite the assessment process, identify current weaknesses, and provide better planning to enhance its level based on its priorities. Additionally, RC2AS proposes four methods to calculate the overall compliance score with ECC. Several scenarios are conducted to assess these methods and compare their performance. The goal is to reflect the accurate compliance score of an organization while considering its domain, needs, resources, and risk level of its security controls. Finally, the outputs of the assessment process are displayed through rich dashboards that comprehensively present the organization's cybersecurity maturity and suggest an improvement plan for its level of compliance.

Keywords: compliance assessment; maturity model; cybersecurity; risk; ECC; Saudi Arabia

Citation: Alfaadhel, A.; Almomani, I.; Ahmed, M. Risk-Based Cybersecurity Compliance Assessment System (RC2AS). *Appl. Sci.* **2023**, *13*, 6145. https://doi.org/10.3390/app13106145

Academic Editors: Peter R.J. Trim and Yang-Im Lee

Received: 7 April 2023
Revised: 10 May 2023
Accepted: 12 May 2023
Published: 17 May 2023

1. Introduction

Currently, many organizations have amalgamated cyberspace solutions within their conventional business processes [1]. The more a business integrates digital solutions and increases its online presence, the more it becomes vulnerable to cybersecurity threats. The COVID-19 pandemic was a motivating factor for many companies to embrace technology-based solutions to aid online learning and virtual communication, support vulnerable supply chains, and avail autonomous systems [2]. Cyberattackers also took advantage of the increased online presence of many businesses to intensify their attacks [3]. Reports indicate that 43% of all cyberattacks targeted small and medium enterprises (SMEs) and their employees by initiating attacks such as SQL injections, distributed denial of services, man-in-the-middle, spam, phishing, and email malware [4]. A significant impediment to depending on cyberspace is the emergence of security complexities that could lead to financial losses and, subsequently, adversely affect organizational reputation and goodwill [5] .Based on the numerous benefits associated with the use of cyberspace in work environments, cybersecurity remains a requirement that businesses must acquire while implementing various types of online technologies to manage their activities.

Several cyber security standards have been established for accountability and obligation to ensure that senior leadership in organizations handles risk and security problems thoughtfully and strategically. The enactment of harmonized international cybersecurity regulations has provided a framework for the development of consistent data protection in many organizations, increasing innovation and interoperability and reducing costs, and minimizing the complexity of implementing security and privacy controls as noted by [6]. The implementation of general cybersecurity practices by organizations has enabled businesses to exercise best practices that reduce the risk of access or loss of data, the disruption of business processes, and the loss of assets due to cyberattacks [7]. The implementation of cybersecurity compliance policies improves the protection of an enterprise's information system and related resources from cyberattacks coming from internal or external cyber attackers. Organizations and entities that fail to comply with already set cybersecurity regulations could subject their assets, information systems, and data in cyberspace to massive losses accrued due to penalties, litigation of cyberattack issues, and loss of reliability of their services, which could impact their performances and competitiveness.

Most recent developments in cybersecurity models are in line with the needs of enterprises using cyberspace to manage their operations and databases. Currently, the most used international cybersecurity standard is the National Institute of Standards and Technology (NIST), which provides security guidelines to companies and individuals in the United States to protect their critical infrastructure from cyberattacks. Such standards are also provided by the International Organization for Standardization/International Electrotechnical Commission (ISO/IEC) 27001 [8] and the Payment Card Industry Data Security Standard (PCI DSS). Muhammad and Alsaleh [9] noted that organizations are required to comply with cybersecurity because failure to do so increases the risk of undesirable cybersecurity habits that can expose an entity's assets to cyberattacks. Despite the awareness of many organizations about threats to cybersecurity, compliance with cybersecurity standards has been ineffective in most organizations.

Various institutions and government agencies ensure enterprises abide by regulations and compliance policies. In the United States, the major agencies are the NIST and the Cybersecurity and Infrastructure Security Agency (CISA). The United Kingdom has similar agencies that enforce cybersecurity compliance policies. The National Cyber Security Centre (NCSC) provides detailed advice, regulation compliance, and management of cybersecurity incidents [10].

On the other hand, Saudi Arabia (SA) currently aims to change its economic patterns, reduce its reliance on oil, and expand its public service industries throughout its 2030 vision plan. An example of this strategy is the implementation of cybersecurity policies as an advancement toward its Vision 2030 [11]. According to Almudaires et al. [11], cybersecurity has become increasingly significant in Saudi Arabia's economy due to its reliance on technology to sustain its economic activities. The Saudi authorities are potentially threatened by cyber criminals as many incidences of cyberattacks have been reported such as phishing and ransomware. With the contemporary cybersecurity challenges and to maintain the organization's information and digital resources, the Royal Decree established the National Cybersecurity Authority (NCA) on 31 October 2017. The primary responsibility of the NCA is to perform administrative and regulatory roles in the field of cybersecurity in Saudi Arabia. In 2020, based on the collective efforts of NCA and the collaboration of national entities, Saudi Arabia ranked (2) globally in the Global Cybersecurity Index issued by the United Nations specialized agency for Information and Communication Technologies [12]. NCA has developed essential cybersecurity control (ECC), an adequate cybersecurity regulation in Saudi Arabia. ECC aims to ensure the development of services in a coordinated, safe, and secure manner. This includes providing security, meeting demands, managing the scarcity of resources, ensuring market development, protecting users, and supporting innovations.

Compliance with the ECC regulation is mandatory for all of Saudi Arabia's organizations with IT systems [13]. However, (a) there are limitations in the existing research works that highlighted and studied the compliance assessment process of the ECC, (b) the compliance assessment processes of different international cybersecurity standards and controls such as (ISO2700x, PCI and NIST) are usually generic for all organizations with different scopes, business functionalities, and criticality levels. A better understanding of the organization's domain and status should be considered to reflect compliance accurately. Incorporating additional factors to differentiate between the compliance of different organizations prioritizes the compliance level and provides a more reliable cybersecurity landscape at the national level. Therefore, this research proposes a risk-based cybersecurity compliance assessment system (RC2AS) that improves the current assessment process by considering the organization's domain and integrating the corresponding risk in the overall compliance score calculations, consequently continually enhancing the cybersecurity compliance assessment process.

Accordingly, the benefits of proposing a risk-based cybersecurity compliance system in terms of a well-developed system can be summarized in the following points:

(a) Measuring the organizations' cybersecurity compliance level using a self-assessment questionnaire (SAQ) approach.
(b) Choosing one of the proposed overall compliance-calculation methods based on the organizations' domains, needs, and resources.
(c) Using color-coding techniques to better reflect the compliance status based on the risk level.
(d) Determining the critical risk controls based on the domain risk level and the impact that is based on the organizations' criticality, scope, and business functionality.
(e) Producing rich dashboards to present the organization's cybersecurity maturity and compliance status.
(f) Setting a clear improvement and action plan to reach the organization's target compliance.
(g) Offering a cybersecurity tool to help the assessor to perform the audit assessment.

Therefore, this paper presents a customized and comprehensive cybersecurity compliance system based on ECC, the national cybersecurity reference regulation in Saudi Arabia. The ECC has well-defined regulations built based on different international standards. Thus, using the ECC for this research will support adopting the RC2AS system for other international standards. The system services are offered through an offline standalone assessment tool for organizations to measure their cybersecurity compliance level efficiently. The assessment results are presented through rich dashboards to reflect the organization's current status. Additionally, the proposed system guides the organization to an action plan to reach its target compliance level. Figure 1 shows our overall methodology to create the proposed RC2AS.

The rest of the paper is organized as follows. Section 2 presents the overall methodology followed to build the proposed RC2AS. Section 3 introduces the current cybersecurity standards and recent related works. Section 4 presents the details of the proposed risk-based cybersecurity compliance assessment system (RC2AS). Section 5 shows the evaluation of RC2AS and discusses its results. Finally, section 6 draws conclusions and suggests possible future works.

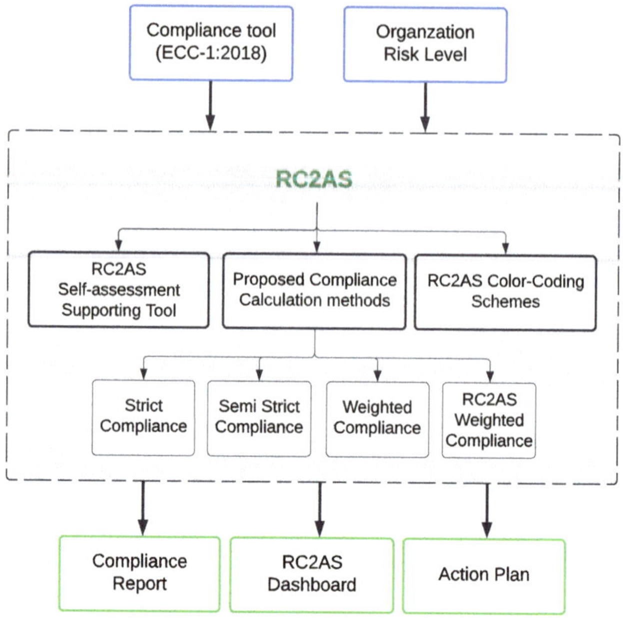

Figure 1. Overall RC2AS building methodology.

2. Methodology

The methodology we followed to solve the research problem addressed in this study is summarized as follows. The main inputs to our proposed solution are the current ECC tool beside the organization risk level of each subdomains. The risk level is identified based on the domain and the nature of the organization, which can be decided by the cybersecurity authority in any country. The proposed solution (RC2AS) includes but is not limited to (a) the RC2AS self-assessment supporting tool, (b) suggested well-studied compliance-calculation methods (four methods), and (c) RC2AS color-coding schemes. Finally, the outcome of our proposed solution and the assessment process are presented in terms of a compliance report, RC2AS's rich dashboards, and future action plans. Figure 1 presents the overall methodology followed to build the proposed RC2AS.

3. Background and Related Work

This section presents the background to national and international cybersecurity frameworks and standards. Additionally, it discusses and compares recent related works.

3.1. International Cybersecurity Framework and Standards

Developing a cybersecurity maturity model aims to help organizations systematically improve their cybersecurity status over time and align it with their overall business objectives. Most cybersecurity maturity models are currently developed according to international standards such as NIST, ISO/IEC 27001, and PCI DSS. Various studies have been proposed focusing on improving enterprises' cybersecurity practices. For instance, Gerl et al. [14] examined the utilization of control objectives for information technologies (COBIT-19) in establishing an IT governance framework for collaboration in higher education settings in Bavaria. The authors hypothesized that the chief information officer (CIO) role is critical to improving collaboration among universities. Based on the findings, implementing COBIT enhanced the trust among collaborating partners. COBIT also creates a consistent model of the role of the CIO in defining the baseline of mutual understanding of competencies and responsibilities. Since the article presented a case study, the generalization of the findings to other problems or settings can be challenging. Cybersecurity is an essential subject in the industrial internet of things because the information equips individuals that use various practices and tools to protect individuals and organizations from occurrences such as data breaches and ensures they comply with cybersecurity policies [15]. Implementing security controls could also involve incorporating blockchain technologies to strengthen cybersecurity. For example, a study exploring a security framework based on blockchain is presented in this study [16]. Thus, adjustments to the security regulations arise because of the constantly shifting information technology environment.

Almuhammadi and Alsaleh [9] defined a five-level maturity model that includes the twenty-two categories of the NIST cyber security framework for critical infrastructure (CSF) to measure the implementation regularly and maintain the security posture. The model presented a comparison between NIST CSF and additional frameworks and standards related to security such as ISO/IEC 27001 and COBIT. According to the authors of [17], Canada can achieve better outcomes in managing health emergencies and maintaining privacy rights by designing laws to comply with European Union (EU) in a way that freedoms relating to privacy can only be limited for shorter periods. Additionally, Aliyu et al. [18] noted that the challenge encountered was a lack of capability maturity models that integrate regulations within the United Kingdom. As a result, they developed a novel framework that includes all of the privacy and security regulations and best practices, such as the general data protection regulation (GDPR), the data security and protection toolkit (DSPT), and PCI DSS. These security standards can be leveraged to enhance the cybersecurity compliance levels of higher education institutions. From a theoretical viewpoint, capability maturity models offer a framework for improving process development operations. The proposed model, which comprises fifteen categories related to security and six maturity levels, can be developed into an online system to support self-assessment and automated compliance reporting. A major weakness of the article is that it is largely theoretical. Indeed, an empirical analysis of the model can provide insight into the effectiveness of its utilization in practical environments.

Although the current maturity models are used to measure the security maturity level of enterprises or specific systems, they cannot be used to create and establish cybersecurity maturity models for protecting cyberspace. The existing maturity models have created static security models and are not flexible to react to new security trends. Zarour et al. [19] explains that the emergence of DevOps is informed by the need to produce fast and high-quality releases by bringing the development and operations teams to work together. DevOps still lack a clear definition in most studies, thereby creating challenges in some quarters. DevOps maturity models are instrumental in providing critical insights regarding what can be done in assessing DevOps-adopted practices. Therefore, there is a need to incorporate perspectives from various levels, such as security experts, practitioners, and management. This can help measure the enterprise's overall security level or the critical system from emerging security threats.

Many cybersecurity compliance and maturity models have been presented. Firstly, Proença and Borbinha [20] introduced a maturity model as an assessment tool for enterprises to provide the current state maturity model of the information security management system (ISMS) based on ISO/IEC 27001. Another approach was proposed by Bolanio et al. [21] to improve the security network of higher education institutions based on ISO27033. [22] Makupi and Masese [23] also created a model to compute the university's information security model based on ISO27001 using related clauses of higher education institutions. Yaokumah and Dawson [24] applied ISO/IEC 21827 [25] to measure the controls related to the security of higher education institutions (HEIs) in Ghana. Another model proposed by [26] examined the maturity level of information systems from a security perspective based on ISO 27001:2013. The idea was to help institutions identify vulnerable areas and implement appropriate interventions to enhance cybersecurity compliance.

Based on the results, most institutions of higher learning in Indonesia have not complied with the requirements of ISO 27001:2013 for cybersecurity; the biggest domain gap has been observed between the current and the expected maturity levels observed in compliance and system acquisition, development, and maintenance. While using a questionnaire as the study methodology helped answer the research questions, the subjectivity associated with this research approach was not addressed. One study proposes a dynamic approach to compliance assessment where organizations consider the return on investment relevant to the savings an organization can realize pertinent to the losses that could arise when security features are not implemented [27]. One of the significant guidelines for implementing cybersecurity governance is ISO/IEC 27001, directing institutions and companies to create specific protocols to mitigate, control, and supervise potential risks. Through the protocols, implementing digital environment rules becomes easier [28]. Suwito et al. [29] applied the assessment security maturity model by combining various models and standards such as ISO/IEC 27001, COBIT 4.1, and ITIL v3 (information technology infrastructure library) for higher education institutions in Indonesia. Hung et al. [30] examined the methods of enhancing the information security governance (ISG) of Taiwanese universities through a questionnaire by building the ISG maturity model by looking at appropriate features. A similar model was designed by Bass [31] as derived from a documentary and the result from selected Ethiopian universities. Ismail et al. [32] proposed a specialized information security framework for Malaysia's higher education institutions.

3.2. Importance of Security in Saudi Arabia

Cybersecurity threats are one of the primary concerns of the Saudi leadership in this digitalized world. Since August 2017, the cyberattack on Saudi Aramco was inflicted with the virus named Shamoon. It considers one of the renowned cyberattacks cases in Saudi Arabia [33]. Due to the contemporary cybersecurity challenges and to maintain the organization's information and digital resources, the government of Saudi Arabia has classified the strategics' priority as cybersecurity and businesses are making it a priority to avoid breaches reputationally and financially. In 2017, a Royal Decree was issued to establish the National Cybersecurity Authority (NCA), which is the national and specialized reference for matters related to cybersecurity in the Kingdom. The primary responsibility of this organization is to realize the idea of a safe and reliable Saudi cyberspace that enables growth and prosperity [34].

Based on the NCA's objectives and in continuation of its part in regulating and protecting Saudi Arabia's cyberspace, NCA has established and developed several cybersecurity frameworks, controls, and guidelines at the national level within its scope to protect its national security vital interests, government services, and critical infrastructure in line with vision 2030 of Saudi [13]. The NCA has set out to establish the cybersecurity minimum standards for national and government agencies at risk of cyberattacks to ensure the safety of their data, for instance, essential cybersecurity controls (ECC), critical systems cybersecurity controls (CSCC), and data cybersecurity controls (DCC).

3.3. Saudi Arabia Security, Frameworks Maturity, and Standards

Enterprise security is very important in Saudi Arabia because many incidences of cyberattacks have been reported compared to other countries. Saudi Arabia's government is committed to developing a powerful and operational cybersecurity framework to overcome these issues. They have designed multiple frameworks, such as NCA creating essential cybersecurity controls (ECC) to help the enterprise follow cybersecurity best practices [13]. The Communications and Information Technology Commission developed a cybersecurity regulatory framework (CRF) for the Information and Communications Technology sector [35]. Moreso, a SAMA cybersecurity framework, was formerly developed by the Saudi Central Bank (SAMA) to secure financial sectors such as banks, financing companies, and financial market infrastructure from cyberattacks [36]. In academic literature, Al Hamed and Alenezi [37] presented a maturity model to mature the ability of business continuity management (BCM) and disaster recovery (DR) for Saudi Arabia's information technology companies. Additionally, Nurunnabi [38] explains that the investigation of the differences between International Financial Reporting Standards (IFRS) and Saudi accounting standards provides an opportunity for the areas that may need to improve further to ensure better standards are realized in the financial sector. Moreover, it ensures that investors have a clear understanding of the financial reporting strategies that they need to embrace in different transactions.

The rationale behind this study was to address cybersecurity issues facing SMEs by presenting an appropriate framework. According to [39], each SME is unique, hence the need to utilize a model that aligns with its needs and wants. To this end, the authors presented specific cybersecurity models for organizations in different industries, including education, health care, and commerce. A holistic model that covers the different models to enhance coordination was also presented. A major weakness of the study was that it merely presented the models and did not evaluate their effectiveness and efficiency. Nevertheless, the adoption of these models can help SMEs in Saudi Arabia improve their cybersecurity implementation processes. Alsahafi et al. [40] stated that there is a need for institutions to implement ISMS such as ISO/IEC 27001 to minimize the risks of cyberattacks on their information assets. The ISO/IEC 27001 acts as a baseline cybersecurity framework. A central hypothesis of the study could be whether universities with ISO/IEC 27001 are fully compliant with NCA-ECC. The assumption here is that it is not fully understood to what extent Saudi institutions with ISO/IEC are compliant with the NCA-ECC. The study design was a qualitative survey. Instrumentation included the use of interviews presented in an interview table from which the answers (data) were collected and analyzed. The sample size used was three universities, whose cybersecurity officers were interviewed on each clause and sub-clause. Research results indicated that ISO/IEC 27001 universities are approximately 64% compliant with the NCA-ECC. Another proposed framework by Almomani et al. [41] found that most cybersecurity models for high-education institutions lacked practical mechanisms for the continual assessment of security levels. Accordingly, they presented a new comprehensive and customized framework, "SCMAF", aligned with international and local security standards. The research method adopted in this study encompassed evaluating current cybersecurity maturity frameworks in Saudi Arabia, mapping local and international frameworks, developing the SCMAF model, implementing it, demonstrating the utilization of SCMAF, and highlighting the approach for keeping the framework updated.

Table 1 presents a comparative analysis of related studies for cybersecurity compliance and the maturity model, for both international and national standards, in terms of the general idea and the technique used; the focus areas were both international or national, the followed standard, and if the proposed solution included ECC . This enabled the organization to measure the maturity level of cybersecurity among the international and Saudi standards using a user self-assessment tool. Part of the presented maturity models was based on international standards, for instance, those [9,18,20,23,24]. To improve cybersecurity in Saudi Arabia other approaches were presented [39,41,42]. The table shows the rest of the comparisons.

Although there are many existing attempts to propose compliance assessment tools and maturity models, there is an apparent absence of studies that highlight the compliance assessment process of the NCA-ECC. Therefore, due to the shortage in the related literature and the importance for organizations to comply with the security regulations in Saudi Arabia, this research takes the existing ECC cybersecurity compliance process as a baseline to build a comprehensive and customized risk-based cybersecurity compliance assessment system (RC2AS). As a result, RC2AS provides an accurate cybersecurity assessment that reflects the organization's current status considering its domain and risk ranges.

Table 1. Comparative analysis of the related works.

Ref.	General Idea	Approach Used	Focus Area	Standards	NCA-ECC?
[9]	Present a five-level maturity model that assesses twenty-three areas, which include the twenty-two categories of NIST CSF and the compliance assessment to measure the implementation regularly and maintain the target security posture.	Compared the scales and domains evaluated by different maturity models to identify the gap in NIST CSF.	International	• NIST CSF • ISO27001 • ISF • COBIT 5	No
[14]	Propose IT governance model for universities in Bavaria. The model defines governance relationships between cooperative IT service providers, CIOs, universities, and all Bavaria stakeholders.	Universities taught applied sciences and CIO boards of higher learning institutions in Bavaria.	International	COBIT 2019	No
[18]	Design framework for maturity assessment (HCYMAF) that higher education institutions in the UK can use to assess their ISMS by using a web-based self-assessment model.	The study combined structured interviews, case study evaluations, feedback, and an online seminar.	International	• GDPR • PCI DSS • DSPT NISD	No
[20]	Propose a maturity model that can be used to plan, implement, review, and enhance an ISMS based on ISO 27001.	Used design science researcher paradigm; an iterative approach; and model adoption techniques such as configuration, specialization, aggregation, and analogy.	International	ISO 27001	No
[21]	Propose a model for assessing and appraising network security in higher education using six components drawn from ISO 27033 framework.	The study relied on the standardized ISO 27033 assessment questionnaire.	International	ISO27033	No
[23]	Find solution that entails creating a maturity model that can be used to assess the information security management systems in universities.	Used design research approach and evaluated cumulative factors statistically to determine their contribution to the proposed model followed ISO 27001 standards.	International	ISO27001	No
[24]	Proposed use of ISO/IEC 21827 maturity model for assessing the IT security posture and security controls.	A questionnaire based on ISO 27033 standards was developed and distributed to network security teams in different learning institutions.	International	ISO21827	No
[26]	Develop a maturity framework that can be used to assess and measure the information security management systems of higher learning institutions in Indonesia for conformity to the ISO 27001 standard.	The research evaluated 35 universities in Indonesia and assessed their compliance with ISO 27001:2013 standards.	International	ISO 27001:2013	No
[29]	Present an approach that combines different frameworks to improve the effectiveness of security maturity management assessments.	A case study on one university in Indonesia.	International	• COBIT 4.1 • ISO27001 • ITIL v3	No
[30]	Propose an information security governance (ISG) model for colleges and universities. The model proposes three maturity levels: low, medium, and high.	Evaluate the maturity of information security governance through a questionnaire survey.	International	None	No
[31]	Present an ICT maturity model for higher education in Ethiopia comprising eight levels.	Adopted action research founded on an iterative approach focused on problem identification, planning, action, and evaluation. The study surveyed education institutions.	International	None	No

Table 1. *Cont.*

Ref.	General Idea	Approach Used	Focus Area	Standards	NCA-ECC?
[32]	Propose an information security management framework comprising five constructs for HEIs in Malaysia.	Interviews and surveys were conducted to gain relevant insights.	International	• COBIT • ISO27001	No
[37]	Propose a model for evaluating the maturity of business continuity and disaster-recovery practices for information technology organizations in SA.	The study adopted an iterative approach to link existing theories to emerging data.	National	ISO22301	No
[39]	Incorporate three models and a combined model for cyber security countermeasures within SMEs in education, healthcare, and commerce in Saudi Arabia.	The study considered organizational special needs and asset sensitivity.	National	NIST	No
[41]	Propose a lightweight cybersecurity maturity assessment framework for HEI in SA.	The study developed a comprehensive policy that bridges local needs and international standards.	Both	• NCA-ECC • CRF • GDPR • NIST • PCI DSS • DSPT	Yes
[40]	Measure the extent to which certified ISO/IEC 27001 Saudi organizations adhere to the NCA-ECC and propose a framework for complying with not fully implemented controls.	The study design is a qualitative survey that included the use of interviews presented in an interview table from which the answers (data) were collected and analyzed.	Both	• NCA-ECC • ISO 27001	Yes
RC2AS	Propose a risk-based cybersecurity compliance assessment system based on ECC.	A comprehensive and customized risk-based cybersecurity compliance assessment system was provided that reflects the current status of the organization, considering its domain and risk ranges.	National	NCA-ECC	Yes

4. Risk-Based Cybersecurity Compliance Assessment System (RC2AS)

This section starts by discussing the existing ECC assessment and compliance tool. Then, it introduces the proposed RC2AS with all its services.

4.1. Existing ECC Assessment and Compliance Tool

The objective of this sub-section is to fully understand and highlight the ECC-1:2018 assessment and compliance tool by studying their domains/controls, scope, objective, and compliance assessment process. A better understanding of the tool will facilitate and pave the way toward establishing the foundation of the RC2AS solution. The list of main functions listed in this section was used as a starting point to build the functions of the proposed system. Before diving into these functions, the key points of the ECC are:

- Description: Minimum cybersecurity standards were customized and developed after reviewing international cybersecurity standards, controls, frameworks, previous cybersecurity attacks incidents, and international practices in cybersecurity to minimize the risk of cyberattack to enterprises' information and technical assets that are created by external and internal threats.
- Scope: It is mandatory for all Saudi Arabian entities within the government and private sectors.
- Objective: The essential objectives must be focused on to protect the information and assets of organizations: confidentiality, availability, and integrity of information, with attention paid to the pillars that cybersecurity focuses on (strategy, people, procedures, and technology).

- ECC Domains and Structure: As shown in Figure 2, ECC consists of 5 main cybersecurity domains, 29 cybersecurity subdomains, and 114 cybersecurity controls.

Currently, each organization should evaluate and assess its compliance with ECC through self assessments and by using the compliance tool [13] . The only and latest release of ECC is (ECC-1:2018). The current ECC self-assessment tool is based on an Excel spreadsheet that is officially posted by the NCA.

(ECC-1:2018 Tool) [43]. The main domains of ECC are placed on separate Excel sheets (ECC.1 Assessment, ECC.2 Assessment, ECC.3 Assessment, ECC.4 Assessment, and ECC.5 Assessment). In addition, the subdomain(s) related to that main domain are also displayed. Table 2 shows the structure of each subdomain and a sample subdomain.

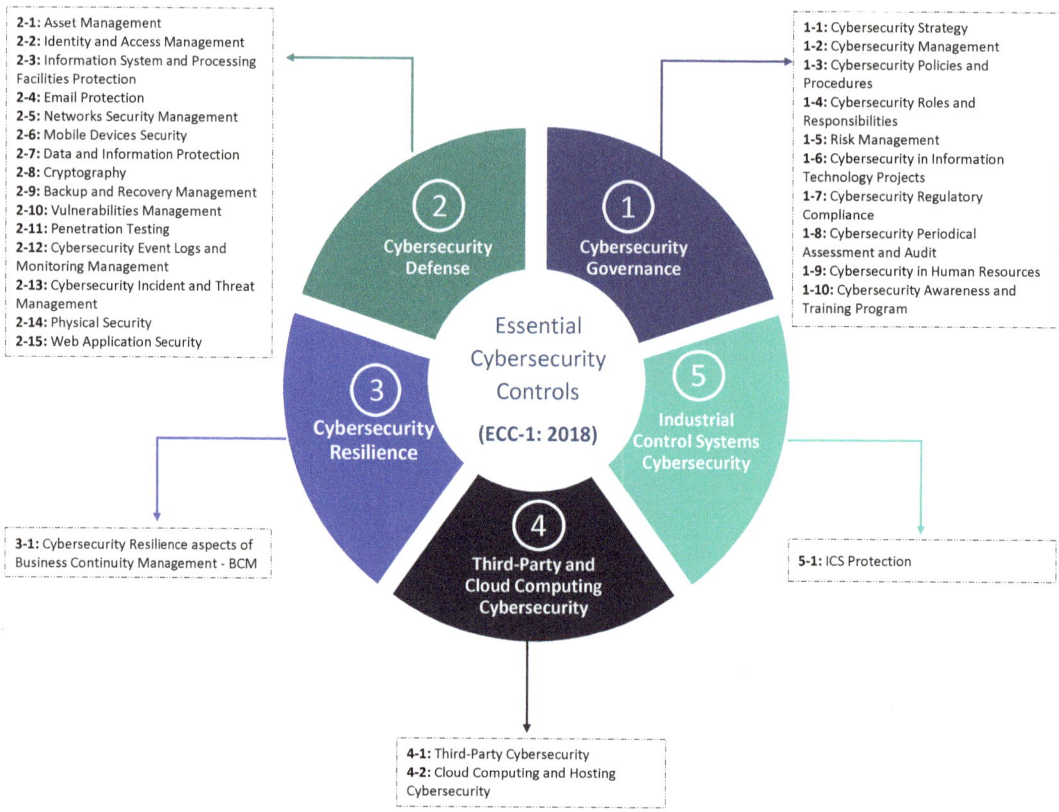

Figure 2. Current ECC-1:2018 domains and subdomains.

Table 2. Current ECC subdomain structure.

Subdomain Structure		
Ref. No.	subdomain's Name	
Control Ref. No.	Control Clause	Compliance Status
Sample of subdomain Structure		
1-1	Cybersecurity Strategy	
1-1-1	A cybersecurity strategy must be defined, documented, and approved. It must be supported by the head of the organization or his/her delegate (referred to in this document as authorizing official). The strategy goals must be in-line with related laws and regulations [13].	Implemented
1-1-2	A roadmap must be executed to implement the cybersecurity strategy [13].	Not Implemented
1-1-3	The cybersecurity strategy must be reviewed periodically according to planned intervals or upon changes to related laws [13].	Not Implemented

The compliance fulfillment of an organization is represented by one of the following statuses: "Implemented", "Partially Implemented", "Not Implemented", or "Not Applicable". Organizations measure and assess how they comply with each control clause's requirement(s). The "Implemented" status means that all of the requirements for this control clause are fully implemented. The "Partially Implemented" status means that some of the control requirements have not been implemented; in other words, the implementation percentage is greater than 0% and less than 100%. If all of the control requirements have not been implemented, then the status will be "Not Implemented". Lastly, the "Not Applicable" status applies to any control that does not apply to the organization. Table 3 shows the compliance status along with the implementation percentage.

Table 3. Current ECC compliance statuses along with implementation percentages.

Compliance Status	Implemented	Partially Implemented	Not Implemented	Not Applicable
Implementation Percentage	100%	>0% to <100%	0%	NA

Currently, the self-assessment tool applies a color-coding technique for the compliance status. Four different colors are used depending on the compliance status. Each status has one color. The "Implemented" status is indicated by the color "Green", "Partially implemented by "Orange" color."Not implemented" status is colored "Red", and the "Gray" color is used for "Not Applicable". Table 4 illustrates the compliance status along with the color coding.

Table 4. Current ECC compliance status along with the color-coding.

Compliance Status	Implemented	Partially Implemented	Not Implemented	Not Applicable
Color-Coding				

Finally, after an organization fills out its compliance status in the self-assessment, a summary of the compliance evaluation results will be generated, as shown in Figure 3. The left side includes **(a)** the total number of security controls under each compliance status, and **(b)** a chart indicating the percentage of controls in each compliance status. There will also be a summary for each domain (five main domains) on the same sheet. For example, Figure 4 illustrates the summary of the overall result for domain 1: "Cybersecurity Governance".

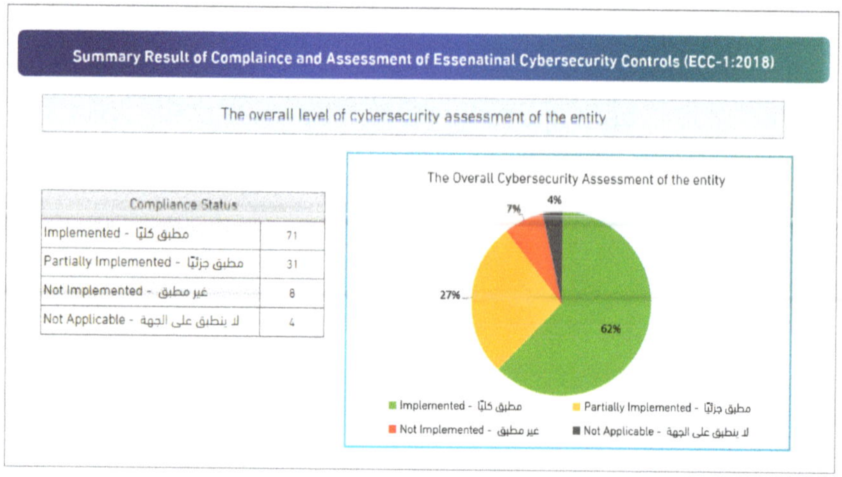

Figure 3. Current ECC—Summary of the results of the overall assessment, Note: the current tool is only available in the Arabic language. For this reason, we translated several sentences into English in Figures 3 and 4 for illustration purposes.

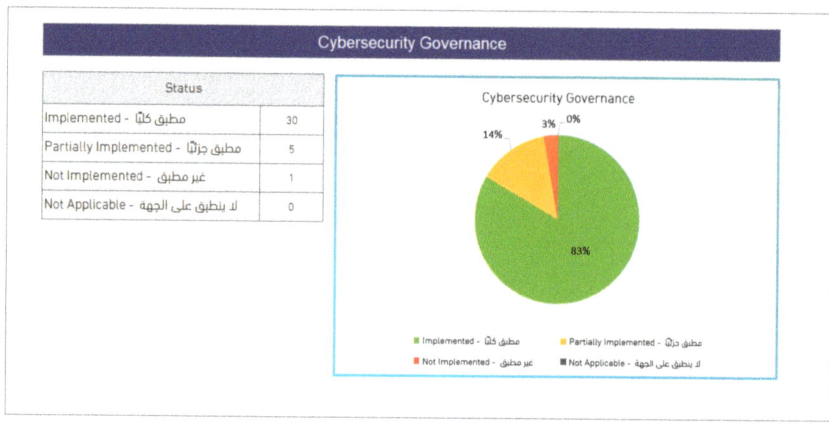

Figure 4. Current ECC—Summary of the results of the main domain "Cybersecurity Governance", Note: the current tool is only available in the Arabic language. For this reason, we translated several sentences into English in Figures 3 and 4 for illustration purposes.

The summary result of the compliance with ECC (Figure 3) shows the percentage of each compliance status of security controls as there is no overall compliance score provided in the current ECC tool. In the pie chart, each compliance status percentage (%) is calculated based on the number of main controls on specific "compliance status" out of the total of main controls (114 controls). For instance, the percentage of "Implemented" in Figure 3 is calculated based on the number of "Implemented" controls (71). So, by using Equation (1), the compliance percentage for "Implemented" compliance status is calculated as follows ($\frac{71}{114} * 100$), corresponding to (62%). The remaining compliance statuses are calculated in the same way. The below Equations (1)–(4) illustrate the formula for how each compliance status percentage is calculated as follows:

$$Implemented = \frac{\sum(F_{Control})}{T_{Controls}}\% \tag{1}$$

$$Partially\ Implemented = \frac{\sum(P_{Control})}{T_{Controls}}\% \quad (2)$$

$$Not\ Implemented = \frac{\sum(N_{Control})}{T_{Controls}}\% \quad (3)$$

$$Not\ Applicable = \frac{\sum(NA_{Control})}{T_{Controls}}\% \quad (4)$$

where; F = fully implemented control(s); P = partially implemented control(s); N = not implemented control(s); NA = not applicable control(s); T = total controls.

4.2. Proposed Risk-Based Cybersecurity Compliance Assessment System (RC2AS)

As mentioned earlier, compliance with ECC is mandatory for all national Saudi organizations (public and private) with IT systems. So, all entities should evaluate and measure their compliance by using the publishing tool (ECC-1:2018 Tool) [43].Many service provider companies help national entities to assess and measure their compliance level through dedicated services and tools. These tools help an entity to demonstrate its commitment to different standards or regulations and enable it to perform gap assessments of its weaknesses and strengths. This will result in enabling the entity to develop its road map based on its priorities.

Therefore, this research takes the existing ECC cybersecurity compliance tool as a baseline to build a comprehensive and customized risk-based cybersecurity compliance assessment system (RC2AS). RC2AS is developed to be compatible with the current ECC tool. This system offers a self-assessment tool that helps the organization evaluate and check its compliance with ECC. Moreover, RC2AS supports weakness identification and provides better planning accordingly to enhance its compliance level based on its priorities for future improvements. Lastly, RC2AS proposes several calculation methods for the overall compliance score of ECC.

This section presents the proposed system that provides various services by highlighting: (a) the RC2AS overall workflow, and (b) the RC2AS supporting tool that provides a well-structured and comprehensive questionnaire derived from ECC controls. Such a tool will provide a practical way to encourage entities to complete the questionnaire and obtain their compliance level, (c) the proposed calculation methods of the overall cybersecurity compliance of ECC, (d) the RC2AS compliance status color-coding scheme, and (e) the rich dashboards RC2AS offer to reflect the current status of compliance with ECC, accurately. Moreover, setting a clear improvement and action plan will allow them to reach their ECC target compliance level.

Additional details of the RC2AS and the services it offers are described in the following:

4.2.1. RC2AS Workflow

The proposed system is offered as an offline version of the self-assessment tool that organizations or auditors can use. To start using the system, there will be high-level questions to establish the applicable domain by answering predefined questions. Accordingly, based on the answers, the appropriate domains/subdomains will be displayed to facilitate and speed up the assessment process by hiding the controls that do not apply to the organization's domain. In addition, the users will choose the recommended and preferable calculation method (one or more) from the options list: (a) strict compliance, (b) semi-strict compliance, (c) weighted compliance, or (d) RC2AS weighted compliance. Table 5 shows a sample of the RC2AS high-level questionnaire.

Table 5. RC2AS High-level questionnaire.

RC2AS High-Level Questionnaire				
Entity Name:				
General Question				
1. Does the entity use cloud computing?	☑	Yes	☐	No
2. Does the entity have a third party?	☑	Yes	☐	No
3. Does the entity use industrial control systems and operational technology (ICS/OT)?	☑	Yes	☑	No
Calculation Compliance Mode				
Select the compliance calculations method:	☑	Strict compliance		
	☐	Semi-strict compliance		
	☐	Weighted compliance		
	☑	RC2AS weighted compliance		

The proposed system assesses the domains one by one. Then, the subdomain(s) of this domain will be fulfilled individually. If the subdomain is applicable to the organization and this subdomain has dependent questions, the associated questions of this subdomain will be shown. If the answers are yes, then the remaining questions will be displayed. The next domain (if any) will appear only in case all questions have been answered, and so forth. Once this domain's subdomain(s) are examined, the following domain will repeat the same steps. Ultimately, when all of the domains' questions are answered, the compliance level will be calculated and displayed through rich dashboards. Figure 5 illustrates the workflow of the proposed system.

4.2.2. RC2AS Supporting Tool

To facilitate the assessment process, the RC2AS supporting tool uses a questionnaire approach. Accordingly, each control will have one or more questions to assess and measure the organization's compliance with a specific control. Table 6 highlights the RC2AS with subdomain structure.

Table 6. RC2AS supporting tool with subdomain structure.

Ref. No. of Subdomain	**Name of Subdomain**			
Control Ref. No.	Control Clauses	Question(1) Question(2) ... Question(n) [1]	RC2AS compliance answer(1) RC2AS compliance answer(2) ... RC2AS compliance answer(n) [1]	Compliance Status

[1] n: means the number of questions related to this control.

The questions on a particular control could depend on each other. An example of questions' dependency is shown in Table 7. Only the first question on control (1-2-1) will be presented. Only if the answer is "Yes" or "Partially Implemented" will Q2 and Q3 be displayed. The following question will not appear if the answer to Q1 is "No". This will expedite the assessment process and make it more convenient. On the other hand, there might be no dependency between the questions, as shown in control (1-2-2), where each question does not depend on the others.

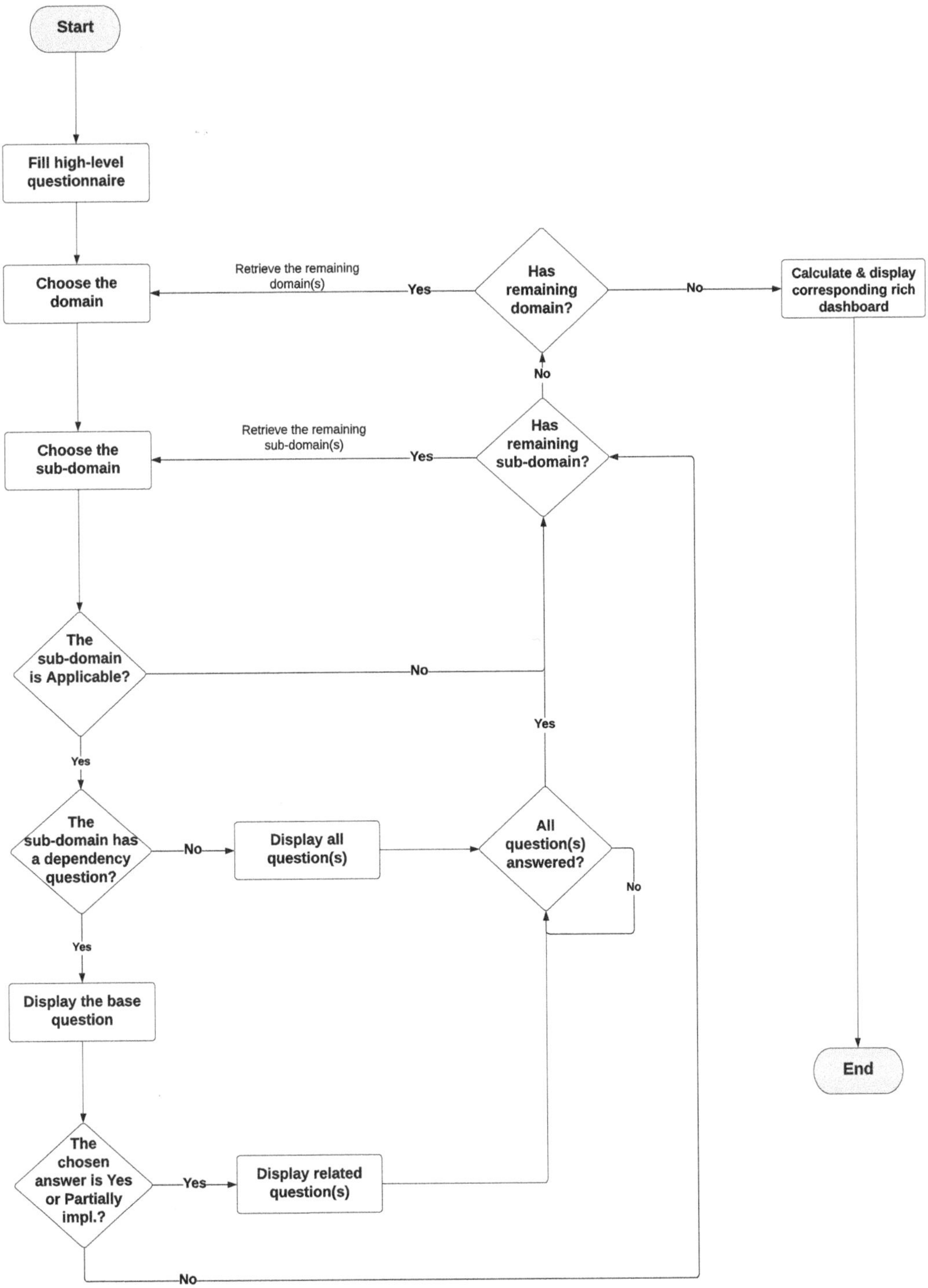

Figure 5. The proposed risk-based cybersecurity compliance assessment system (RC2AS) workflow.

Table 7. Sample of RC2AS subdomain structure (with and without dependency).

1-2	Cybersecurity Management	
	Sample of Control's Question(s) with Dependency	
1-2-1	A dedicated cybersecurity function (e.g., division, department) must be established within the organization. This function must be independent of the information technology/information communication and technology (IT/ICT) functions (as per the Royal Decree number 37140 dated 14/8/1438H). It is highly recommended that this cybersecurity function reports directly to the head of the organization or his/her delegate while ensuring that this does not result in a conflict of interest [13].	**Q1:** Does the entity have a cybersecurity department? **Q2:** Is the cybersecurity department independent of information technology management in the entity? **Q3:** Does the cybersecurity department report directly to the organization's head or his/her delegate while ensuring that this does not result in a conflict of interest?
	Sample of Control's Question(s) without Dependency	
1-2-2	The position of cybersecurity function head (e.g., CISO), and related supervisory and critical positions within the function, must be filled with full-time and experienced Saudi cybersecurity professionals [13].	**Q1:** Is the cybersecurity department function headed by full-time and experienced Saudi cybersecurity professionals? **Q2:** Do the related supervisory and critical positions in the cybersecurity department function with full-time and experienced Saudi cybersecurity professionals?

Accordingly, the organization answers each question to measure its compliance level. The answer could be one of the following options. (a) If the organization accomplished all of the requirements of an associated question, then the chosen answer is 'Yes'. (b) If the organization partially implements the requirements, then the organization selects the percentage of this implementation (>0% to $\leq$ 35%, >35% to $\leq$ 85%, or >85% to <100%). Otherwise, (c) the answer selection will be 'No'. Lastly, (d) if it is not applicable, the answer 'NA' will be selected. These RC2AS compliance answers are derived and inspired by the current compliance status of the ECC assessment tool shown in Table 3. The different RC2AS answer options are shown in Table 8.

Table 8. RC2AS compliance answer options.

RC2AS Compliance Answer	**Yes**	**Partially Implemented with**			**No**	**NA**
Implementation Percentage	100%	>0% to $\leq$35%	>35% to $\leq$85%	>85% to <100%	0%	NA

4.2.3. Proposed Calculation Methods for the Overall Compliance Score of ECC

As we have mentioned in Section 4.1, what has been published by NCA and presented in the current compliance and assessment tool of the ECC tool does not provide an overall compliance score. Therefore, RC2AS proposes several possible calculation methods to provide the overall compliance score:

- Method (1): Strict compliance.
- Method (2): Semi-strict compliance.
- Method (3): Weighted compliance.
- Method (4): RC2AS Weighted compliance.

Now, we will describe each method and the attributes and factors that are needed to calculate the overall compliance score.

(A) **Strict Compliance**

This is the most strict method as it only considers the fully implemented controls. The "Partiality Implemented" and "Not Implemented" controls are discarded. The fully implemented control will have a weight value of (1), and the rest will have a (zero) value.

This means that only fully implemented controls will be counted (100% implementation), and other controls will not be counted (from > 0% to < 100% implementation). Equation (5) shows how the compliance score of this method is calculated:

$$Compliance\,Score = \frac{\sum(F_{Control})}{TA_{Controls}}\% \tag{5}$$

where;

F = No. Fully Implemented Control(s); TA = No. of Applicable Control(s);

(B) **Semi-Strict Compliance**

This method is less strict than the above method. Here, the compliance status "Implemented" will have the same weighted value, which is (1), whereas "Partially Implemented" has a weight value equal to (0.5). Otherwise, no weight value is given (zero value). Accordingly, Equation (6) calculates the overall compliance score for this method:

$$Compliance\,Score = \frac{\sum(F_{Control} + 1/2P_{Control})}{TA_{Controls}}\% \tag{6}$$

where;

F = No. Fully Implemented Control(s); P = No. Partially Implemented Control(s);
TA = No. of Applicable Control(s);

(C) **Weighted Compliance**

Before we present the attributes of this compliance score equation, referring to the RC2AS compliance answer options mentioned above in Table 8, where there are three different levels of compliance status based on the implementations percentage of the control, which are (a) >0% to <=35%, (b) >35% to <=85%, and (c) >85% to <100%. Thus, weighted "Partiality Implemented" is embedded in this equation to impact the calculation of the overall compliance score. We map each of these RC2AS compliance answer options to dedicated weight values depending on the percentage of implementation of this control as detailed in Table 9:

Table 9. Mapping RC2AS compliance answer options with weight values.

RC2AS Compliance Answer	Yes	Partially Implemented with			No	NA
Implementation Percentage	100%	>0% to <=35%	>35% to <=85%	>85% to <100%	0%	NA
Weight Value	1	0.75	0.5	0.25	0	-

Therefore, the overall compliance score is calculated in Equation (7):

$$Compliance\,Score = \frac{\sum_{i=1}^{n}(Weight\,Value_i)}{TA_{Controls}}\% \tag{7}$$

where;

n = No. of total Control(s); TA = No. of Applicable Control(s);

(D) **RC2AS Weighted Compliance**

A better understanding of the organization's domain and status is needed to reflect compliance levels accurately. Therefore, this method differentiates organizations by different scopes, business functionality, and criticality level. For this reason, the overall compliance score should not be measured similarly.

This method includes the risk level of the subdomains to calculate the overall compliance score. This means that the regulator predates a risk level for each subdomain in ECC (29 subdomains) depending on different criteria and conditions. To clarify more, each organization will have a risk level of one of the following (high, medium, or low) according to the risk impact in case the subdomain controls are not implemented. However, the organization can officially request the regulator to change the risk level of the subdomains (if required).

The following example elaborates more on how this method is used. There are two organizations: the first is a university, and the second is a CNI (Critical National Infrastructure) organization. Both organizations are not fully implementing control in a subdomain (2-10). Accordingly, the impact of not fully implementing this control is definitely not the same for both organizations. Thus, both organizations will be assigned to a different risk level for the subdomain; subsequently, the overall compliance score will be affected by the risk value of that subdomain. The above formula (Equation (7)) will be modified to consider the risk value as shown in Equation (8):

$$Compliance\ Score = \frac{\sum_{i=0}^{n}(Risk\ Based\ Weight\ Value_i)}{TA_{Controls}}\% \qquad (8)$$

where

n = No. total Control(s); TA = No. of Applicable Control(s);

As mentioned before, there are three different risk levels (high, medium, and low), each with a risk value (1, 0.75, and 0.5), respectively. To calculate the "Risk-based Weight Value", the compliance status weight value and the risk value are multiplied as shown in Equation (9):

$$Risk\ Based\ Weight\ Value = Weight\ Value * Risk\ Value \qquad (9)$$

Table 10 describes the difference in the weight value, in case the risk is considered or not.

4.2.4. RC2AS Color-Coding Scheme

Influenced by the current color coding highlighted in Table 4, RC2AS enhances the color-coding scheme to reflect the risk level of the compliance status. So, in the case of low risk, lighter degrees of the color are used, whereas, in the case of high risk, darker degrees of the color are used. This new color-coding scheme is applied only to the "Implemented" and "Partially Implemented" compliance statuses. There are no changes for the "Not Implemented" and "Not Applicable" statuses. To elaborate, Table 11 shows the used color depending on two factors: the compliance status and the risk-based weight value.

Table 10. RC2AS compliance answers along with the corresponding weight value while considering the risk level or not.

RC2AS Compliance Answer	Weight Value	Risk Level	Risk Value	Risk-Based Weight Value
Yes (100%)	1	High Medium Low	1 0.75 0.5	1.00 0.75 0.5
Partially (>85% to <100%)	0.75	High Medium Low	1 0.75 0.5	0.75 0.56 0.38
Partially (>35% to <=85%)	0.50	High Medium Low	1 0.75 0.5	0.50 0.38 0.25
Partially (>0% to <=35%)	0.25	High Medium Low	1 0.75 0.5	0.25 0.19 0.13
No (0%)	0	High Medium Low	0	0.00
Not Applicable (NA)	NA	High Medium Low	NA	NA

Table 11. RC2AS color-coding scheme.

Compliance Status	Risk-Based Weight Value	Color
Implemented	100%	
	75%	
	50%	
Partially Implemented	>85% to <100%	
	>35% to ≤ 85%	
	>0% to ≤ 35%	
Not Implemented	0	
Not Applicable	N/A	

4.2.5. RC2AS Rich Dashboards

Lastly, the dashboard page will be displayed after filling out the self-assessment and answering all of the questions related to all subdomains. First, the user needs to select the model from the following options: (strict compliance, semi-strict compliance, weighted compliance, and RC2AS weighted compliance) along with the date range. Accordingly, comprehensive dashboards will be generated and displayed. These dashboards include (a) the overall compliance score with ECC based on the selected model. Moreover, they include the overall compliance score across all models (Figure 6), and (b) the compliance score for each main domain per compliance statute is also based on the selected model (Figure 7). Accordingly, they provide the organization with insight into its cybersecurity posture per domain, and they provide a detailed view of each main domain and the compliance level for each control. This helps the organization to gain valuable insight into the most critical compliance concerns, which will support it in building its activities and actions to enhance its cybersecurity controls and prepare action plans accordingly. (c) The RC2AS evaluation of the organization among dates ranges across the main domains based on the selected model (Figure 8), which allows for the close monitoring of the changes within the domain controls along with the selected duration, and (d) the proposed action plan for the organization based on its current compliance status.

As mentioned before, the ECC has 29 subdomains. Each subdomain has a risk level (high, medium, and low) that the regulator redefines. Accordingly, based on compliance level, the subdomains will be divided among these risk levels using the proposed color-coding scheme. The visualized action plan helps the organization understand its current compliance status with ECC. A complete summary is provided for all domains and subdomains, including their compliance and risk levels. Additionally, this plan helps the organization to identify its weaknesses and design a well-structured strategy to enhance its level based on its priorities and resources. To illustrate more, the chart in Figure 9 allows the organization to prioritize its actions. For example, the subdomains that are not fully implemented and have a high risk level should be considered first. Therefore, this will be used as an action plan to draw future implementations based on the risk and compliance levels. Moreover, such a summary encourages the organization to continuously implement the subdomains that are categorized as high-risk and fully implemented.

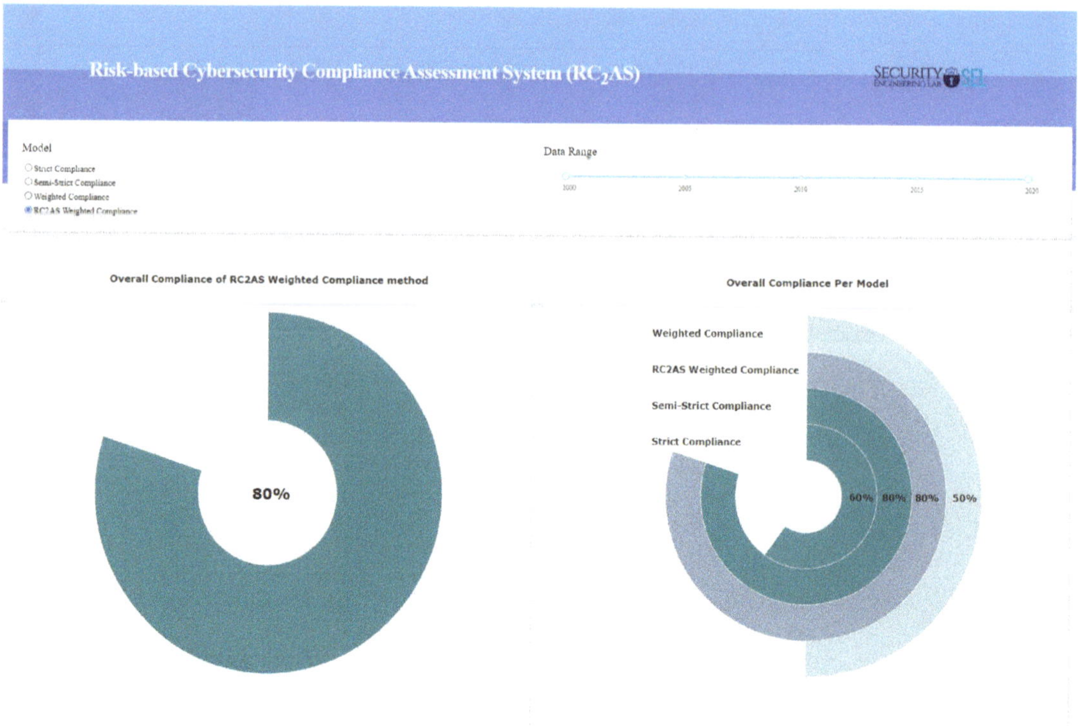

Figure 6. RC2AS dashboard of ECC—(a).

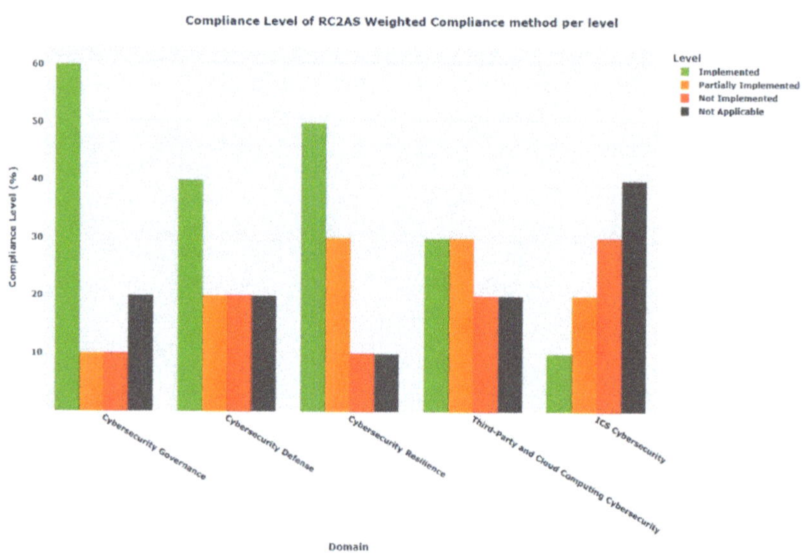

Figure 7. RC2AS dashboard of ECC—(b).

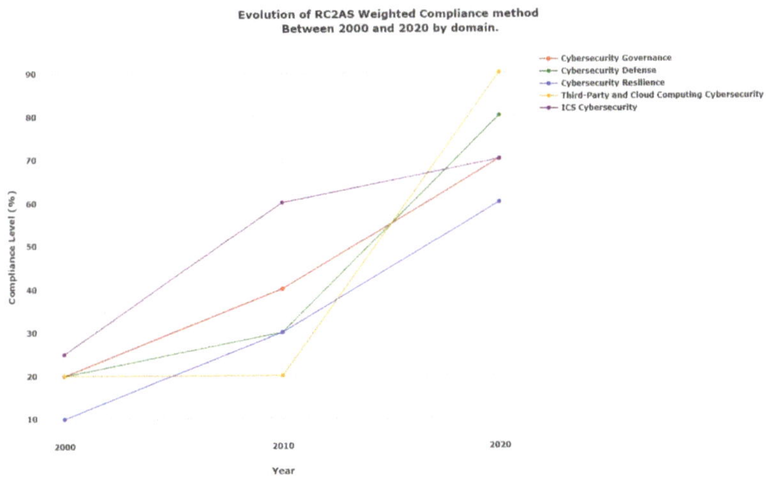

Figure 8. RC2AS dashboard of ECC—(c).

Figure 9. RC2AS dashboard of ECC—(d).

5. Proposed System Evaluation and Results Discussion

As highlighted before, the current ECC did not publish any compliance score calculation methods. Therefore, to evaluate the effectiveness of the proposed risk-based cybersecurity compliance assessment system (RC2AS) and examine all of its proposed calculation methods, several scenarios have been considered. These scenarios aim to conduct a deep comparative analysis OF these methods. Two entities, X and Y, will be utilized in implementing the different calculation models. For simplicity, one of the ECC subdomains is chosen to run these scenarios, which is subdomain (1-2) -"Cybersecurity Management", as shown in Table 12.

Table 12. Subdomain (1-2)-"Cybersecurity Management".

1-2	Cybersecurity Management
Control No.	**Control Clauses**
1-2-1	A dedicated cybersecurity function (e.g., division, department) must be established within the organization. This function must be independent of the Information technology/information communication and technology (IT/ICT) functions (as per the Royal Decree number 37140 dated 14/8/1438H). It is highly recommended that this cybersecurity function reports directly to the head of the organization or his/her delegate while ensuring that this does not result in a conflict of interest [13].
1-2-2	The position of cybersecurity function head (e.g., CISO), and related supervisory and critical positions within the function, must be filled with full-time and experienced Saudi cybersecurity professionals [13].
1-2-3	A cybersecurity steering committee must be established by the authorizing official to ensure the support and implementation of the cybersecurity programs and initiatives within the organization. Committee members, roles and responsibilities, and governance framework must be defined, documented, and approved. The committee must include the head of the cybersecurity function as one of its members. It is highly recommended that the committee reports directly to the head of the organization or his/her delegate while ensuring that this does not result in a conflict of interest [13].

5.1. Comparison Using the "Strict Compliance" Method

Assume that there are two entities (entity X and entity Y). Both entities implement controls (1-2-1 and 1-2-2) in subdomains (1-2). On the other hand, control (1-2-3) is "Partially Implemented" by entity X, while in entity Y it is "Not Implemented". Table 13 compares the number of each compliance status for subdomain (1-2) between the entities X and Y.

Table 13. Comparison of the number of each compliance status between two entities (X and Y).

Compliance Status	Implemented	Partially Implemented	Not Implemented	Not Applicable
Entity X	2	1	0	0
Entity Y	2	0	1	0

Accordingly, applying (Equation (5)) of the compliance score for both entities will result in achieving the same score value. Table 14 compares the overall compliance score for entities X and Y using the "Strict Compliance" method. This equation considers only the number of fully implemented controls out of the total controls. In this scenario, the number of fully implemented controls for both entities is (2); then, the compliance score is calculated as follows: $(\frac{2}{3} * 100) = (66.67\%)$.

Consequently, it is observed that this method does not reflect the accurate, current state of compliance with subdomain (1-2). The implementation percentage on this subdomain is supposed to measure the overall compliance score. However, this method counts only the fully implemented controls and ignores everything else. So, "Partially Implemented" and "Not Implemented" statuses are not included.

Table 14. Comparison of the overall compliance score between the two entities, using "Strict Compliance" method.

Method Name	Entity X	Entity Y
Strict Compliance Method	66.67%	66.67%

5.2. Comparison Using the "Semi-Strict Compliance" Method

In this scenario, both entities have the same compliance status for all controls (1-2-1, 1-2-2, and 1-2-3), which are ("Implemented", "Implemented", and "Partially Implemented"). The main difference is in the implementation percentage for control (1-2-3). As illustrated in Table 15, entity X implemented around 70% of this control, whereas entity Y implemented 30%.

Table 15. Comparison of the number of each compliance status between two entities (X and Y).

Compliance Status	Implemented	Partially Implemented	Not Implemented	Not Applicable
Entity X	2	1 (70%)	0	0
Entity Y	2	1 (30%)	0	0

Nevertheless, by using (Equation (6)), the overall compliance score for entity X is the same as entity Y. To be specific, the compliance calculated for both entities will be $(\frac{2+0.5}{3} * 100) = (83.33\%)$. In summary, both entities' "Partially Implemented" status will be treated the same (weight value 0.5) even though the implementation percentage is different. Therefore, this method still does not accurately reflect the overall compliance score, as described in Table 16.

Table 16. Comparison of overall compliance score between two entities by using "Semi-Strict Compliance Method".

Method Name	Entity X	Entity Y
Strict Compliance Method	66.67%	66.67%
Semi-Strict Compliance Method	83.33%	83.33%

5.3. Comparison Using the "Weighted Compliance" Method

In this method, we include the weight of the "Partially Implemented" status as part of the calculation. We will continue with the same scenario in the above method (Section 5.3), but in this case, we will evaluate the weights for the level of "Partially Implemented" (3 levels), which are ((a) >0% to <=35%, (b) >35% to<=85%, (c) >85% to <100%) taken into account in the overall calculation compliance score. The RC2AS self-assessment for entity X is shown in Table 17, while entity Y is shown in Table 18.

Table 17. Sample of RC2AS self-assessment of entity X.

1-2	Cybersecurity Management			
Control No.	Control Clauses	Question	RC2AS Compliance Answer	Compliance Status
1-2-1	A dedicated cybersecurity function (e.g., division, department) must be established within the organization. etc. [13].	**Q1:** Does the entity have a cybersecurity department?	Yes (100%)	Implemented
		Q2: Is the cybersecurity department independent of information technology management in the entity?	Yes (100%)	
		Q3: Does the cybersecurity department report directly to the organization's head or his/her delegate while ensuring that this does not result in a conflict of interest?	Yes (100%)	
1-2-2	The position of cybersecurity function head (e.g., CISO), and related supervisory and critical positions within the function, etc. [13].	**Q1:** Is the cybersecurity department function headed by full-time and experienced Saudi cybersecurity professionals?	Yes (100%)	Implemented
		Q2: Have the related supervisory and critical positions in the cybersecurity department functioned with full-time and experienced Saudi cybersecurity professionals?	Yes (100%)	
1-2-3	A cybersecurity steering committee must be established by the Authorizing Official to ensure the support and implementation of the cybersecurity programs and initiatives within the organization. Committee members, roles, etc. [13].	**Q1:** Does the entity have a cybersecurity steering committee to ensure the support and implementation of cybersecurity programs and initiatives within the organization?	Yes (100 %)	Partially Implemented
		Q2: Does the entity have defined, documented, and approved the committee members, roles and responsibilities, and governance framework?	Partially Imp. (>85% to <100%)	
		Q3: Does the committee in the entity include the head of the cybersecurity function as one of its members?	Yes (100%)	
		Q4: Does the committee report directly to the head of the organization or his/her delegate while ensuring that this does not result in a conflict of interest?	No (0%)	

Table 18. Sample of RC2AS self-assessment of entity Y.

1-2		Cybersecurity Management		
Control No.	Control Clauses	Question	RC2AS Compliance Answer	Compliance Status
1-2-1	A dedicated cybersecurity function (e.g., division, department) must be established within the organization. This etc. [13].	**Q1:** Does the entity have a cybersecurity department?	Yes (100%)	Implemented
		Q2: Is the cybersecurity department independent of information technology management in the entity?	Yes (100%)	
		Q3: Does the cybersecurity department report directly to the organization's head or his/her delegate while ensuring that this does not result in a conflict of interest?	Yes (100%)	
1-2-2	The position of cybersecurity function head (e.g., CISO), and related supervisory and critical positions within the function, etc. [13].	**Q1:** Is the cybersecurity department function headed by full-time and experienced Saudi cybersecurity professionals?	Yes (100%)	Implemented
		Q2: Have the related supervisory and critical positions in the cybersecurity department functioned with full-time and experienced Saudi cybersecurity professionals?	Yes (100%)	
1-2-3	A cybersecurity steering committee must be established by the authorizing official to ensure the support and implementation of the cybersecurity programs and initiatives within the organization. Committee members, roles and responsibilities, etc. [13].	**Q1:** Does the entity have a cybersecurity steering committee to ensure the support and implementation of the cybersecurity programs and initiatives within the organization?	Yes (100%)	Partially Implemented
		Q2: Does the entity have defined, documented, and approved the committee members, roles and responsibilities, and governance framework?	Partially Imp. (>0% to <35%)	
		Q3: Does the committee in the entity include the head of the cybersecurity function as one of its members?	No (0%)	
		Q4: Does the committee report directly to the head of the organization or his/her delegate while ensuring that this does not result in a conflict of interest?	No (0%)	

In summary, the difference in the implementation percentage for control (1-2-3) will be presented in this method. To illustrate this further, the implementation percentage for entity X on this control is around (70%), while for entity Y it is around (30%). Hence, by using (Equation (7)), the weight value for entity X in control (1-2-3) will be $(\frac{1+0.75+1+0}{4})$ = (0.69%), so the overall compliance score for entity X in subdomain (1-2) will be followed, $(\frac{2+0.69}{3} * 100)$ = (89.67%). In the same way, the weight value for entity Y in control (1-2-3) will be $(\frac{1+0.25+0+0}{4})$ = (0.31%), so the overall compliance score for entity X in subdomain (1-2) will be $(\frac{2+0.31}{3} * 100)$ = (77.00%).

Therefore, this method has two main advantages: (a) the reflective color coding for the compliance status; and (b) the overall compliance score of this subdomain, which better integrates the partially implemented control in the score calculation. Table 19 compares a case of including the weight value for "Partially Implemented" status on subdomain (1-2): "Cybersecurity Management".

Table 19. Comparison of overall compliance score between two entities by using "Weighted Compliance" Method.

Method Name	Entity X	Entity Y
Strict Compliance Method	66.67%	66.67%
Semi-Strict Compliance Method	83.33%	83.33%
Weighted Compliance Method	89.67%	77.00%

5.4. Comparison Using "RC2AS Weighted Compliance" Method

In the same way, we will continue with the example mentioned before in Section 5.2. However, here we will evaluate the overall compliance score while considering (a) the weight for the level of "Partially Implemented" status, and (b) the risk level of the subdomain that the regulator has predefined. Assume the risk level of subdomain (1-2) is "Low" for both entities. The self-assessments for both entities are shown in Tables 20 and 21.

This method presented the difference in compliance for both entities, specifically for control (1-2-3) with the risk values included. The overall compliance score of the subdomain (1-2) for both entities has been reduced due to the risk value for this subdomain being

"Low". For more illustrations, the risk-based weight value for both entities in controls (1-2-1) and (1-2-2) using Equations (8) and (9) will be reduced from value 1 without risk to 0.5 after a low-risk value. At the same time, the overall compliance score for entity X in control (1-2-3) will be ($\frac{0.5+0.38+0.5+0}{4}$) = (0.343%). So, the overall compliance score for subdomain (1-2) will be ($\frac{0.5+0.5+0.343}{3} * 100$) = (44.79%). Similarly, the risk-based weight value for entity Y in control (1-2-3) will be ($\frac{0.5+0.34+0+0}{4}$) = (0.16%). So, the overall compliance score for entity X in subdomain (1-2) will be followed, ($\frac{1+0.16}{3} * 100$) = (38.54%).

Table 20. Sample of RC2AS self-assessment of entity X.

	1-2	Cybersecurity Management		
	Risk Level:	Low		
Control No.	Control Clauses	Question	RC2AS Compliance Answer	Compliance Status
1-2-1	A dedicated cybersecurity function (e.g., division, department) must be established within the organization, etc. [13].	**Q1:** Does the entity have a cybersecurity department?	Yes (100%)	Implemented
		Q2: Is the cybersecurity department independent of information technology management in the entity?	Yes (100%)	
		Q3: Does the cybersecurity department report directly to the head of the organization or his/her delegate while ensuring that this does not result in a conflict of interest?	Yes (100%)	
1-2-2	The position of cybersecurity function head (e.g., CISO), and related supervisory and critical positions within the function, etc. [13].	**Q1:** Is the cybersecurity department function headed by full-time and experienced Saudi cybersecurity professionals?	Yes (100%)	Implemented
		Q2: Do the related supervisory and critical positions in cybersecurity department function with full-time and experienced Saudi cybersecurity professionals?	Yes (100%)	
1-2-3	A cybersecurity steering committee must be established by the authorizing official to ensure the support and implementation of the cybersecurity programs and initiatives within the organization. Committee members, roles, and etc.[13].	**Q1:** Does the entity have a cybersecurity steering committee to support and implement cybersecurity programs and initiatives within the organization?	Yes (100%)	Partially Implemented
		Q2: Does the entity have defined, documented, and approved the committee members, roles and responsibilities, and governance framework?	Partially Imp. (>85% to <100%)	
		Q3: Does the committee in the entity include the head of the cybersecurity function as one of its members?	Yes (100%)	
		Q4: Does the committee report directly to the head of the organization or his/her delegate while ensuring that this does not result in a conflict of interest?	No (0%)	

Accordingly, by using this method it has been noticed that **(b)** the overall compliance score has been reduced here as this is an indicator that the entity needs to focus first on the subdomain with the highest level of impact. Furthermore, **(a)** the color coding for the "Implemented" and "Partially Implemented" statuses are lighter than the normal ones, as been affected by risk value as illustrated in Table 11. The comparison between the compliance level by including "Risk Value" is shown in Table 22.

As another example, we will provide the advantage of adding the risk level to the compliance score calculation method. Let us assume that we have two entities whose compliance statuses are similar for subdomain (1-2). However, the risk level is different as classified by the regulator. The risk level for subdomain (1-2) for entity X is "High", while for entity Y it is "Medium". The self-assessment for entity X is presented in Table 23:

Table 21. Sample of RC2AS self-assessment of entity Y.

	1-2	Cybersecurity Management		
	Risk Level:	Low		
Control No.	Control Clauses	Question	RC2AS Compliance Answer	Compliance Status
1-2-1	A dedicated cybersecurity function (e.g., division, department) must be established within the organization. This etc. [13].	**Q1:** Does the entity have a cybersecurity department?	Yes (100%)	Implemented
		Q2: Is the cybersecurity department independent of information technology management in the entity?	Yes (100%)	
		Q3: Does the cybersecurity department report directly to the head of the organization or his/her delegate while ensuring that this does not result in a conflict of interest?	Yes (100%)	
1-2-2	The position of cybersecurity function head (e.g., CISO), and related supervisory and critical positions within the function, etc. [13].	**Q1:** Is the cybersecurity department function headed by full-time and experienced Saudi cybersecurity professionals?	Yes (100%)	Implemented
		Q2: Have the related supervisory and critical positions in the cybersecurity department function with full-time and experienced Saudi cybersecurity professionals?	Yes (100%)	
1-2-3	A cybersecurity steering committee must be established by the authorizing official to ensure the support and implementation of the cybersecurity programs and initiatives within the organization. Committee members, roles, etc. [13].	**Q1:** Does the entity have a cybersecurity steering committee to support and implement cybersecurity programs and initiatives within the organization?	Yes (100%)	Partially Implemented
		Q2: Does the entity have defined, documented, and approved the committee members, roles and responsibilities, and governance framework?	Partially Imp. (>0% to <35%)	
		Q3: Does the committee in the entity include the head of the cybersecurity function as one of its members?	No (0%)	
		Q4: Does the committee report directly to the head of the organization or his/her delegate while ensuring that this does not result in a conflict of interest?	No (0%)	

Table 22. Comparison of overall compliance score between two entities by using the "RC2AS Weighted Compliance" method in case of low-risk control.

Method Name	Entity X	Entity Y
Strict Compliance Method	66.67%	66.67%
Semi-Strict Compliance Method	83.33%	83.33%
Weighted Compliance Method	89.67%	77.00%
RC2AS Weighted Compliance Method	44.79%	38.54%

Although both entities have the same implementation percentage and the only difference between the two entities is the risk level of this control, the compliance score will be different since one entity has a "High" risk and the other has a "Low" risk level. Using Equation (8) to calculate the overall compliance score for both entities, the compliance score for entity X in subdomain (1-2) is calculated as $(\frac{1+1+0.69}{3} * 100) = (89.58\%)$, whereas the compliance score for entity Y is $(\frac{0.5+0.5+0.343}{3} * 100) = (44.79\%)$. Entity X obtains a higher compliance level for subdomain (1-2) than entity Y, even though entity Y implemented the controls for this subdomain (1-2) similar to entity X. The compliance level of this subdomain is affected by the risk level. The comparison of the overall compliance score is shown in Table 24.

Table 23. Sample of self-assessment of entity X (risk level: high).

Control No.	Control Clauses	Question	Answer	Compliance Status
1-2		Cybersecurity Management		
Risk Level:		High		
1-2-1	A dedicated cybersecurity function (e.g., division, department) must be established within the organization, etc. [13].	**Q1:** Does the entity have a cybersecurity department?	Yes (100%)	Implemented
		Q2: Is the cybersecurity department independent of information technology management in the entity?	Yes (100%)	
		Q3: Does the cybersecurity department report directly to the head of the organization or his/her delegate while ensuring that this does not result in a conflict of interest?	Yes (100%)	
1-2-2	The position of cybersecurity function head (e.g., CISO), and related supervisory and critical positions within the function, etc. [13].	**Q1:** Is the cybersecurity department function headed by full-time and experienced Saudi cybersecurity professionals?	Yes (100%)	Implemented
		Q2: Do the related supervisory and critical positions in cybersecurity department function with full-time and experienced Saudi cybersecurity professionals?	Yes (100%)	
1-2-3	A cybersecurity steering committee must be established by the Authorizing Official to ensure the support and implementation of the cybersecurity programs and initiatives within the organization. Committee members, roles and, etc. [13].	**Q1:** Does the entity have a cybersecurity steering committee to support and implement cybersecurity programs and initiatives within the organization?	Yes (100%)	Partially Implemented
		Q2: Does the entity have defined, documented, and approved the committee members, roles and responsibilities, and governance framework?	Partially Imp. (>35% to <=85%)	
		Q3: Does the committee in the entity include the head of the cybersecurity function as one of its members?	Yes (100%)	
		Q4: Does the committee report directly to the head of the organization or his/her delegate while ensuring that this does not result in a conflict of interest?	No (0%)	

Table 24. Comparison between two entities with different risks.

Method Name	Entity X Risk Level: High	Entity Y Risk Level: Low
RC2AS Weighted Compliance Method	89.58%	44.79%

5.5. Real Case Study

As part of the validation of the new proposed system RC2AS, we conducted a real case study to provide a convincing argument for the efficacy of the proposed risk-based cybersecurity compliance assessment system and demonstrate its services through a real-life environment. We conducted a real study with a figure organization in Saudi Arabia in the academic sector. The selected organization's name is kept private here for confidentiality and privacy (anonymity) reasons and be referred to as (Alpha). The Alpha has won several accolades and distinctions for its outstanding academic and scientific achievements. An effective way to learn about the advantages and disadvantages of the proposed solution is to complete a case study with experts who are acquainted with the compliance assessment process. This strategy ensures that the appropriate individuals review the tool, provide inputs and insightful information about the suggested solution, and enhance its efficacy, leading to a more thorough and informative assessment. The evaluation process has been conducted with specialized experts at the Alpha organization. The phases of the evaluation process are as follows:

(A) **RC2AS Overview Description:** In this phase, before the specialized experts start using and experimenting with the proposed RC2AS self-assessment supporting tool, we arranged several workshops with the organization introducing the RC2AS tool and exploring its main features and functions. The proposed RC2AS tool has been offered as an offline version through a dedicated Excel file.

(B) **RC2AS Experiment:** This phase offered a live experiment, with the Alpa organization given a chance to perform a live examination of the RC2AS supporting self-assessment tool by itself and start answering the question(s) on each subdomain to assess its cybersecurity compliance based on its domain. The RC2AS supporting tool uses a self-assessment questionnaire (SAQ) approach.

(C) **RC2AS Evaluation:** After finishing the assessment process, we conducted a final workshop with them to obatin their insights, recommendations, and feedback.

In conclusion, based on the feedback from the participants, the RC2AS is useful and valuable for the cybersecurity compliance assessment process, not only by expediting the assessment process but also by letting the organization choose between one of the proposed overall compliance-calculation methods based on their needs and resources. Finally, we have incorporated some of their recommendations to enhance the RC2AS tool functions and their user experience. Moreover, some of the suggestions will be considered for improvement in the proposed RC2AS solution in future work.

5.6. Results Discussion

Based on the comparisons conducted among the proposed calculation methods, it can be observed that the "RC2AS weighted compliance" method can provide organizations with a fair and accurate assessment that measures the current cybersecurity compliance level with ECC.

Additionally, there is no right or wrong method as each calculation method has a different purpose and usage based on the organization's needs and objectives and its business functions. Therefore, the main comparisons among the calculation methods with regard to purpose and usage are summarized in Table 25:

Table 25. The comparison among RC2AS proposed methods in terms of purpose and usage .

Method Name	Purpose and Usage
Strict Method	This method helps the organization to fully comply with the controls and requirements. Having a weak hole within the control will lead to exposing the entity and prevent achieving the objective of that control.
Semi-Strict Method	This method takes into consideration the efforts that the organization has made to comply with the requirements by increasing the compliance score of the control's requirements.
Weighted Compliance	This method differentiates between the level of implementation and gives a more specific score on the requirements that have been implemented. Thus, this will give a better idea of the progress of the entities' current status.
RC2AS Weighted Compliance	This method provides a better understanding of the organization's domain and status by differentiating between organizations with different scopes, business functionalities, and criticality levels. This method includes the risk level of the subdomains.

6. Conclusions and Future Work

Complying with the cybersecurity regulatory framework becomes essential to reduce the risk of security attacks and protect the nation's individuals, organizations, and economies. Such compliance is encouraged regionally and internationally. Therefore, Saudi Arabia is leading in defining a comprehensive cybersecurity regulatory standard that is followed and assessed through the introduced essential cybersecurity control (ECC).

The compliance assessments of international cybersecurity standards are general for all organizations regardless of their domains, business functionality, and criticality level. In addition, their current assessment approach does not consider that the risk level in case the security control is implemented or not in reference to the organization's scope. Having a unified compliance assessment process for all organizations may affect the national cybersecurity landscape. To ensure the protection of organizations with critical infrastructure, further factors need to be injected into the compliance assessment process.

Therefore, this research has been motivated to build a comprehensive and customized risk-based cybersecurity compliance assessment system (RC2AS) based on ECC, which is well-defined and inspired by many international standards. All national organizations having IT systems as part of their infrastructure need to orient themselves toward ECC using its assessment tool. RC2AS introduces a self-assessment tool that allows an organization to measure its compliance with the ECC and calculate the compliance score using different compliance-calculation methods that meet the organization's needs, criticality, and resources. This will provide a realistic, fair, and accurate assessment of the organization's compliance with the ECC. The offered assessment tool by RC2AS provides enhancements not only in regard to compliance score calculations but also to the assessment methodology, carried out in a very convenient way with expressive color-coding schemes. The assessment results are visualized in rich dashboards that illustrate the organization's current status in fully complying with the ECC. The RC2AS tool guides the organization by addressing its weaknesses, setting a proper plan to maintain what has been achieved and suggesting possible solutions and ways to improve.

The proposed RC2AS has been evaluated by conducting several case studies that examine all of the suggested compliance-calculation methods, including "Strict Compliance", "Semi-Strict Compliance", "Weighted Compliance", and "RC2AS Weighted Compliance" methods, where each method has different features and equations. The selection of the method depends on the nature of the organization. A deep comparative analysis was conducted to differentiate between these methods and recommend their application scopes.

Even though the proposed RC2AS solution considers many factors that facilitate and help organizations assess their current cybersecurity compliance posture, it can be customized to meet the organization's needs, for instance, integrating the RC2AS with the existing internal mechanisms in the organization to avoid duplication of efforts. The employment of this feature will not only be effective but will also ensure a seamless experience for users.

For future work, an online version of RC2AS could be offered. In addition, RC2AS will be offered to be used by different types of organizations to take their inputs and suggestions for improvements. Finally, RC2AS can be adopted by other international standards and controls.

Author Contributions: Conceptualization, I.A. and A.A.; methodology, I.A.; software, A.A.; validation, A.A., I.A. and M.A.; formal analysis, A.A. and I.A; investigation, A.A., I.A. and M.A; resources, A.A. and I.A; data curation, A.A. and I.A; writing—original draft preparation, A.A, I.A and M.A.; writing—review and editing, I.A.; visualization, A.A., I.A. and M.A.; supervision, I.A.; project administration, I.A.; and funding acquisition, I.A. All authors have read and agreed to the published version of the manuscript.

Funding: The APC was funded by Prince Sultan University, Riyadh, Saudi Arabia.

Institutional Review Board Statement: Not applicable.

Informed Consent Statement: Not applicable.

Data Availability Statement: Not applicable.

Acknowledgments: The authors would like to thank the Security Engineering Lab members at Prince Sultan University for their technical support.

Conflicts of Interest: The authors declare no conflict of interest.

Abbreviations

The following abbreviations are used in this manuscript:

NCA	National Cybersecurity Authority
ECC	Essential cybersecurity controls
SAMA	Saudi Arabian Monetary Authority

References

1. Li, Y.; Liu, Q. A comprehensive review study of cyberattacks and cyber security; Emerging trends and recent developments. *Energy Rep.* **2021**, *7*, 8176–8186. [CrossRef]
2. He, W.; Zhang, Z.J.; Li, W. Information technology solutions, challenges, and suggestions for tackling the COVID-19 pandemic. *Int. J. Inf. Manag.* **2021**, *57*, 102287. [CrossRef] [PubMed]
3. AlDaajeh, S.; Saleous, H.; Alrabaee, S.; Barka, E.; Breitinger, F.; Raymond Choo, K.K. The role of national cybersecurity strategies on the improvement of cybersecurity education. *Comput. Secur.* **2022**, *119*, 102754. [CrossRef]
4. Dalal, R.S.; Howard, D.J.; Bennett, R.J.; Posey, C.; Zaccaro, S.J.; Brummel, B.J. Organizational science and cybersecurity: Abundant opportunities for research at the interface. *J. Bus. Psychol.* **2021**, *37*, 1–29. [CrossRef]

5. Perera, S.; Jin, X.; Maurushat, A.; Opoku, D.G.J. Factors Affecting Reputational Damage to Organisations Due to Cyberattacks. *Informatics* **2022**, *9*, 28. [CrossRef]
6. Fathi, S.; Hikal, N. A Review of Cyber-security Measuring and Assessment Methods for Modern Enterprises. *JOIV Int. J. Inform. Vis.* **2019**, 3, 157–172. [CrossRef]
7. Bailey, T.; Greis, J.; Watters, M.; Welle, J. Cybersecurity Legislation: Preparing for Increased Reporting and Transparency. 2022. Available online: https://www.mckinsey.com/capabilities/risk-and-resilience/ourinsights/cybersecurity/cybersecurity-legislation-preparing-for-increased-reporting-and-transparency (accessedon 26 July 2022).
8. ISO/IEC 27001:2013; Information Technology—Security Techniques—Information Security Management Systems—Requirements, The International Organization for Standardization (ISO): Geneva, Switzerland, 2013.
9. Almuhammadi, S.; Alsaleh, M. Information Security Maturity Model for Nist Cyber Security Framework. In Proceedings of the Sixth International Conference on Information Technology Convergence and Services. Academy and Industry Research Collaboration Center (AIRCC), Sydney, Australia, 25–26 February 2017; pp. 51–62. [CrossRef]
10. Lee, Y.C. Financial Sector's Cybersecurity. 2021. Available online: https://docslib.org/doc/12762763/financial-sectors-cybersecurity-a-regulatory-digest (accessed on 20 September 2020).
11. Almudaires, F.; Rahman, M.H.; Almudaires, M. An Overview of Cybersecurity, Data Size and Cloud Computing in light of Saudi Arabia 2030 Vision. In Proceedings of the 2021 International Conference on Information Technology (ICIT), Amman, Jordan, 14–15 July 2021; pp. 268–273.
12. NCA. Global Cybersecurity Index 2020—International Telecommunication Union. 2020. Available online: https://www.itu.int/dms_pub/itu-d/opb/str/D-STR-GCI.01-2021-PDF-E.pdf (accessed on 20 February 2023).
13. NCA. Essential Cybersecurity Controls (ECC-1: 2018). 2018. Available online: https://nca.gov.sa/files/ecc-en.pdf (accessed on 20 July 2022).
14. von der Heyde, M.; Gerl, A.; Seck, R.; Groß, R.; Watkowski, L. Applying COBIT 2019 to IT Governance in Higher Education—Establishing IT governance for the collaboration of all universities and universities of applied sciences in Bavaria. In Proceedings of the Conference: INFORMATIK 2020, Karlsruhe, Germany, 2 October 2021. [CrossRef]
15. Corallo, A.; Lazoi, M.; Lezzi, M.; Luperto, A. Cybersecurity awareness in the context of the Industrial Internet of Things: A systematic literature review. *Comput. Ind.* **2022**, *137*, 103614. [CrossRef]
16. Asaithambi, S.; Ravi, L.; Kotb, H.; Milyani, A.H.; Azhari, A.A.; Nallusamy, S.; Varadarajan, V.; Vairavasundaram, S. An Energy-Efficient and Blockchain-Integrated Software Defined Network for the Industrial Internet of Things. *Sensors* **2022**, *22*, 7917. [CrossRef]
17. Sarabdeen, J.; Chikhaoui, E.; Ishak, M.M.M. Creating standards for Canadian health data protection during health emergency—An analysis of privacy regulations and laws. *Heliyon* **2022**, *8*, e09458. [CrossRef]
18. Aliyu, A.; Maglaras, L.; He, Y.; Yevseyeva, I.; Boiten, E.; Cook, A.; Janicke, H. A holistic cybersecurity maturity assessment framework for higher education institutions in the United Kingdom. *Appl. Sci.* **2020**, *10*, 3660. [CrossRef]
19. Zarour, M.; Alhammad, N.; Alenezi, M.; Alsarayrah, K. A Research on DevOps Maturity Models. *Int. J. Recent Technol. Eng.* **2019**, *8*, 4854–4862. [CrossRef]
20. Proença, D.; Borbinha, J. Information security management systems—A maturity model based on ISO/IEC 27001. In *Proceedings of the Lecture Notes in Business Information Processing*; Springer: Berlin/Heidelberg, Germany, 2018; Volume 320, pp. 102–114. [CrossRef]
21. Bolanio, J.B.; Paredes, R.K.; Yoldan, A.L., Jr.; Acapulco, R.E., II. Network Security Policy for Higher Education Institutions based on ISO Standards. *Mediterr. J. Basic Appl. Sci.* **2021**, *5*, 1–17. [CrossRef]
22. ISO/IEC 27033-1:2010;Information Technology—Security Techniques—Network Security—Part 1: Overview and Concepts. The International Organization for Standardization (ISO): Geneva, Switzerland, 2010.
23. Makupi, D.; Masese, N. Determining Information Security Maturity Level of an organization based on ISO 27001. *Int. J. Comput. Sci. Eng.* **2019**, *6*, 5–11. [CrossRef]
24. Yaokumah, W.; Dawson, A.A. *Network and Data Transfer Security Management in Higher Educational Institutions*; IGI Global: Hershey, PA, USA, 2019; pp. 1–19.
25. ISO/IEC 21827:2008; Information Technology—Security Techniques–Systems Security Engineering—Capability Maturity Model (SSE-CMM). The International Organization for Standardization (ISO): Geneva, Switzerland, 2008.
26. Mantra, I.; Rahman, A.A.; Saragih, H. Maturity Framework Analysis ISO 27001: 2013 on Indonesian Higher Education. *Int. J. Eng. Technol.* **2020**, *9*, 429–436. [CrossRef]
27. Tejay, G.; Goel, S. Editorial: Time to move away from compliance—Cybersecurity in the context of needs and investments of organizations. *Organ. Cybersecur. J. Pract. Process. People* **2022**, *2*, 1–2. [CrossRef]
28. Mijwil, M.; Filali, Y.; Aljanabi, M.; Bounabi, M.; Al-Shahwani, H.; ChatGPT. The Purpose of Cybersecurity Governance in the Digital Transformation of Public Services and Protecting the Digital Environment. *Mesopotamian J. Cybersecur.* **2023**, *2023*, 2–4.
29. Suwito, M.H.; Matsumoto, S.; Kawamoto, J.; Gollmann, D.; Sakurai, K. An analysis of IT assessment security maturity in higher education institution. In *Proceedings of the Information Science and Applications (ICISA) 2016*; Springer: Berlin/Heidelberg, Germany, 2016; Volume 376, pp. 701–713. [CrossRef]
30. Hung, C.; Hwang, M.; Liu, Y. Building a Maturity Model of Information Security Governance for Technological Colleges and Universities in Taiwan. *Appl. Mech. Mater.* **2013**, *284–287*, 3657–3661. [CrossRef]

31. Bass, J.M. An Early-Stage ICT Maturity Model derived from Ethiopian education institutions. *Int. J. Educ. Dev. Using Inf. Commun. Technol. IJEDICT* **2011**, *7*, 5–25.
32. Ismail, Z.; Masrom, M.; Sidek, Z.; Hamzah, D. Framework to Manage Information Security for Malaysian Academic Environment. *J. Inf. Assur. Cybersecur.* **2010**, 2010, 1–16.
33. Dehlawi, Z.; Abokhodair, N. Saudi Arabia's response to cyber conflict: A case study of the Shamoon malware incident. In Proceedings of the 2013 IEEE International Conference on Intelligence and Security Informatics, Seattle, WA, USA, 4–7 June 2013; pp. 73–75. [CrossRef]
34. Saudi GAZETTE Report. King Orders Setting Up of National Cyber Security Authority. 2017. Available online: https://saudigazette.com.sa/article/520782/SAUDI-ARABIA/King-orders-setting-up-of-National-Cyber-Security-Authority (accessed on 26 August 2022).
35. CITC. Cybersecurity Regulatory Framework. 2020. Available online: https://www.citc.gov.sa/en/RulesandSystems/CyberSecurity/Documents/CRF-en.pdf (accessed on 20 August 2022).
36. SAMA. Cyber Security Framework Saudi Arabian Monetary Authority. 2017. Available online: https://www.sama.gov.sa/enUS/Laws/BankingRules/SAMA20Cyber/20Security/20Framework.pdf (accessed on 20 July 2022).
37. Hamed, T.A.; Alenezi, M. Business Continuity Management & Disaster Recovery Capabilities in Saudi Arabia ICT Businesses. *Int. J. Hybrid Inf. Technol.* **2016**, *9*, 99–126. [CrossRef]
38. Nurunnabi, M. IFRS and Saudi accounting standards: A critical investigation. *Int. J. Discl. Gov.* **2017**, *14*, 4854–4862. [CrossRef]
39. Ajmi, L.; Hadeel; Alqahtani, N.; Rahman, A.U.; Mahmud, M. A Novel Cybersecurity Framework for Countermeasure of SME's in Saudi Arabia. In Proceedings of the 2nd International Conference on Computer Applications and Information Security, ICCAIS 2019, Riyadh, Saudi Arabia, 1–3 May 2019. [CrossRef]
40. Alsahafi, T.; Halboob, W.; Almuhtadi, J. Compliance with Saudi NCA-ECC based on ISO/IEC 27001. *Tech. Gaz.* **2022**, *29*, 2090–2097. [CrossRef]
41. Almomani, I.; Ahmed, M.; Maglaras, L. Cybersecurity maturity assessment framework for higher education institutions in Saudi Arabia. *PeerJ Comput. Sci.* **2021**, *7*, e703. [CrossRef] [PubMed]
42. Singh, H.P.; Alshammari, T.S. An Institutional Theory Perspective on Developing a Cyber Security Legal Framework: A Case of Saudi Arabia. *Beijing Law Rev.* **2020**, *11*, 637–650. [CrossRef]
43. NCA ECC-1:2018 Assessment and Compliance Tool. Available online: https://nca.gov.sa/legislation?item=176&slug=controls-list (accessed on 20 July 2022).

MDPI AG
Grosspeteranlage 5
4052 Basel
Switzerland
Tel.: +41 61 683 77 34

Applied Sciences Editorial Office
E-mail: applsci@mdpi.com
www.mdpi.com/journal/applsci